MICROWAVE ENGINEERING

WILEY SURVIVAL GUIDES IN ENGINEERING AND SCIENCE

Emmanuel Desurvire, Editor

Wiley Survival Guide in Global Telecommunications: Signaling Principles, Network Protocols, and Wireless Systems *Emmanuel Desurvire*

Wiley Survival Guide in Global Telecommunications: Broadband Access, Optical Components and Networks, and Cryptography *Emmanuel Desurvire*

Fiber to the Home: The New Empowerment *Paul E. Green, Jr.*

Electric Power Systems: A Conceptual Introduction *Alexandra von Meier*

Fiber Optic Essentials *K. Thyagarajan and Ajoy Ghatak*

Fiber to the Home: The New Empowerment *Paul E. Green, Jr.*

Microwave Engineering: Land & Space Radiocommunications *Gérard Barué*

MICROWAVE ENGINEERING
LAND & SPACE RADIOCOMMUNICATIONS

Gérard Barué

WILEY

A JOHN WILEY & SONS, INC., PUBLICATION

Published by John Wiley & Sons, Inc.

Published simultaneously in Canada

Library of Congress Cataloging-in-Publication Data

Barue, Gerard, 1945–
Microwave engineering : land & space radiocommunications / by Gerard Barue.
p. cm.
Includes bibliographical references and index.
ISBN 978-0-470-08996-5 (cloth)
1. Microwave communication systems. 2. Microwaves. I. Title.
TK7876.B37 2008
621.381′3—dc22

2008019299

Printed in the United States of America

10 9 8 7 6 5 4 3 2 1

To my son Vincent, a brilliant architect, first prizewinner at the French Academy of Architecture

To my wife, Dominique, who had the privilege of reading the manuscript entirely

This book is dedicated to the memory of Philippe Magne, THOMSON-CSF former Scientific and Technical Manager, who developed extensively the microwave radiocommunications in all domains.

CONTENTS

FOREWORD

It is always a great pleasure to see the introduction into the literature of a synthesis work which answers a great number of questions about a given topic; questions which often remain unanswered, either because they are hidden in bulky and difficult works, or because they were not published. The book of Gérard Barué has the immense merit to treat the various fundamental aspects of land and space radiocommunications. The constant concern of Gérard Barué to start by considering the basic physical phenomena makes the reading very pleasant. The reader never feels himself lost since the new concepts introduced take root in the fertile and well-known ground of classical physics. The student will find there an illustration of the effectiveness of the physical laws by concrete applications of these laws to the resolution of *a priori* complex problems. Thus, the establishment of a budget link, which needs to take into account several fields of physics (often taught in a disjoined way), reveals the approach that an engineer must control. The young engineer will see with pleasure how much its fundamental knowledge appears essential to the very concrete dimensioning of a telecom application. The senior engineer will find in the work of Gérard Barué an incomparable source of accurate data, validated by experiment (and often difficult to find) in particular thanks to many abacuses which enrich the text by values essential to the quantification of the established formulas. However, this book is not restricted to the presentation of the architecture of various systems of telecommunication and to the establishment of associated budget links, it makes us discover the world of engineering and infrastructure of site, without the control of which the best theoretical designs can prove to be of a remarkable inefficiency! The most critical points are reviewed and the proven solutions are described with a constant preoccupation of an application easily adaptable to the specific cases which the various readers can encounter. Gérard Barué recalls us that the physical laws which support the developments that one finds in his work belong essentially to the field of classical physics. That shows how much rich and fertile is this physics resulting essentially from work of the XIXth century and which still constitutes, even at the beginning of XXIth century, the foundation of the teaching in our schools of engineering. As Gérard Barué is working for nearly forty years in the field of the telecommunication within the Thales group, we would like to conclude this foreword by a short evocation of the origin of the group where the tradition is precisely the realization of innovating systems founded on the richness of the physical laws.

As each one knows, Thales (625–548 BC) is the founder of the philosophy of the "physicists" and was the first to notice that the dry rubbed yellow amber acquires the property to attract the light bodies: he is thus universally regarded as the father of electricity. With the passing time, telecommunications became an important activity attached to the field of the science of electricity. The birth of telecommunications is contemporary with Thales and goes back to ancient Greece. Later, the Romans developed their communication network and thus Cesar could communicate with his armies until the ends of his Empire. Nowadays, the civil and military stakes of telecommunications grow unceasingly, and the phenomenon of globalization is largely due to these techniques. The communications by optical fiber allow an incomparable rate and along with telecommunications by satellite it contributes to make our world a small planetary garden. One will find precisely described in the book of Gérard Barué the techniques of land and space radiocommunications which have still beautiful days in front of them. But let us return to the history. The previous *Thomson-CSF* owes its name with fusion, in 1968, of the *C.S.F* with the *Thomson Houston* company, founded in 1893 by the physicist Elihu Thomson. If name C.S.F ("Compagnie Sans Fil" or "*Wireless Company*") clarifies sufficiently the link with radio telecommunications, it is interesting to say few words on the physicist Elihu Thomson. For that it is necessary to go up in 1875. This very year, Thomas Edison notices that the electric relays of his telegraph make shoot out, without contact, the sparks from close metal objects. He concludes to an: "etheric force hitherto hidden in the depths of ignorance" and publishes his "discovery" in *Scientific American*. After reading this article, Elihu Thomson, then young professor of the *Technical Academy of Philadelphia*, remembered to have observed four years earlier the same phenomenon. He did again his experiments and succeeded in making shoot out sparks with more than 30m of the Ruhmkorff coil which he used as generator. Contrary to Edison, he attributes an electric origin to this phenomenon, which leads him to a public controversy with the famous inventor. Unfortunately, Elihu Thomson will not go further in the interpretation of this phenomenon, leaving the glory of the electromagnetic waves discovery to Heinrich Hertz (1888). Thus three years only after the discovery by James Clerk Maxwell of the fundamental equations of electromagnetism (1868), Elihu Thomson had observed the whole phenomena; emission, propagation and detection of electromagnetic waves, which validated the forecasts of the great Scottish physicist. But in the United States the work of Maxwell was not yet known and the father founder of Thomson group could not establish the link between his observations and the predictions of the incipient electromagnetic theory.

Few years after the celebration of the centenary of the first transatlantic radio connection (December 1901) we measure, upon reading the book of Gérard Barué, the way covered in a little more than one century in the field of telecommunications. It is thanks to the physical understanding, ingeniousness, entrepreneurship and tenacity of several generations of engineers that this tremendous development could be done. It is in this very spirit of implementation of the physical laws to concrete applications that the work of Gérard Barué has been done, following the example of the fathers founders of Thales company. The followed approach, all along this book, facilitates its reading and the progressive acquisition knowledge of this vast field of telecommunications. We wish for the many readers to feel the same pleasure as we had by reading this work.

JEAN-CLAUDE BOUDENOT
Thales senior expert

PREFACE

To face the tremendous expansion of the telecommunication market, the field of wireless communication has greatly developed in the past 50 years, especially in the domains of line of sight, satellites, and troposcatter. Therefore, the worldwide implementation of digital telecommunication equipment, continuously growing, requires incessantly systems which can provide an increasing capacity for an exploding number of users and creates the emergence of new types of services and new technologies, essentially in the areas of data processing, signal coding, and multiplexing. In addition, increasing link capacity within fixed spectral constraints requires more sophistication in signal processing and electronics at terminals that make more efficient use of the transmission medium by using the minimal radio frequency spectrum that is a shared and finite resource.

As applications have become increasingly sophisticated, more complex models and improved diversity techniques are needed. For instance, multipath propagation for terrestrial links is a fundamental and capacity-limiting problem for spectrally efficient information transmission; the corresponding propagation mechanisms therefore need to be well understood and the resulting channel behavior limitations in its modeling also require thorough study in order to design efficient microwave radio links.

The evaluation of the performance of digital communication over the various types of fading channels for single and multichannel diversity reception is consequently of prime importance, and the book develops copious methods, formulas, and numerical evaluations illustrated in the form of parametric performance curves. Such calculations assume that the physical channels follow well-defined statistical models based on commonly encountered probability distribution functions of random signals that are reviewed in the book. The most used configurations for each type of main communication system are also described as well as optimum transceiver structures and the optimum diversity process corresponding to a wide variety of combinations in relation with fading parameters such as amplitude, phase, and delay. Emphasis is placed on evaluation of system error rate performance and availability of practical applications using various digital signaling techniques and basic system parameters and requirements, such as signal-to-noise ratio, probability of error, and channel bandwidth, and methods of enhancing performance are also given.

There are a large number of excellent works about telecommunication and information theory, but most of them are dedicated to the physical layers of digital networks and to their numerous applications, so that wireless communication systems generally appear as "clouds in the sky" in their relevant diagrams. However, these mysterious communication systems, which appear so "nebulous" to most specialists of telecommunication networks and signal design, have great importance in terms of overall quality and availability performances, and their costs may be not negligible as they often require important structures in terms of shelters, towers, and antenna size.

The primary objective of this book is thus to introduce the microwave student as well as the practicing engineer to most of the propagation phenomena encountered in practice and that must be considered in planning and evaluation of a communication system of any given type in a wide frequency range of operation. Fundamental theory and principles are stressed and complete derivations are provided for all significant formulas and relationships as well as the methods of analysis and design that are the foundations for microwave engineering. Attention is paid to the effects of terrain and atmospheric conditions on microwave propagation with a detailed account of the many varied phenomena that have to be considered. Various examples of link evaluations and real systems performance recordings are included to illustrate the application of the theory to microwave line-of-sight, troposcatter, and satellite systems. The second objective is to provide the engineer with up-to-date technical background coupled with applications that are essential for the design of communication systems. An attempt is made to provide a sound understanding of the physical channel limitations as well as propagation models with signal processing concepts and design criteria in order to satisfy the minimum system performance standards in the presence of the hazards of propagation. The book is composed of five chapters that present the various phenomena of propagation, their respective effects on the transmitted signal and the equipment, and the appropriate methods for the definition of the links in each domain with the maximum of clarity.

Chapter 1 contains an introduction to the fundamental theory of radiation and propagation in order to provide a comprehensive summary of basic electromagnetic theory and principles that are needed as background for proper understanding of the following topics: microwave free-space attenuation, scattering mechanism losses due to diffraction, reflection on plane surface or rough terrain, multipath and atmospheric constituents, and typical antenna concepts such as gain, directivity, radiation patterns, effective area, aperture efficiency, cross-polarization properties, and performance of passive repeaters or radio reflectors. As the properties of the antenna radiation pattern are difficult to separate from the effect of the distribution of the direct and reflected components arriving at the antennas, special attention is given to the physics of reflection and the appropriate use of antennas to launch and receive the radio waves most efficiently. The chapter also deals with the statistical properties of a signal resulting from a combination of the various causes of perturbation given above. The distribution functions commonly encountered in the study of communication systems are thus exposed because the mathematical modeling, such as the prediction of path loss and fade depth, and the characterization of physical channels require some familiarity with random variables and basic theory of probability.

Chapter 2 gives an overview of digital communication systems by exposing the main spectral characteristics of digitally modulated signals in relation to the various available modulation techniques and signal processing that enhance the bandwidth efficiency and error control to improve performance. It explains the origin of thermal noise and introduces notions such as noise figure, noise temperature, antenna system noise, transmission line noise, receiver threshold, and mechanisms by which external noise combines with the inherent thermal noise of the receiver to contribute to the overall system noise. Other causes of degradation of communication systems, such as radio channel interference, polarization mismatch, impedance mismatch, echo phenomena, and intermodulation in nonlinear circuits, are also developed from basic principles.

Chapter 3 concerns the fundamentals of microwave line-of-sight systems operating over various types of path profiles under all possible atmospheric conditions, regarding the vertical refractivity gradient and the scattering by hydrometeors, which affect the radio channels by either slow flat fading or high-speed frequency-selective fading. With the view of minimizing the effects of reflection and multipath, an interferometric method based on classical interference equations in a two-ray model allows us to define the appropriate diversity techniques which have to be settled according to the size of the interference fringes at the respective terminals. Antennas, propagation, and radio channel interference are thus interrelated in systems that can increase performance and spectral efficiency. It shows the importance of diversity to solve the problems linked to reflection, ducting, and multipath for all conditions of propagation, from atmospheric subrefraction to superrefraction, and the way to use the interferometric method to get the optimum diversity gain and improvement factor. It therefore provides a detailed study of digital microwave radio systems design based on various static and dynamic characteristics such as receiver signature, fading speed, and notch speed. As the communication systems have to conform to international standards for reasons of compatibility, the chapter recalls the performance requirements of radio relay systems based on the International Telecommunication Union (ITU) recommendations for low-, medium-, and high-quality sections for plesiochronous and synchronous digital hierarchy (PDH, SDH) in integrated services digital networks (ISDNs), and it gives a description of the various measurements that are necessary to characterize the radio channel and evaluate its operational performance. The chapter considers the constraints imposed by the frequency and polarization plan which have to be fully defined in order to avoid any interference between radio channels in local and distant modes, and it shows how to optimize the frequency resource by using high-performance antennas and available frequency reuse techniques.

Chapter 4 is dedicated to the design of troposcatter communication systems with special emphasis on the theories and the experimental results issued in

National Bureau of Standards technical note 101 and ITU-R. The basic aspects of this mode of propagation beyond the horizon are described for the purpose of establishing design criteria for a specified system reliability. Objectives and design parameters are discussed in order to lead to the evaluation of channel performance in terms of quality and availability. The chapter covers the propagation by diffraction on obstacles and by scattering mechanisms in the troposphere, in particular multipath statistical relations regarding time delay, long-term variation, and short-term deep and rapid fading. It is shown how to establish the path profile between the terminals and how to determine the meteorological parameters of great importance in order to proceed to the modeling of the channel characteristics in terms of bandwidth capability and availability. The properties of scattering propagation, based on experimental analysis, are fully described as well as the most common diversity schemes, such as space, frequency, and angle diversity, and the various combining methods used to improve transmission capability. A complete description of an impressive system using a 6th-order frequency and three-beams angle diversity is also given including real performance recording. The phenomenon of transverse directivity, that the author experienced in the 1970s, is also described as it has a great influence on the pointing of the antennas as well as on the transverse angle diversity performance.

Chapter 5 is devoted to the domain of satellite communication systems with the purpose of presenting the organization of the various applications in relation to their respective orbital parameters from low Earth orbit to geostationary orbit. It shows how to compute the uplink and downlink budgets for both space and ground segments and how to define the different components of the repeater and Earth terminals. The chapter develops the aspects of Earth–satellite geometry to determine azimuth, distance, and elevation and the important notions of coverage, repeater gain, equivalent isotropic radiated power of the transmitter, power flux density, figure of merit of the receiver, linear and nonlinear repeater, multiple-access principles, and carrier-to-noise ratio. The influence of the propagation medium, such as the effects due to hydrometeors in the atmosphere, in terms of attenuation and increase of the system noise and those of the scintillation and rotation of polarization in the ionosphere, is also treated.

ACKNOWLEDGMENTS

The author thanks the company for which he worked and that gave him the opportunity to develop his experience in this comprehensive field for nearly 40 years. Thanks are also due to the International Telecommunication Union for the recommendations cited in the text. The author is also grateful to Jean-Claude Boudenot and Emmanuel Desurvire for their encouragement to write this book. A special thanks goes to Yves Le Nevez, who provided invaluable information about microwave line-of-sight propagation. A final thanks goes to Jacques Ricroque and Jean-Luc Linder for their important contribution in the design and preparation of tests and recordings on a number of line-of-sight and troposcatter radio links, which allowed us to put into practice the famous saying of Auguste Comte: "To experiment means to know and to know means to can." (1830).

CHAPTER 1

Electromagnetic Wave Propagation

1.1 PROPERTIES OF PLANE ELECTROMAGNETIC WAVE

1.1.1 Equation of Wave or Propagation

Electromagnetic waves are propagated in a vacuum, in dielectrics and conductors; here we will be interested in the propagation of radiated waves of periodic type that are characterized by a wavelength defined by a wave velocity which depends on the permittivity and permeability of the crossed medium.

Thus we consider a plane wave, as represented in Figure 1.1, propagating in direction x of an orthogonal reference system (x, y, z) while transporting the electric field **E** polarized in the direction y and the magnetic induction **B** in the direction z. The properties of such a wave can be deduced from Maxwell's equations, which link together the electric field **E**, the magnetic induction **B**, and the current density **J**:

$$\text{Curl } E = -\frac{\partial B}{\partial t} \quad \text{(Faraday's law)} \tag{1.1}$$

$$\text{Curl } B = \mu J + \varepsilon\mu\frac{\partial E}{\partial t} \quad \text{(Ampere's generalized theorem)} \tag{1.2}$$

where μ = absolute permeability of medium of propagation (H m^{-1})
ε = its absolute permittivity (F m^{-1})

These equations can also be written in differential form by considering, for example, a nonconducting medium ($J = 0$):

$$\frac{\partial E}{\partial x} = -\frac{\partial B}{\partial t} \tag{1.3}$$

$$\frac{\partial B}{\partial x} = -\varepsilon\mu\frac{\partial E}{\partial t} \tag{1.4}$$

Microwave Engineering: Land & Space Radiocommunications, by Gérard Barué

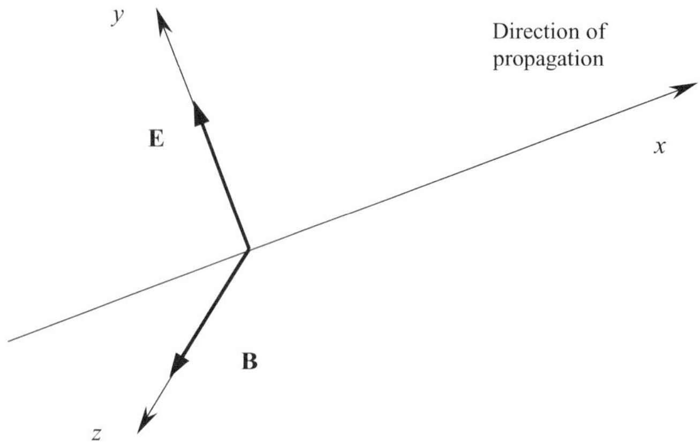

Figure 1.1 Plane electromagnetic wave.

After a second derivation according to x and t, these equations become

$$\frac{\partial^2 E}{\partial x^2} = -\frac{\partial}{\partial x}\left(\frac{\partial B}{\partial t}\right) = -\frac{\partial}{\partial t}\left(\frac{\partial B}{\partial x}\right) = \varepsilon\mu\frac{\partial^2 E}{\partial t^2}$$

$$\frac{\partial^2 B}{\partial x^2} = \varepsilon\mu\frac{\partial^2 B}{\partial t^2}$$

A simple solution is a wave varying sinusoidally over time whose electric field and magnetic induction amplitudes are given by the relations

$$E = E_{max}\cos(\beta x - \omega t) \tag{1.5}$$

$$B = B_{max}\cos(\beta x - \omega t) \tag{1.6}$$

where

$$\beta = \frac{2\pi}{\lambda}$$

$$\omega = 2\pi f = \frac{2\pi}{T} \quad \text{angular velocity (rad s}^{-1}\text{)}$$

$$\lambda = \frac{v}{f} = vT \quad \text{wavelength (m)}$$

where v = wave velocity (m s^{-1})
f = frequency (Hz)
T = period (s)

By deriving these two last expressions, we obtain

$$\frac{\partial E}{\partial x} = -\beta E_{\max} \sin(\beta x - \omega t)$$

$$\frac{\partial B}{\partial t} = \omega B_{\max} \sin(\beta x - \omega t)$$

Then, starting from expression (1.3), we can write

$$\beta E_{\max} = \omega B_{\max}$$

$$\frac{E_{\max}}{B_{\max}} = \frac{\omega}{\beta} = \lambda f = \frac{\lambda}{T} = v \tag{1.7}$$

The general expression of a sinusoidal periodic electromagnetic plane wave propagating in the direction x can thus be written in its complex form:

$$E = E_{\max} e^{-\alpha x} e^{j(\omega t - \beta x)} \tag{1.8}$$

where α is the attenuation constant of the medium ($\alpha = 0$ in a vacuum).

1.1.2 Wave Velocity

The velocity of an electromagnetic wave in a vacuum is a universal constant (speed of light) used to define the meter in the International System of Units (SI) and has as a value

$$c = \frac{1}{\sqrt{\varepsilon_0 \mu_0}} = 299{,}792{,}458 \,\mathrm{m\,s^{-1}} \tag{1.9}$$

where

$\mu_0 = 4\pi \times 10^{-7}\,\mathrm{H\,m^{-1}}$ absolute magnetic permeability of vacuum

$\varepsilon_0 = \dfrac{1}{\mu_0 c^2} = 8.8542 \times 10^{-12}\,\mathrm{F m^{-1}}$ absolute permittivity of vacuum

The wave velocity in a medium of absolute permittivity ε and permeability μ corresponding to a relative permittivity ε_r and a relative permeability μ_r compared to those of the vacuum is given by the relation

$$v = \frac{1}{\sqrt{\varepsilon \mu}} = \frac{c}{\sqrt{\varepsilon_r \mu_r}}$$

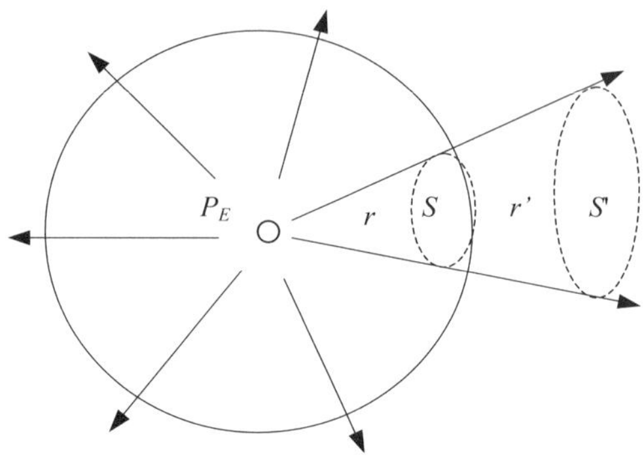

Figure 1.2 Radiation of isotropic source.

As the magnetic permeability of the medium can be regarded as equal to that of the vacuum, the wave velocity becomes

$$v \approx \frac{c}{\sqrt{\varepsilon_r}} \tag{1.10}$$

1.1.3 Power Flux Density

A very small surface that emits a uniform radiation in all directions of space inside a transparent medium is shown in Figure 1.2. The quantity of energy E radiated by such a fictitious element, called a point source or an isotropic antenna,[1] that crosses per second a certain surface S located on a sphere at a distance r is called the flux, and the ratio of this quantity to this surface is the power flux density. If we consider a second surface S' at a distance r' based on the contour of S, we see that the power flux density which crosses S' varies by a factor S/S' compared to that which crosses S, that is, by the inverse ratio of the square of the distance. Consequently, a transmitter of power P_E supplying an isotropic antenna generates a power flux density P_D on the surface of the sphere of radius r equal to

$$P_D = \frac{P_E}{4\pi r^2} \tag{1.11}$$

[1]An isotropic antenna radiating in an identical way in all directions of space could not have physical existence because of the transverse character of the electromagnetic wave propagation, but this concept has been proven to be very useful and it is common practice to define the characteristics of an antenna compared to an isotropic source, whose theoretical diagram of radiation is a sphere, rather than to the elementary doublet of Hertz or the half-wave doublet.

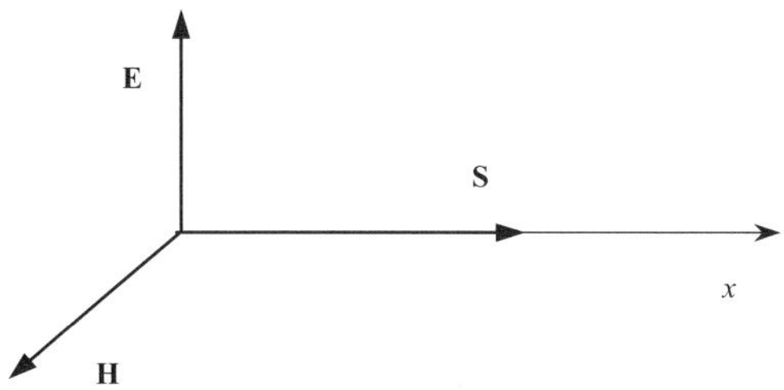

Figure 1.3 Poynting's vector.

1.1.4 Field Created in Free Space by Isotropic Radiator

The power flux density, created at a sufficiently large distance d so that the spherical wave can be regarded as a plane wave,[2] can also be expressed by using Poynting's vector, illustrated in Figure 1.3, which is defined by the relation

$$\mathbf{S} = \mathbf{E} \wedge \mathbf{H} \qquad (\mathrm{W\,m^{-2}}) \tag{1.12}$$

where $\mathbf{H} = \mathbf{B}/\mu_0$ in a vacuum and the electric field $\mathbf{E}$ in volts per meter, magnetic field $\mathbf{H}$ amperes per meter, and Poynting vector $\mathbf{S}$ watts per square meter orientated in the direction of propagation form a direct trirectangular trihedron.

The electric field and magnetic field are connected in the vacuum by the scalar relation

$$\frac{E}{H} = \sqrt{\frac{\mu_0}{\varepsilon_0}} = Z_0 = 120\pi \approx 377\,\Omega \tag{1.13}$$

where Z_0 is the impedance of the vacuum and the modulus of Poynting's vector—or power flux density—is completely defined by the electric field in the relation

$$S = \frac{E^2}{Z_0} \tag{1.14}$$

1.1.5 Wave Polarization

An important property of an electromagnetic wave is its polarization, which describes the orientation of the electric field E. In general, the electric field of

[2]A plane wave is a mathematical and nonphysical solution of Maxwell's equations because it supposes an infinite energy, but a spherical wave all the more approaches the structure of the plane wave since it is distant from its point of emission.

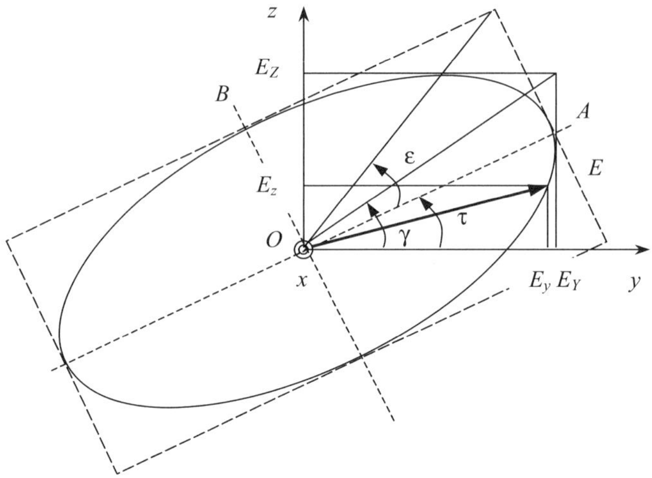

Figure 1.4 Electric field resultant vector.

a wave traveling in one direction may have components in both other orthogonal directions and the wave is said to be elliptically polarized.

Consider a plane wave traveling out through the page in the positive x direction, as shown in Figure 1.4, with electric field components in the y and z directions as given by

$$E_y = E_Y \sin(\omega t - \beta x) \tag{1.15}$$

$$E_z = E_Z \sin(\omega t - \beta x + \delta) \tag{1.16}$$

where E_Y = modulus of component according to y (constant)
E_Z = modulus of component according to z (constant)
δ = phase difference between E_y and E_z

Equations (1.15) and (1.16) describe two linearly polarized waves which combine vectorially to give the resultant field:

$$\mathbf{E} = \mathbf{E}_y + \mathbf{E}_z$$

It follows that

$$\mathbf{E} = \mathbf{E}_Y \sin(\omega t - \beta x) + \mathbf{E}_Z \sin(\omega t - \beta x + \delta)$$

At the origin ($x = 0$),

$$E_y = E_Y \sin \omega t$$

$$E_z = E_Z \sin(\omega t + \delta)$$

$$E_z = E_Z(\sin \omega t \cos \delta + \cos \omega t \sin \delta)$$

and thus

$$\sin \omega t = \frac{E_y}{E_Y}$$

$$\cos \omega t = \sqrt{1 - \left(\frac{E_y}{E_Y}\right)^2}$$

Rearranging these equations in order to eliminate time yields

$$\frac{E_y^2}{E_Y^2} - \frac{2E_y E_z \cos \delta}{E_Y E_Z} + \frac{E_z^2}{E_Z^2} \tag{1.17}$$

or $aE_y^2 - bE_yE_z + cE_z^2$ where

$$a = \frac{1}{E_Y^2 \sin^2 \delta} \qquad b = \frac{2\cos \delta}{E_Y E_Z \sin^2 \delta} \qquad c = \frac{1}{E_Z^2 \sin^2 \delta}$$

Equation (1.17) describes the polarization ellipse with tilt angle τ and whose semimajor axis OA and semiminor axis OB determine the axial ratio AR $\ni (1, \infty)$:

$$\text{AR} = \frac{OA}{OB} \tag{1.18}$$

If $E_Y = 0$, the wave is linearly polarized in the z direction.
If $E_Z = 0$, the wave is linearly polarized in the y direction.
If $E_Y = E_Z$, the wave is linearly polarized in a plane at angle of 45°.
If $E_Y = E_Z$ with $\delta = \pm 90°$, the wave is circularly polarized (left circularly polarized for $\delta = +90°$ and right circularly polarized for $\delta = -90°$).

The polarization is right circular if the rotation of the electric field is clockwise with the wave receding and left circular if it is counterclockwise. It is also possible to describe the polarization of a wave in terms of two circularly polarized waves of unequal amplitude, one right (E_r) and the other left (E_l). Thus, at $x = 0$

$$E_r = E_R e^{j\omega t} \qquad E_l = E_L e^{-j(\omega t + \delta')}$$

where δ' is the phase difference. Then,

$$E_y = E_R \cos \omega t + E_L \cos(\omega t + \delta') \qquad E_z = E_R \sin \omega t + E_L \sin(\omega t + \delta') \tag{1.19}$$

The other parameters can be determined from the properties of Poincaré's sphere:

$$\cos 2\gamma = \cos 2\varepsilon \cos 2\tau \qquad \tan \delta = \frac{\tan 2\varepsilon}{\sin 2\tau}$$

$$\tan 2\tau = \tan 2\gamma \cos \delta \qquad \sin 2\varepsilon = \sin 2\gamma \sin \delta$$

As a function of time and position, the electric field of a plane wave traveling in the positive x direction is given by

$$E_y = E_Y e^{j(\omega t - \beta x)} \qquad E_z = E_Z e^{-j(\omega t - \beta x + \delta)}$$
$$\mathbf{E} = \mathbf{E}_Y e^{j(\omega t - \beta x)} + \mathbf{E}_Z e^{-j(\omega t - \beta x + \delta)}$$

The magnetic field components H associated respectively to E_y and E_z are

$$H_z = H_Y e^{j(\omega t - \beta x - \theta)} \qquad H_y = -H_Z e^{j(\omega t - \beta x + \delta - \theta)}$$

where θ is the phase lag of H in the absorbing medium ($\theta = 0$ in a lossless medium).

The total H field vector is thus

$$\mathbf{H} = \mathbf{H}_Y e^{j(\omega t - \beta x + \theta)} - \mathbf{H}_Z e^{j(\omega t - \beta x + \delta - \theta)}$$

The average power of the wave per unit area, or power flux density, is given by the Poynting's vector formula:

$$S = \frac{1}{2}[E_Y H_Y + E_Z H_Z] \cos \theta$$

which in a lossless medium becomes[3]

$$S = \frac{1}{2} \frac{E_Y^2 + E_Z^2}{Z_0} \tag{1.20}$$

as

$$E = \sqrt{E_Y^2 + E_Z^2} \qquad H = \sqrt{H_Y^2 + H_Z^2}$$

[3]This equation takes into account the absolute values of the electric field instead of the root-mean-square (rms) values for relation (1.14).

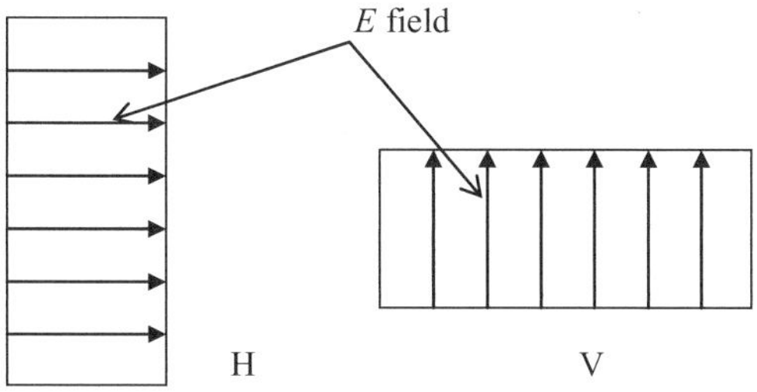

Figure 1.5 Representation of linear polarization H and V.

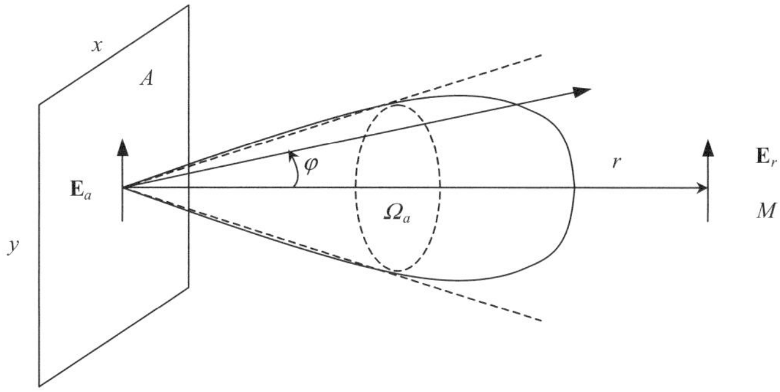

Figure 1.6 Radiation pattern of planar rectangular aperture.

As linear polarization is mostly used, Figure 1.5 shows the direction of the electric field relative to the position of a waveguide for horizontal (H) and vertical (V) polarizations.

1.2 RADIANT CONTINUOUS APERTURE

1.2.1 Expression of Directivity and Gain

Consider planar rectangular aperture of surface A uniformly illuminated by an electromagnetic wave whose field conserves an intensity E_a and a direction constant, as illustrated in Figure 1.6, and a point M located in the normal direction and at a sufficiently large distance r so that the wave reradiated by the aperture and received at this point can be regarded as a plane wave. Moreover, the dimensions of the aperture are large in front of the wavelength λ, we suppose that is, x, $y > 10\lambda$. By summing the contributions at distance r of all the elements of current uniformly distributed in the aperture of dimensions (x, y), we obtain the relation

$$E_r(\varphi) = -\frac{j\omega\mu y e^{-j\beta r}}{4\pi r Z} \int E(x) e^{j\beta x \sin\varphi} dx \tag{1.21}$$

where $E(x) = E_a$
x = transverse dimension of aperture
y = vertical dimension of aperture
φ = angle compared to perpendicular to aperture
A = area of aperture, $= x\ y$
Z = intrinsic wave impedance of propagation medium

The field E_r has as a modulus

$$|E_r| = \frac{|E_a|}{r\lambda} A \tag{1.22}$$

We can associate a radiation pattern to the radiant aperture characterized by the following:

- A gain G compared to isotropic radiation:

$$G = \frac{P}{P'} \tag{1.23}$$

  where P, P' are the powers of the transmitter which would generate the same field at a given distance by means respectively of an isotropic antenna and the radiant aperture.
- A beam solid angle Ω_a, in which would be concentrated all the energy if the power flux density remained constant and equal to the maximum value inside this angle, represented by φ and ϕ in two orthogonal planes:

$$\Omega_a = \iint_{4\pi} g(\varphi, \phi) d\Omega$$

  where

$$g(\varphi, \phi) = \frac{G(\varphi, \phi)}{G}$$

Under these conditions, the power radiated by the aperture can be written as

$$P = \frac{|E_a|^2}{Z} A = \frac{|E_r|^2}{Z} r^2 \Omega_a$$

which yields

$$\Omega_a = \frac{|E_a|^2}{|E_r|^2}\frac{A}{r^2}$$
$$= \frac{\lambda^2}{A}$$

Since $G = 4\pi/\Omega_a$, we find the general expression of the directivity of an antenna,

$$G = \frac{4\pi A}{\lambda^2} \tag{1.24}$$

and, taking account of the efficiency, its gain,

$$G = \frac{\eta 4\pi A}{\lambda^2} = \frac{4\pi A_e}{\lambda^2} \tag{1.25}$$

where η = efficiency of antenna
A_e = its effective area

or, according to the diameter of the antenna when it is about a circular aperture,

$$G = \eta\left[\frac{\pi D}{\lambda}\right]^2 \tag{1.26}$$

with $A = \pi D^2/4$. An isotropic antenna having a theoretical gain equal to unity we can deduce its equivalent area, which still does not have physical significance, while posing from the relation (1.24):

$$\frac{4\pi A_{\text{eq}}}{\lambda^2} = 1$$

that is,

$$A_{\text{eq}} = \frac{\lambda^2}{4\pi} \tag{1.27}$$

According to the principle of reciprocity, the characteristics of an antenna are the same at the emission as at the reception.

Using the power received by an isotropic antenna, which is equal to

$$P = A_{\text{eq}} S$$

we can deduce the power flux density:

$$S = \frac{4\pi P}{\lambda^2} \tag{1.28}$$

1.2.2 Radiation Pattern

Here we will consider aperture-type antennas whose dimensions are at least 10 times higher than the wavelength used. When a conducting element of infinite size is placed in an electromagnetic field, it undergoes a certain induction; a current of induction then circulates on its surface and this element in turn radiates energy. The spatial distribution of this radiation can be defined by integration of the distribution of these elements of surface current by using the principle of superposition of energy. As the dimensions of the antennas are necessarily finished, we observe additional radiation located in angular zones not envisaged by calculations relating to the radiation and which are due to diffraction on the edges of the aperture.

The distribution of the field $E(\sin\varphi)$ at long distances corresponds to Fourier's transform of the distribution of the field $E(X)$ in the aperture, which can be written as

$$E(\sin\varphi) = \int_{-\infty}^{+\infty} E(X) e^{j2\pi X \sin\varphi} dX$$

where $X = x/\lambda$ and reciprocally as

$$E(X) = \int_{-\infty}^{+\infty} E(\sin\varphi) e^{-j2\pi X \sin\varphi} d(\sin\varphi)$$

When the aperture has a finished dimension L, these expressions become

$$\begin{aligned} E(\sin\varphi) &= \int_{-L/2}^{+L/2} E(X) e^{j2\pi X \sin\varphi} dX \\ E(X) &= \int_{-L/2}^{+L/2} E(\sin\varphi) e^{-j2\pi X \sin\varphi} d(\sin\varphi) \end{aligned} \tag{1.29}$$

where $L = l/\lambda$. The radiation pattern thus depends on both the distribution of the field in the aperture, or the law of illumination, and its dimensions. Expression (1.29) is similar to expression (1.21), previously obtained by integration of the distribution of surface current, the first giving the relative value of the remote field and the second its absolute value.

For example, we will consider a rectangular aperture of sides a and b subjected to a uniform primary illumination, as shown in Figure 1.7. The radiation pattern can be expressed by the relation

$$\frac{E}{E_0} = \frac{\sin u}{u} \tag{1.30}$$

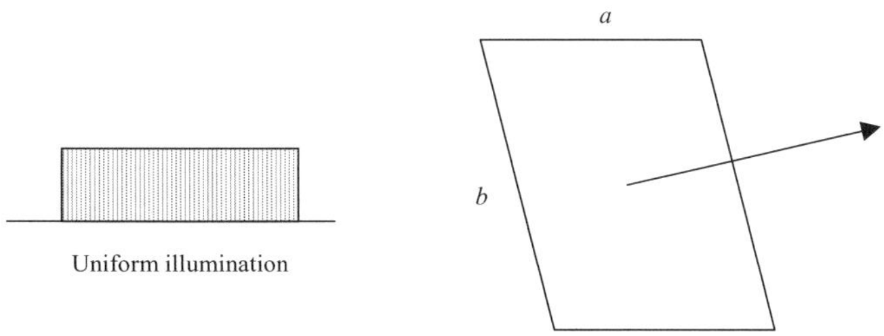

Figure 1.7 Rectangular aperture subjected to uniform illumination.

where

$$u = \begin{cases} \dfrac{\pi a}{\lambda}\sin\varphi \quad (\text{rad}) & \text{according to dimension } a \\ \dfrac{\pi b}{\lambda}\sin\varphi \quad (\text{rad}) & \text{according to dimension } b \end{cases}$$

from which

$$\frac{P}{P_0} = \left(\frac{\sin u}{u}\right)^2$$

The radiation pattern is presented versus u in Figure 1.8. As shown in the figure, this radiation pattern has a main beam, or main lobe, associated with minor lobes, called side lobes, whose relative levels are –13.3 dB for the first side lobe, –17.9 dB for the second, –20.8 dB for the third, and so on.

There are a number of factors that affect the side-lobe performance of an antenna, such as the aperture illumination function, the edge diffraction, and the feed spillover. By using a decreasing primary illumination toward the edges of the aperture (e.g., parabolic law, triangular, Gaussian, cosine squared), it is possible to reduce and even remove the minor lobes but at the cost of a widening of the main lobe and a reduction of the nominal gain.

The total beamwidth at half power for each dimension of the aperture corresponds to the values

$$\theta_{T3\text{dB}} = \begin{cases} 0.886\dfrac{\lambda}{a} & \text{according to dimension } a \\ 0.886\dfrac{\lambda}{b} & \text{according to dimension } b \end{cases}$$

In the case of a circular aperture of diameter D subjected to a uniform illumination, the diagram of radiation is given by the relation

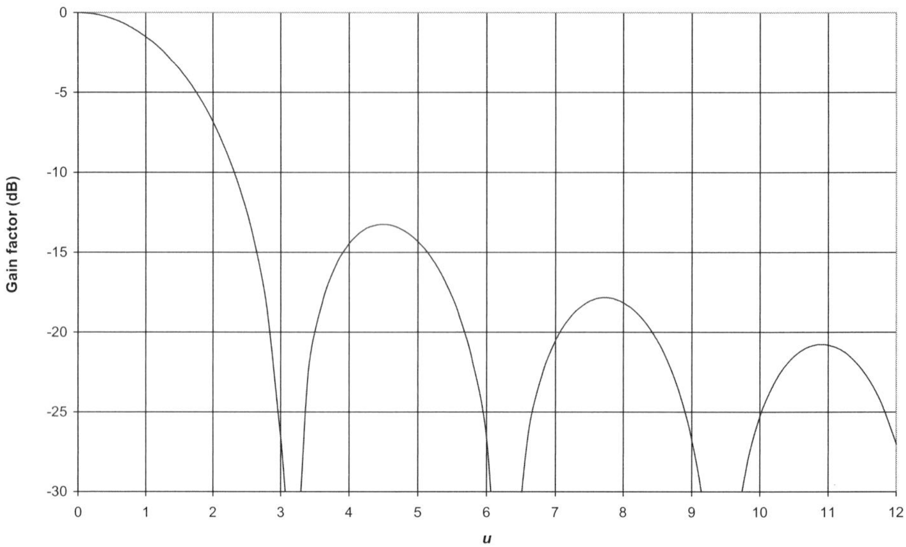

Figure 1.8 Radiation pattern of rectangular aperture subjected to uniform illumination.

$$\frac{E}{E_0} = C\frac{J_1(u)}{u} \tag{1.31}$$

where $J_1(u)$ is Bessel's function of order 1 and

$$u = \frac{\pi D}{\lambda}\sin\varphi \qquad \text{(rad)}$$

that is,

$$\frac{P}{P_0} = \left(C\frac{J_1(u)}{u}\right)^2$$

The levels of the minor lobes relative to the main lobes, as illustrated by Figure 1.9, are located respectively at –17.6 dB, –22.7 dB, –25.3 dB, and so on.

In the same way, it is possible to reduce the side-lobe level with decreasing illumination toward the edges. Using, for example, a tapered parabolic distribution of the form $(1 - r^2)^p$ with

$$r = \frac{\rho}{R}$$

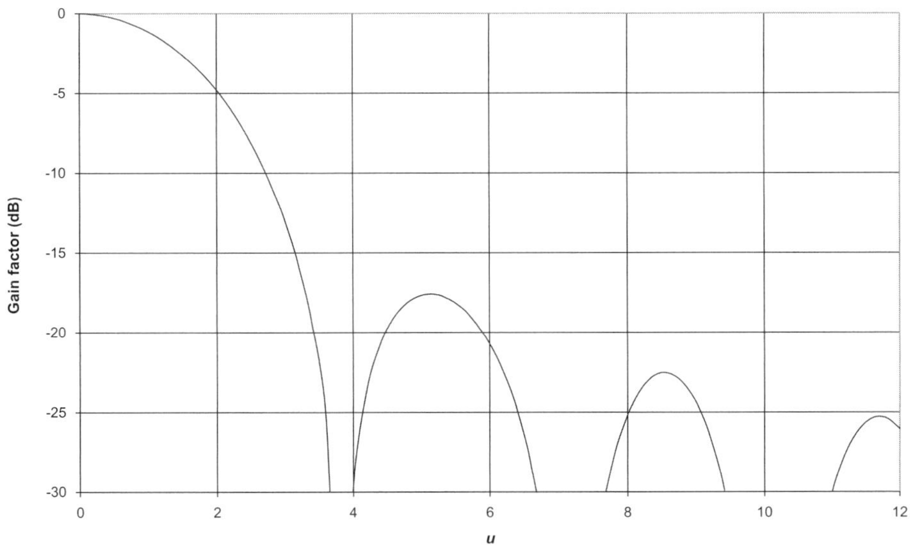

Figure 1.9 Radiation pattern of circular aperture subjected to uniform illumination.

where R is the radius of the aperture ($0 < \rho < R$), we obtain for various decreasing scale parameter p the characteristics of the radiation pattern presented in the table below (Silver, 1986):

p	Gain Reduction	Total Half-Power Beamwidth (rad)	First Zero	First Lobe (dB)
0 (uniform)	1	$1.02\dfrac{\lambda}{D}$	$\arcsin\left(\dfrac{1.22\lambda}{D}\right)$	17.6
1	0.75	$1.27\dfrac{\lambda}{D}$	$\arcsin\left(\dfrac{1.63\lambda}{D}\right)$	24.6
2	0.56	$1.47\dfrac{\lambda}{D}$	$\arcsin\left(\dfrac{2.03\lambda}{D}\right)$	30.6

1.2.3 Near Field and Far Field

Three zones in a field diffracted by a radiating aperture versus distance are illustrated in Figure 1.10, with each features well defined:

1. *Rayleigh's Zone* This is the near-field zone, which extends at the distance

$$Z_{\text{Rayleigh}} = \frac{D^2}{2\lambda} \tag{1.32}$$

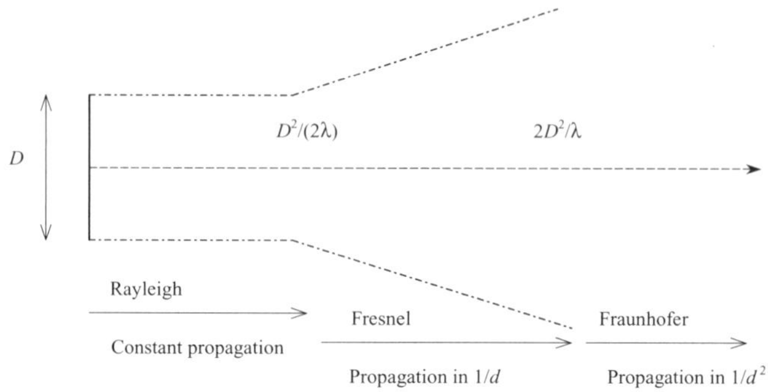

Figure 1.10 Zones of radiation of an aperture.

in a square projected aperture of size D (in meters) uniformly illuminated at a wavelength λ (in meters). Inside this zone, the energy is propagated in a tube delimited by the aperture D by presenting an equal-phase wave front and some oscillations of the amplitude versus the longitudinal axis. The extent of Rayleigh's zone depends on the form of the aperture and the distribution of the field inside the zone; for instance:

(a) For a circular projected aperture of diameter D and uniform illumination

$$Z_{\text{Rayleigh}} = 0.82\frac{D^2}{2\lambda}$$

(b) For a circular aperture and illumination of parabolic type

$$Z_{\text{Rayleigh}} = 0.61\frac{D^2}{2\lambda}$$

2. *Fresnel's Zone* This is the intermediate zone, where the diffracted field starts and the wave front tends to become spherical, which extends with a law for attenuation in $1/d$ at the distance

$$Z_{\text{Fresnel}} = 2\frac{D^2}{\lambda} \tag{1.33}$$

3. *Fraunhofer's Zone* Beyond Fresnel's zone, the energy is propagated in an inverse ratio to the square of the distance (law in $1/d^2$) and the radiant characteristics of the aperture are well defined (radiation pattern, positions and levels of the minor lobes, gain in the axis, spherical wave front).

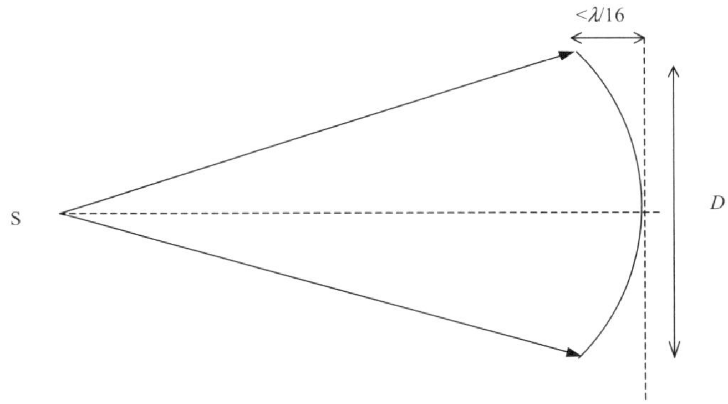

Figure 1.11 Plane wave in far field.

The far-field zone also corresponds to the distance beyond which the difference between the spherical wave and the plane wave becomes lower than $\lambda/16$, as shown in the Figure 1.11.

1.2.4 Effective Aperture

The effective aperture of an antenna for an electromagnetic plane wave linearly polarized at the emission as at the reception is defined by the relation

$$A_e = \frac{P_R}{P_D} \tag{1.34}$$

where P_R = power available at antenna port
P_D = power flux density given by relation (1.11)

1.2.5 Skin Effect

This effect appears when a conductor is traversed by a sinusoidal current and results in an increase of high-frequency resistance; in particular, the efficiency of the feeders as well as the reflectors of the antennas is thus affected.

By considering a conductor of conductivity σ and permeability μ limited by an insulator by which arrives the plane wave at frequency f whose direction of propagation is perpendicular to the surface of the conductor, we show that the current density amplitude decreases exponentially inside the conductor starting from the surface of separation according to the relation

$$i = i_0 \exp\left(-\frac{x}{\delta}\right) \tag{1.35}$$

where x is the penetration depth in meters and the critical depth δ (in meters) is given by the relation

$$\delta = \frac{1}{\sqrt{\pi\mu\sigma f}} = \frac{503.3}{\sqrt{\mu_r \sigma f}}$$

where μ = absolute permeability ($\mathrm{H\,m^{-1}}$)
μ_r = relative permeability, = μ/μ_0
σ = conductivity (S)
f = frequency (Hz)

For example, for steel and copper, we obtain the following δ values by frequency f:

	50 Hz	10 kHz	1 MHz	100 MHz	10 GHz
Steel	0.33 mm	0.024 mm	2.4 μm	0.24 μm	0.024 μm
Copper	8.5 mm	0.6 mm	60 μm	6 μm	0.6 μm

In the case of propagation in a round conductor, the current density decreases in the same way starting from its periphery.

1.3 GENERAL CHARACTERISTICS OF ANTENNAS

1.3.1 Expression of Gain and Beamwidth

The nominal or maximum isotropic gain in the far field of circular aperture-type antennas, expressed in dBi, can be calculated by using the relation in Section 1.2.1:

$$G_{\max} = \eta \left[\frac{\pi D}{\lambda} \right]^2 \qquad (1.36)$$

where η = total efficiency[4] of antenna, in general between 0.5 and 0.7
D = diameter (m)
λ = wavelength (m)

The total half-power beamwidth (3 dB) of the antenna, expressed in degrees, is given by the approximate formula

$$\alpha_T \approx 69.3 \frac{\lambda}{D} \qquad (1.37)$$

[4]The total efficiency of the antenna depends on numerous factors, such as the tapering illumination function, the spillover loss, the phase error, and the surface accuracy.

or ±34.65(λ/D). The gain $G(\varphi)$ relative to the isotropic antenna, expressed in dBi, in the direction φ relative to the axis is given by the following approximate relation which is valid for the main lobe of radiation:

$$G(\varphi) = G_{\max} - 12\left(\frac{\varphi}{\alpha_T}\right)^2 \tag{1.38}$$

where φ and α_T are in degrees. These formulas are also used to determine the following:

- Antennas gain loss due to their possible misalignment (e.g., action of the wind) or variation of the launch and arrival angles of the rays according to the conditions of refractivity of the atmosphere
- Discrimination of reflected ray compared to direct ray
- Level of signals coming from other transmitters of a network that interfere with the useful signal during setup of the frequency and polarization plan

Figure 1.12 illustrates the nominal isotropic gain and total half-power beamwidth versus frequency, calculated using relations (1.36) and (1.37), for various sizes of antennas usually employed in terrestrial microwave links and satellite communications.

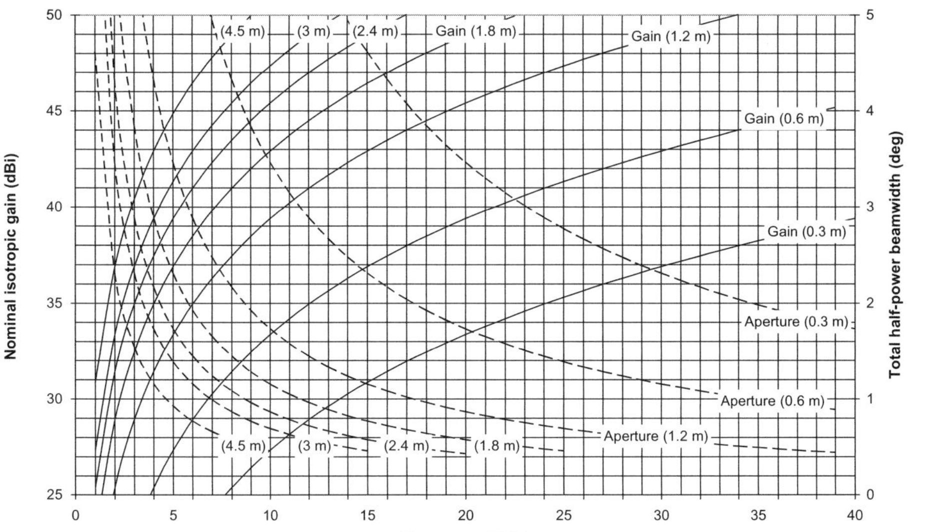

Figure 1.12 Nominal isotropic gain and total half-power beamwidth versus frequency.

When the apertures are not of revolution type, we can employ the general relation

$$G_{\max} = \eta \frac{4\pi(\mathrm{sr})}{\alpha_T \beta_T} = \eta \frac{4\pi(57.3°)^2}{\alpha_T \beta_T} = \eta \frac{41{,}253}{\alpha_T \beta_T} \tag{1.39}$$

or

$$G_{\max} = \eta \frac{4\pi D_\alpha D_\beta}{\lambda^2} = \frac{4\pi A_{\mathrm{eff}}}{\lambda^2}$$

where α_T = total half-power beamwidth corresponding to D_α dimension
β_T = total half-power beamwidth corresponding to D_β dimension
A_{eff} = effective aperture area

By taking account of the efficiency, we can employ the approximate formulas

$$G_{\max} \approx \begin{cases} \dfrac{25{,}000}{\alpha_T \beta_T} & \text{at 1–6 GHz} \\ \dfrac{30{,}000}{\alpha_T \beta_T} & \text{at 6–18 GHz} \end{cases}$$

Generally, the size of the antennas is determined by the microwave radio link budget, which is necessary to achieve the performances goals of the connection.

1.3.2 Reference Radiation Patterns

Radiation patterns must comply with rules of coordination defined by the International Telecommunication Union (ITU)[5] in order to reduce mutual interference as much as possible not only between microwave line-of-sight radio relay systems but also between radio relay systems and services of satellite communications. In the absence of particular features concerning the radiation pattern of antennas used in line-of-sight radio relay systems, ITU-R F.699 recommends for coordination aspects adopting the reference radiation pattern given in dBi by the following formulas, which are valid between 1 and 40 GHz in the far field:

[5]The complete texts of the ITU recommendations cited can be obtained from Union Internationale des Télécommunications, Secrétariat Général—Service des Ventes et Marketing, Place des Nations CH-1211, Genera 20, Switzerland.

• If $D/\lambda \leq 100$:

$$G(\varphi) = \begin{cases} G_{MAX} - 2.5 \times 10^{-3} \left[\frac{D}{\lambda} \varphi \right]^2 & \text{for } 0 < \varphi < \varphi_m \\ G_1 & \text{for } \varphi_m \leq \varphi < \frac{100\lambda}{D} \\ 52 - 10 \log \left[\frac{D}{\lambda} \right] - 25 \log \varphi & \text{for } \frac{100\lambda}{D} \leq \varphi < 48^\circ \\ 10 - 10 \log \left[\frac{D}{\lambda} \right] & \text{for } 48^\circ \leq \varphi \leq 180^\circ \end{cases} \tag{1.40}$$

• If $D/\lambda > 100$:

$$G(\varphi) = \begin{cases} G_{\max} - 2.5 \times 10^{-3} \left[\frac{D}{\lambda} \varphi \right]^2 & \text{for } 0 < \varphi < \varphi_m \\ G_1 & \text{for } \varphi_m \leq \varphi < \varphi_r \\ 32 - 25 \log \varphi & \text{for } \varphi_r \leq \varphi < 48^\circ \\ -10 & \text{for } 48^\circ \leq \varphi \leq 180^\circ \end{cases} \tag{1.41}$$

where $G(\varphi)$ = gain referred to isotropic antenna (dBi)
G_{Max} = maximum isotropic gain of main lobe (dBi)
φ = angle off the axis (deg)
D = diameter of antenna (m)
λ = wavelength (m)
G_1 = gain of first minor lobe (dBi), $= 2 + 15 \log(D/\lambda)$
φ_m $= (20\lambda/D)\sqrt{G_{\max} - G_1}$
ϕ_r $= 15.85[D/\lambda]^{-0.6}$

Figure 1.13 presents reference antenna radiation patterns related to $G_{\max}$ for the main-lobe and side-lobe envelopes which correspond to standard antennas. Antennas with high-performance radiation have a notably weaker side-lobe level as well as a higher front-to-back ratio. If only the maximum gain of the antenna is known, the ratio D/λ can be evaluated by using

$$20 \log \left[\frac{D}{\lambda} \right] \approx G_{MAX} - 7.7$$

When the total half-power beamwidth is known, $G_{\max}$ can be obtained by the relation

$$G_{\max} = 44.5 - 20 \log \alpha_T$$

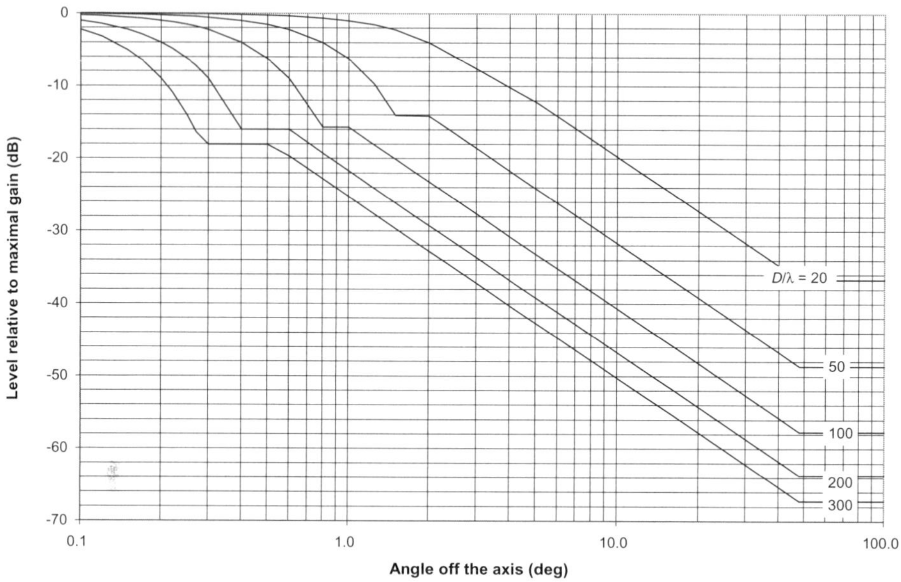

Figure 1.13 Reference antenna radiation patterns compared to $G_{\max}$.

In horn-type and "offset" antennas, we can employ the following relation, which is valid apart from the main lobe and for $\varphi < 90°$:

$$G = 88 - 30 \log \frac{D}{\lambda} - 40 \log \varphi \tag{1.42}$$

It is of course preferable to take into account the real or guaranteed radiation patterns of the antennas in the budget links and calculations related to jamming in order to avoid problems when using frequency sharing and frequency reuse techniques.

More specifically, concerning the coordination between fixed services by satellite, recommendation ITU-R S.465 gives the following values for antenna radiation patterns, expressed in dBi, in the frequency bands ranging between 2 and 30 GHz:

$$G = \begin{cases} 32 - 25 \log \varphi & \text{for } \varphi_{\min} \le \varphi \le 48° \\ -10 & \text{for } 48° \le \varphi \le 180° \end{cases} \tag{1.43}$$

where $\varphi_{\min}$ equals 1° or $100\lambda/D$ by taking the highest value. In systems used before 1993 and for $D/\lambda \le 100$

$$G = \begin{cases} 52 - 10\log\left(\dfrac{D}{\lambda}\right) - 25\log\varphi & \text{for } \left(\dfrac{100\lambda}{D}\right) \le \varphi \le 48^\circ \\ 10 - 10\log\left(\dfrac{D}{\lambda}\right) & \text{for } 48^\circ \le \varphi \le 180^\circ \end{cases}$$

The gain variation due to reflector surface errors can be obtained by the following approximate formula, by considering the effective value of the irregularities ε:

$$\Delta G = 10\log\left[\exp\left(-\frac{4\pi\varepsilon}{\lambda}\right)^2\right] \approx -686\left(\frac{\varepsilon}{\lambda}\right)^2 \tag{1.44}$$

It is advisable to make sure that the radiation pattern of the selected antenna makes it possible to respect the maximum levels of equivalent isotropic radiated power [EIRP, in decibels related to 1 W (dBW)] or maximum equivalent isotropic radiated spectral densities (dBW per hertz) which are authorized by the ITU rules of radiocommunications concerning both useful signal and nonessential radiations.

The values to be taken into account depend on the frequency band as well as the type of service to protect (fixed or mobile service using, e.g., radio relay systems or satellites); for example:

- The direction of the maximum radiation of an antenna delivering higher EIRP than +35 dBW must deviate by at least 2° the orbit of the geostationary satellites in the frequency bands ranging between 10 and 15 GHz.
- If it is not possible to conform to this recommendation, the EIRP should not exceed +47 dBW in any direction deviating by less 0.5° the geostationary orbit or from +47 to +55 dBW (8 dB per degree) in any direction ranging between 0.5° and 1.5°.
- The power provided to the antenna should not exceed +13 dBW in the frequency bands ranging between 1 and 10 GHz and +10 dBW in the higher-frequency bands.

These arrangements must be taken at both emission and reception sites in order to ensure the highest possible protection against jamming.

1.3.3 Characteristics of Polarization

In practice, the majority of antennas radiate in linear or circular polarization, which are particular cases of elliptic polarization as described in Section 1.1.5.

For any elliptic polarization, we can define an orthogonal polarization whose direction of rotation is opposite; two waves with orthogonal polarization in theory being insulated, the same antenna can simultaneously receive and/or emit two carriers at the same frequency with polarizations horizontal and vertical or circularly right and left. We can also show that any radio wave with elliptic polarization can be regarded as the sum of two orthogonal components, for example, of two waves with perpendicular linear polarization or two waves circularly right and left polarized.

In ordinary radio communication systems, waves are usually completely polarized and the electric field corresponding to the useful signal is great compared to the cross-polarized unwanted signal; under such conditions, the moduli of the orthogonal components are close to the semimajor and semiminor axes of the polarization ellipse presented in Figure 1.4.

Elliptic polarization may thus be characterized by the maximum and minimum levels according to the nominal polarization and its opposite in the following parameters:

- Axial ratio:

$$\mathrm{AR} = \frac{E_{\max}}{E_{\min}} \tag{1.45}$$

- Ellipticity ratio:

$$\mathrm{ER} = \frac{\mathrm{AR}+1}{\mathrm{AR}-1} = \frac{E_{\max}+E_{\min}}{E_{\max}-E_{\min}} \tag{1.46}$$

  or, in decibels,

$$\mathrm{ER} = 20\log\left[\frac{E_{\max}+E_{\min}}{E_{\max}-E_{\min}}\right]$$

An important feature of the radio wave consists in its purity of polarization, that is, the relationship between the copolarized component, which represents the useful signal, and the cross-polarized component, which constitutes the unwanted signal. The relationship between the useful signal and the unwanted one is called cross-polarization discrimination[6] XPD, expressed in decibels and given by the relation

[6]Between orthogonally polarized signals, a cross-polarization discrimination on the order of 30–40 dB can be expected.

$$\mathrm{XPD} = 20\log[\mathrm{AR}] = 20\log\left(\frac{\mathrm{ER}+1}{\mathrm{ER}-1}\right) \tag{1.47}$$

For an ellipticity ratio lower or equal to 3 dB, we can also employ the following approximate formulas with XPD and ER expressed in decibels:

$$\mathrm{XPD} = 24.8 - 20\log[\mathrm{ER}] \qquad \mathrm{ER} = 17.37\ 10^{-\mathrm{XPD}/20}$$

Due to its imperfections, an antenna will thus generate some useful signal on the regular polarization at the same time as an unwanted signal on the orthogonal polarization, which may parasitize another carrier; this feature is related to the purity of polarization at the transmission side. In the same way, an antenna may receive some useful signal on one polarization at the same time as some part of another signal transmitted on the opposite polarization; this feature concerns the cross-polarization isolation at the reception side.

1.4 FREE-SPACE LOSS AND ELECTROMAGNETIC FIELD STRENGTH

1.4.1 Attenuation of Propagation

Consider an isotropic source supplied with a transmitter of power P_E and an isotropic reception antenna located at a distance d. The received power P_R is the product of the power flux density created at the distance d by the effective aperture area A_e of the reception antenna. Starting from relations (1.11) and (1.34), we can write

$$P_R = \frac{P_E A_e}{4\pi d^2} \tag{1.48}$$

where A_e = effective area of isotropic radiator (m^2) = $\lambda^2/(4\pi)$
d = distance (m)
λ = wavelength (m)

The free-space attenuation A_{FS} between two isotropic antennas is thus equal to

$$\frac{P_R}{P_E} = \frac{1}{4\pi d^2}\frac{\lambda^2}{4\pi}$$

that is,

$$\frac{P_R}{P_E} = \left[\frac{\lambda}{4\pi d}\right]^2 = \frac{1}{A_{\mathrm{FS}}} \tag{1.49}$$

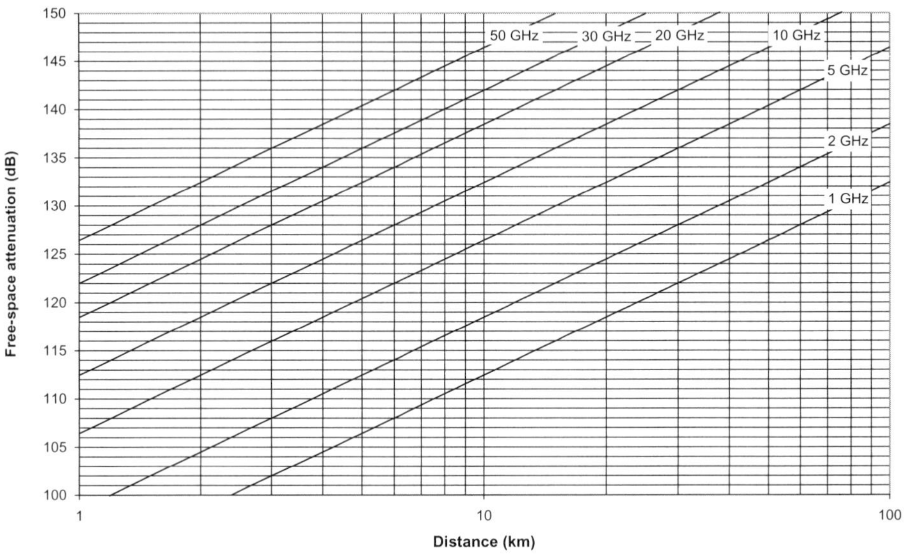

Figure 1.14 Free-space attenuation versus distance.

from which, expressed in decibels,

$$A_{\mathrm{FS}} = 20\log\left(\frac{4\pi d}{\lambda}\right)$$

Supposing, now, that the transmitting antenna has a gain G_E compared to an isotropic[7] radiator and the reception antenna a gain G_R, we obtain the general relation of free-space loss according to the wavelength and the distance expressed in meters:

$$\frac{P_R}{P_E} = G_E G_R\left[\frac{\lambda}{4\pi d}\right]^2 \tag{1.50}$$

Figure 1.14 shows the free-space attenuation versus distance at various frequencies.

1.4.2 Electromagnetic Field Strength

The product of the emitted power and the gain of the transmitting antenna corresponds to the EIRP, expressed in Watts, and the received power can still be expressed by the relation

[7]It is shown that in free space the absolute gain of a Hertz doublet is 1.75 dB and that of a half-wave doublet is 2.15 dB.

$$P_R = \text{EIRP}\,G_R\left[\frac{\lambda}{4\pi d}\right]^2 \tag{1.51}$$

where

$$\text{EIRP} = P_E G_E \tag{1.52}$$

In the transmission omnidirectional-type a system or one that is common to several receivers, such as broadcasting or telecommunications by satellite, it may be preferable to express the field strength in terms of power flux density or of electric field or even of magnetic field by using the following relations for free space:

$$\text{Power flux density:}\quad S(\text{W m}^{-2}) = \frac{\text{EIRP}}{4\pi d^2} = \frac{E^2(\text{V m}^{-1})}{120\pi} = \frac{4\pi p_{\text{riso}}}{\lambda^2}$$

$$\text{Electric field:}\quad E(\text{V m}^{-1}) = \frac{\sqrt{30\text{EIRP}}}{d}$$

$$\text{Magnetic field:}\quad H(\text{A m}^{-1}) = \frac{E(\text{V m}^{-1})}{120\pi} = \sqrt{\frac{\text{EIRP}}{480\pi^2 d^2}}$$

where P_{riso} is the power, expressed in watts, collected by an isotropic radiator. Certain authors also use the unit dBμV, which is equivalent to the power received by comparison to that which would be developed by a field of 1 μV m^{-1} expressed in decibels. Then

$$1\,\mu\text{V m}^{-1} = 10^{-6}\,\text{V m}^{-1}$$

from which

$$P(\mu\text{V m}^{-1}) = 10^{-12}\,P(\text{V m}^{-1})$$

or

$$0\,\text{dB}(\mu\text{V m}^{-1}) \Leftrightarrow -120\,\text{dB}(\text{V m}^{-1})$$

$$0\,\text{dB}(\mu\text{V m}^{-1}) \Leftrightarrow -145.8\,\text{dB}(\text{W m}^{-2})$$

$$0\,\text{dB}(\text{W m}^{-2}) \Leftrightarrow 0\,\text{dBW} + 10\log\left(\frac{\lambda^2}{4\pi}\right)$$

Figure 1.15 shows, for example, the relationship between the electric field expressed in Volts per meter and the power flux density in decibels (watts per square meter). In addition, the unit used to characterize transmitter–antenna

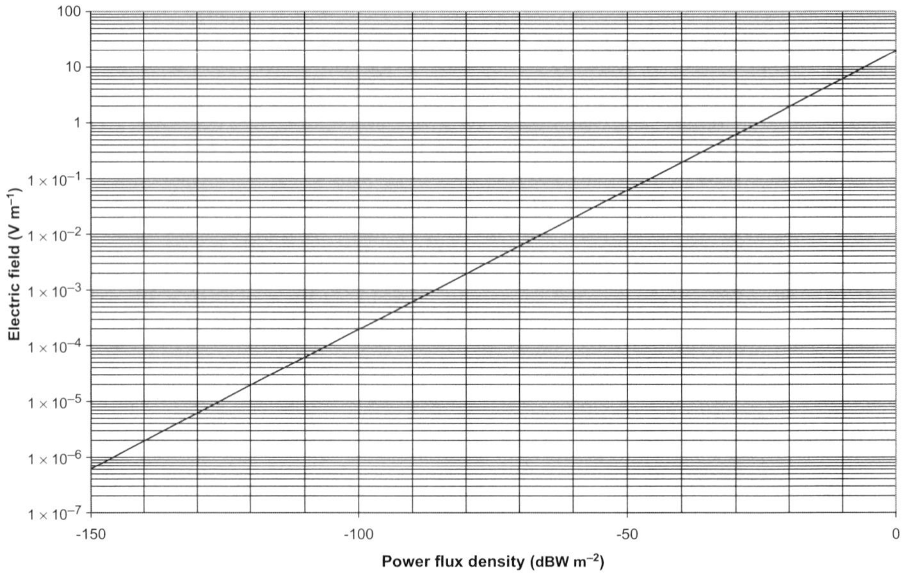

Figure 1.15 Relationship between electric field and power flux density.

systems for broadcasting purposes, called cymomotive force, is the product of the electric field and the distance and is usually expressed in volts and, according to the type of reference antenna, has the expression

$$E(\text{mV m}^{-1})d(\text{km}) = k\sqrt{P(\text{kW})}$$

where

$$K = \begin{cases} 173 & \text{for isotropic radiator} \\ 212 & \text{for Hertz doublet} \\ 222 & \text{for half-wave doublet} \end{cases}$$

Finally, a unit of power often employed, dBm, refers to decibels relative to 1 mW of power, that is,

$$0\,\text{dBm} \leftrightarrow -30\,\text{dBW} \qquad 0\,\text{dBW} \leftrightarrow 30\,\text{dBm}$$

1.5 REFLECTOR AND PASSIVE REPEATER

1.5.1 Reflector in Far Field

A reflector of surface S, as illustrated in Figure 1.16, reflects, under an angle of incidence α referred to the normal to the reflector, an electromagnetic wave

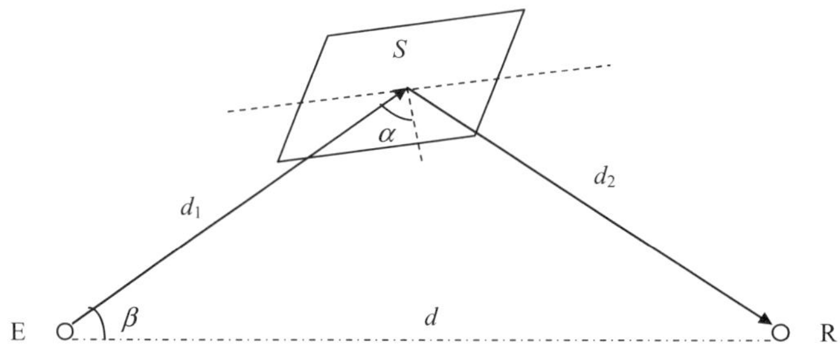

Figure 1.16 Reflector in far field.

issued from a transmitter located at a distance d_1 toward a receiver located at a distance d_2 with an efficiency of reflection η. In the case of a plane reflector, the reflected wave presents the same coherence of phase as the incidental wave and, because of the symmetry inherent in the reflection, the apparent area S_a (projected surface) seen by the transmitter is identical to that which is seen by the receiver and can be written as

$$S_a = S\cos\alpha \tag{1.53}$$

Indicating by G_{rp} the directivity of the reflector in the direction of the receiver, we can write, according to relation (1.25), that

$$G_{rp} = \frac{\eta 4\pi S_a}{\lambda^2} \tag{1.54}$$

Assuming that the antennas used at the emission and at the reception are isotropic and that the reflector is placed in the far field of both antennas, we can write that the ratio of the received power to the emitted power is equal to

$$\left(\frac{P_R}{P_E}\right)_{\text{iso}} = \frac{\eta S_a}{4\pi d_1^2} G_{rp} \left[\frac{\lambda}{4\pi d_2}\right]^2$$

that is, the product of flux density per unit power created at the level of the reflector at distance d_1 according to relation (1.11) by

- the effective area of the reflector $\eta\ Sa$,
- the gain G_{rp} of the reflector toward the receiver compared to an isotropic antenna according to relation (1.54), and
- the attenuation between isotropic antennas at distance d_2 according to relation (1.49), from which we get the general relation for isotropic radiating–receiving antennas which is independent of wavelength:

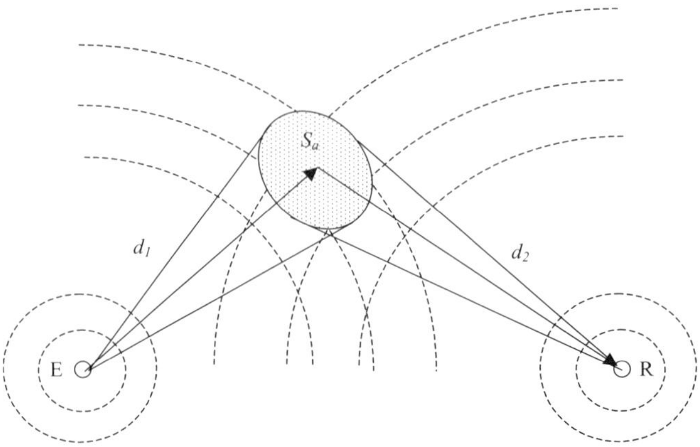

Figure 1.17 Transmitter and receiver solid angles.

$$\left(\frac{P_R}{P_E}\right)_{\text{iso}} = \left[\frac{\eta S_a}{4\pi d_1 d_2}\right]^2 \tag{1.55}$$

We easily find the same result when considering, as illustrated in Figure 1.17, solid angles seen by the transmitter and the receiver of the same effective surface ηS_a; the relationship between the received power and the emitted power is then equal to the product of the ratios:

- $\eta S_a/(4\pi d_1^2)$ (effective surface of reflector/surface of sphere of radius d_1)
- $\eta S_a/(4\pi d_2^2)$ (effective surface of reflector/surface of sphere of radius d_2)

The total reflection efficiency η^2 is a function of the dimensions of the reflector, its surface condition, and the incidence angle. By supposing that the radiating and receiving antennas are not isotropic but have respectively gains G_E and G_R, we obtain the general relation

$$\frac{P_R}{P_E} = G_E G_R \left[\frac{\eta S_a}{4\pi d_1 d_2}\right]^2 \tag{1.56}$$

or

$$\frac{P_R}{P_E} = G_E G_R \left[\frac{\eta S \cos\alpha}{4\pi d_1 d_2}\right]^2$$

The reflector gain is defined from

$$\left(\frac{P_R}{P_E}\right)_{\text{iso}} = G_P \left[\frac{\lambda}{4\pi d_1}\right]^2 \left[\frac{\lambda}{4\pi d_2}\right]^2 = \left[\frac{\eta S \cos\alpha}{4\pi d_1 d_2}\right]^2$$

from which

$$G_P = \left[\frac{4\pi\eta S\cos\alpha}{\lambda^2}\right]^2 \tag{1.57}$$

We can also deduce the total equivalent gain G_{Pe} of the link comprising a far-field reflector compared to the direct link which would be carried out in free space between the transmitter and the receiver using isotropic antennas:

$$\left(\frac{P_R}{P_E}\right)_{\text{iso}} = \left[\frac{\eta S_a}{4\pi d_1 d_2}\right]^2 \approx G_{Pe}\left[\frac{\lambda}{4\pi d}\right]^2$$

that is,

$$G_{Pe} \approx \left[\frac{\eta S_a}{4\pi d_1 d_2}\right]^2 \left[\frac{4\pi d}{\lambda}\right]^2$$

where the first term represents the total attenuation of propagation in two hops by reflection on the reflector according to relation (1.55) and the second term corresponds to the attenuation of propagation between isotropic antennas over the overall length of the link in free space according to the relation (1.49) while supposing $d \approx d_1 + d_2$, resulting in

$$G_{Pe} \approx \left[\frac{\eta S_a d}{\lambda d_1 d_2}\right]^2 \tag{1.58}$$

Alternatively, if

$$d = (d_1 + d_2)\frac{\sin\alpha}{\sin(\alpha+\beta)}$$

then

$$G_{Pe} \approx \left[\frac{\eta S\cos\alpha(d_1 + d_2)\sin\alpha}{\lambda d_1 d_2 \sin(\alpha+\beta)}\right]^2 \tag{1.59}$$

using the meter as the unit.

Apart from the zones close to the source of emission or the receiving antenna, the gain of the system with the reflector compared to the direct link is lower than unity and is a minimum value when $d_1 = d_2$, that is, when the reflector is in the middle of the connection, which is the case for the target radar.

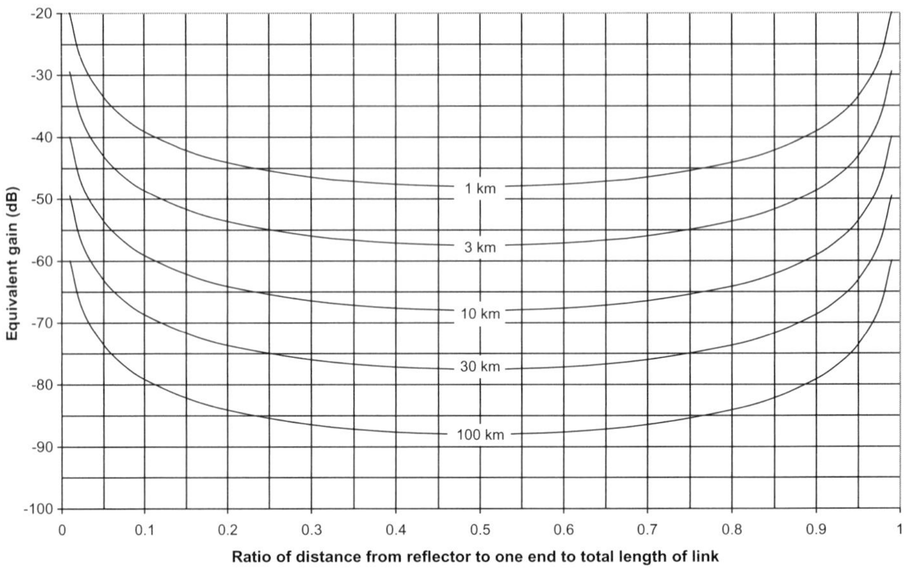

Figure 1.18 Equivalent gain of reflector of reduced surface $S_a/\lambda = 1$ in far field versus length of connection and relative position to one of the ends.

Figure 1.18 illustrates the equivalent gain G_{Pe} of a system using a reflector of reduced surface $S_a/\lambda = 1$ versus the relationship between the distance from the reflector to one of the ends and the overall length of the link, d_1/d or d_2/d, assuming $d \approx d_1 + d_2$. To determine the equivalent gain of a given system using a reflector, of which the dimensions and the relative position to the ends are known, it is enough to add to the results obtained on the graph 1.18 the quantity $[\eta S_a/\lambda]^2$ expressed in decibels with S_a in square meters and λ in meters.

The use of a far-field reflector is thus of interest only if it is set up close to one end of the link and its performance has to be compared with the attenuation resulting from diffraction on the obstacle to ensure there is no risk of self-interference. We will avoid, consequently, placing the reflector in the same plane as the two stations in order to benefit from the discrimination brought on by the antennas. Moreover, since the reflection polarizes the electromagnetic waves, the plane of reflection will have to coincide as much as possible with the plane of polarization for, by supposing that they differ by an angle θ, it would result in a significant reduction in gain, which can be evaluated using the general relation for antennas:

$$G(\theta) = G_0 \cos^2 \theta \tag{1.60}$$

For a radar-type reflector at a distance d which returns the wave toward the source, we indicate the product $G_{rp}\eta S_a$ by the equivalent target cross section S_{eq} and obtain the equation of the radar for an isotropic source:

$$\frac{P_R}{P_E} = S_{\text{eq}} \frac{\lambda^2}{(4\pi)^3 d^4}$$

where $S_{\text{eq}} = G_{rp}\eta S_a$. If the radar antenna has a gain G, we then obtain the general equation for a radar system:

$$\frac{P_R}{P_E} = G^2 S_{\text{eq}} \frac{\lambda^2}{(4\pi)^3 d^4} \tag{1.61}$$

1.5.2 Passive Repeater in Far Field

A passive repeater consists of two antennas placed at a distance d_1 from the transmitter and d_2 from the receiver and connected back to back as presented in Figure 1.19. By supposing that the transmitting antenna has a gain G_E compared to an isotropic radiator and the receiving antenna a gain G_R, just as the antennas constituting the passive repeater have a gain G'_E and G'_R, and by considering a loss in the connecting feeder P_f, we can write the general relation of global loss, which includes two sections of length d_1 and d_2 in cascade, according to relation (1.50):

$$\frac{P_R}{P_E} = \frac{G_E G'_R}{P_f}\left[\frac{\lambda}{4\pi d_1}\right]^2 G'_E G_R \left[\frac{\lambda}{4\pi d_2}\right]^2$$

That is,

$$\frac{P_R}{P_E} = \frac{G_E G_R G'_E G'_R}{P_f}\left[\frac{\lambda^2}{(4\pi)^2 d_1 d_2}\right]^2 \tag{1.62}$$

This expression can also be written if both repeater antennas are identical:

$$\frac{P_R}{P_E} = \frac{G_E G_R}{P_f}\left[\frac{\eta A}{4\pi d_1 d_2}\right]^2 \qquad G'_E = G'_R = \frac{\eta 4\pi A}{\lambda^2}$$

where A is the area of the antennas back to back. We find the same formula, independent of the wavelength, than that one for the far-field reflector, (1.56),

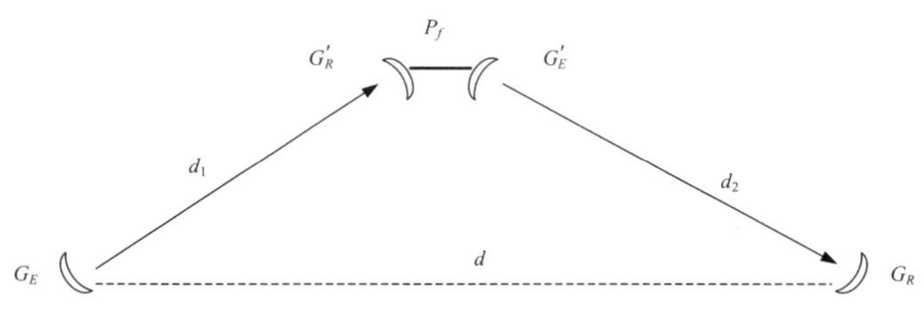

Figure 1.19 Passive far-field repeater composed of back-to-back antennas.

by replacing the apparent surface of the reflector by the effective area of the antennas connected back to back.

The gain of the passive repeater, G_P, is then equal to

$$G_P = \frac{G'_E G'_R}{P_f} \tag{1.63}$$

or

$$G_P = \left[\frac{4\pi}{\lambda^2}\right]^2 \frac{\eta_E A_E \eta_R A_R}{P_f}$$

and can also be expressed according to the antenna diameters D'_E and D'_R:

$$G_P = \frac{\eta_E D'^2_E \eta_R D'^2_R}{P_f}\left(\frac{\pi}{\lambda}\right)^4$$

with

$$A_E = \frac{1}{4}\pi D'^2_E \qquad A_R = \frac{1}{4}\pi D'^2_R$$

that is, by using identical antennas of diameter D' and efficiency η':

$$G_P = \frac{1}{P_f}\left[\frac{\eta' D'^2 \pi^2}{\lambda^2}\right]^2$$

As in the case of the far-field reflector, we can determine the total equivalent gain G_{Pe} of the link comprising a passive far-field repeater compared to the direct link which would be carried out in free space between the transmitter and the receiver using isotropic antennas while supposing $d \approx d_1 + d_2$:

$$\frac{P_R}{P_E} = G_P\left[\frac{\lambda^2}{(4\pi)^2 d_1 d_2}\right]^2 \approx G_{Pe}\left[\frac{\lambda}{4\pi d}\right]^2$$

from which

$$G_{Pe} \approx G_P\left[\frac{\lambda d}{4\pi d_1 d_2}\right]^2 \tag{1.64}$$

or

$$G_{Pe} \approx \frac{1}{P_f}\left[\frac{\eta' D'^2 \pi d}{4\lambda d_1 d_2}\right]^2$$

It appears that the only advantage of the passive repeater on the reflector lies in the fact that the back-to-back antennas are pointed respectively toward the transmitter and the receiver, which makes the device independent of the angle of deflection and makes it possible to reduce its size for an equivalent gain; however, the efficiency of a reflector can be higher than that of an antenna, which results in a higher side-lobe level.

As above, the equivalent gain G_{Pe} of the passive repeater can easily be deduced from Figure 1.18 by replacing the apparent surface of the reflector by that of the repeater antennas, that is, the quantity $\eta S_a/\lambda$ by $\eta'\pi D'^2/4\lambda$.

1.5.3 Reflector in Near Field (Periscope)

The gain of the far-field reflector, as defined previously, tends toward infinity when we bring it closer to one of the two ends of the link; this is of course not the case in reality. When the reflector, also called a "mirror," is placed in the near field of an antenna, the device constitutes a periscope, as illustrated in Figure 1.20.

The equivalent gain is defined as

$$G_{Per} = \left(\frac{G_{\text{passive}}}{G_{\text{antenna}}}\right)\left(\frac{\eta_{\text{illumination}}}{P_{\text{spillover}}}\right)$$

where the gain of the passive repeater corresponds to that of the projected aperture of area S_a given by the relation (1.25) and that of the antenna by relation (1.26):

$$G_{\text{passive}} = \frac{4\pi S_a}{\lambda^2} \qquad G_{\text{antenna}} = \eta\left[\frac{\pi D}{\lambda}\right]^2$$

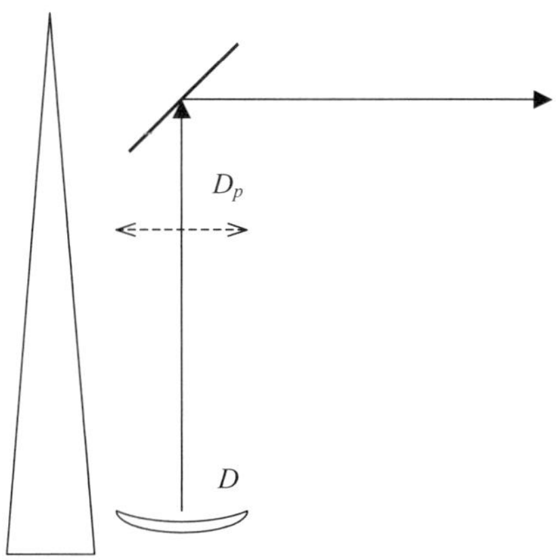

Figure 1.20 Periscope.

and the efficiency of illumination ($\eta_{\text{Illumination}}$) and the loss by spillover ($p_{\text{Spill-over}}$) depend on the height of the reflector, the relative dimensions of the antenna and reflector, and the form of the latter. Generally, the gain of this device is close to unity, which makes it possible for the unit to be freed from the usual branching losses by coaxial cables or waveguides on the height of the tower, or even greater than unity. Preferably the periscope must be placed in the meridian plane containing the antennas of both stations in order to not modify the plane of polarization. The reflector can be rectangular, octogonal, elliptic, plane, or curved, according to whether we want to improve the gain or the radiation pattern or both at the same time, the worse results being obtained with the rectangular form. For example, Figure 1.21 presents relative gain compared to the nominal gain of the antenna calculated for an elliptic plane reflector for various ratios D/D_p of respective size according to the parameter P defined by

$$P = \frac{\lambda H}{D_p^2}$$

where λ = wavelength (m)
D = antenna diameter (m)
D_p = projected diameter of reflector (m)
H = relative height of reflector (m)

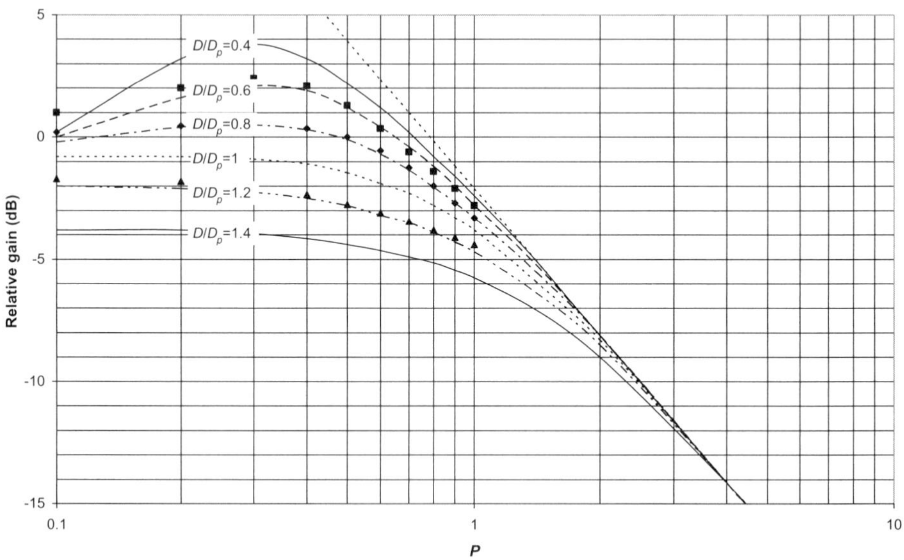

Figure 1.21 Gain of periscope comprising elliptic plane reflector versus parameter *P*.

These curves converge toward the asymptote, defined by the relation[8]:

$$20\log\left(\frac{\pi D_P^2}{4\lambda H}\right)$$

Also reproduced in the figure are experimental values obtained using a 46-cm antenna operating at the frequency of 8.4 GHz in combination with three elliptic plane reflectors of different sizes with the following projected diameters and relevant size ratios:

77 cm ($D/Dp = 0.6$)
57 cm ($D/Dp = 0.8$)
38 cm ($D/Dp = 1.2$)

For $0.6 \le D/D_p \le 1.2$ we can use the empirical formula

$$G_{Per} = \begin{cases} [0.3125\times 10^{0.9(Dp/D)} + 0.8](1+\log P) - 0.75\left(\dfrac{D}{D_p}\right)^{5.5} & P<2 \\ \quad -2.703\times 10^{0.41(Dp/D)}(1+\log P)^2 & \\ 20\log\left(\dfrac{\pi}{4P}\right) - \dfrac{D}{D_p(2+\log P)} & P\ge 2 \end{cases}$$

In addition to the significant gain corresponding to the suppression of the usual branching losses on the tower height, the disadvantages of this system are as follows:

- The dimensions of the reflector are in general higher than the diameter of the antenna, which can result in an increased catch of wind.
- The loss of gain due to the misalignment of the reflector varies by twice that of a parabolic antenna placed under the same conditions, since it is about a reflection, requiring a more rigid tower.
- The characteristics of radiation with reference to the side-lobe level and especially the front-to-back ratio are in general worse than those of a parabolic antenna and the presence of the tower can cause parasitic reradiations.

1.6 MODEL OF PROPAGATION

1.6.1 Spherical Diffraction

Previously a radiated wave was shown to propagate in all directions of space and that, to connect two points in space called transmitter and receiver, we

[8]This relation can be related to the relation (1.58), which corresponds to the equivalent gain of the device comprising a far-field reflector.

introduced the concept of a radioelectric ray that is characterized by an electromagnetic field and a wavelength.

According to wave motion theory, electromagnetic waves are propagated gradually due to the phenomenon of spherical diffraction, which consists in considering that each point of a wave front reemits in its turn in all directions; Augustin Fresnel showed that backward reradiation of all these point sources was destroyed and that their forward contribution depended on their respective position on the wave front.

1.6.2 Fresnel's Ellipsoid

Consider a source of emission E, a receiver R, and any plane (P) perpendicular to the line joining E and R, as shown in Figure 1.22, and determine the properties of the field received in R. James Clerck Maxwell's theory makes it possible to calculate the field received in R starting from the field created by a source E in all the points of the space that separates E and R. We can see that all points of the same circle centerd on the axis ER contribute by the same share since they are at the same distance from E and R and the waves they produce are consequently in phase. By neglecting the aspects related to polarization, the point at R thus receives energy coming from all the points of the plane (P) which one can break up into concentric rings by taking account of the phase which rises from the difference in pathlength, ΔL, between the axis ER, which constitutes the shortest path, and any path EPR given by the relation

$$\Delta L = EP + PR - ER$$

The contribution is positive if ΔL is smaller than a half wavelength or an add number of it and negative if not; the space between the source E and the receiver R can thus be divided into concentric ellipsoids of focus at E and R, called Fresnel's ellipsoids, such as

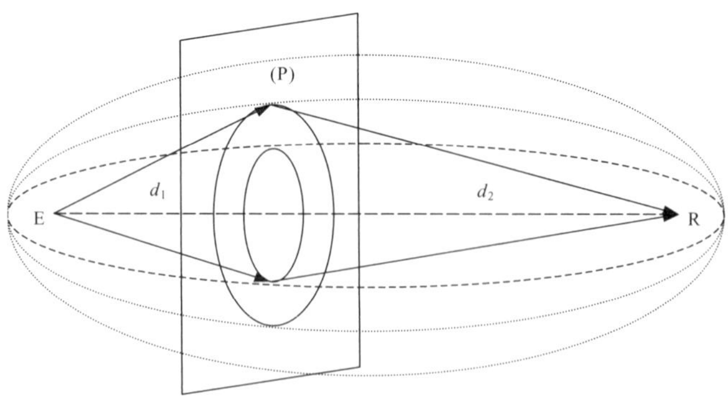

Figure 1.22 Fresnel's ellipsoid.

$$\Delta L = n\frac{\lambda}{2}$$

The first ellipsoid, obtained for $n = 1$, contains most of energy and delimits the free space. To determine the radius of each ellipsoid, we can write

$$n\frac{\lambda}{2} = \sqrt{d_1^2 + R_F^2} + \sqrt{d_2^2 + R_F^2} - d \qquad d \approx d_1 + d_2$$

from which, by limited development,

$$n\frac{\lambda}{2} \approx \frac{R_F^2}{2}\left(\frac{1}{d_1} + \frac{1}{d_2}\right) \qquad R_F \approx \sqrt{\frac{d_1 d_2 n\lambda}{d_1 + d_2}} \tag{1.65}$$

where d_1, d_2, and λ are in meters. The equatorial radius, obtained for $d_1 = d_2$, has as a value

$$R_{FM} \approx \frac{1}{2}\sqrt{dn\lambda}$$

1.7 REFLECTION AND REFRACTION

1.7.1 General Laws

An electromagnetic wave undergoes the laws of reflection and refraction at the passage of a surface separating two different media; it will be assumed that this surface is large with respect to the wavelength, just as its radius of curvature.

Figure 1.23 illustrates an electromagnetic wave reaching the surface of separation between two media of respective refractive indexes n_1 and n_2.

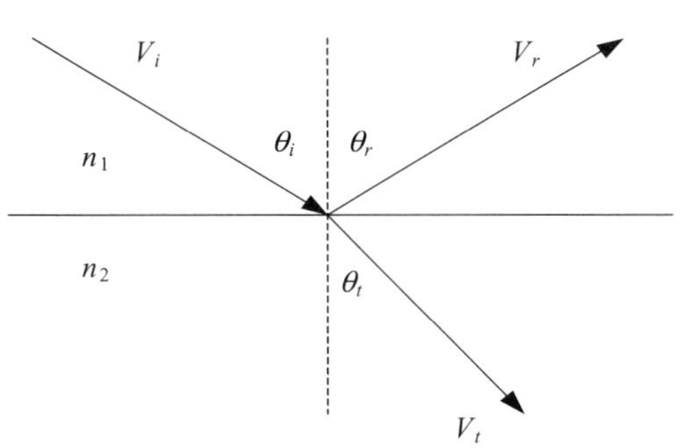

Figure 1.23 Reflection and refraction of plane electromagnetic wave.

Calling V_i, V_r, and V_t the respective velocities of propagation of the rays incident, reflected, and refracted, we can write

$$\frac{V_i}{\sin\theta_i} = \frac{V_r}{\sin\theta_r} = \frac{V_t}{\sin\theta_t} \tag{1.66}$$

where

$$V_i = \frac{c}{n_1} \qquad V_r = \frac{c}{n_1} \qquad V_t = \frac{c}{n_2}$$

from which $V_i = V_r$ and $\theta_i = \theta_r$ and

$$n_1 \sin\theta_i = n_2 \sin\theta_t \qquad \text{(Descartes' law of refraction)}$$

From relation (1.13), for each medium, we can also write

$$\frac{E}{H} = \sqrt{\frac{\mu}{\varepsilon}} = Z_0\sqrt{\frac{\mu_r}{\varepsilon_r}}$$

where Z_0 is the impedance of the vacuum and

$$c = \frac{1}{\sqrt{\varepsilon_0\mu_0}} \qquad v = \frac{c}{\sqrt{\varepsilon_r\mu_r}}$$

As μ is in general close to unity, one has $\sqrt{\varepsilon_r} \approx n$.

The coefficients of reflection E_r/E_i for horizontal polarization R_H and vertical polarization R_V are given by the fundamental relations of Fresnel:

$$R_H = \frac{\sin(\theta_i - \theta_t)}{\sin(\theta_i + \theta_t)} \qquad R_V = \frac{\tan(\theta_i - \theta_t)}{\tan(\theta_i + \theta_t)}$$

It appears that the coefficient of reflection in vertical polarization is canceled for

$$\theta_i + \theta_t = \frac{1}{2}\pi \quad \text{to which corresponds Brewster's angle}$$

$$\tan\theta_i = \frac{n_2}{n_1} \quad \text{for which the wave is completely refracted}$$

Using the angle of inclination φ, complementary to the angle of incidence θ_i, we obtains

$$R_H = \frac{\sin\varphi - \sqrt{(n_2/n_1)^2 - \cos^2\varphi}}{\sin\varphi + \sqrt{(n_2/n_1)^2 - \cos^2\varphi}}$$
$$R_V = \frac{(n_2/n_1)^2\sin\varphi - \sqrt{[(n_2/n_1)^2 - \cos^2\varphi]/\eta^2}}{(n_2/n_1)^2\sin\varphi + \sqrt{[(n_2/n_1)^2 - \cos^2\varphi]/\eta^2}} \tag{1.67}$$

For two media with similar indexes of refraction, for example two layers of the atmosphere separated by a plane surface through which the refractive index undergoes a small discontinuity *dn* positive or negative, the preceding relations become, with $(n_2/n_2)^2 \approx 1 + 2\,dn$,

$$R_H = \frac{\sin\varphi - \sqrt{\sin^2\varphi + 2dn}}{\sin\varphi + \sqrt{\sin^2\varphi + 2dn}}$$
$$R_V = \frac{(1+2dn)\sin\varphi - \sqrt{\sin^2\varphi + 2dn}}{(1+2dn)\sin\varphi + \sqrt{\sin^2\varphi + 2dn}} \tag{1.68}$$

Figure 1.24 presents the magnitude of the reflection coefficient according to the inclination angle and the discontinuity *dn* positive and negative for horizontal polarization. The reflection coefficient has practically the same value for both polarizations and reflection is total when the discontinuity is negative for all angles lower than the limit angle given by

$$\varphi_{\text{lim}} = \sqrt{2|dn|} \tag{1.69}$$

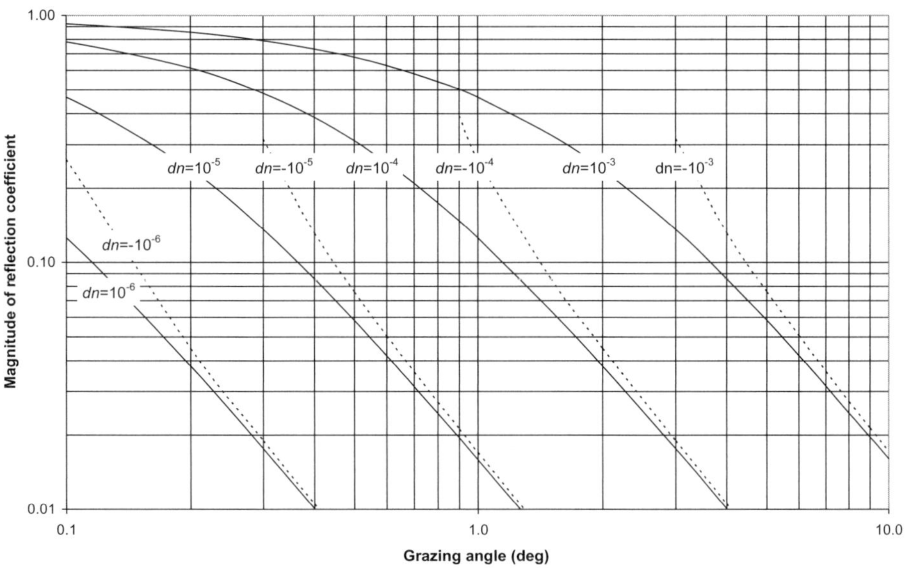

Figure 1.24 Reflection due to discontinuity of refractive index between two dielectric media.

When the surface of separation is between a dielectric medium and a conducting medium of conductivity σ, the reflection coefficient is given by the following formulas (ITU-R Rep.1008):

$$R_H = \frac{\sin\varphi - \sqrt{\eta - \cos^2\varphi}}{\sin\varphi + \sqrt{\eta - \cos^2\varphi}}$$
$$R_V = \frac{\sin\varphi - \sqrt{(\eta - \cos^2\varphi)/\eta^2}}{\sin\varphi + \sqrt{(\eta - \cos^2\varphi)/\eta^2}} \quad (1.70)$$

with the complex permittivity $\eta = \varepsilon - j60\sigma\lambda$.

The amplitude and phase difference of the reflected wave vary according to the nature of the terrain, the frequency, the grazing angle, and the polarization, as shown in Figures 1.25*a* and 1.25*b*, which illustrate reflections on the sea and on average ground. We can see that the magnitude of the reflection coefficient is close to unity for radio waves that are horizontally polarized either on the sea or on the ground.

For the majority of microwave links where the angles of reflection on the ground or in the low layers of the atmosphere are small, the magnitude of the reflection coefficient and the phase difference of the reflected wave are respectively equal to unity and 180° for horizontal polarization and close to these values for vertical polarization at frequencies greater than gigahertz and angles of reflection less than a few degrees. The reflection coefficient also depends on other factors such as divergence, related to the terrestrial curvature, roughness of the ground, and size of the reflection zone.

We have seen that most of the energy of an electromagnetic wave was contained in the first Fresnel's ellipsoid; it is necessary thus to take account of the Fresnel's ellipsoid of the reflected wave, which corresponds to that generated by a fictitious transmitter E′ symmetrical to transmitter E relative to the surface of reflection, as shown in Figure 1.26.

1.7.2 Divergence Factor

The divergence factor is given by the relation

$$D = \frac{1}{\sqrt{1 + 2d_1d_2/(Rd\sin\varphi)}} \quad (1.71)$$

where R = radius of curvature of reflective surface (km)

$\varphi \approx \dfrac{H_1 + H_2}{d} - \dfrac{d}{4R}\left[1 + \left(\dfrac{d_1 - d_2}{d_1 + d_2}\right)^2\right]$ (mrad)

H_1, H_2 = heights of antennas above reflection plane (m)

d, d_1, d_2 = distances (km)

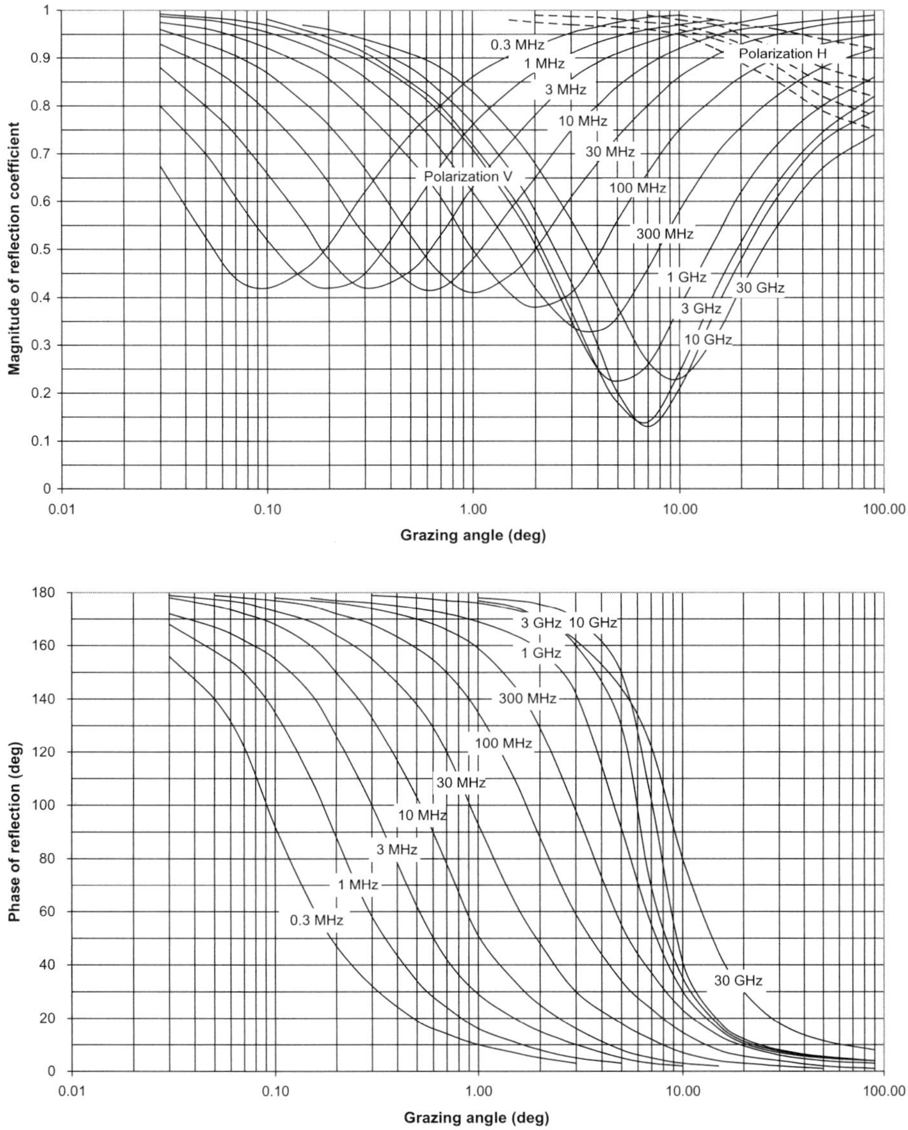

Figure 1.25 (*a*) Reflection on sea: magnitude and phase of coefficient of reflection of plane surface versus grazing angle in vertical (V) and horizontal (H) polarization. (*b*) Reflection on ground (average ground): Magnitude and phase coefficient of reflection of plane surface versus grazing angle in vertical (V) and horizontal (H) polarization.

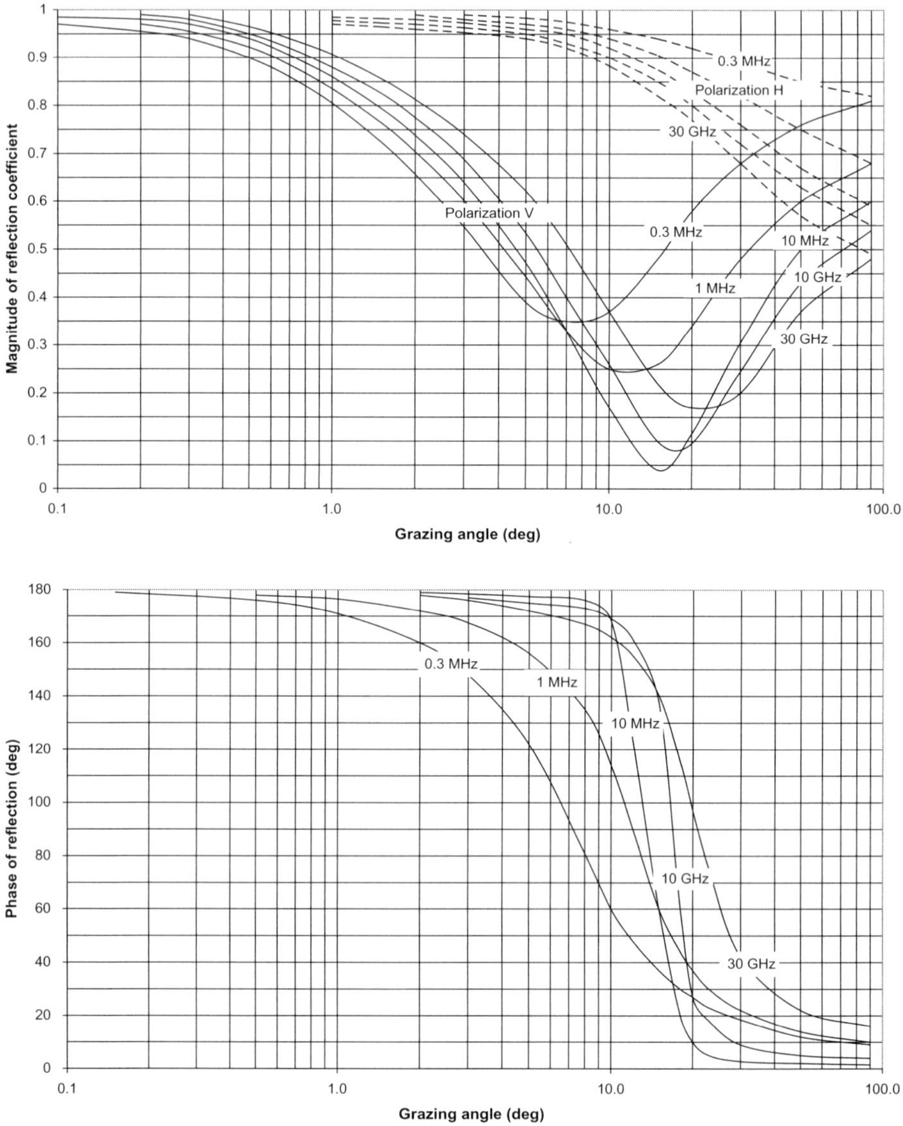

(b)

Figure 1.25 *Continued*

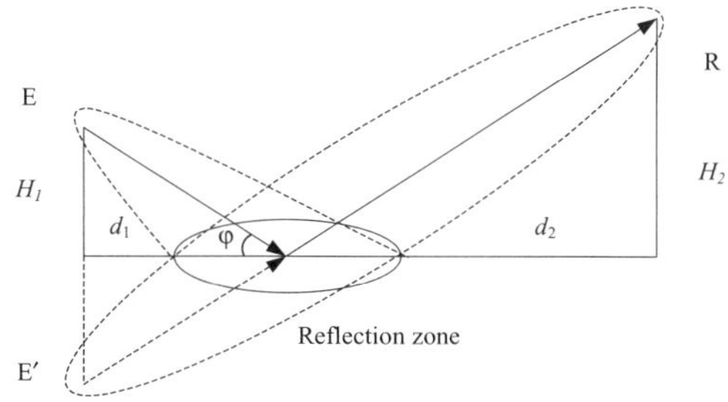

Figure 1.26 Fresnel's zone of reflected wave.

for a reflection angle higher than the limit angle:

$$\varphi_{\text{lim}} = \sqrt[3]{10\lambda} \qquad \text{(mrad)}$$

In the case of Earth–space paths, this formula becomes

$$D = \frac{1}{\sqrt{1 + 2H \tan\theta / (Rd \sin\varphi)}}$$

where $\varphi \approx E + \dfrac{H \operatorname{cotan}\theta}{R + H}$
θ = elevation angle toward satellite
H = height of Earth station

1.7.3 Roughness Factor

The roughness factor of the reflection zone is given by a relation in ITU-R Rep.1008:

$$\rho = \exp\left[-0.5\left(\frac{4\pi\, dh \sin\varphi}{\lambda}\right)^2\right] \tag{1.72}$$

where dh is the rms height of the irregularities and $4\pi\, dh \sin\varphi/\lambda$ constitutes the Rayleigh criterion, illustrated in Figure 1.27. Figure 1.28 presents the magnitude of the roughness factor according to the ratio dh/λ for various values of the reflection angle.

1.7.4 Factor of Limitation of Reflection Zone

It is shown (Boithias 1983) that the limitation factor of the zone of reflection, if its edges are not too far from the reflection point, is given by the approximate relation

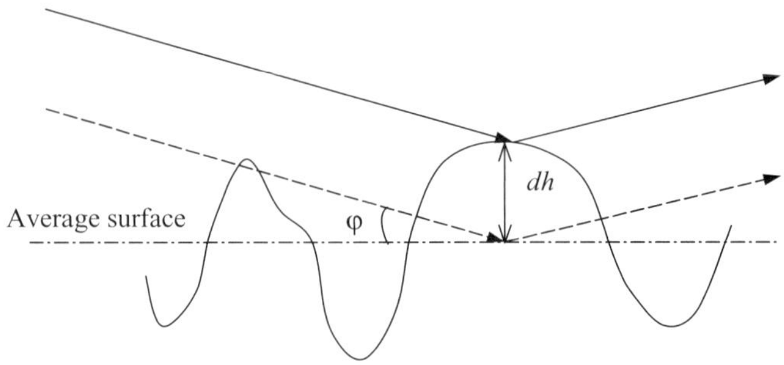

Figure 1.27 Rayleigh criterion.

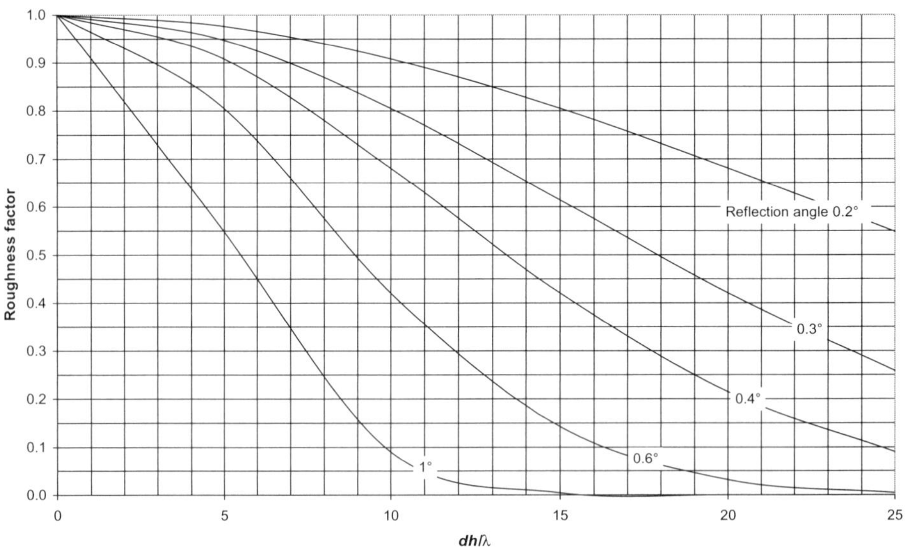

Figure 1.28 Roughness factor due to irregularities of reflective surface.

$$\frac{P}{P_0} \approx \frac{(H_1 + H_2)^4 \Delta x^2}{H_1 H_2 \lambda d^3} \tag{1.73}$$

where P = power received by reflection (W)
P_0 = power received in free space (W)
H_1, H_2 = heights of antennas above reflective surface (m)
λ = wavelength (m)
d = hop length (m)
Δx = size (m) of reflection zone extending on both sides of reflection point, represented in Figure 1.29

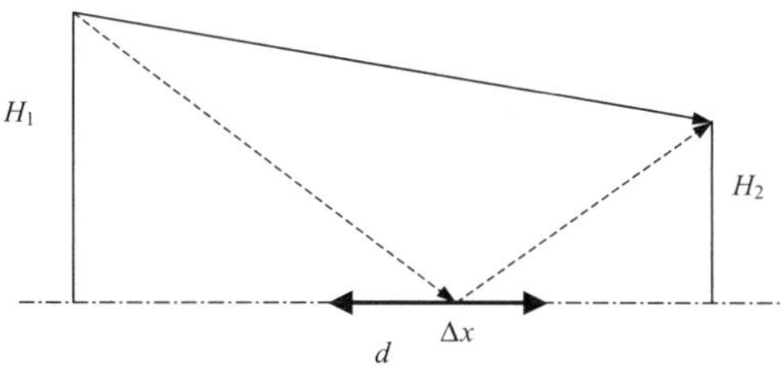

Figure 1.29 Limitation of the reflection zone.

1.8 INFLUENCE OF ATMOSPHERE

1.8.1 Refractivity

We saw that the influence of the propagation medium is entirely determined by its refractive index; the refractive index n of the air being very close to unity, we substitute the value N to it, called refractivity and expressed in N-units, such as

$$N = (n - 1) \times 10^6 \tag{1.74}$$

The atmospheric pressure, temperature, and water vapor concentration influence the refractivity according to the following relation, which is valid for all frequencies up to 100 GHz:

$$N = N_{\text{dry}} + N_{\text{wet}} = 77.6\frac{P}{T} + (3.732 \times 10^5)\frac{p_v}{T^2} \tag{1.75}$$

where T = temperature (K)
P = atmospheric pressure (hPa)
p_v = water vapor pressure (hPa)

The surface refractivity N_S varies with altitude according to the relation

$$N_S = N_0 \exp(-0.136h) \tag{1.76}$$

where N_0 = mean value of atmospheric refractivity reduced to sea level (N-units)
h = altitude (km)

1.8.2 Vertical Gradient of Refractive Index

World charts of monthly averages of N_0 reduced to sea level are presented in Figures 1.30 and 1.31 for the months of February and August, respectively. Failing these data, one can consider the average exponential atmosphere:

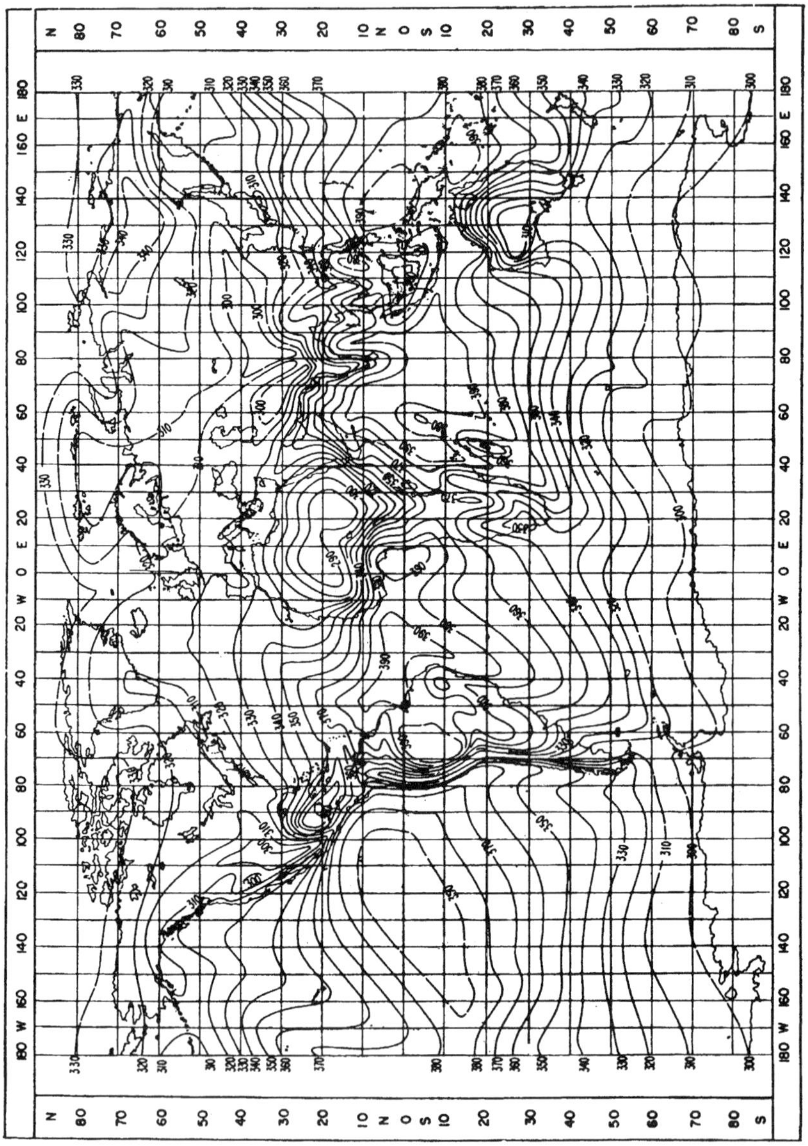

Figure 1.30 Monthly mean value of atmospheric refractivity N_0 in February (ITU-R P.453).

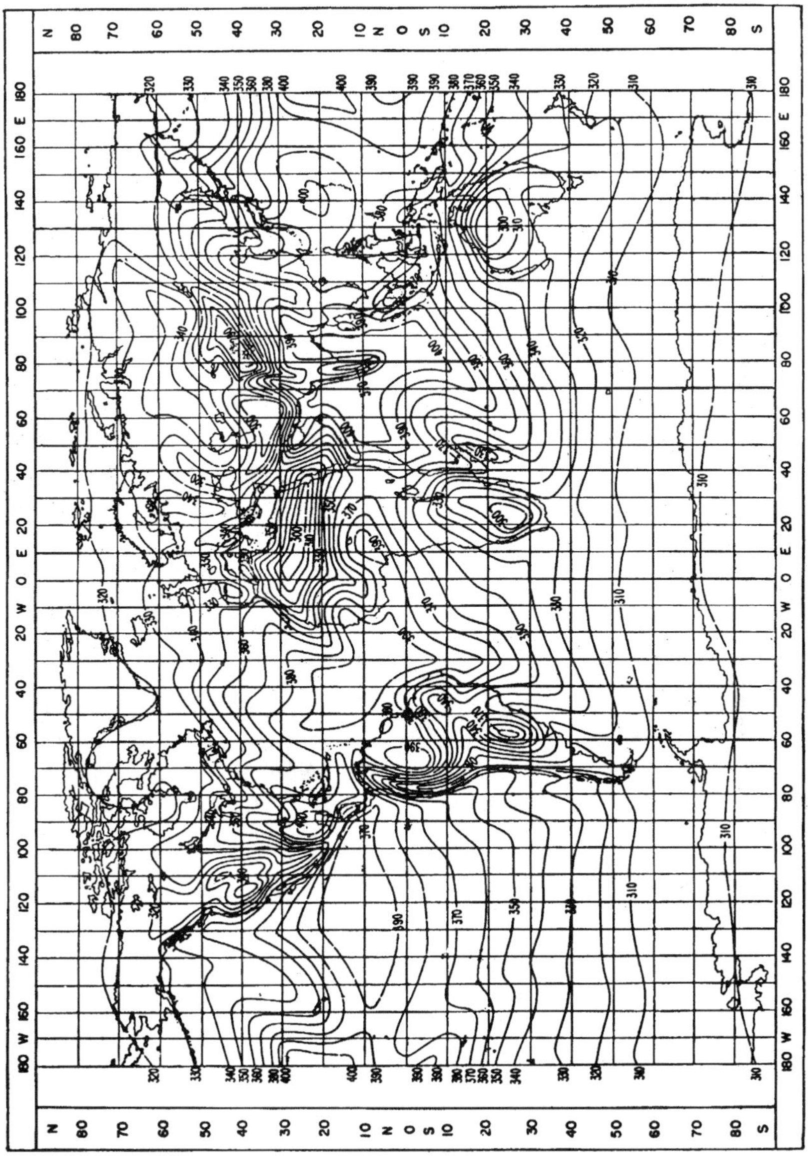

Figure 1.31 Monthly mean value of atmospheric refractivity N_0 in August (ITU-R P.453).

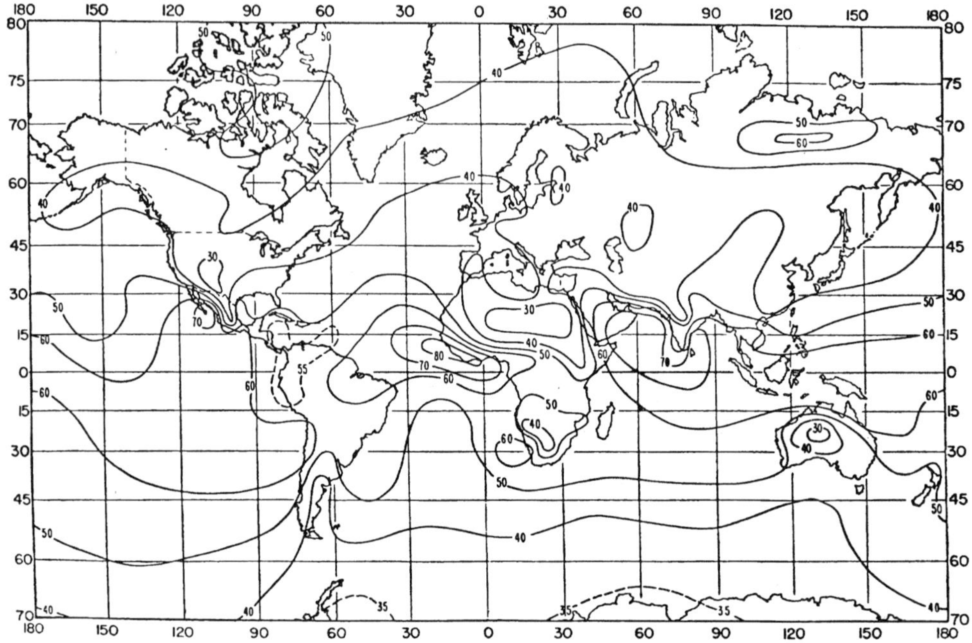

Figure 1.32 Monthly average of ΔN in February (ITU-R P.453).

$$N_S = N_A \exp(-0.136h) \qquad (1.77)$$

where $N_A = 315$ N-units is the average value on the surface of Earth that corresponds to the standard refractivity gradient of $-40\,\mathrm{N\,km^{-1}}$ on the first kilometer. The vertical refractivity gradient dN/dh in the low layer of the atmosphere is an important parameter for the estimate of the effects of refraction on the electromagnetic wave propagation (ray curvature, multipath, atmospheric ducts).

Figures 1.32–1.35 show the monthly average decreases of the refractivity ΔN in a layer 1 km above the Earth surface for February, May, August, and November, respectively: that is,

$$\Delta N = N_S - N_1$$

where N_1 is the value of the refractivity at a height of 1 km above the surface, ΔN not being reduced to the reference surface. The vertical refractivity gradient varies not only according to the geographical localization but also statistically in the course of time; we thus found that, in Florida, in the first 100 m, the refractivity gradient was between 230 and $-370\,\mathrm{N\,km^{-1}}$ in values exceeded for percentages of time corresponding to 0.05 and 99.9%, respectively.

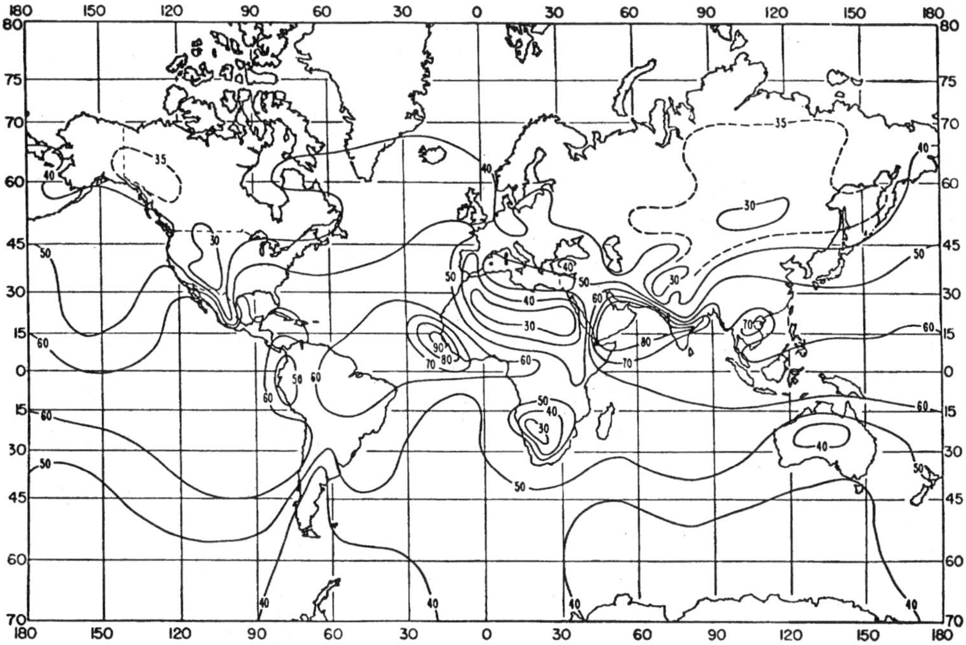

Figure 1.33 Monthly average of ΔN in May (ITU-R P.453).

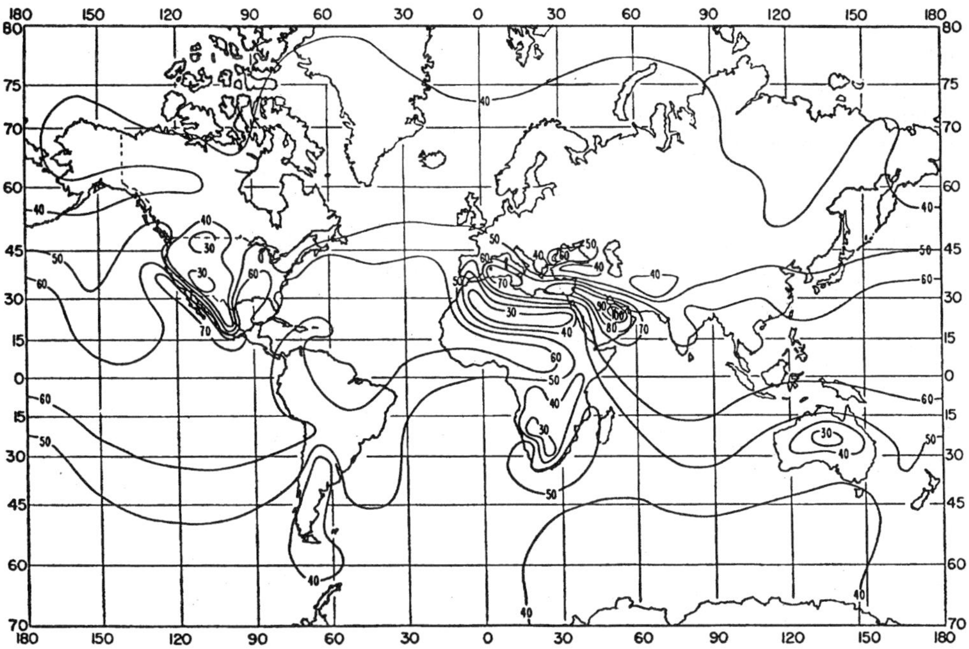

Figure 1.34 Monthly average of ΔN in August (ITU-R P.453).

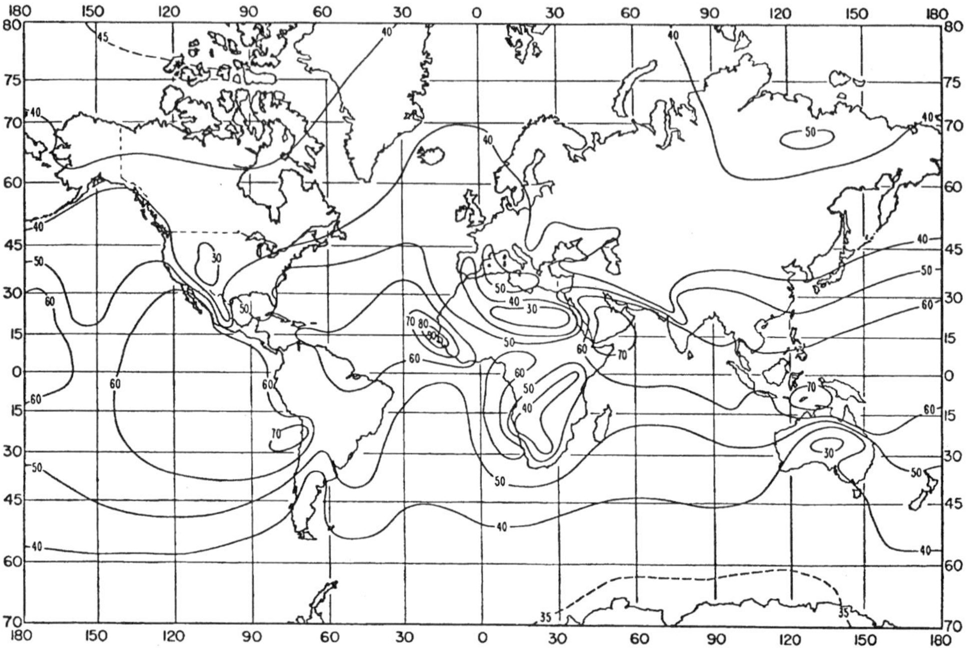

Figure 1.35 Monthly average of ΔN in November (ITU-R P.453).

Figures 1.36–1.39 indicate the percentage of time during which the vertical refractivity gradient is lower or equal to $-100\,\mathrm{N\,km^{-1}}$ in the first 100 m during February, May, August, and November, respectively; this value, called climatic variable P_L, corresponds to the particular conditions of trapping or ducting of the electromagnetic waves which are at the origin of severe fading with deep depressions of the transmitted signal (radioelectric holes).

To use the weather data given in relation (1.75), the following relations exist between water vapor concentration ($\mathrm{g\,m^{-3}}$), water vapor pressure (hPa), and humidity ratio (%) which are valid between –20 and +50 °C:

- Water vapor concentration ($\mathrm{g\,m^{-3}}$):

$$v = 216.7\frac{p_v}{T} \tag{1.78}$$

- Humidity ratio (%):

$$\mathrm{HR} = 100\frac{p_v}{p_s}$$

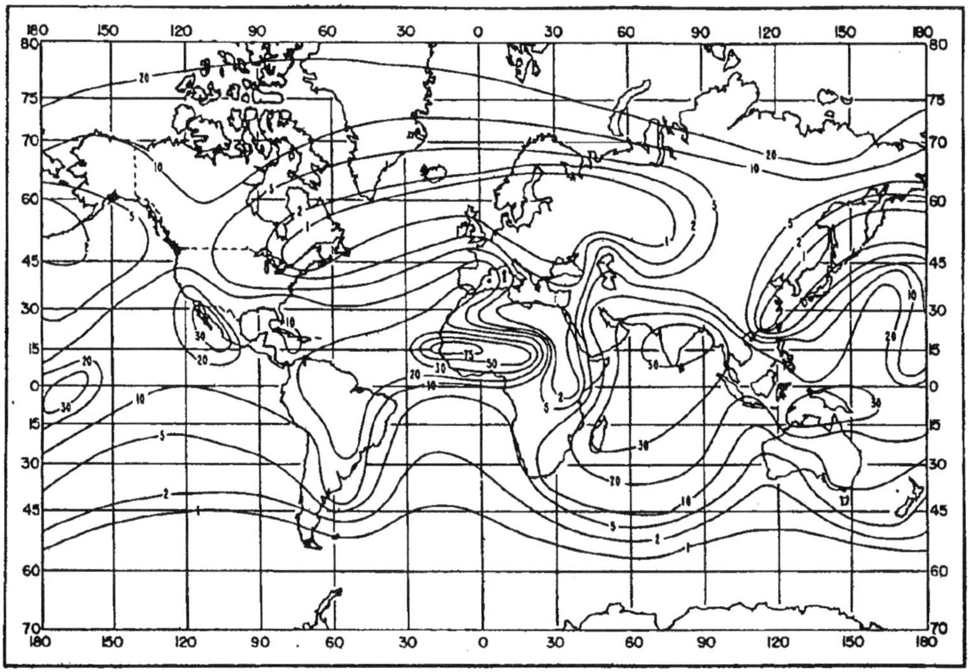

Figure 1.36 Percentage of time $dN/dh < -100\,\mathrm{N\,km^{-1}}$ in February (ITU-R P.453).

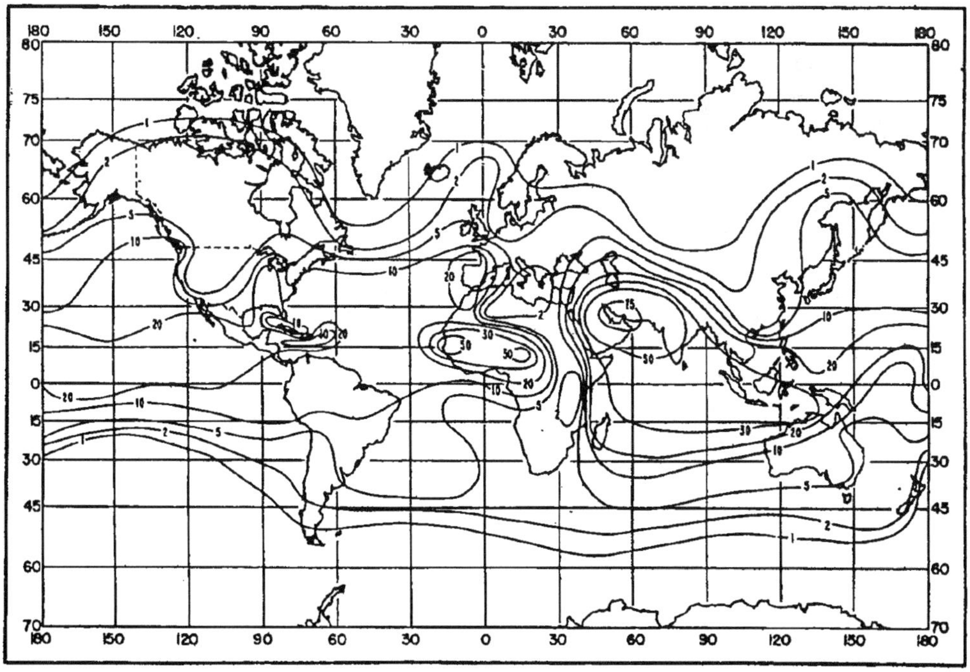

Figure 1.37 Percentage of time $dN/dh < -100\,\mathrm{N\,km^{-1}}$ in May (ITU-R P.453).

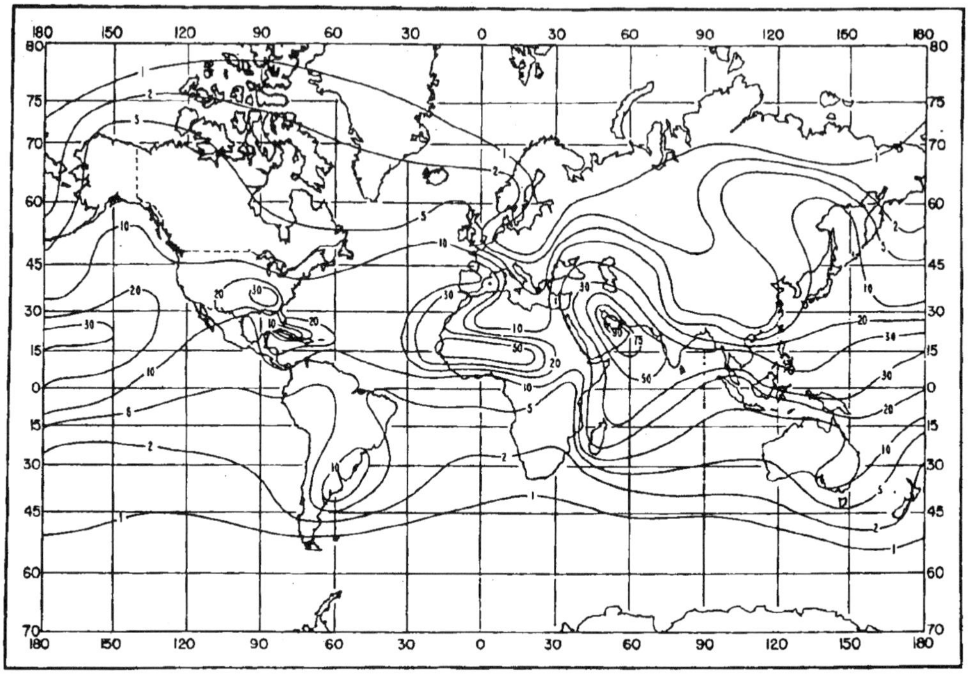

Figure 1.38 Percentage of time $dN/dh < -100\,\mathrm{N\,km^{-1}}$ in August (ITU-R P.453).

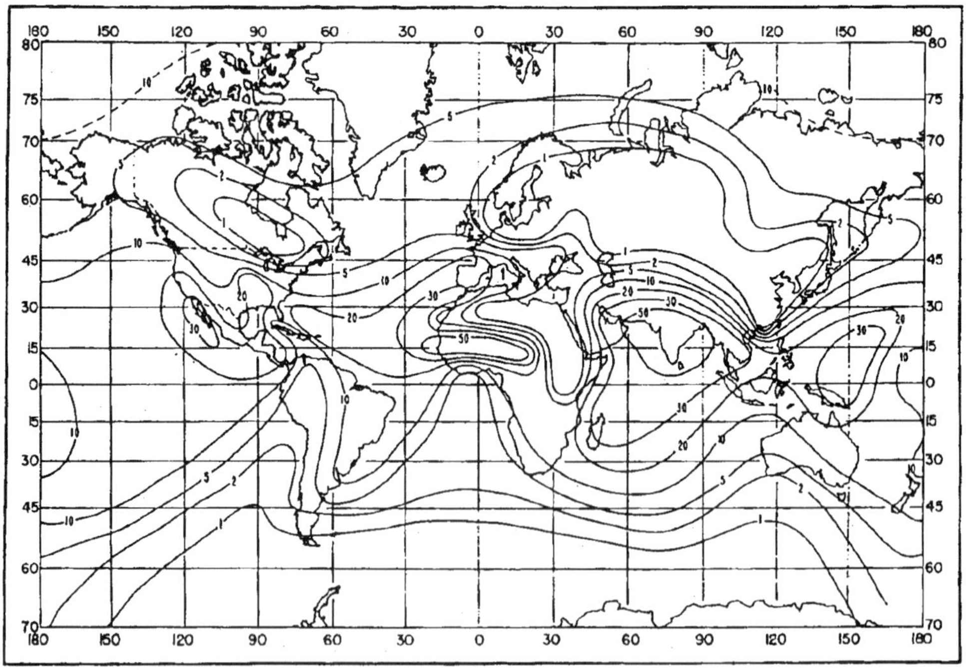

Figure 1.39 Percentage of time $dN/dh < -100\,\mathrm{N\,km^{-1}}$ in November (ITU-R P.453).

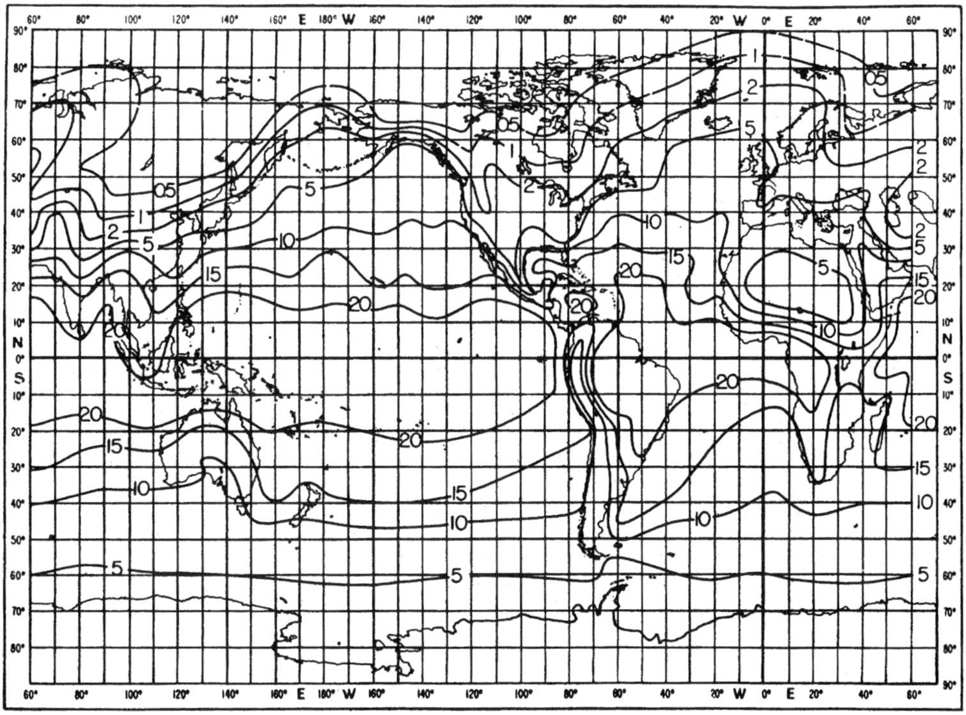

Figure 1.40 Water vapor concentration ($g\,m^{-3}$) in February (ITU-R Rep.563).

- Water vapor pressure at saturation (hPa):

$$p_s = 6.1121 \exp\left[17.502 \frac{t}{t + 240.97}\right]$$

where p_v = water vapor pressure (hPa)
T = absolute temperature (K)
t = température (°C) (0 °C = 273.16 K)

Figure 1.40 presents the distribution of water vapor concentration for February. Figure 1.41 presents the distribution of water vapour concentration for August. Figure 1.42 presents the psychrometric chart[9] which links water vapor concentration and humidity ratio (HR) to temperature. Figure 1.43 illustrates the distribution versus percentage of time of vertical refractivity gradient for various standard climates.

[9]The psychrometric chart can be used to solve numerous process problems with moist air; the concepts of internal energy, enthalpy, and entropy are linked to it.

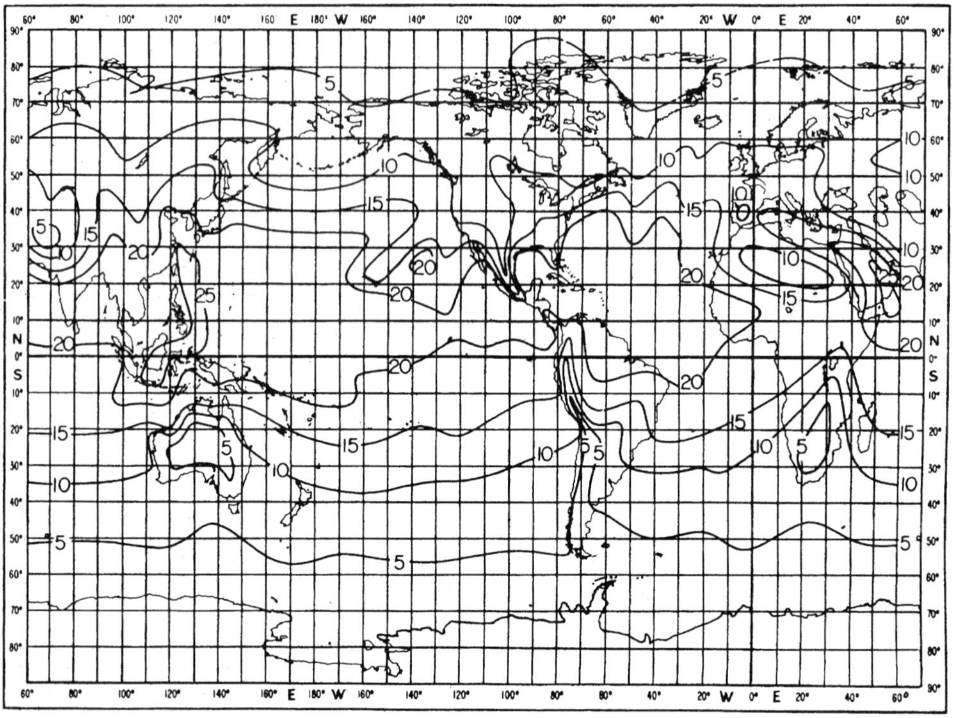

Figure 1.41 Water vapor concentration ($g\,m^{-3}$) in August (ITU-R Rep. 563).

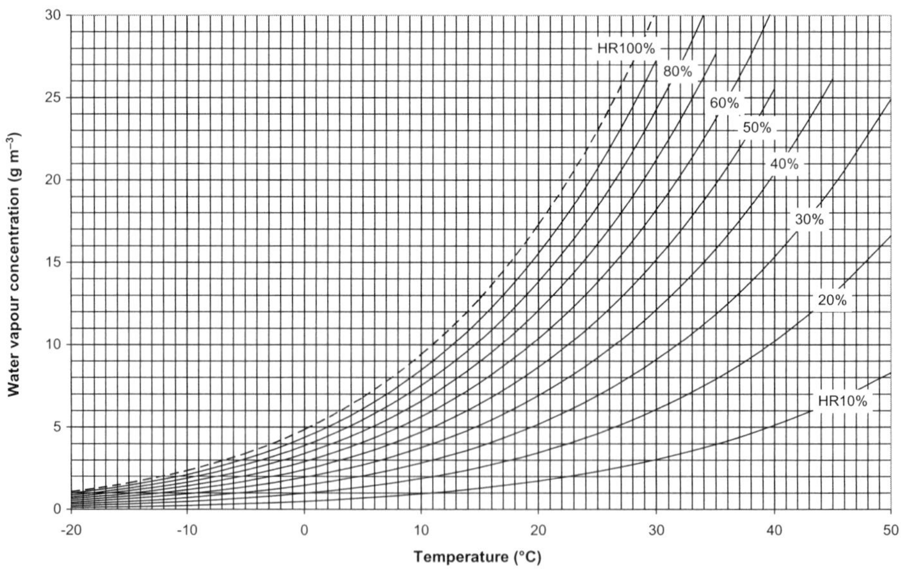

Figure 1.42 Psychrometric chart.

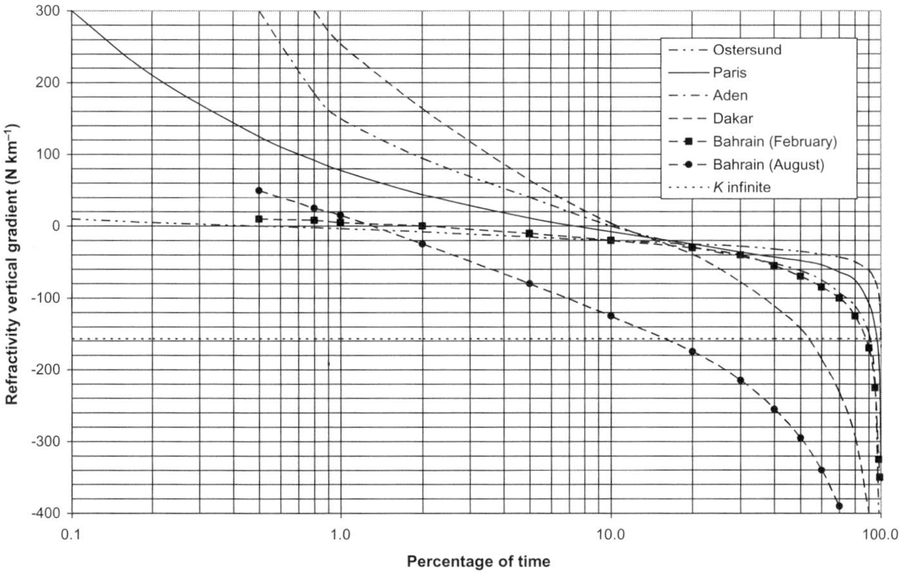

Figure 1.43 Typical distributions of vertical refractivity gradient.

The pressure, temperature, and humidity ratio vary with altitude. It is the same for the vertical refractivity gradient; one can apply, as a first approximation, the following differential equation:

$$\frac{dN}{dh} = 0.35\frac{dP}{dh} - 1.3\frac{dt}{dh} + 7\frac{dv}{dh} \tag{1.79}$$

where P = atmospheric pressure (hPa)
t = temperature (°C)
v = water vapor concentration (g kg^{-1} dry air) for an air density

$$\rho \approx (1.2\,\text{kg}\,\text{m}^{-3})\frac{293}{t+273}$$

Figure 1.44 shows the evolution of the vertical refractivity gradient in a layer 100 m thick, expressed in N-units × km^{-1} or N km^{-1}, according to the relevant gradients of temperature and water vapor concentration assuming the pressure remains constant.

Generally, the atmosphere above the ground is made homogeneous by the wind or during the day by the convection caused by the heating action of the sun. However, in certain circumstances, a stratification of the atmosphere can occur in the shape of layers at different temperature, pressure, and hygrometry which present very different refractive indexes as well as different vertical

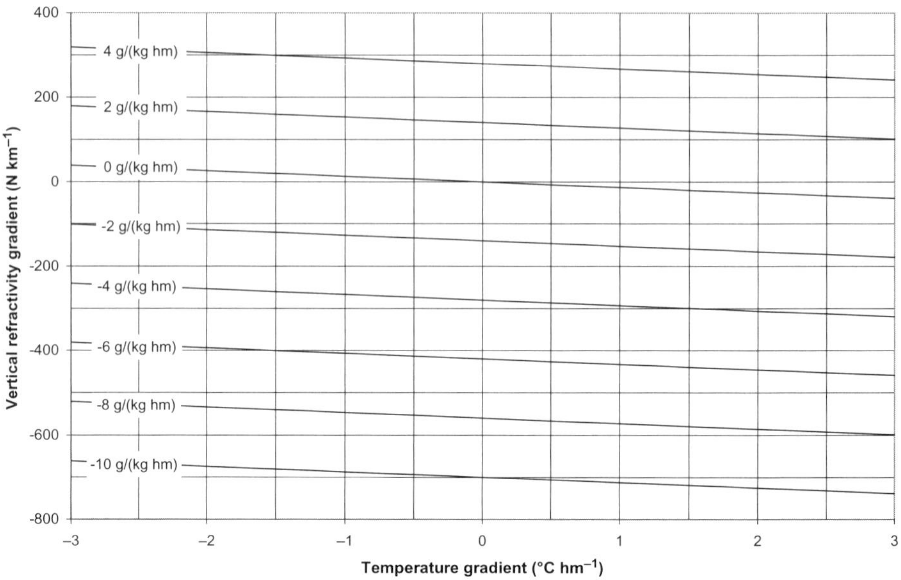

Figure 1.44 Evolution of vertical refractivity gradient (dN/dh) at constant pressure versus gradients of temperature and water vapor concentration in layer 100 m high.

refractive gradients. Then tropospheric radio ducts from that extend a few tens to a few hundred meters vertically and several tens to several hundred kilometers horizontally inside which the radio waves are guided. For example, the psychrometric chart for moist air shows that, for high temperatures and a humidity ratio of about 40%, the water vapor concentration can exceed $30\,\mathrm{g\,m^{-3}}$, which corresponds to current weather conditions in the area of the Persian Gulf in summer when the phenomena of ducting propagation dominate. Then we can calculate the air density variation according to temperature and water vapor concentration as follows:

- Temperature

$$\rho = \rho_0 \frac{273 + t_0}{273 + t} \approx (1.2\,\mathrm{kg\,m^{-3}}) \frac{293}{273 + t} \tag{1.80}$$

from which $d\rho/dt \approx -4.1 \times 10^{-3}\,\mathrm{kg\,m^{-3}\,K^{-1}}$

- Water vapor concentration:

$$\rho \approx \rho_0(\mathrm{kg\,m^{-3}}) + v(\mathrm{g\,m^{-3}}) \times 10^{-3}$$

from which $d\rho/dv \approx 10^{-3}\,\mathrm{kg/m^{-3}}\,(\mathrm{g/m^{-3}})^{-1}$

It is seen that an increase in water vapor concentration of some $4\,\mathrm{g\,m^{-3}}$ compensates for the reduction in air density due to an increase in temperature of 1 °C, which explains why a layer of hot and wet air can be formed under a layer of cold and dry air (phenomenon of inversion of temperature) with a very negative refractivity gradient, as shown in Figure 1.44. A positive vertical temperature gradient and/or a negative water vapor concentration gradient thus result in a negative refractivity gradient, whereas a negative temperature gradient and/or a positive water vapor concentration gradient result in a positive refractivity gradient.

1.8.3 Curvature of Radioelectric Rays

The propagation of radio waves is rectilinear in a dielectric medium whose index remains constant at any point of crossed space. On the other hand, it is not the same when this index varies and it is shown, using a spherical model of atmosphere, that the trajectory of the rays undergoes a continuous deflection when the index varies continuously in the vertical direction.

The phenomenon of ray curvature may be understood by dividing the atmosphere into layers with discrete values for the refractive index in each layer, as shown in Figure 1.45. Since each successive value of n_n is smaller than the preceding value n_{n-1}, as shown in the figure, the angle φ_n must increase and the ray curves in the downward direction as

$$n_1 \sin\varphi_1 = n_2 \sin\varphi_2 = n_3 \sin\varphi_3 = \cdots = n_n \sin\varphi_n$$

This results in a curvature radius r of the trajectory of the radioelectric rays expressed in the form

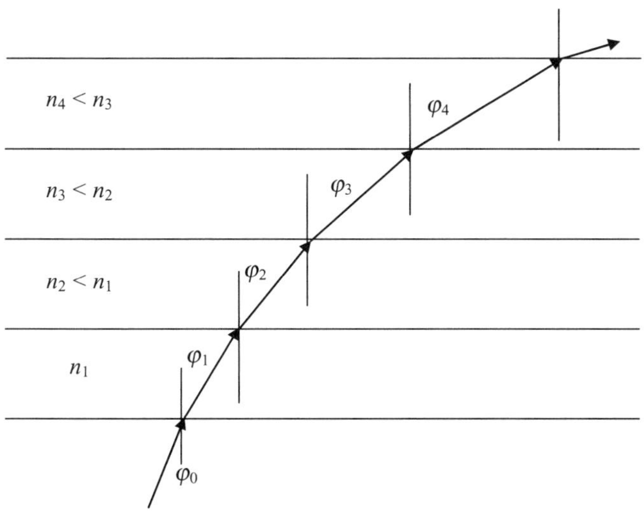

Figure 1.45 Ray curvature.

$$\frac{1}{r} = -\frac{dn}{dh}$$

1.8.4 Effective Earth Radius

To obtain a rectilinear representation of the rays trajectory of the rays in the vicinity of Earth, the actual terrestrial radius R is modified by a factor K, called the effective Earth radius factor:

$$\frac{dn}{dh} + \frac{1}{R} = \frac{1}{KR}$$

that is,

$$K = \frac{1}{1 + R(dn/dh)}$$

where $R = 6.38 \times 10^3$ km and

$$\frac{dn}{dh} = \frac{dN}{dh} \times 10^{-6}$$

We obtain

$$K = \frac{1}{1 + (6.38 \times 10^3)(dN/dh) \times 10^{-6}}$$

Then

$$K = \frac{157}{157 + dN/dh} \tag{1.81}$$

The factor K can also be obtained using the following empirical formula from technical note NBS-101 [National Bureau of Standards (NBS), 1966] which connects dN/dh to the surface refractivity N_S for a layer 1 km high:

$$\frac{dN}{dh} = -7.32 \exp(0.005577 N_S)$$

Thus

$$K = \frac{1}{1 - 0.04665 \exp(0.005577 N_S)} \tag{1.82}$$

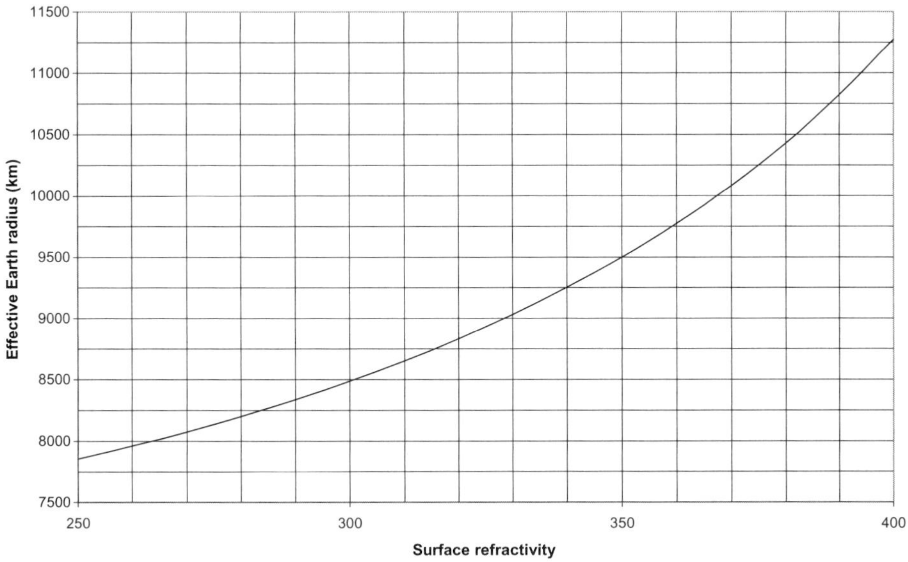

Figure 1.46 Effective terrestrial radius versus surface refractivity N_S.

The effective Earth radius a is then

$$a = KR = \frac{157R}{157 + dN/dh} \tag{1.83}$$

Figure 1.46 presents the effective Earth radius versus surface refractivity N_S. The effective Earth radius becomes

- infinite when $dN/dh = -157\,\mathrm{N\,km^{-1}}$ ($K \pm \infty$), in which case the trajectory of the rays follows the actual curvature of Earth;
- equal to the actual Earth radius for $dN/dh = 0\,\mathrm{N\,km^{-1}}$ ($K = 1$);
- negative for $dN/dh < -157\,\mathrm{N\,km^{-1}}$ (concave Earth for $K < 0$); or
- lower than the real Earth-radius for $dN/dh > 0\,\mathrm{N\,km^{-1}}$ ($K < 1$).

Figure 1.47 illustrates the trajectory of the rays compared to the actual Earth radius and to a model utilizing an effective Earth radius versus factor K.

The standard radio atmosphere corresponds to a spherical atmosphere whose vertical refractivity gradient would be constant and equal to the median value (50% of time) in a moderate climate:

$$\frac{dN}{dh} = -40\,\mathrm{N\,km^{-1}}$$

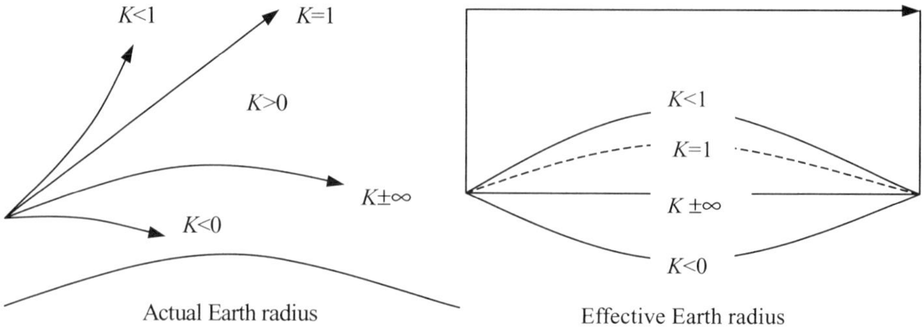

Figure 1.47 Trajectory of rays and effective Earth radius versus K.

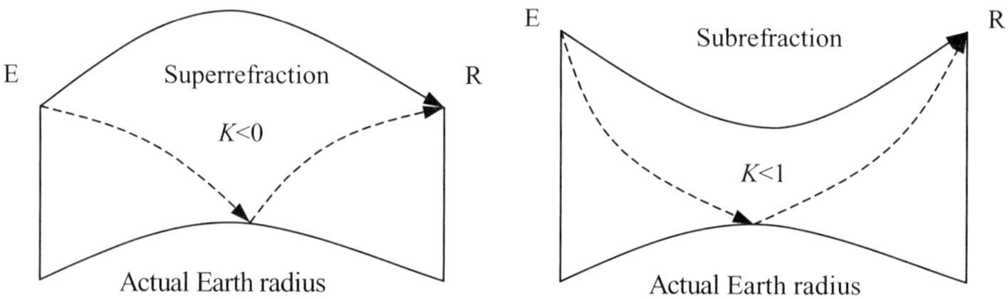

Figure 1.48 Relative position of rays compared to ground.

that is, $K \approx \frac{4}{3} \rightarrow a \approx 8470\,\text{km}$. When dN/dh is higher than this value, it corresponds to subrefraction conditions and in the contrary case to superrefraction. When K is negative, which corresponds to superrefraction, the rays are curved toward the ground; consequently, the rays from the transmitter reach the receiver either directly while deviating from Earth or after reflection on the ground, as shown in Figure 1.48. On the other hand, when K is positive and lower than 1, which corresponds to subrefraction, the rays are deflected upward; it follows that the lower rays issued from the transmitter reach the receiver while approaching the ground.

In the event of subrefraction, the ground can thus become an obstacle to the propagation of the rays, which then are affected by an additional loss by diffraction, while in superrefraction, the direct rays pass largely above the ground, but these conditions also favor the reflected rays, which can interfere with the direct rays, as we will see. Figure 1.49 summarizes the behavior of electromagnetic waves which cross layers of air of different indexes; this phenomenon explains the mirages which occur in the desert when the image of the sky is reflected on the ground (curvature of the rays upward) as well as the vision at very long distance above the sea (curvature of the rays downward). The main cause of the subrefraction corresponds to an unstable arrange-

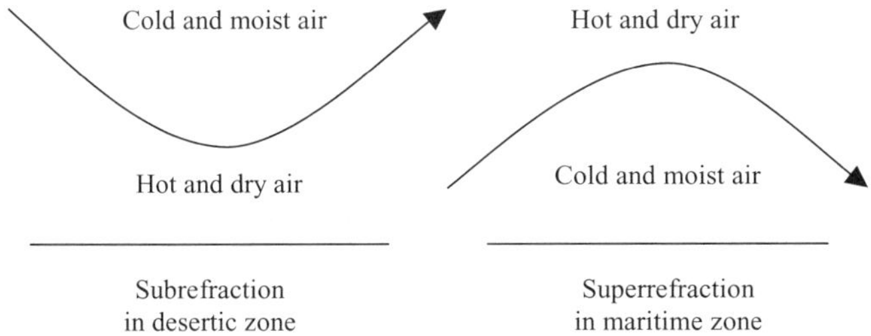

Figure 1.49 Typical examples of ray curvature.

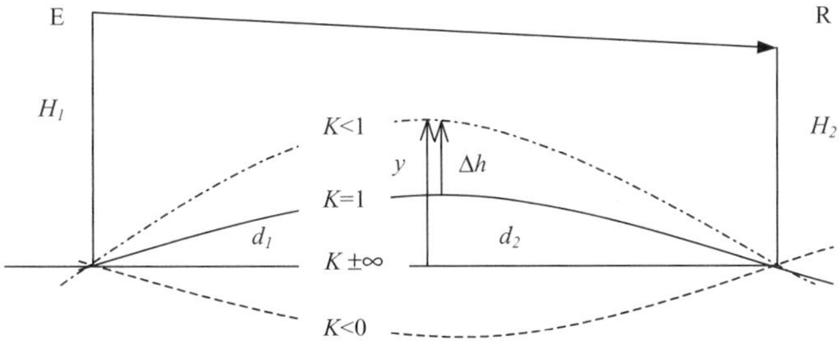

Figure 1.50 Effective Earth radius.

ment of the layers of the atmosphere, contrary to that which is at the origin of the superrefraction, and consequently to a small percentage of time.

The effective Earth radius compared to the infinite apparent Earth radius, as presented in Figure 1.50, is given according to the factor K by the following relation using the same unit for heights and distances (kilometers):

$$y = \frac{d_1 d_2}{2a} = \frac{d_1 d_2}{2KR} \tag{1.84}$$

The Earth curvature y from a flat Earth ($K = \pm\infty$) at midpath is represented in Figure 1.51 as well as the height difference Δh from the actual Earth ($K = 1$), which is given by the relation

$$\Delta h = \frac{d_1 d_2}{2R}\left(\frac{1}{K} - 1\right) \tag{1.85}$$

1.8.5 Variation of Launch and Arrival Angles of Rays

Ray curvature results in variation of the launch and arrival angles of rays, both direct and reflected, at the antennas. The approximate value of this variation can be calculated as follows on the basis of the geometric elements presented in Figure 1.52. For greater convenience, we will consider a ray issued from the source S, located at a height h, and tangent to the Earth at point T, which is

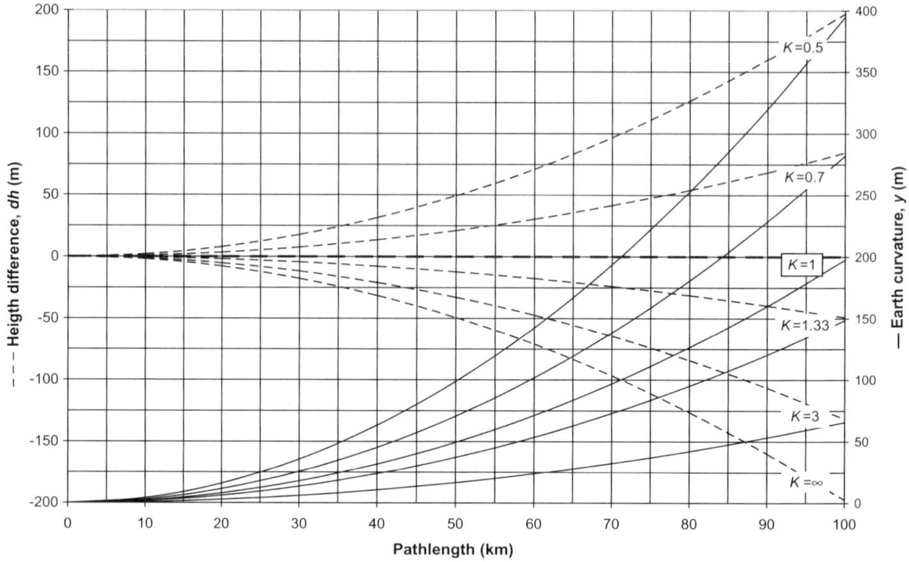

Figure 1.51 Earth curvature and height difference at midpath.

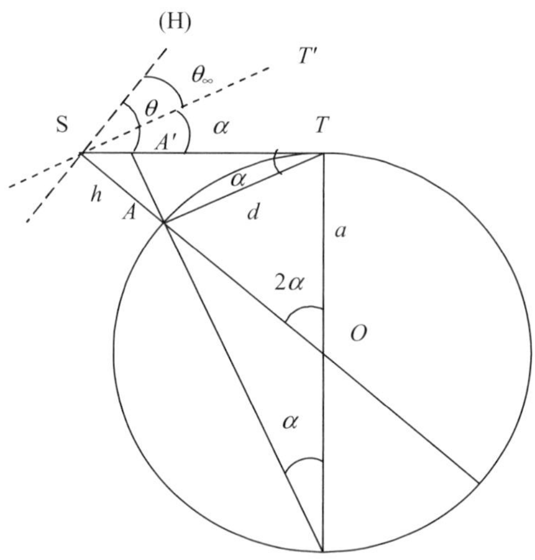

Figure 1.52 Apparent launch/arrival angles.

the same as a path between two points placed at altitudes differing by the height h. The launch angle θ of the ray is related to the horizontal line (H) passing by the source S. Moreover, the distance d and the height h are small with respect to the effective Earth radius a. The similarity of triangles ATA' and ABT makes it possible to write

$$\tan\alpha = \frac{AA'}{AT} = \frac{AT}{AB}$$

from which we get the downward slope from the horizon line, such as $AB \approx 2a$ and $h << d << a$:

$$\alpha \approx \frac{d}{2a}$$

The angle θ is the sum of the angle θ_∞, which corresponds to the elevation angle for $a = \infty$, and the angle α, which is equal to half of the angle at the center of Earth, that is,

$$\theta = \theta_\infty + \alpha \approx \frac{h}{d} + \frac{d}{2a} \tag{1.86}$$

The launch and arrival angles of the direct ray, $\Delta\theta_{DL}$ and $\Delta\theta_{DA}$, thus vary with the factor K according to the expressions[10]

$$\Delta\theta_{DL} = \Delta\theta_{DA} \approx -\frac{d}{2a} = -\frac{d}{2KR}$$

The launch angle $\Delta\theta_{RL}$ and arrival angle $\Delta\theta_{RA}$ of the reflected ray vary at first approximation and using relation (1.84) according to the expressions[11]

$$\Delta\theta_{RL} \approx -\frac{y}{d_1} = -\frac{d_1 d_2}{2ad_1} = -\frac{d_2}{2a} = -\frac{d_2}{2KR} \tag{1.87}$$

$$\Delta\theta_{RA} \approx -\frac{y}{d_2} = -\frac{d_1 d_2}{2ad_2} = -\frac{d_2}{2a} = -\frac{d_1}{2KR} \tag{1.88}$$

1.8.6 Positive Minimal Value of Effective Earth-radius Factor

The factor K varies throughout a given path; therefore, one defines an equivalent factor K_e for this path. The positive minimal value $K_{\min}$ can then be obtained in the following way:

- Determine the distribution of the vertical refractivity gradient at a point of the path $(dN/dh)_0$ and evaluate its median value μ_0 as well as its

[10]The minus sign refers to the fact that the terrestrial curvature varies in the opposite direction to that of the curvature of the rays.

[11]It is seen that the launch and arrival angles of the reflected ray vary less than those of the direct ray, that is, in the ratios d_1/d and d_2/d.

standard deviation σ_0 for the positive gradients by supposing that it is about a normal distribution.

- Assume that the distribution is the same throughout the path and, since the instantaneous value varies from one point to another, utilize an equivalent vertical refractivity gradient $(dN/dh)_e$, which makes it possible to obtain the equivalent factor K_e by the relation

$$K_e = \frac{157}{157 + [dN/dh]_e}$$

- The median value of $(dN/dh)_e$, which is the average vertical refractivity gradient along the path, and its standard deviation are respectively given by the relations

$$\mu_e \approx \mu_0 \qquad \sigma_e \approx \frac{\sigma_0}{\sqrt{1 + d/d_0}}$$

where: $d_0 = 13.5\,\text{km}$.

- Calculate the values of $(dN/dh)_e$, and thus those of K_e, which are exceeded during any percentage of time using the following Gaussian coefficients C_G:

Percentage of Time	$(dN/dh)_e$	Percentage of Time	$(dN/dh)_e$
50	$(dN/dh)_e \approx \mu_e$	99.9	$(dN/dh)_e \approx \mu_e + 3.09\sigma_e$
84.14	$(dN/dh)_e \approx \mu_e + \sigma_e$	99.99	$(dN/dh)_e \approx \mu_e + 3.719\sigma_e$
90	$(dN/dh)_e \approx \mu_e + 1.28\sigma_e$	99.999	$(dN/dh)_e \approx \mu_e + 4.265\sigma_e$
99	$(dN/dh)_e \approx \mu_e + 2.327\sigma_e$	99.9999	$(dN/dh)_e \approx \mu_e + 4.753\sigma_e$

The positive minimal value $K_{\min}$ can thus be defined for any given percentage of time by the expression

$$K_{\min} = \frac{157}{157 + [dN/dh]_{50\%} + C_G\sigma_0/\sqrt{1 + d/d_0}} \tag{1.89}$$

Figure 1.53 presents the positive minimal value $K_{\min}$ in temperate climate for percentages of time of the average year ranging between 99 and 99.9999% using the following parameters[12]:

- Median value of vertical gradient: $[dN/dh]_{50\%} = -40\,\text{N}\,\text{km}^{-1}$
- Standard deviation: $\sigma_0 = 50\,\text{N}\,\text{km}$
- Reference distance: $d_0 = 13.5\,\text{km}$

[12]These values are in agreement with the distribution for Paris in Figure 1.43.

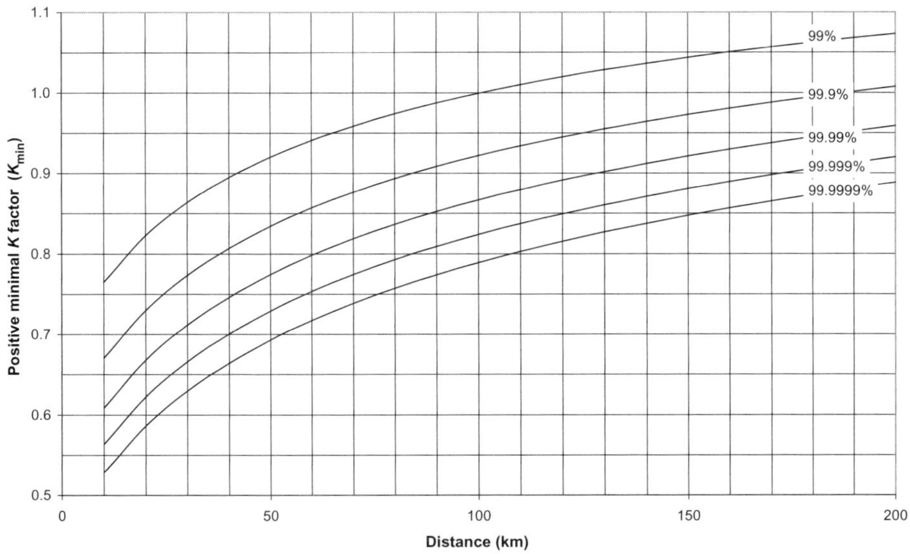

Figure 1.53 Positive minimal value K_{min} for various percentages of average year.

Figure 1.54 presents the positive minimal value K_{min} of the equivalent factor K_e versus pathlength that is exceeded during roughly 99.9% for any month in continental temperate climate.

1.8.7 Tropospheric Radio Ducts

1.8.7.1 Conditions of Appearance We saw previously that the atmosphere can be divided into quasi-horizontal layers with different refractive indexes and whose vertical refractivity gradients can take very negative values, resulting in superrefraction conditions of propagation which may trap the waves inside a layer called a radio duct. The meteorological conditions which cause such stratification of the troposphere are the following:

1. Evaporation caused by the sun above a wetland or a broad stretch of water
2. Advection, that is, the horizontal movement of a mass of hot and dry air passing above a mass of cold and wet air
3. Cooling of the ground by radiation at the same time as the air at its contact while the upper layers of the atmosphere cool more slowly
4. Subsidence, which corresponds to downward movement, contrary to thermal ascent, of a mass of air which cools slowly above a mass of colder and wetter air

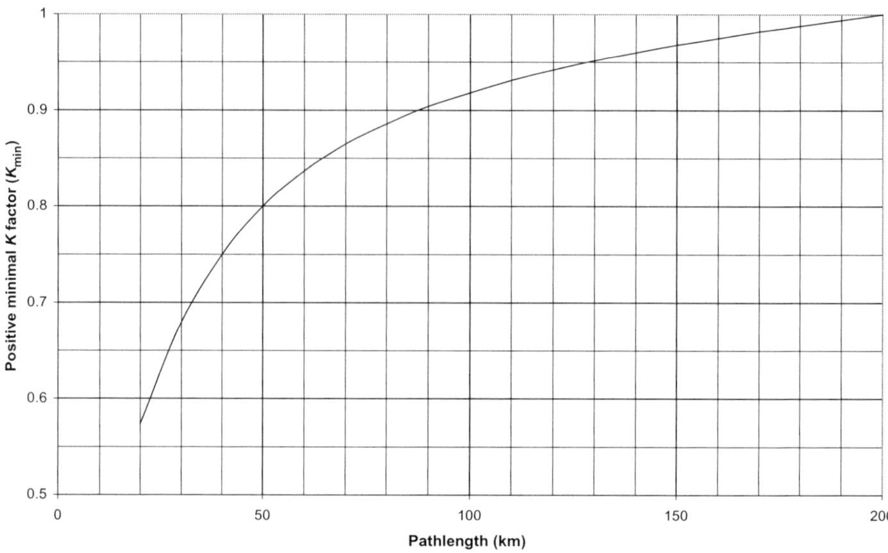

Figure 1.54 Positive minimal value K_{min} for 99.9% of any month (ITU-R P.530).

These weather conditions correspond to a stable arrangement of the air masses since the densest ones occupy the low part of the atmosphere and can consequently last several hours and extend over very vast zones.

Tropospheric radio ducts generate a major and slow fading of the signal which can reach several tens of decibels, called depression or a radioelectric hole, which is accompanied by a very fast and deep fading due to multipath, having a speed of variation of several tens to several hundreds of decibels per second, alternating with periods of important reinforcement of the field strength. Recommendation ITU-R P.453-9 provides world statistics on the frequency of occurrence, average intensity, and average height of the following ducts types:

(a) Surface ducts
(b) Surface elevated ducts
(c) Elevated ducts

For example, Figures 1.55 and 1.56 illustrate duct types (a) and (b), commonly encountered near the ground in microwave radio links, which are characterized by a modified refractive index profile according to the height by considering the refractive modulus M defined by

$$M = \left(n + \frac{h}{a} - 1\right) \times 10^6 \Rightarrow M(h) = N(h) + 157h$$

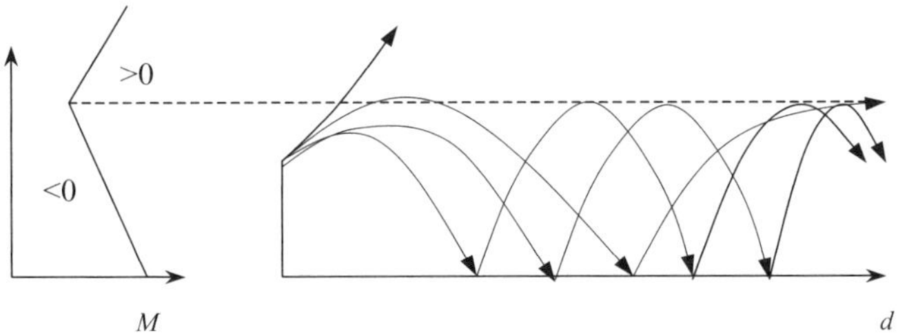

Figure 1.55 Propagation in surface duct.

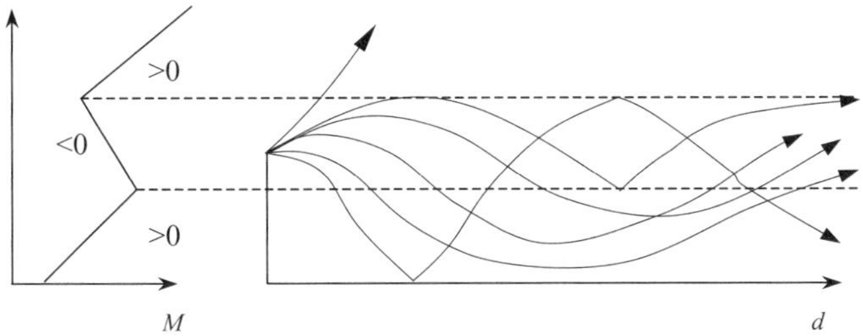

Figure 1.56 Propagation in elevated surface duct.

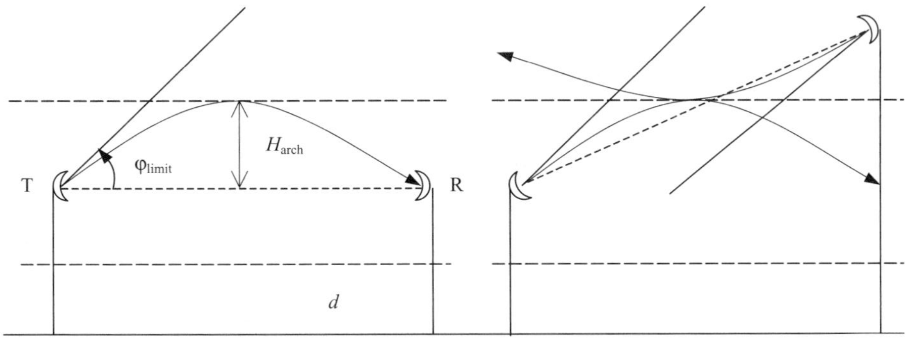

Figure 1.57 Height of direct arch and angle limit of rays in duct with very negative vertical refractivity gradient.

1.8.7.2 Trajectories of Rays in Radio Duct We saw previously that radioelectric rays are curved downward when the vertical refractivity gradient is strongly negative and are curved upward if it is positive; Figures 1.55 and Figure 1.56 illustrate these trajectories in superrefractive conditions, which characterize the guided propagation.

Figure 1.57 presents the parameters of the parabolic trajectory of a direct ray between a transmitter and a receiver that are placed at the same height inside a radio duct with a very negative vertical refractivity gradient. It is shown that the trajectory of waves in a duct with a very negative vertical refractivity gradient is parabolic and one can then define the height of the

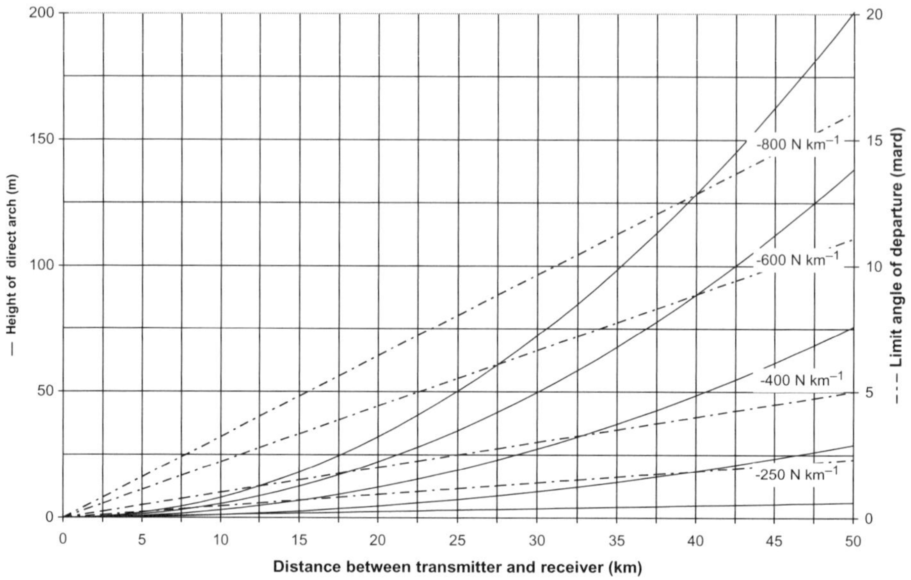

Figure 1.58 Height of direct arch and limit angle.

direct arch, H_{arch}, expressed in meters, as well as the limit angle φ_{limit}, expressed in milliradians, which are necessary to reach a receiver placed at the same height as the transmitter and at a distance d using the following relations:

$$H_{\text{arch}} = -\frac{157 + dN/dh}{8000} d^2 \tag{1.90}$$

$$\varphi_{\text{limit}} = -\frac{157 + dN/dh}{2000} d \tag{1.91}$$

In addition to the curvature of the rays according to the refractivity of the atmosphere, these can thus be reflected by the interface between two layers of different indexes or by the ground, according to the parameters in Figures 1.24 and 1.25, or leave the duct when the incident angle is higher than the limit angle.

Figure 1.58 presents the height of the direct arch and the value of the limit launch angle of the rays versus distance between the transmitter and the receiver for various values of the vertical refractivity gradient.

1.9 PROPAGATION BY DIFFRACTION

Figure 1.59 illustrates an obstruction on a path that divides two zones known as zone of visibility and zone of diffraction. According to geometric optics, the

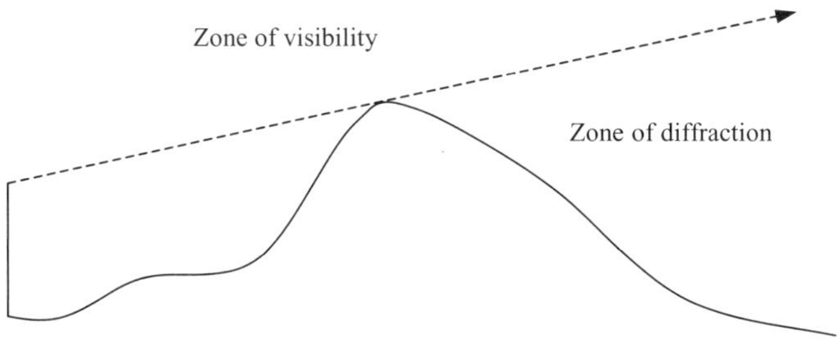

Figure 1.59 Diffraction due to an obstacle.

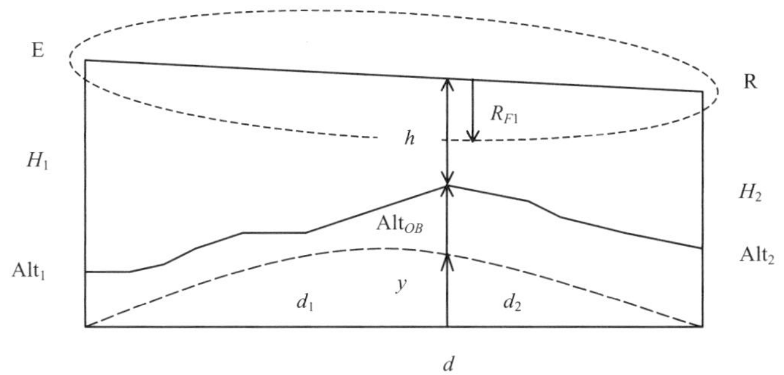

Figure 1.60 Clearance of microwave line-of-sight radio link.

field strength below the ray to the horizon or the tangent ray is null; however, because of propagation by diffraction, the radiated field goes around the obstacle and penetrates into the shadow zone by decreasing rapidly as the receiver moves deeper inside. Thus, when an obstacle interposes between the transmitter and the receiver, there is an additional transmission loss which depends on a great number of parameters related to its nature, the wavelength, the polarization, the height of the antennas, and the geometry of the link. The nature of the ground and the polarization play a great part at low frequencies, but, as microwave links in the majority of the cases are established at frequencies higher than 300 MHz, we will see in what follows that their influence can be neglected; however, although the ITU recommendation does not indicate higher limit of validity in frequency, a certain prudence is essential when this exceeds some gigahertz.

It is advisable to first define the parameter of clearance of the connection, h/R_{F1}, which is the ratio of the clearance h under the line joining the two ends of the path and the first Fresnel's ellipsoid radius R_{F1}, as illustrated in Figure 1.60, which presents a microwave line-of-sight radio link established between a transmitter E and a receiver R, respectively located at heights H_1 and H_2 on a path profile of length d. The relief is reported to the arc of the Earth great circle corresponding to the level of reference whose plane passes by E and R, with respective altitudes Alt_1 and Alt_2, and we consider the prevailing obstacle of altitude Alt_{OB} placed at distance d_1 from E and at distance d_2 from R. We define the clearance h geometrically as

$$h = \mathrm{Alt}_1 + H_1 + [\mathrm{Alt}_2 + H_2 - (\mathrm{Alt}_1 + H_1)]\frac{d_1}{d} - (y + \mathrm{Alt}_{OB}) \qquad (1.92)$$

Then, starting from relation (1.65) and giving the radius of Fresnel's first ellipsoid R_{F1}, we deduce the clearance criterion[13] h/R_{F1}. Generally, there are five obstacle types:

Smooth spherical Earth
Knife edge
Rounded edge
Multiple edges
Irregular terrain

1.9.1 Diffraction over Smooth Spherical Earth

The diffraction field strength E relative to free-space field strength E_0, expressed in decibels, is given by the following relation (ITU-R P.526):

$$20\log\left(\frac{E}{E_0}\right) = F(X) + G(Y_1) + G(Y_2) \qquad (1.93)$$

where X = pathlength
Y_1 = height of transmission antenna
Y_2 = height of reception antenna

and

$$X = 2.2\beta f^{1/3} a^{-2/3} d \qquad Y = 9.6\times 10^{-3}\beta f^{2/3} a^{-1/3} H$$

where d = pathlength (km)
a = equivalent terrestrial radius (km)
H = antenna height (m)
f = frequency (MHz)
β = Boithias's parameter depending on nature of ground

with

$$\beta = \frac{1 + 1.6\kappa^2 + 0.75\kappa^4}{1 + 4.5\kappa^2 + 1.35\kappa^4}$$

[13]By convention, we will consider that the clearance h is negative when the chord joining E and R passes over the top of the obstacle and positive in the contrary case.

where the standardized surface admittance factor κ depends on polarization:

$$\kappa_H = 0.36(af)^{-1/3}\left[(\varepsilon_r - 1)^2 + \left(1.8\times 10^4 \frac{\sigma}{f}\right)^2\right]^{-1/4}$$

$$\kappa_V = \kappa_H\left[\varepsilon_r^2 + \left(1.8\times 10^4 \frac{\sigma}{f}\right)^2\right]^{1/2}$$

where ε_r = relative permittivity
σ = conductivity

Then

$$F(X) = 11 + 10\log X - 17.6X$$

$$G(Y) \approx \begin{cases} 17.6\sqrt{Y-1.1} - 5\log(Y-1.1) - 8 & \text{for } Y > 2 \\ 20\log(Y + 0.1Y^3) & \text{for } 10\kappa < Y < 2 \\ 2 + 20\log\kappa + 9\log\left(\frac{Y}{\kappa}\right)\left[\log\left(\frac{Y}{\kappa}\right) + 1\right] & \text{for } \frac{\kappa}{10} < Y < 10\kappa \\ 2 + 20\log\kappa & \text{for } Y < \frac{\kappa}{10} \end{cases}$$

When $\kappa < 10^{-3}$, the electric characteristics of the ground and polarization are without influence. For horizontal polarization at all frequencies and for vertical polarization above 20 MHz over land or 300 MHz over sea, β can be taken equal to 1. For frequencies below these values, which are not of interest for usual microwave links, β must be calculated as a function of κ.

Figure 1.61 gives the relative permittivity and conductivity of various ground types versus frequency. The values for obstacle diffraction loss over a smooth spherical Earth obtained thus are close to those given by Bullington's method, according to the clearance criterion h/R_{F1} established above, by using the following relation expressed in decibels:

$$A_{OB} = \left(14 + 22.24\frac{h}{R_{F1}}\right)\frac{2.456}{M^{0.15}} \quad (1.94)$$

where

$$M = \frac{3.28H_1}{4K^{1/3}}\left(1 + \sqrt{\frac{H_2}{H_1}}\right)^2\left(\frac{F}{4000}\right)^{2/3}$$

and F is the frequency (in megahertz), K is the effective Earth radius factor, and h, R_{F1}, H_1, and H_2 are in meters.

Figure 1.62 presents curves of diffraction loss over a smooth spherical Earth that have been established according to Bullington for various values of the coefficient M from $h/R_{F1} \geq -1$ with a coefficient of reflection of –1.

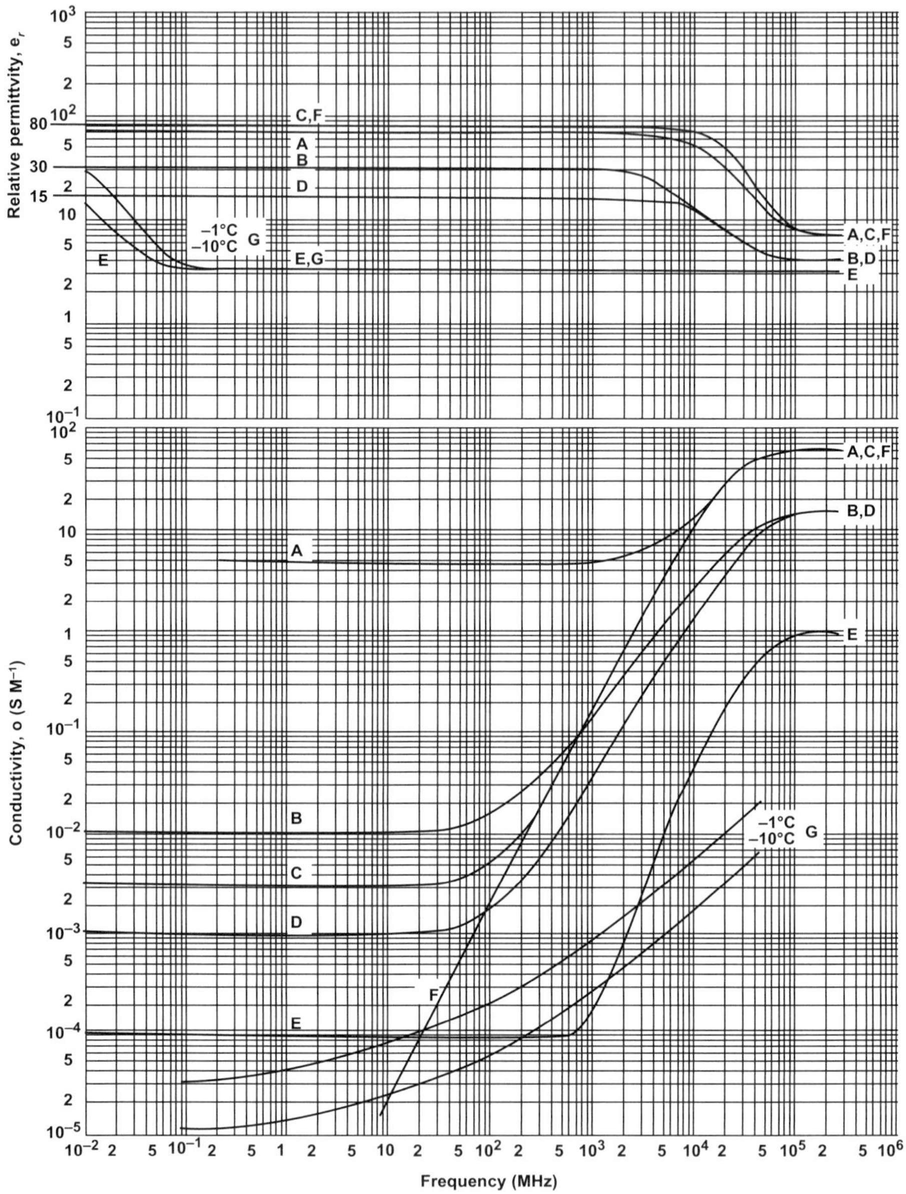

Figure 1.61 Relative permittivity and conductivity versus frequency for various ground types (ITU-R Rec.527): A, sea water of average salinity at 20 °C; B, wet ground; C, fresh water at 20 °C; D, medium dry ground; E, very dry ground; F, pure water at 20 °C; G, ice (fresh water).

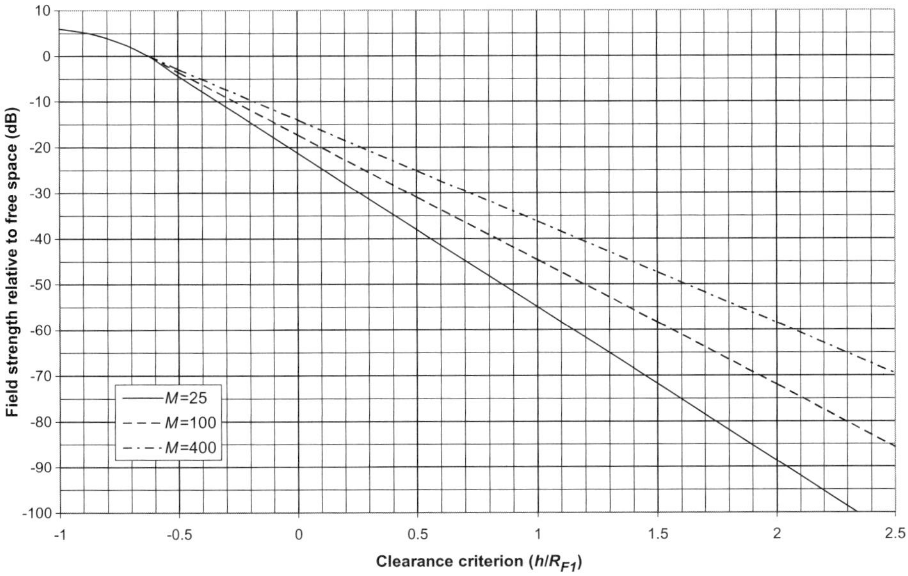

Figure 1.62 Diffraction loss over smooth Earth according to Bullington.

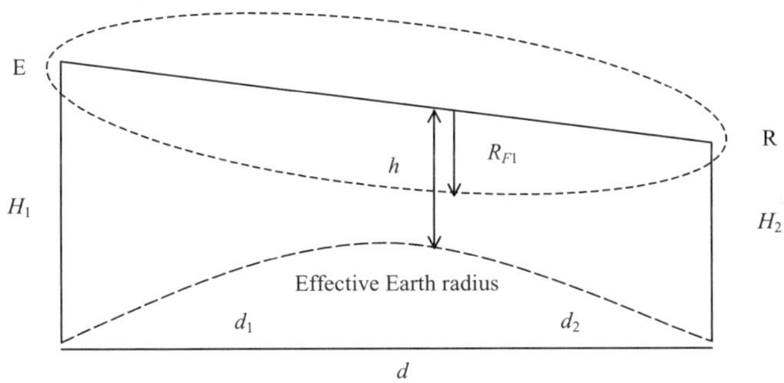

Figure 1.63 Graphic determination of $(h/R_{F1})_{\min}$.

It is however necessary to seek the minimal value of the h/R_{F1} criterion, which cannot be obtained directly from expressions (1.65), (1.84), and (1.92); the value of h is given by the following relation using the same unit for heights and distances (kilometers):

$$h = H_1 + (H_2 - H_1)\frac{d_1}{d} - \frac{d_1 d_2}{2a} \tag{1.95}$$

Using this method, h/R_{F1} is calculated by successive approximations or determined graphically on the real path profile, as shown in Figure 1.63, for the considered effective Earth radius. The attenuation due to the spherical

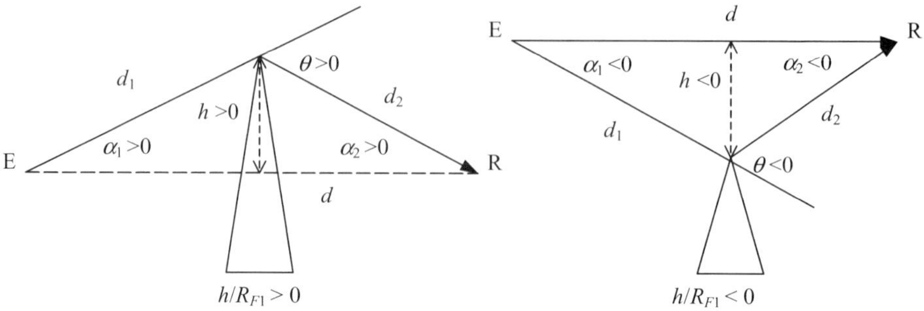

Figure 1.64 Diffraction over knife edge.

diffraction varies considerably according to the vertical refractivity gradient *dN/dh*; one can calculate this loss for various percentages of time while varying the effective Earth radius in expressions (1.83) and (1.95). The minimal value of the clearance criterion $(h/R_{F1})_{min}$ corresponds to the minimal positive value of the equivalent factor K_e presented in Figures 1.53 and 1.54 or calculated for the considered effective Earth radius.

1.9.2 Diffraction over Single-Knife-Edge Obstacle

Assume that the wavelength is small compared to the size of the obstacle, which is in general the case at frequencies higher than 30 MHz. The Fresnel–Kirchhoff obstacle loss due to midpath single-knife-edge diffraction as presented in Figure 1.64 is given by the following relation expressed in decibels (ITU-R P.526):

$$J(v) = 6.9 + 20\log\left[\sqrt{(v-0.1)^2+1} + v - 0.1\right]$$

$$v = h\sqrt{\frac{2(d_1+d_2)}{\lambda d_1 d_2}} = \theta\sqrt{\frac{2d_1 d_2}{\lambda(d_1+d_2)}} = \sqrt{\frac{2h\theta}{\lambda}} = \sqrt{\frac{2d\alpha_1\alpha_2}{\lambda}} \qquad (1.96)$$

where h = height of top compared to chord ER (m)
d, d_1, d_2 = distances, $d \approx d_1 + d_2$ (m)
α_1, α_2 = obstruction angles (rad)
θ = diffraction angle of same sign as h (rad) given by $\alpha_1 + \alpha_2$ = $\arcsin(h/d_1) + \arcsin(h/d_2) \approx h/d_1 + h/d_2$

Figure 1.65 presents the diffraction loss due to knife-edge-type obstacle with a negligible thickness which depends on the clearance parameter h/R_{F1} according to the following expression derived from relation (1.96):

$$A_{OB} = 6.9 + 20\log\left[\sqrt{\left(\frac{h}{R_{F1}}\sqrt{2}-0.1\right)^2+1} + \frac{h}{R_{F1}}\sqrt{2} - 0.1\right]$$

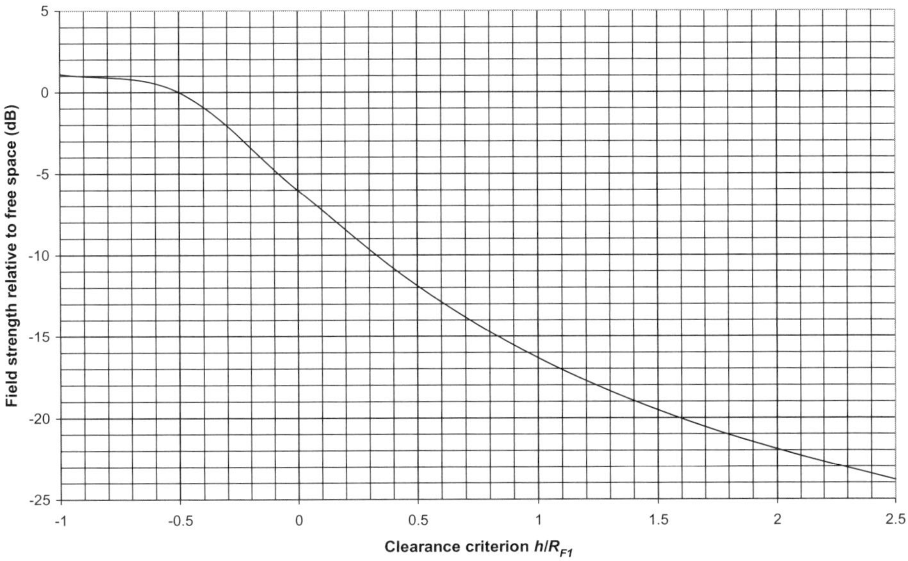

Figure 1.65 Diffraction loss single over knife edge.

where

$$\nu = \frac{h}{R_{F1}}\sqrt{2}$$

1.9.3 Diffraction over Single Rounded Obstacle

Figure 1.66 gives the geometric elements needed for determination of the obstacle loss that results from a single rounded obstruction. The diffraction loss over a rounded obstacle can be expressed by the relation (ITU-R P.526)

$$A_{OB} = J(\nu) + T(m, n) \tag{1.97}$$

The attenuation $J(\nu)$, which corresponds to the equivalent sharp edge of height h, is given by relation (1.96) and the additional loss $T(m,n)$, due to the obstacle curvature, by the expression

$$T(m, n) = km^b$$

with

$$k = 8.2 + 12n \qquad m = \frac{R[(d_1 + d_2)/d_1 d_2]}{(\pi R/\lambda)^{1/3}}$$

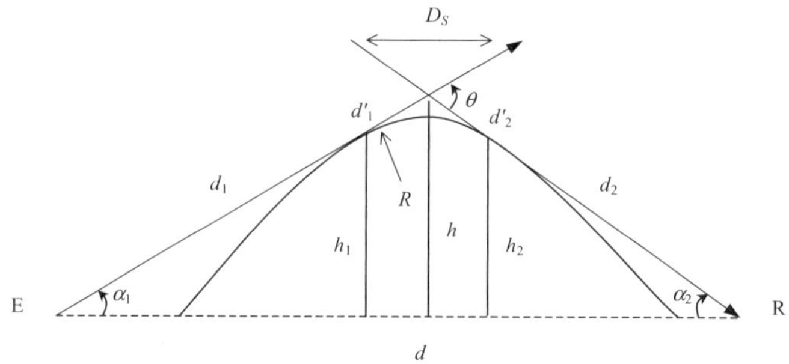

Figure 1.66 Diffraction over a rounded edge.

$$b = 0.73 + 0.27[1 - \exp(-1.43n)] \qquad n = \frac{h}{R}\left(\frac{\pi R}{\lambda}\right)^{2/3}$$

where R = equivalent radius of curvature of obstacle (m)
h = height of vertex (m)
λ = wavelength (m)
d_1, d_2 = distances from ends to vertex (m)

The equivalent radius of curvature of the obstacle can be calculated as

$$R = \frac{D_S}{\theta} \qquad \text{or} \qquad R = K\frac{D_S}{\theta}$$

where Ds = occultation distance along rounded edge (m)
K = effective Earth radius factor according to whether K is considered in calculation of h_1, h_2, and h
θ = angle of diffraction, such as $\theta < 0.2$ (rad)

with

$$\theta = \arcsin\left(\frac{h_1}{d_1}\right) + \arcsin\left(\frac{h_2}{d_2}\right) = \arcsin\left(\frac{h}{d_1' + D_S/2}\right) + \arcsin\left(\frac{h}{d_2' + D_S/2}\right)$$

where d_1' and d_2' are the distances between the terminals and their horizons to the obstacle in meters. Note that when R tends towards zero, expression (1.97) is reduced to the first term, which corresponds to the knife-edge loss.

1.9.4 Diffraction over Double Isolated Edges

When the two edges have similar altitudes, we use the method of Epstein and Peterson, whose geometric elements are presented in Figure 1.67. The method of Epstein and Peterson consists in successively applying to the two obstacles

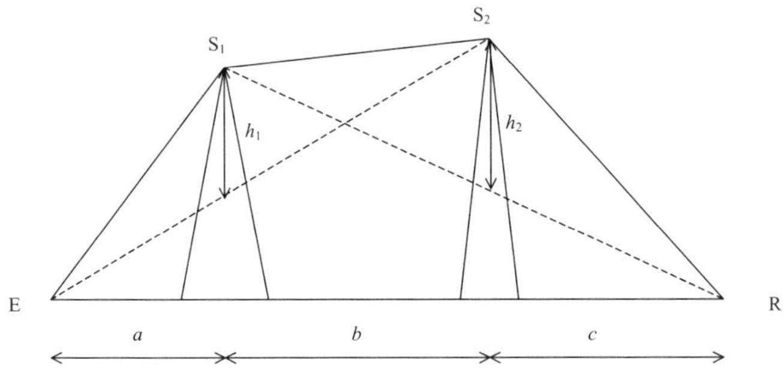

Figure 1.67 Method of Epstein and Peterson.

the theory of diffraction by a single knife edge, the top of the first obstacle being used as the source of diffraction over the second. The first path, defined by the distance $a + b$ and the height h_1, gives the loss A_1 and the second path, defined by the distance $b + c$ and the height h_2, gives the loss A_2, to which one adds a term of correction A_C. The obstacle loss is thus given by the relation

$$A_{OB} = A_1 + A_2 + A_C \tag{1.98}$$

where

$$A_C = 10\log\left(\frac{(a+b)(b+c)}{b(a+b+c)}\right)$$

which is valid when A_1 and A_2 each exceed approximately 15 dB.

When one of the edges is prevalent, Deygout's method is applied using the geometric elements presented in Figure 1.68.

In Deygout's method, the first path of diffraction corresponds to the edge which is prevalent and is defined by the distances a and $b + c$ and by the height h_1; in the same way, the second path is determined by the distances b and c and the height h_2. The obstacle loss is then given by the relation

$$A_{OB} = A_1 + A_2 \tag{1.99}$$

The diffraction loss over a knife edge, expressed in decibels, can be calculated using the following simplified relation when its value is higher than about 15 dB:

$$A = 16 + 20\log\left(\frac{h}{R_{F1}}\right) \tag{1.100}$$

In the case of rounded top obstacles, the same method can be applied by using relation (1.97).

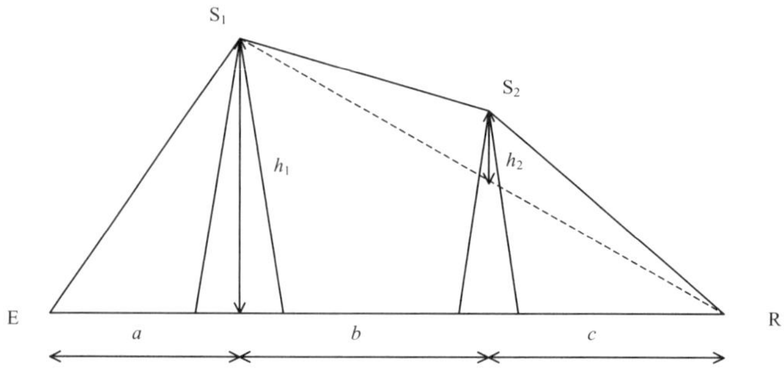

Figure 1.68 Deygout's method.

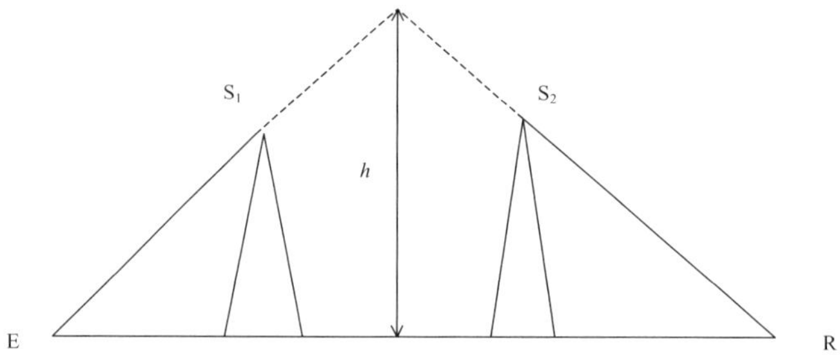

Figure 1.69 Millington's method.

Millington's method consists in replacing the obstacles met on the whole path by a single obstacle at the location of the vertex as shown in Figure 1.69.

1.9.5 Diffraction Loss General Curves

In the case of an irregular terrain, we can employ the following approximate formula, expressed in decibels, which is intermediate between diffraction over a smooth spherical Earth and a knife edge, where h is the height of the most significant obstacle:

$$A = 9 + 20\frac{h}{R_{F1}} \quad \text{for } \frac{h}{R_{F1}} > -0.5 \tag{1.101}$$

In the same way, for diffraction over a smooth spherical Earth, we can use the following approximate formula:

$$A = \begin{cases} 14 + 28\dfrac{h}{R_{F1}} & \text{for } \dfrac{h}{R_{F1}} > 0 \\ 14 + 23\dfrac{h}{R_{F1}} & \text{for } -0.5 < \dfrac{h}{R_{F1}} < 0 \end{cases} \tag{1.102}$$

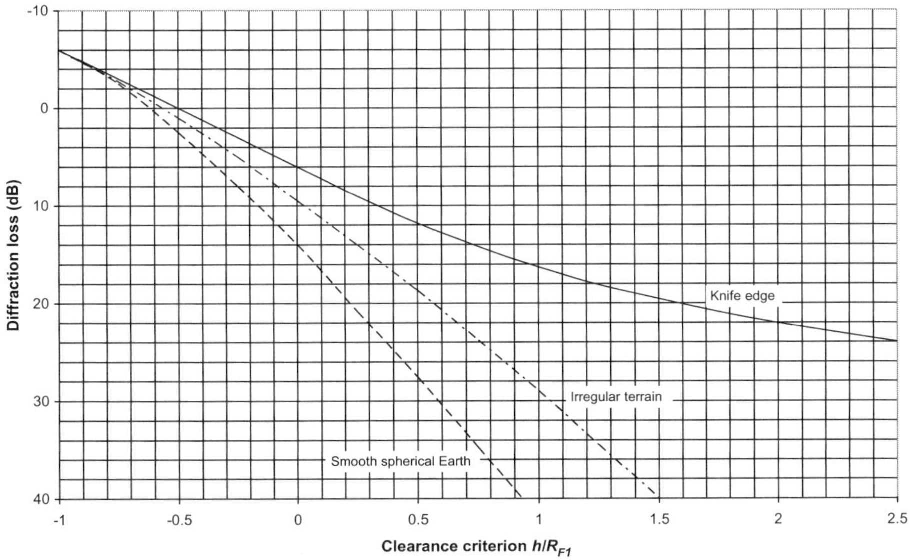

Figure 1.70 Diffraction loss for main obstacle types.

Figure 1.70 presents a synthesis of the diffraction loss for the main obstacle types which can be useful for an approximate calculation in the majority of cases. The curves corresponding to the knife edge and the smooth spherical Earth are very close to the values obtained using the general relation of diffraction loss (1.97).

1.9.6 Attenuation by Vegetation

Figure 1.71 presents specific attenuation through vegetation versus frequency for horizontal and vertical polarizations given by ITU-R Rep.1145 using an empirical method.

1.10 ATTENUATION BY ATMOSPHERIC GASES

Attenuation by atmospheric gases is due primarily to molecular absorption by oxygen and water vapor; oxygen has an absorption line at 118.74 GHz and a series of resonance lines between 50 and 70 GHz while water vapor presents three absorption lines at frequencies of 22.2, 183.3, and 325.4 GHz. Recommendation ITU-R P.676 gives the procedures to be used to accurately calculate gaseous attenuation at frequencies up to 1000 GHz. However, as most radio links are realized at frequencies lower than 50 GHz, for which the influence of dry air and water vapor is predominant, we will restrict interest to this limit in the following analysis which is based on reports such ITU-R Rep.719 and Rep.564 as well as on technical note 101 (NBS, 1966).

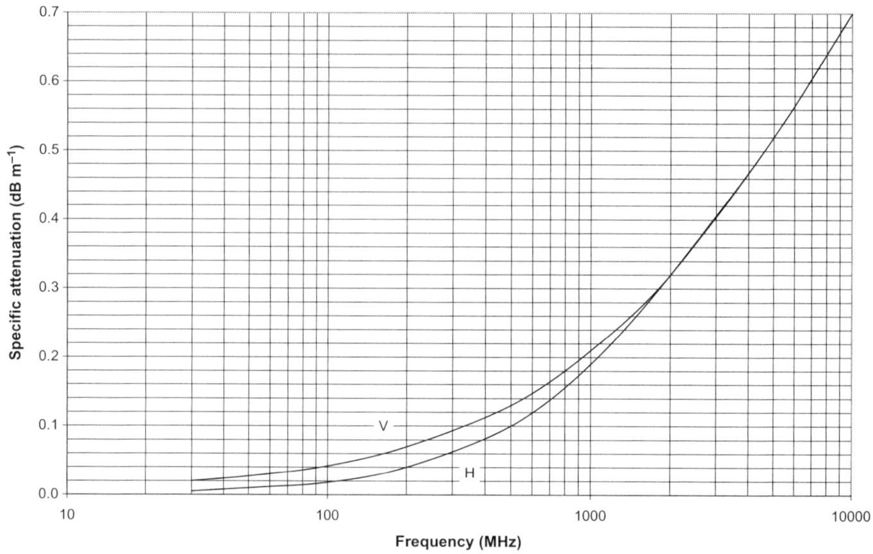

Figure 1.71 Attenuation by vegetation.

1.10.1 Specific Attenuation

Figure 1.72 presents specific attenuation by dry air and water vapor under normal conditions at ground level (1013 hPa; 15 °C; 7.5 g m^{-3}) as well as total specific attenuation versus frequency.

The specific absorption produced by either dry air or water vapor at ground level (1013 hPa, 15 °C), expressed in decibels per kilometer, breaks up into two terms: γ_0, which represents the contribution of dry air, and γ_W, which represents that of water vapor:

$$\gamma_0 = \begin{cases} \left[7.19\times10^{-3} + \dfrac{6.09}{f^2+0.227} + \dfrac{4{,}81}{(f-57)^2+1.5}\right] f^2 \times 10^{-3} & \text{for } f < 57\,\text{GHz} \\ \left[(3.79\times10^{-7})f + \dfrac{0.265}{(f-63)^2+1.59} + \dfrac{0.028}{(f-118)^2+1.47}\right](f+198)^2 \times 10^{-3} & \\ & \text{for } f > 63\,\text{GHz} \end{cases} \tag{1.103}$$

$$\gamma_W = \left[0.05 + 0.0021\nu + \frac{3.6}{(f-22.2)^2+8.5} + \frac{10.6}{(f-183.3)^2+9} + \frac{8.9}{(f-325.4)^2+26.3}\right] f^2 \nu \times 10^{-4} \quad \text{for } f < 350\,\text{GHz}$$

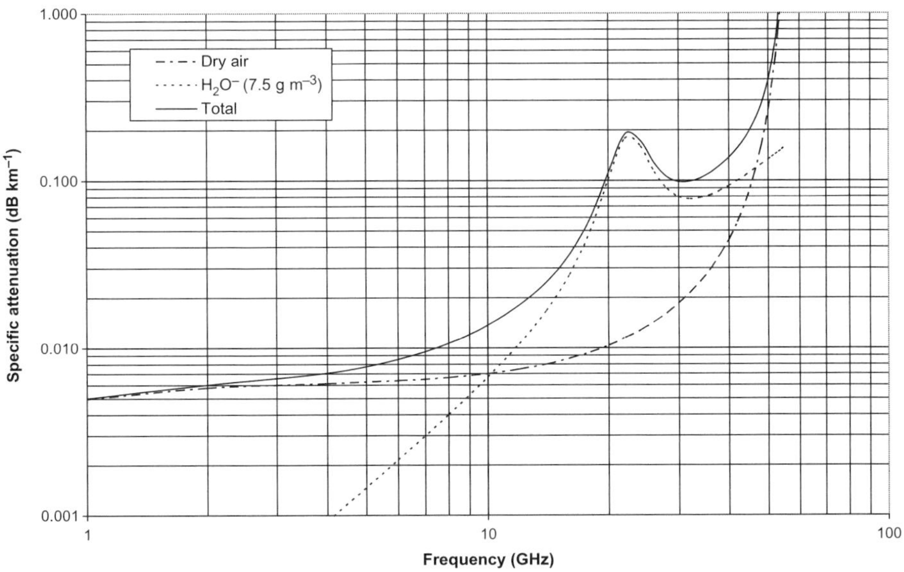

Figure 1.72 Specific attenuation by atmospheric gases.

where v is water vapor concentration in grams per cubic meter. Figures 1.40 and 1.41 give the geographical distribution for various periods of the year of water vapor concentration, and this can also be given from usual weather data using relations (1.78).

1.10.2 Terrestrial Paths

Total absorption by atmospheric gases on a terrestrial radio link of length d becomes

$$A_a = (\gamma_0 + \gamma_W)d \tag{1.104}$$

Figure 1.73 presents another way to evaluate the median absorption according to distance and frequency that is largely used in the calculation of very long microwave radio links budgets such as those which use the troposcatter mode. The median absorption by water vapor and oxygen according to frequency and distance, expressed in decibels, can be calculated using the relation

$$A_a = 0.01485\sqrt{f}\left(0.004075d + 1.45 - \sqrt{0.00001173d^2 - 0.0097875d + 2.1}\right) \tag{1.105}$$

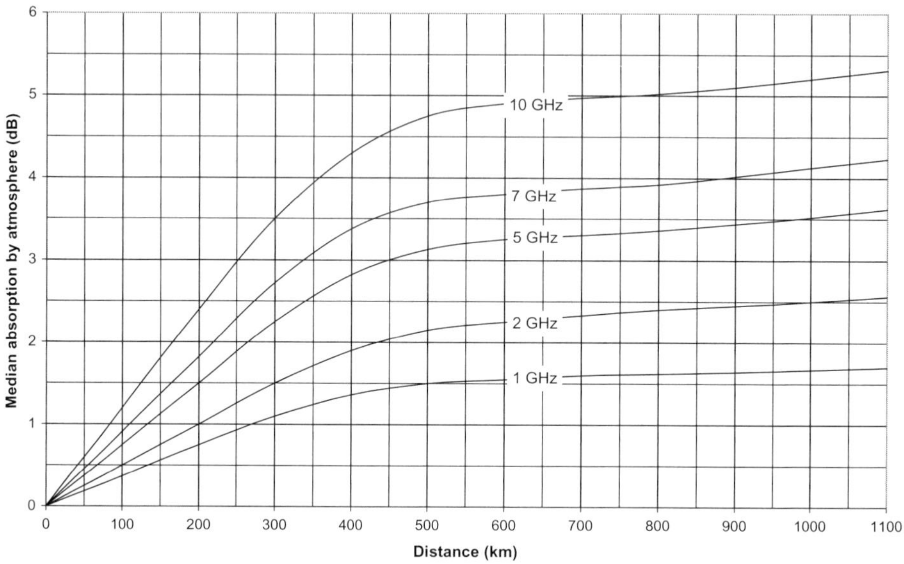

Figure 1.73 Median absorption versus frequency and distance (NBS, 1966).

where f = frequency (MHz)
d = distance (km)

1.10.3 Slant Paths

In the case of oblique paths, we suppose an exponential decrease of air density with altitude and utilize an equivalent height h_0 for dry air and h_W for water vapor, expressed in kilometers, for frequencies outside the range 57–63 GHz, such as

$$h_0 = \begin{cases} 6\,\text{km} & \text{for } f < 57\,\text{GHz} \\ 6 + \dfrac{40}{(f-118.7)^2+1} & \text{for } 63\,\text{GHz} < f < 350\,\text{GHz} \end{cases}$$

$$h_W = h_{W0}\left[1 + \frac{3}{(f-22.2)^2+5} + \frac{5}{(f-183.3)^2+6} + \frac{2.5}{(f-325.4)^2+4}\right] \quad \text{for } f < 350\,\text{GHz}$$

where h_{W0} equals 1.6 km in clear weather and 2.1 km in rainy weather. Total attenuation in the direction of the zenith starting from sea level then becomes

$$A_a = h_0\gamma_0 + h_W\gamma_W \tag{1.106}$$

1.10.3.1 Elevation Angle Greater Than 10° Total attenuation in the case of inclined paths between a satellite and an Earth station situated at sea level is given by the relation

$$A_a = \frac{\gamma_0 h_0 + \gamma_W h_W}{\sin\theta} \tag{1.107}$$

Figure 1.74 presents the total attenuation at sea level for water vapor concentrations of respectively 7.5 g m^{-3} (continental climate) and 20 g m^{-3} (oceanic and tropical climates) versus frequency for various values of elevation angle. To determine the attenuation on an oblique path, in which elevation angle θ is generally higher than 10°, from a station located at an altitude h_S, this last expression becomes

$$A_a = \frac{\gamma_0 h_0 \exp(-h_S/h_0) + \gamma_W h_W}{\sin\theta} \qquad \theta > 10°$$

Relation (1.107) is also applicable to terrestrial inclined paths between a station situated at an altitude h_1 and another one at a higher altitude h_2 by replacing the values h_0 and h_W by the following values:

$$h_0' = h_0\left[\exp\left(-\frac{h_1}{h_0}\right) - \exp\left(-\frac{h_2}{h_0}\right)\right] \qquad h_W' = h_W\left[1 - \exp\left(\frac{h_1 - h_2}{h_W}\right)\right]$$

1.10.3.2 Elevation Angle between 0° and 10° In the case of inclined paths between a satellite and an Earth station situated at sea level, the total attenuation becomes

$$A_a = \frac{\sqrt{a}}{\cos\theta}\left[\gamma_0\sqrt{h_0}\,F\left(\tan\theta\sqrt{\frac{a}{h_0}}\right) + \gamma_W\sqrt{h_W}\,F\left(\tan\theta\sqrt{\frac{a}{h_W}}\right)\right]$$

where a = effective Earth radius including refraction (km)
θ = elevation angle
F = function defined as $F(x) - 1/\left(0.661x + 0.339\sqrt{x^2 + 5.51}\right)$

1.11 ATTENUATION AND DEPOLARIZATION BY HYDROMETEORS

An attenuation can be experienced on terrestrial paths, either on microwave radio links or between an Earth station and a satellite, as a result of absorption and scattering by rain, snow, hail, or fog. Normally, for the very small percentage of time that is of interest in system design at frequencies above 5 GHz, we need only consider the attenuation due to rain. Attenuation by rain depends

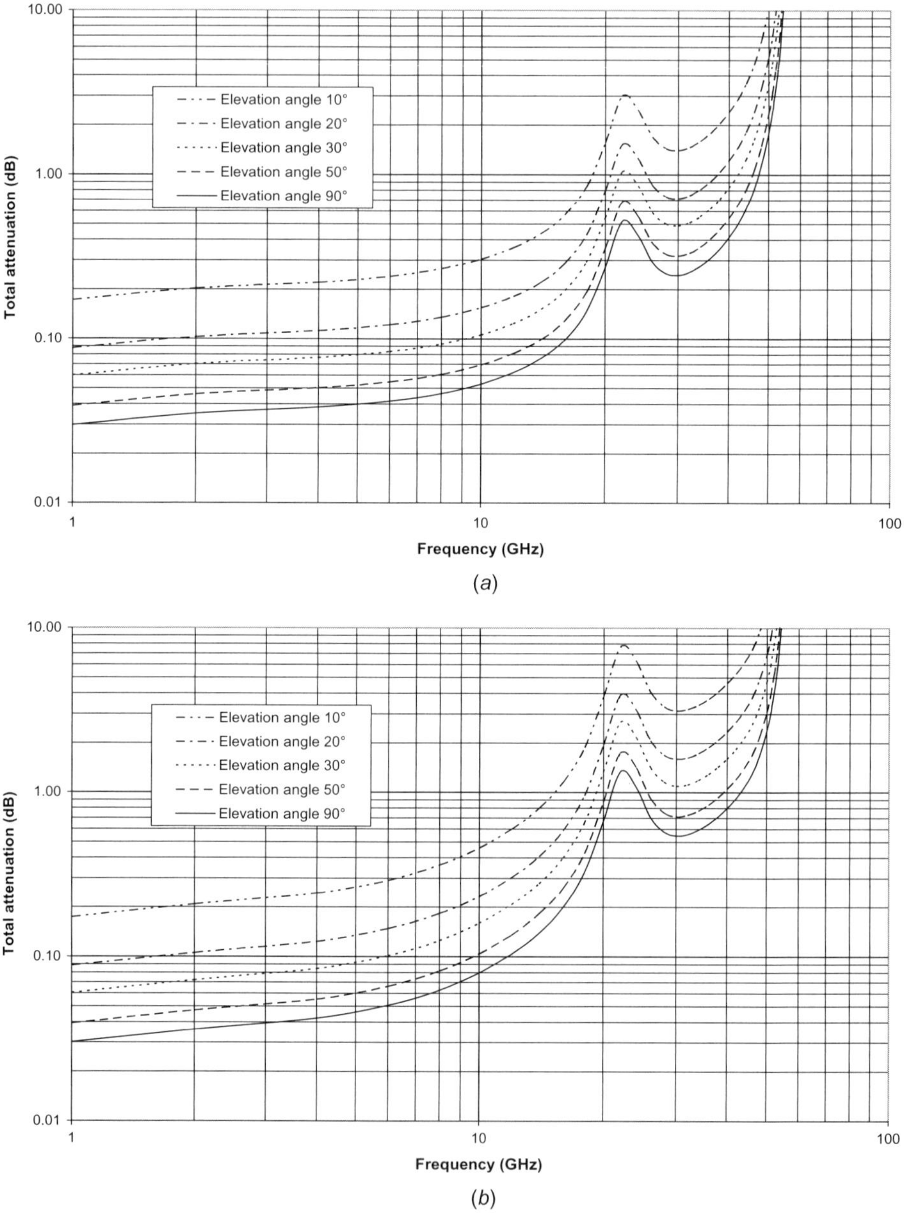

Figure 1.74 Total attenuation at sea level for water vapor concentration of (a) 7.5 $g\,m^{-3}$ and (b) 20 $g\,m^{-3}$ on oblique paths.

on the frequency and intensity of precipitation as well as the distance because of the nonuniformity of the rainfall rate throughout the path. Moreover, as the drops of rain have a lentil form during the fall, their axis of revolution being vertical, the attenuation differs according to whether the wave is horizontally or vertically polarized; it appears, in the course of propagation, as a component whose polarization is orthogonal to the expected polarization due to a phenomenon called cross-polarisation.

1.11.1 Rain Attenuation

Figure 1.75 shows the attenuation coefficient due to the rain, γ_R, versus frequency and rainfall rate. The relation between specific attenuation γ_R (in decibels per kilometer) and rainfall rate R (millimeters per hour) given by ITU-R P.838 is

$$\gamma_R = kR^{\alpha} \tag{1.108}$$

with

$$k = \frac{k_H + k_V + (k_H - k_V)\cos^2(\theta)\cos(2\tau)}{2}$$

$$\alpha = \frac{k_H\alpha_H + k_V\alpha_V + (k_H\alpha_H - k_V\alpha_V)\cos^2(\theta)\cos(2\tau)}{2k}$$

where θ = path elevation (deg)
τ = tilt angle of linearly polarized electric field vector with respect to horizontal or 45° in circular polarization (deg)

The coefficients k and α are given by the relations

$$\log k = \sum_{j=1}^{4} a_j \exp\left[-\left(\frac{\log f - b_j}{c_j}\right)^2\right] + m_k \log f + c_k$$

$$\alpha = \sum_{j=1}^{4} a_j \exp\left[-\left(\frac{\log f - b_j}{c_j}\right)^2\right] + m_\alpha \log f + c_\alpha$$

where the regression coefficients k_H, k_V, α_H, and α_V are given below:

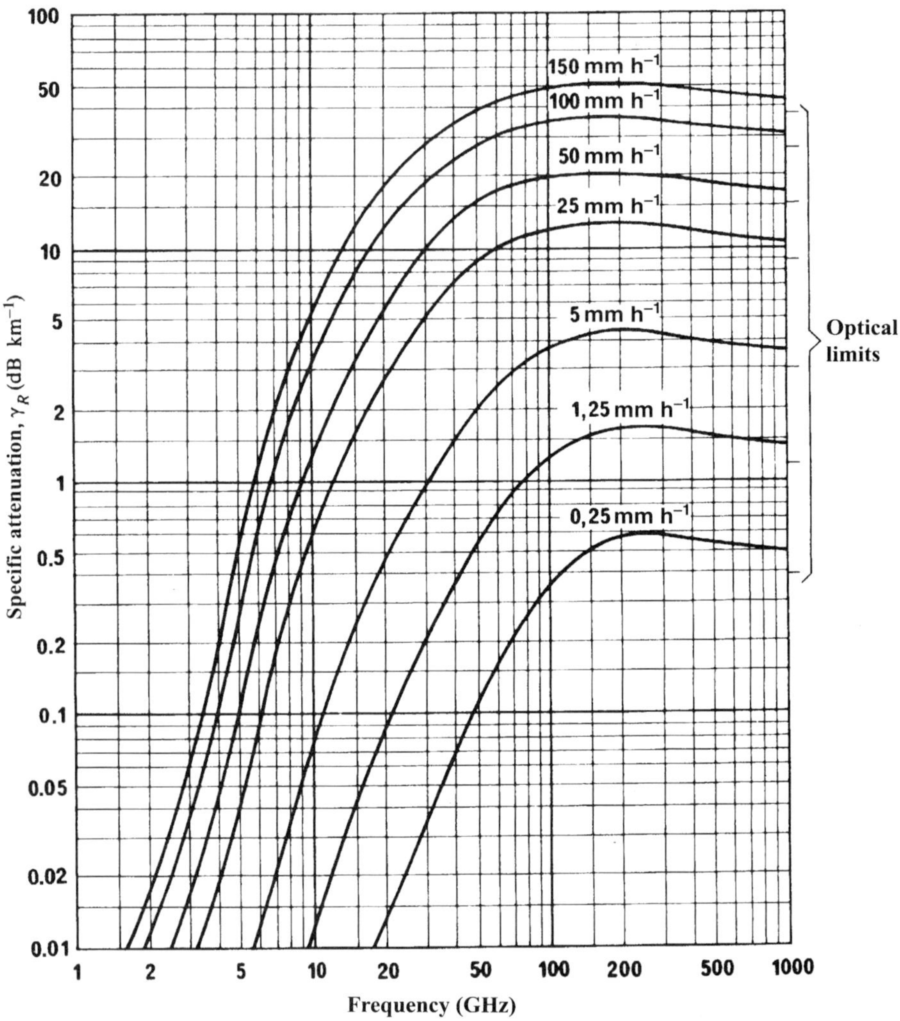

Figure 1.75 Rain attenuation (ITU-R Rep.721).

REGRESSION COEFFICIENTS FOR K_H

j	a_j	b_j	c_j	m_k	c_k
1	−5.33980	−0.10008	1.13098	−0.18961	0.71147
2	−0.35351	1.26970	0.45400	−0.18961	0.71147
3	−0.23789	0.86036	0.15354	−0.18961	0.71147
4	−0.94158	0.64552	0.16817	−0.18961	0.71147

REGRESSION COEFFICIENTS FOR κ_V

j	a_j	b_j	c_j	m_k	c_k
1	–3.80595	0.56934	0.81061	–0.16398	0.63297
2	–3.44965	–0.22911	0.51059	–0.16398	0.63297
3	–0.39902	0.73042	0.11899	–0.16398	0.63297
4	0.50167	1.07319	0.27195	–0.16398	0.63297

REGRESSION COEFFICIENTS FOR α_H

j	a_j	b_j	c_j	m_k	c_k
1	–0.14318	1.82442	–0.55187	0.67849	–1.95537
2	0.29591	0.77564	0.19822	0.67849	–1.95537
3	0.32177	0.63773	0.13164	0.67849	–1.95537
4	–5.37610	–0.96230	1.47828	0.67849	–1.95537
5	16.17210	–3.29980	3.43990	0.67849	–1.95537

REGRESSION COEFFICIENTS FOR α_V

j	a_j	b_j	c_j	m_k	c_k
1	–0.07771	2.33840	–0.76284	–0.053739	0.83433
2	0.56727	0.95545	0.54039	–0.053739	0.83433
3	–0.20238	1.14520	0.26809	–0.053739	0.83433
4	–48.2991	0.791669	1.47828	–0.053739	0.83433
5	48.5833	0.791459	3.43990	–0.053739	0.83433

Recommendation ITU-R P.838 is bound to ITU-R P.837, which call for precise data files for all locations on the globe and a wide range of probabilities with an integration time of 1 min. For a quick calculation, the following tables, from ITU-R Rep.563, Rep.721, Rep.564, and Rep.724, provide the cumulative distribution over the average year, by rainfall climatic zones, of the exceeded rate of precipitation in millimeters per hour as well as the regression coefficients that correspond to the main frequency bands.

P (%)	A	B	C	D	E	F	G	H	J	K	L	M	N	P	Q
1	<0.1	0.5	0.7	2.1	0.6	1.7	3	2	8	1.5	2	4	5	12	24
0.3	0.8	2	2.8	4.5	2.4	4.5	7	4	13	4.2	7	11	15	34	49
0.1	2	3	5	8	6	8	12	10	20	12	15	22	35	65	72
0.03	5	6	9	13	12	15	20	18	28	23	33	40	65	105	96
0.01	8	12	15	19	22	28	30	32	35	42	60	63	95	145	115
0.003	14	21	26	29	41	54	45	55	45	70	105	95	140	200	142
0.001	22	32	42	42	70	78	65	83	55	100	150	120	180	250	170

Frequency (GHz)	k_H	k_V	α_H	α_V
1	0.0000387	0.0000352	0.912	0.880
2	0.000154	0.000138	0.963	0.923
4	0.000650	0.000591	1.121	1.075
6	0.00175	0.00155	1.308	1.265
7	0.00301	0.00265	1.332	1.312
8	0.00454	0.00395	1.327	1.310
10	0.0101	0.00887	1.276	1.264
12	0.0188	0.0168	1.217	1.200
15	0.0367	0.0335	1.154	1.128
20	0.0751	0.0691	1.099	1.065
25	0.124	0.113	1.061	1.030
30	0.187	0.167	1.021	1.000
35	0.263	0.233	0.979	0.963
40	0.350	0.310	0.931	0.929
45	0.442	0.393	0.903	0.897
50	0.536	0.479	0.873	0.868
60	0.707	0.642	0.826	0.824
70	0.851	0.784	0.793	0.793
80	0.975	0.906	0.769	0.769
90	1.06	0.999	0.753	0.754
100	1.12	1.06	0.743	0.744
120	1.18	1.13	0.731	0.732
150	1.31	1.27	0.710	0.711
200	1.45	1.42	0.689	0.690
300	1.36	1.35	0.688	0.689
400	1.32	1.31	0.683	0.684

1.11.1.1 Terrestrial Paths The intensity of precipitation varies from one point to another one and over time; as a result, the size of the rain cells is smaller as the intensity increases and the radio wave meets different rain conditions along the path. The estimate of the value of the attenuation A_R due to rain exceeded during a percentage of time p ranging between 0.001 and 1% can be obtained using the following relations (ITU-R P.530):

- For latitudes[14] ≥30°:

$$A_R = 0.12 A_{0.01} p^{-(0.546+0.043 \log p)} \tag{1.109}$$

- For latitudes[15] <30°:

$$A_R = 0.07 A_{0.01} p^{-(0.855+0.139 \log p)}$$

[14]This formula was established in order to obtain values equal to 0.12, 0.39, 1, and 2.14 for respectively 1, 0.1, 0.01, and 0.001% of time.

[15]This formula was established in order to obtain values equal to 0.07, 0.36, 1, and 1.44 for respectively 1, 0.1, 0.01 and 0.001% of time.

with

$$A_{0.01} = \gamma_{R0.01} d r_{0.01} \qquad \gamma_{R0.01} = k R_{0.01}^{\alpha}$$

$$r_{0.01} = \frac{1}{1 + d/d_0} \qquad d_0 = 35 \exp(-0.015 R_{0.01})$$

where $r_{0.01}$ = pathlength[16] reduction factor
$A_{0.01}$ = exceeded attenuation for 0.01% of time
d = pathlength (km)
$R_{0.01}$ = exceeded rainfall rate for 0.01% of time[17] (integration time 1 min)

Where long-term attenuation statistics exist at one polarization on a given link, the attenuation for the other polarization over the same link may be estimated as

$$A_V = \frac{300 A_H}{335 + A_H} \qquad A_H = \frac{335 A_V}{300 - A_V}$$

Figure 1.76 presents the 15 rainfall zones. Recommendation ITU-RP.841 gives further detailed information and data about the conversion of the statistics over the average year into statistics over any month (called "worst month" in the past); concerning the effects due to the rain, the percentage of annual time, p_A, and the percentage of time for any month, p_M, are generally related by the following formulas, which are illustrated in Figure 1.77:

$$p_A = 0.3 p_M^{1.15} \qquad P_M = 2.85 p_A^{0.87} \tag{1.110}$$

1.11.1.2 Earth–Space Paths The approximate method of calculation of the long-term statistics of attenuation due to the rain on an oblique way (ITU-R Rep.564) rises from the preceding one by considering the following parameters, which are presented in Figure 1.78:

H_S: altitude of Earth station (km)
θ: elevation angle (deg)
φ: latitude of Earth station (deg)

We calculate the effective rain height H_R, in kilometers, starting from the latitude of the Earth station:

$$H_R = \begin{cases} 3 + 0.028\varphi & \text{for } 0 \le \varphi \le 36° \\ 4 - 0.075(\varphi - 36) & \text{for } \varphi \ge 36° \end{cases} \tag{1.111}$$

[16]The attenuation by rain covers the atmospheric attenuation for the $r_{0.01}$ fraction of the path.
[17]If $R_{0.01} > 100$ mm/h, use the value 100 mm/h for the determination of d_0.

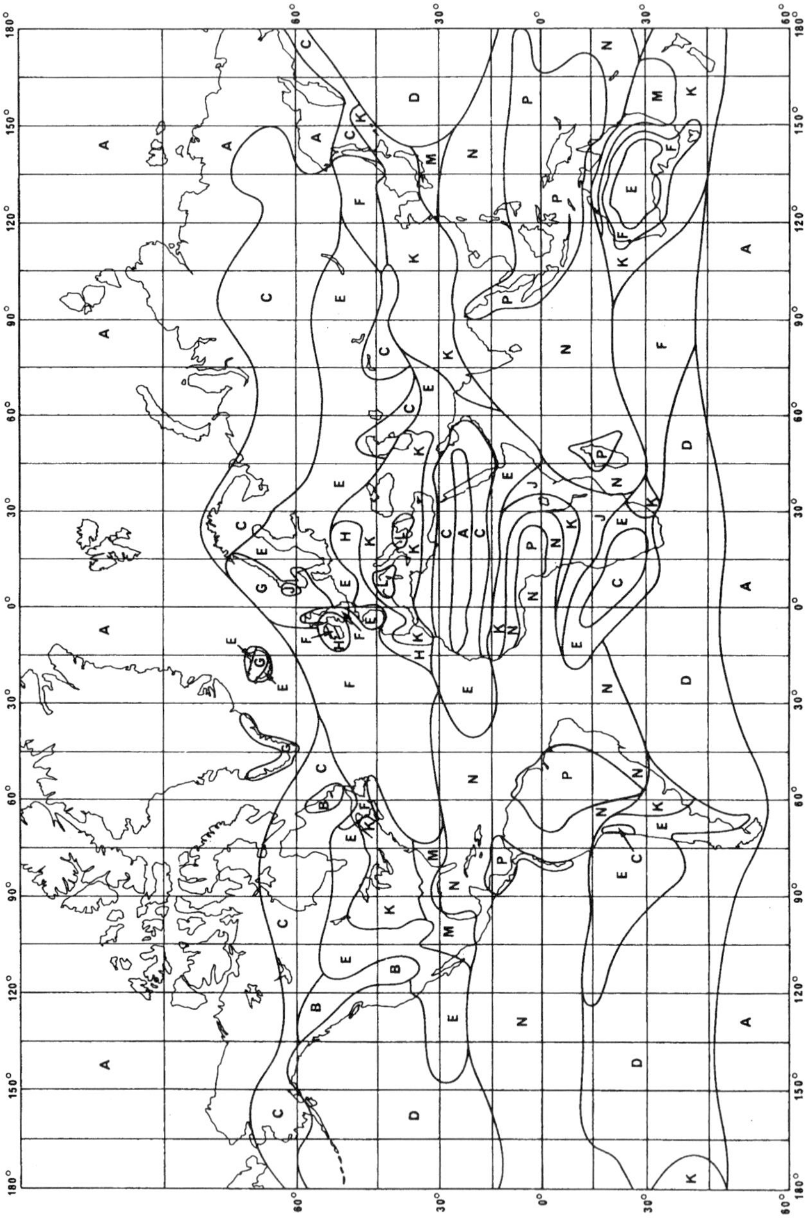

Figure 1.76 Rainfall zones (ITU-R Rep.724).

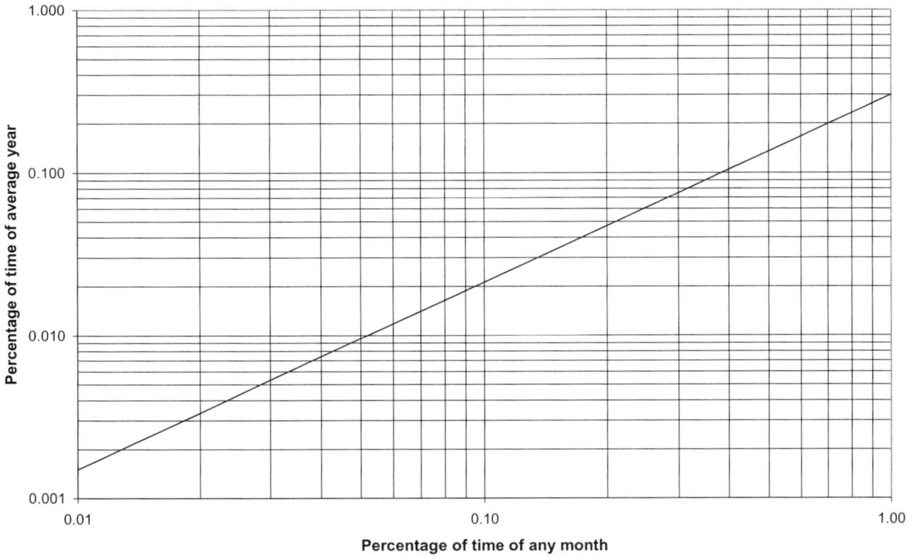

Figure 1.77 Relation between percentage of time for any month and for average year.

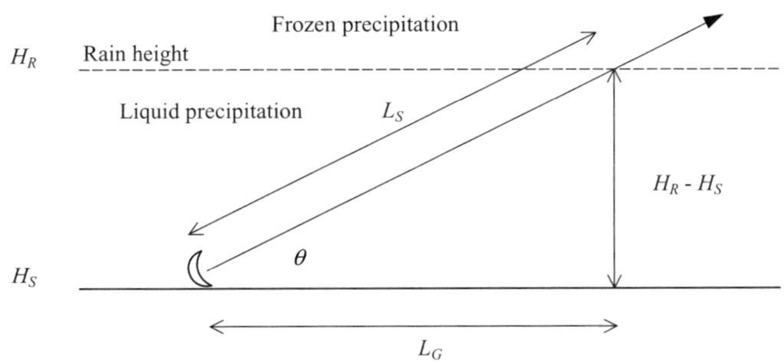

Figure 1.78 Earth–space path.

next, the length of the oblique path, L_S, below the effective rain height:

$$L_S = \frac{H_R - H_S}{\sin\theta} \quad \text{for } \theta \geq 5^\circ \tag{1.112}$$

then, the horizontal projection length L_G of the oblique path:

$$L_G = L_S \cos\theta$$

and, finally, the attenuation A_P given by relation[18] (1.109) with

[18]The Attenuation by rain covers the atmospheric loss for the $r_{0.01}$ fraction of the path defined below.

$$A_{0.01} = \gamma_{R0.01} L_S r_{0.01} \tag{1.113}$$

$$\gamma_{R0.01} = k(R_{0.01})^{\alpha} \qquad r_{0.01} = \frac{1}{1 + L_G/L_0} \qquad L_0 = 35\exp(-0.015 R_{0.01})$$

1.11.1.3 Site Diversity (Rayleigh's Equations for Correlation between Random Variables) As rain cells are of finite size and distributed in a random way, it is possible to improve considerably the availability of communications by satellite that use high frequencies by employing site diversity. We consider two stations located at two distinct sites which are each affected by the same individual probability P_i for a given attenuation A_i and bound by a correlation coefficient K. The combined probability P_c so that both stations simultaneously undergo the same attenuation A_i corresponding to their individual probability P_i is given by Rayleigh's relation:

$$P_c = \frac{P_i^2}{K^2 P_i + 1 - K^2} \tag{1.114}$$

such that for all (K, P_i)

$$P_i^2 \le P_c \le P_i$$

$$P_i = 1 \Rightarrow P_c = 1$$

$$P_i \to 0 \Rightarrow P_c \to P_i^2$$

$$K = 0 \Rightarrow P_c = P_i^2$$

$$K = 1 \Rightarrow P_c = P_i$$

We define the improvement factor by the expression

$$F_a = \begin{cases} \dfrac{P_i}{P_c} \\[2ex] \dfrac{K^2 P_i + 1 - K^2}{P_i} \end{cases} \tag{1.115}$$

Figure 1.79 shows the various terms which come into play; we consider there the probabilities P_1, P_2, and P_3 and the individual and combined attenuations which correspond to them in order to determine the diversity gain as well as the equivalent increase of margin. Figure 1.80 illustrates the relation between the combined probability P_c and the individual probability P_i for various values of the correlation coefficient. Figure 1.81 presents the improvement factor F_a according to the individual probability P_i for various values of the correlation coefficient. The diversity gain is given by the relation

$$G_d = A_i(P_2) - A_c(P_2) = A_i(P_2) - A_i(P_1) \tag{1.116}$$

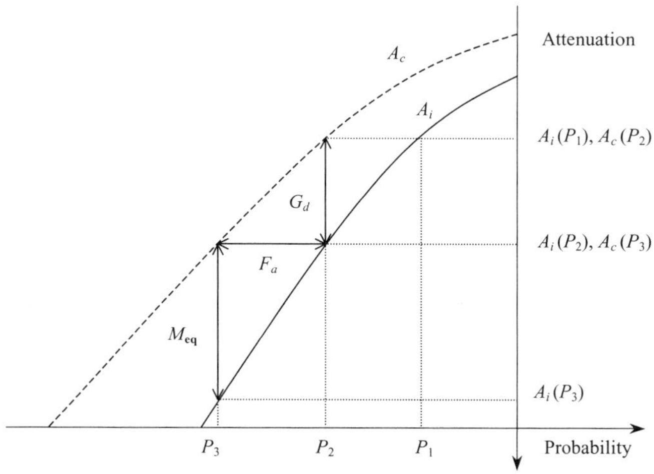

Figure 1.79 Representation of individual and combined distributions.

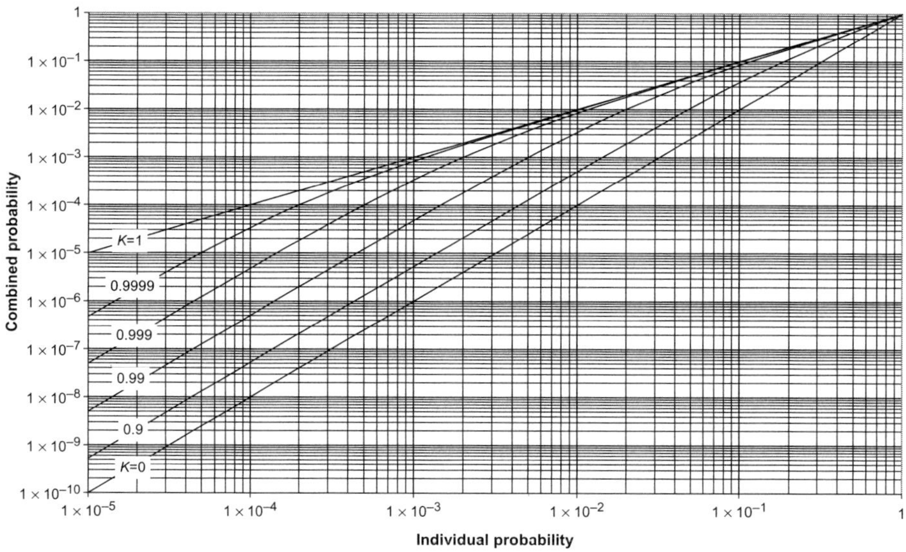

Figure 1.80 Relation between combined probability and individual probability versus correlation coefficient.

where

$$P_2 = \frac{P_1^2}{K^2 P_1 + 1 - K^2}$$

from which we obtain

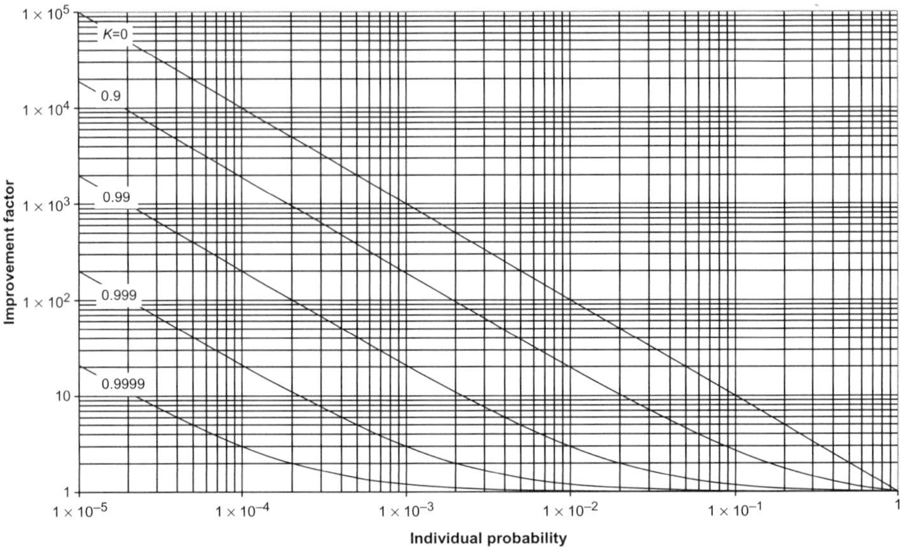

Figure 1.81 Relation between improvement factor and individual probability versus correlation coefficient.

$$P_1 = \frac{1}{2}\left(K^2 P_2 + \sqrt{K^4 P_2^2 + 4P_2(1-K^2)}\right) \tag{1.117}$$

and finally the equivalent increase of margin:

$$M_{\text{eq}} = A_i(P_3) - A_c(P_3) = A_i(P_3) - A_i(P_2) \tag{1.118}$$

where

$$P_3 = \frac{P_2^2}{K^2 P_2 + 1 - K^2}$$

When the two sites are sufficiently distant from one another, the individual probabilities of exceeding a given value can be different. Setting $P_{i1} < P_{i2}$, relation (1.114) becomes

$$P_c = \frac{P_{i1} P_{i2}}{K^2 P_{i2} + 1 - K^2}$$

In practical applications, we consider the probability of exceeding a given value for a single site P_s and the combined probability for the same exceeded value with two sites P_c while posing

$$\frac{1}{K^2} = 1 + \beta^2$$

from which

$$P_c = \frac{P_s^2(1+\beta^2)}{P_s+\beta^2} \tag{1.119}$$

$$F_a = \frac{P_s}{P_c} = \frac{1+\beta^2/P_s}{1+\beta^2} \approx 1 + \frac{\beta^2}{P_s} \tag{1.120}$$

with the following empirical values given by ITU-R P.618:

$$\beta^2 = \begin{cases} 10^{-4} d^{1.33} & \text{in general case} \\ 10^{-4}(\sin\theta)^{0.5} d^{1.5} & \text{considering elevation angle } \theta \end{cases} \tag{1.121}$$

As the rain corresponds to a small percentage of time (~1%), one replaces the probabilities corresponding to the total time by the percentage of time affected by the rain, which yields:

$$p_{c\%} = \frac{p_{s\%}^2(1+\beta^2)}{p_{s\%}+100\beta^2} \tag{1.122}$$

and

$$F_a = \frac{1+100\beta^2/p_{s\%}}{1+\beta^2} \approx 1 + \frac{100\beta^2}{p_{s\%}} \tag{1.123}$$

Figure 1.82 shows the relation between the percentage of time with and without diversity for the same attenuation.

Introducing the distribution function of the attenuation due to rain (1.109), we obtain

$$A_s = A_{0.01}\, 0.12\, p_{s\%}^{-(0.546+0.043\log p_{s\%})} \tag{1.124}$$

$$A_c = A_{0.01}\, 0.12\, p_{c\%}^{-(0.546+0.043\log p_{c\%})} \tag{1.125}$$

where

$$p_{c\%} = \frac{p_{s\%}}{2}\left[1+\sqrt{1+0.04\frac{d^{1.33}}{p_{s\%}}}\right] \tag{1.126}$$

Figure 1.83 shows, for example, by considering an individual attenuation $A_{0.01}$ of 25 dB, the improvement which would introduce the site diversity versus

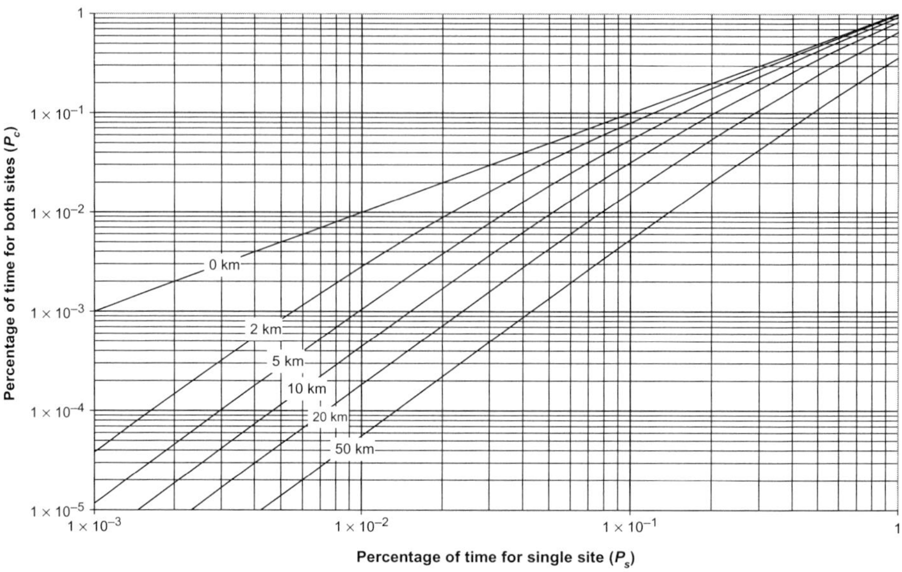

Figure 1.82 Individual and combined probabilities versus spacing.

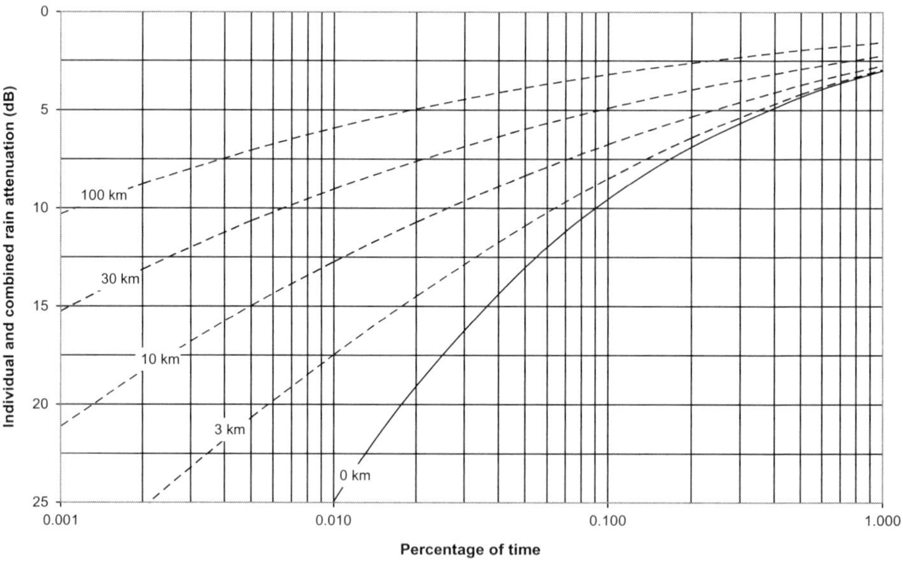

Figure 1.83 Site diversity improvement for $A_{0.01} = 25\,\text{dB}$.

percentage of time and spacing, that is, the distance between both Earth stations. An empirical expression for the diversity gain G is also given by the relation (Hodge, 1982)

$$G = G_d G_f G_\theta G_\psi \tag{1.127}$$

with

$$G_d = a(1 - e^{-bd}) \qquad G_f = e^{-0.025f}$$

$$G_\theta = 1 + 0.006\theta \qquad G_\psi = 1 + 0.002\psi$$

where $a = 0.78A - 1.94(1 - e^{-0.11A})$
$b = 0.59(1 - e^{-0.1A})$
d = distance between two sites (km)
A = attenuation by rain for only one site (dB)
f = frequency (GHz)
θ = elevation angle (deg)
ψ = angle in degrees formed between base connecting two sites and azimuth of satellite chosen so that $\psi \leq 90°$

1.11.1.4 Attenuation Due to Gases, Clouds, and Fog The method of pediction of attenuation due to rain was defined so that the calculated results coincide as much as possible with the experiments; consequently, it includes other losses due to gases, clouds, and fog for frequencies up to 40 GHz.

1.11.2 Depolarization by Hydrometeors

1.11.2.1 Polarization of Electromagnetic Wave As indicated in Section 1.1.5 and 1.3.3, the polarization of an electromagnetic wave is defined by the orientation of the electric field that is perpendicular to the direction of propagation; the projection of the point E of the electric field on a plane (P) perpendicular to the direction of propagation describes an ellipse for one period and this is that one calls the elliptic polarization represented on Figure 1.84. In short, elliptic polarization is characterized by the following parameters:

Direction of rotation

Axial ratio, ellipticity ratio, and angle of inclination ϕ of ellipse

In practice, most of the antennas radiate in linear polarization or in circular polarization which are particular cases of the elliptic polarization:

Polarization is linear when the axial ratio is infinite, that is, when the ellipse is completely flat and the vector E varies only in intensity.

Polarization is circular when the axial ratio is equal to 1.

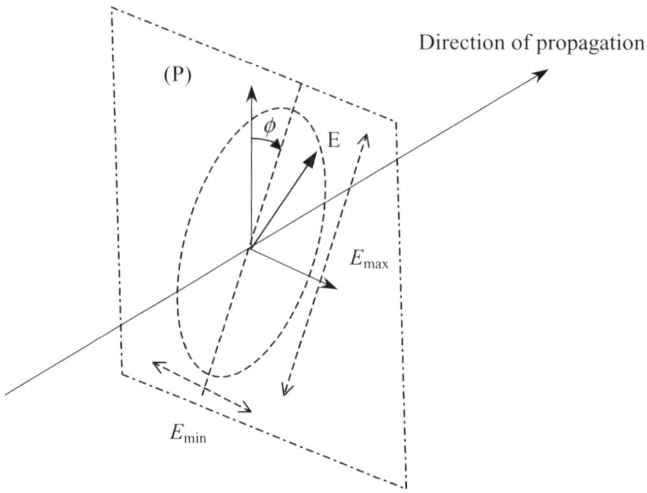

Figure 1.84 Elliptic polarization of radiated wave.

The elliptic polarization can also be regarded as the sum of two orthogonal components whose directions of rotation are opposite; two waves with orthogonal polarizations being in theory insulated, the same antenna can receive and/or emit simultaneously two carriers at the same frequency with polarizations horizontal and vertical or right-circular and left-circular.

1.11.2.2 Cross-Polarization Effects An important characteristic of the radio wave is its purity of polarization, i.e. the ratio between the component known as co-polar, which represents the useful signal, and the component known as cross-polar which constitutes the jamming power.

The cross-polarization results from the mechanism of depolarization by which appears, in the course of propagation, a polarization component which is orthogonal to the one expected; beyond 6 GHz, the principal effects are produced by hydrometeors, that is, the rain and the ice crystals. Figure 1.85 presents, on the one hand, the cross-polarization of two orthogonal waves propagated through a depolarizing medium and, on the other hand, the discrimination of both polarizations at the reception.

For a radio wave transmitted with a given polarization, the ratio of the power received with the expected polarization to the power received with the orthogonal polarization is called cross-polarization discrimination XPD, given by relation (1.47). For two radio waves transmitted with the same power and orthogonal polarization, the ratio of the copolarized power to the cross-polarized power in a given receiver is called cross-polarization isolation[19] (XPI).

[19]The XPI corresponds as well to the isolation between ports, that is, to the ratio of the transmitted power to the power collected on the orthogonal polarization when using the same feed horn for emission and reception.

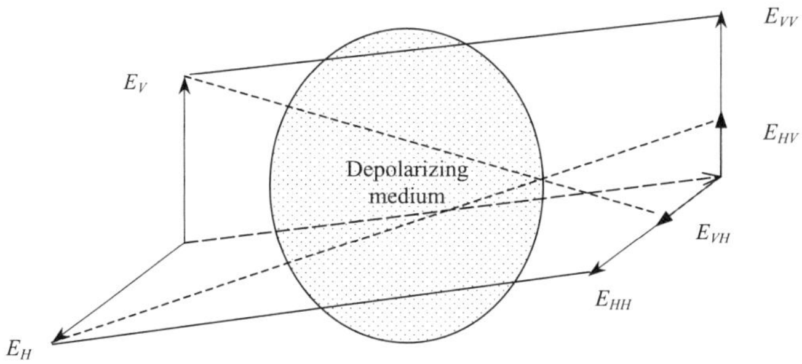

Figure 1.85 Cross-polarization.

Cross-polarization discrimination depends on the depolarization by the propagation medium and on the characteristics of purity of polarization of the emitting and receiving antennas, as it corresponds to the ratio of the copolarized power to the cross-polarised power of the same wave. We see that the wave emitted on vertical polarization E_V is received with level E_{VV} on the vertical polarization, which is the copolarized component constituting the useful signal, and E_{VH} on horizontal polarization, which is the cross-polarized or jamming component—reciprocally, levels E_{HH} and E_{HV} correspond to the wave emitted on horizontal polarization E_H.

Cross-polarization discrimination is given by the relations

$$(\text{XPD})_V = 20\log\left[\frac{E_{VV}}{E_{VH}}\right] \qquad (\text{XPD})_H = 20\log\left[\frac{E_{HH}}{E_{HV}}\right] \tag{1.128}$$

Cross-polarization isolation is given by the relations

$$(\text{XPI})_V = 20\log\left[\frac{E_{VV}}{E_{HV}}\right] \qquad (\text{XPI})_H = 20\log\left[\frac{E_{HH}}{E_{VH}}\right] \tag{1.129}$$

Cross-polarization discrimination due to hydrometeors $(\text{XPD})_{\text{hydro}}$ is strongly correlated with the copolar attenuation by rain, A_R; the method of prediction, given in ITU-R P.618, consists in calculating the terms separately depending on the rain and ice expressed in decibels:

$$(\text{XPD})_{\text{hydro}} = (\text{XPD})_{\text{rain}} - C_{\text{ice}} \tag{1.130}$$

To calculate the long-term statistics of depolarization starting from the statistics of the attenuation due to the rain, we take into account the following parameters:

A_R: attenuation exceeded during percentage of time $p_{\%}$ (dB)
τ: tilt angle of linearly polarized electric field vector with respect to horizontal or 45° in circular polarization (deg)
f: frequency (GHz)
θ: elevation angle (deg)

The cross-polarization discrimination due to rain not exceeded for $p_{\%}$ of the time is given by

$$(\text{XPD})_{\text{rain}} = C_f - C_A + C_\tau + C_\theta + C_\sigma \qquad (\text{dB}) \tag{1.131}$$

with

$$C_f = 30\log(f) \quad \text{for } 8 \le f \le 35\,\text{GHz}$$
$$C_A = \text{V}(f)\log(A_R)$$

where

$$\text{V}(f) = \begin{cases} 12.8 f^{0.19} & \text{for } 8 \le f \le 20\,\text{GHz} \\ 22.6 & \text{for } 20 \le f \le 35\,\text{GHz} \end{cases}$$

and

$$C_\tau = -10\log[1 - 0.484(1 + \cos(4\tau))]$$
$$C_\theta = -40\log[\cos\theta] \quad \text{for } \theta \le 60°$$
$$C_\sigma = 0.0052\sigma^2$$

where σ is the equivalent standard deviation of the distribution of the oblique angle of the rain drops, expressed in degrees, which takes the values of 0°, 5°, 10°, and 15° for 1, 0.1, 0.01, and 0.001% of the time. The term depending on ice crystals is given by the relation

$$C_{\text{ice}} = (\text{XPD})_{\text{rain}}\left[\frac{0.3 + 0.1\log(p_{\%})}{2}\right] \tag{1.132}$$

In the case of terrestrial paths, the depolarization due to ice crystals can be neglected because their presence is above the isotherm 0°, but reduction of the cross-polarization discrimination can occur in clear atmosphere under conditions of multipath propagation or in heterogeneous turbulence along the path.

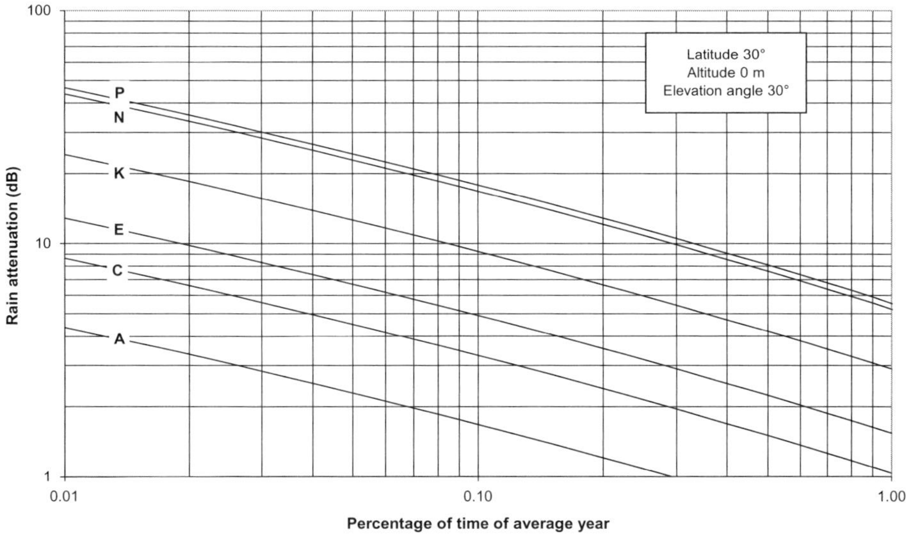

Figure 1.86 Attenuation by rain at 20 GHz.

The cross-polarization due to hydrometeors, $(XPD)_{hydro}$, is added to the total cross-polarization due to equipment, $(XPD)_E$, which is primarily related to the characteristics of the sources and the antennas concerning their purity of polarization in the following way:

$$(XPD)_{total} = -10\log(10^{-(XPD)_E/10} + 10^{-(XPD)_{hydro}/10}) \qquad (1.133)$$

Figures 1.86 and 1.87 show, for example, the attenuation and depolarization due to rain according to the percentage of time at frequency 20 GHz for various standard climates on an Earth–space path with:

- Axial ratio and cross-polarization isolation of 30 dB for satellite (ER = 0.5 dB)
- Axial ratio and cross-polarization isolation of 20 dB (ER = 1.7 dB) for Earth station

1.12 INFLUENCE OF IONOSPHERE

The table below, from ITU-R P.618, indicates the estimated maximum values of the ionospheric effects for an elevation angle of approximately 30°; these effects, which occur at the time of the crossing of the ionosphere, vary according to the reverse of the square of the frequency.

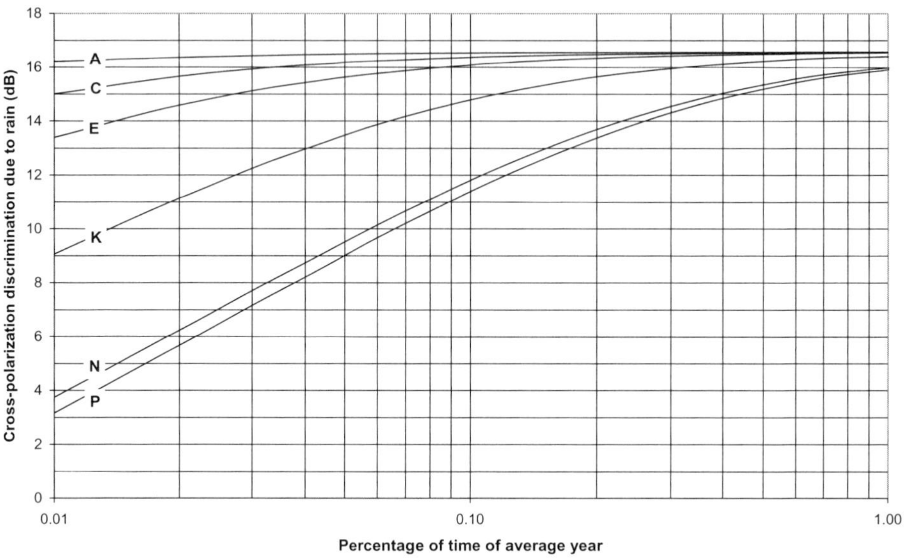

Figure 1.87 Depolarization by rain at 20 GHz.

Effects	Variation	0.1 GHz	0.25 GHz	0.5 GHz	1 GHz	3 GHz	10 GHz
Faraday rotation[a]	$1/f^2$	30 rotations	4.8 rotations	1.2 rotation	108°	12°	1.1°
Propagation delay	$1/f^2$	25 μs	4 μs	1 μs	0.25 μs	0.028 μs	0.0025 μs
Refraction	$1/f^2$	<1°	<0.16°	<2.4′	<0.6′	<4.2″	<0.36″
Variation in direction of arrival (rms)	$1/f^2$	20′	3.2′	48″	12″	1.32″	0.12″
Absorption (midlatitude)	$1/f^2$	<1 dB	<0.16 dB	<0.04 dB	<0.01 dB	<0.001 dB	$<10^{-4}$ dB
Dispersion (ps Hz^{-1})	$1/f^3$	0.4	0.026	0.0032	0.0004	$1.5.10^{-5}$	4.10^{-7}
Scintillation[b]	—	—	—	—	>20 dBcc	≈10 dBcc	≈4 dBcc

[a]Ionospheric effects above 10 GHz are negligible.

[b]Values observed near geomagnetic equator during early nighttime hours at equinox under conditions of high sunspot number.

1.12.1 Scintillation

Scintillation refers to the variations of amplitude, phase, polarization, and arrival angle which appear when radio waves cross zones of turbulent irregularities of the refractive index in the troposphere or the electron density in the ionosphere. Figure 1.88 presents the distribution of the fade depth due to

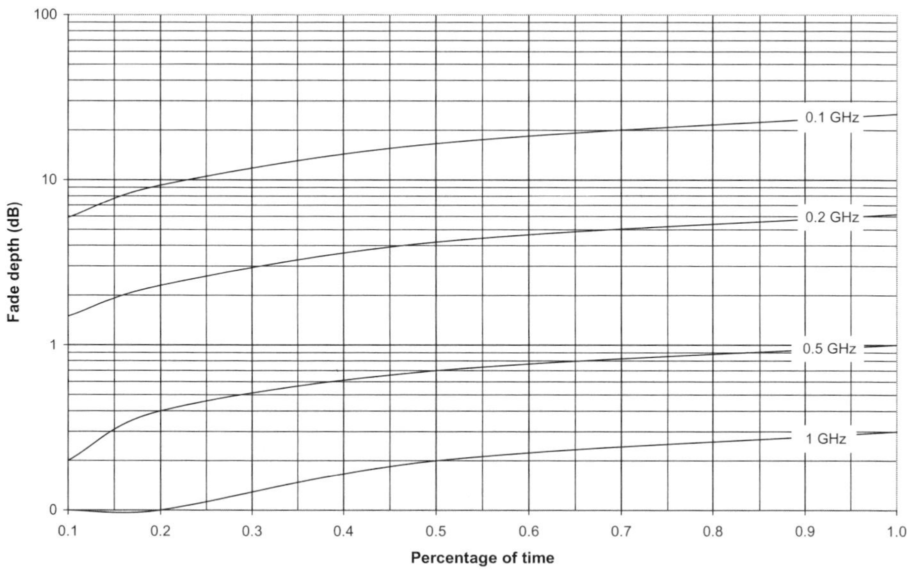

Figure 1.88 Distribution of fade depth due to ionospheric scintillation.

ionospheric scintillation at average latitudes for various frequency bands. In addition, the amplitude of the fast fluctuations of the received signal due to crossing the troposphere is all the more important as the frequency is high, the elevation angle is small, and the moisture is high and decreases when the beamwidth of the antenna decreases; for elevation angles higher than 3°, the standard deviation of the scintillation of amplitude would lie between 0.1 and 1 dB according to the elevation angle at the frequency 7 GHz and between 0.4 and 4 dB at the frequency 100 GHz.

1.12.2 Faraday's Rotation

Recommendation ITU-R P.531 expresses Faraday's rotation of the plane of polarization of a linearly polarized wave due to the Earth magnetic field and the ionosphere by the relation

$$\theta = 2.36\,10^2 B_{av} N_T f^{-2} \qquad (1.134)$$

where θ = angle of rotation (rad)
B_{av} = average Earth magnetic field (Wb m^{-2})
f = frequency (GHz) and the electron density N_T of the ionosphere is given as

$$N_T = \int_s n_e(s)\, ds$$

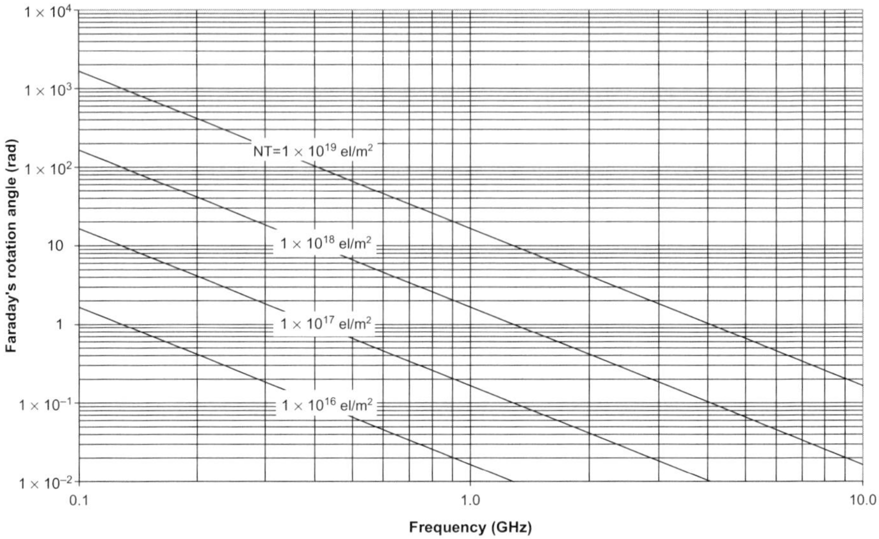

Figure 1.89 Faraday's rotation.

where n_e = electron concentration (el/m^3)
s = traversed path

Figure 1.89 shows the variation of Faraday's rotation angle versus frequency for various values of electron density.

1.12.3 Propagation Time Delay

The presence of charged particles in the ionosphere causes a delay in electromagnetic wave propagation; the propagation time delay is given by the relation

$$t = 1.345 \times 10^{-7} \frac{N_T}{f^2} \tag{1.135}$$

Figure 1.90 presents the propagation delay due to crossing the ionosphere versus frequency for various values of electron density.

1.13 THERMAL RADIATION

1.13.1 Origin of Thermal Radiation

Any hot body emits a thermal radiation whose intensity increases with the temperature; this radiation transports energy that is propagated in the form

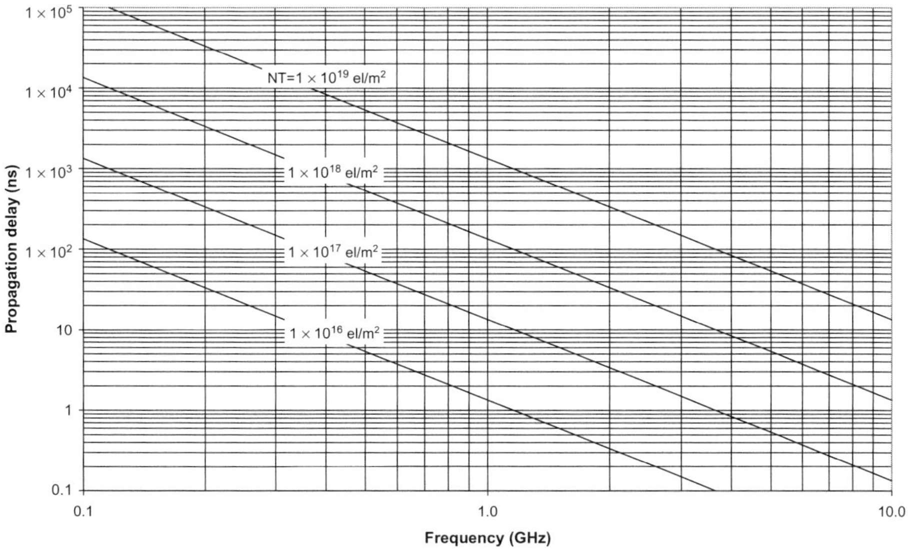

Figure 1.90 Propagation delay due to crossing of ionosphere.

of electromagnetic waves and has a continuous spectrum; that is, that it comprises components at all the frequencies with different intensities. The macroscopic concept of temperature rests on the thermal agitation of the elementary corpuscles which constitute the matter. The first theories, which lead to Rayleigh's law for blackbody, postulated the presence of a great number of elementary oscillators radiating all possible frequencies; these oscillators are the electrons bound to the atoms of the body that emit radiation while oscillating and are excited by absorbing other radiation. Since Rayleigh's law is valid only for radio frequencies, it is by introducing the concept of minimal quantity of energy, or quantum of energy, that Planck could establish the law which bears his name and that represents perfectly the phenomena observed at all frequencies; that is, electrons can radiate or absorb energy only by jump(s) or quanta of value $h\nu$, where h is the Planck's constant and ν is the radiated frequency.

1.13.2 Propagation of Thermal Radiation

Consider a surface S sufficiently small to constitute a point source of radiation in all directions of space, as shown in Figure 1.91. The quantity of energy which crosses in 1 s the surface S' located on a sphere at a certain distance from S is propagated inside the cone C and is called the power flux; bringing back the surface to $1\,m^2$, we obtain the power flux density, which is the quantity of energy which crosses this unit of area per second and is expressed in watts per square meter.

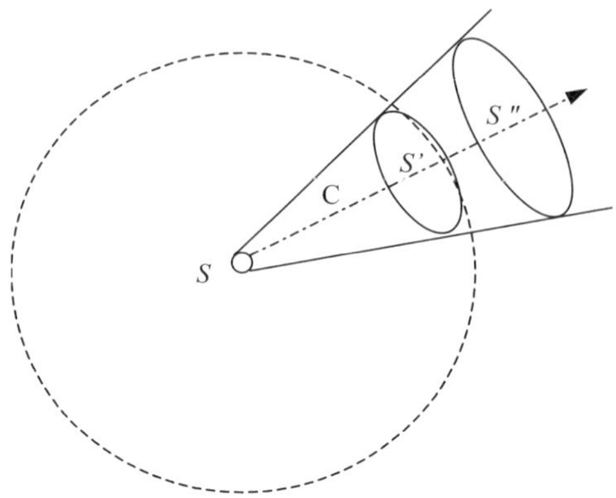

Figure 1.91 Propagation of thermal radiation.

In the same way, another surface S'' located on a more distant sphere would receive a power flux density whose intensity would be in the ratio S'/S''; the power flux density thus varies as the inverse ratio of the square of the distance to the radiating source and depends only on the properties of the surface S and the inner volume of cone C. The inner portion of space of the cone C constitutes a solid angle; the unit of solid angle, the steradian (sr), is that under which one sees a unit of area from the center of the sphere. When the radiating source is sufficiently extended to comprise zones from which the characteristics of radiation are different, we use the unit of brightness, that is, the power flux density that is received from an area of the source which is seen under a given solid angle, expressed as watts per square meter per steradian.

The power flux density and the brightness of a body depend on the spectrum part which has been chosen for the observation; by considering an interval of frequencies $d\nu$ in the vicinity of the frequency ν sufficiently small so that the characteristics of radiation are constant, we report these terms to an interval of 1 cycle/s or 1 Hz and thus obtain:

- Monochromatic power flux density in vicinity of frequency ν ($W\,m^{-2}\,Hz^{-1}$)
- Spectral brightness or monochromatic brightness ($W\,m^{-2}\,Hz^{-1}\,sr^{-1}$)

The spectral brightness $B(\nu)$ of a body in the vicinity of the frequency ν depends only on its temperature and its capacity to absorb radiations at the frequency ν.

1.13.3 Blackbody

A blackbody is a body which absorbs all radiation that it receives; it is shown that, for such a body, the spectral brightness depends only on its temperature

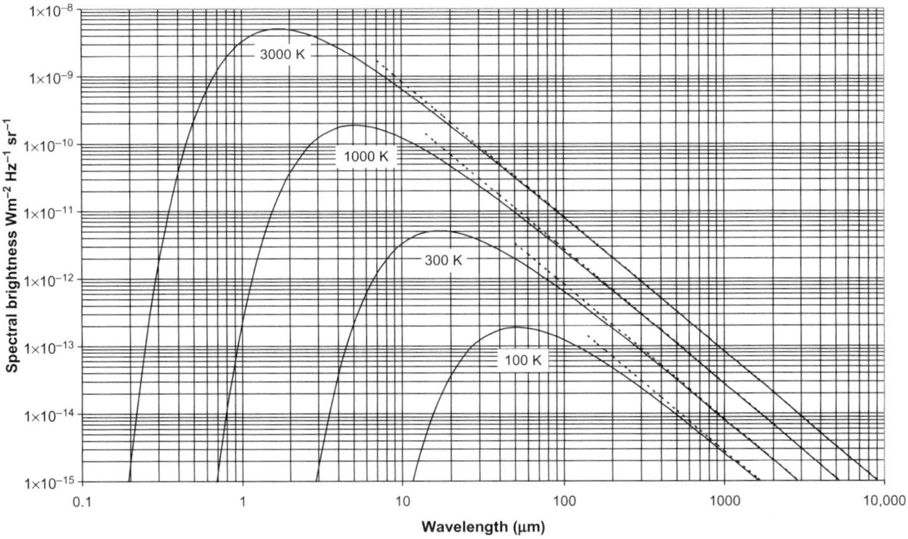

Figure 1.92 Spectral distribution of brightness of blackbody (Planck's law and Rayleigh's law).

and on the frequency considered, and the law of radiation of the blackbody, or Planck's law, is given by the relation

$$B(\nu) = \frac{2h\nu^3 c^{-2}}{\exp(h\nu/kT) - 1} \qquad (\mathrm{W\,m^{-2}\,Hz^{-1}\,sr^{-1}}) \qquad (1.136)$$

where Planck's constant $h = 6.62618 \times 10^{-34}\,\mathrm{J\,s}$
Speed of light $c = 299{,}792{,}458\,\mathrm{m\,s^{-1}}$
Frequency ν is in hertz
Boltzmann's constant $k = 1.380664 \times 10^{-23}\,\mathrm{J\,K^{-1}}$
Absolute temperature T is in kelvin ($0\,\mathrm{K} = -273.16\,°\mathrm{C}$)

Figure 1.92 presents the spectral brightness of blackbodies brought up to usual temperatures versus wavelength in micrometers. The monochromatic brightness is maximum at a given temperature when

$$\frac{c}{\nu_m} T = 5.0996 \times 10^{-3}\,\mathrm{m\,K} \quad \text{for } \frac{dB_\nu}{d\nu} = 0$$

$$\lambda_m T = 3.6698 \times 10^{-3}\,\mathrm{m\,K} \quad \text{for } \frac{dB_\lambda}{d\lambda} = 0$$

In the field of radio waves, where the wavelengths are large, $h\nu$ becomes much smaller than kT and one obtains the law by Rayleigh before that of Planck, which is illustrated by the dotted lines in Figure 1.92:

$$B(\nu) = \frac{2kT\nu^2}{c^2} = \frac{2kT}{\lambda^2} \tag{1.137}$$

with

$$\lambda = \frac{c}{\nu} \quad \text{for } h\nu \ll kT$$

The brightness temperature of a body is the temperature of the blackbody which would have the same spectral brightness for the frequency of observation. By integrating Planck's law on the whole spectrum, we obtain the relation of Stefan–Boltzmann, which represents the total power radiated per unit area of a blackbody:

$$B_{\text{total}} = \frac{2h}{c^2}\int_0^\infty \frac{\nu^3}{\exp(h\nu/kT)-1}\,d\nu$$

which becomes

$$B_{\text{total}} = \sigma T^4 \tag{1.138}$$

with Stefan's constant

$$\sigma = \frac{2\pi^5 k^4}{15h^3c^2} = 5.66956 \times 10^{-8}\ \text{W m}^{-2}\ \text{K}^{-4}$$

Figure 1.93 presents the total power radiated per unit area according to the law of Stefan–Boltzmann.

1.13.4 Gray Body

A gray body is a body which is not perfectly absorbent; when it receives energy, it absorbs a certain fraction $p(\nu)$ in the vicinity of the frequency ν which is the absorption coefficient. According to Kirchhoff's law, the spectral brightness of a gray body is equal to that of the blackbody at the same temperature and the same frequency multiplied by $P(\nu)$. Kirchhoff's law makes it possible to interpret and determine all the phenomena of absorption of radiation.

Let one consider, for example, a body of brightness temperature T_S which is seen by a sensor through an absorbing medium at temperature T_a, as represented on Figure 1.94, and suppose that, in the vicinity of frequency ν, this absorbing medium retains a fraction p of the incidental energy.

The power flux coming from the source is thus proportional to T_S (Rayleigh's law), then a flux proportional to $(1 - p)Ts$ crosses the absorbing medium which radiates in turn a flux proportional to $p\ T_a$; in this manner, the

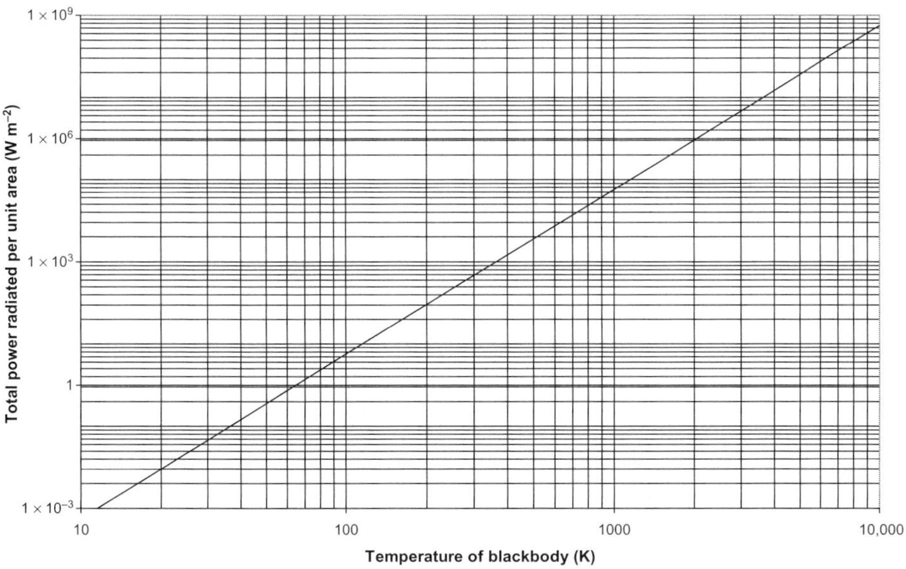

Figure 1.93 Total radiated power by blackbody (Stefan–Boltzmann).

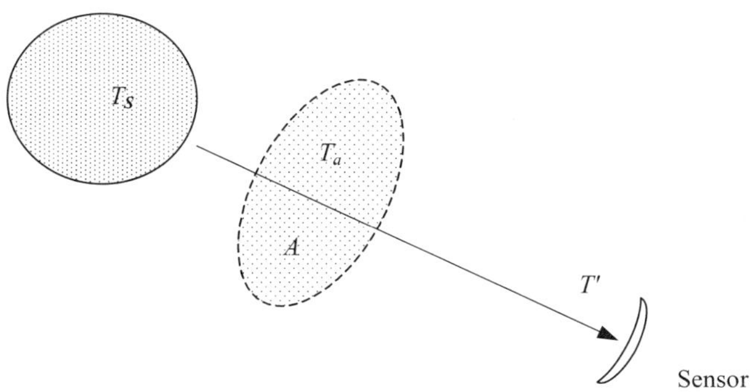

Figure 1.94 Partial absorption by a grey body.

received power flux density is that of a black body which would have the apparent temperature T', such as:

$$T' = (1 - p)T_S + pT_a \tag{1.139}$$

By calling A the attenuation caused by the grey body, one obtains:

$$p = 1 - \frac{1}{A}$$

and the relation (1.139) becomes:

$$T' = \frac{T_S}{A} + T_a\left[1 - \frac{1}{A}\right] \tag{1.140}$$

1.14 PROBABILITY DISTRIBUTIONS

1.14.1 Introduction

The electromagnetic wave propagation in the vicinity of the Earth is carried out in a random medium; it is thus necessary to analyze the phenomena of propagation by using the statistical methods in order to lead to a model which would be able to describe satisfactorily the variations of the transmitted signals in the course of the time and across the space.

The statistical distributions are described either by a density of probability or by a function of distribution:

- The probability density function $p(X)$ is such as the probability that a variable X takes a value ranging between x and $x + dx$ is $p(x)dx$,
- The cumulative distribution function $F(x)$ indicates the probability that the variable takes a value lower than x, just as $[1 - F(x)]$ is the probability for a value higher than x.

The relations between these two functions are thus the following:

$$p(x) = \frac{d}{dx}[F(x)] \tag{1.141}$$

$$F(x) = \int_c^x p(t)dt \tag{1.142}$$

where: c is the lower limit that can take the variable X.

There are many laws of distribution among which the Gauss's law, known as the normal law, and the Rayleigh's law, called diffusion law, are used to represent most of the phenomena of propagation of the electromagnetic waves.

1.14.2 Gauss's Law of Distribution

By considering a random variable X, of mean value m and standard deviation σ, the probability density and the distribution function are written:

$$p(x) = \frac{1}{\sigma\sqrt{2\pi}}\exp\left[-\frac{1}{2}\left(\frac{x - m}{\sigma}\right)^2\right] \tag{1.143}$$

$$F(x) = \frac{1}{\sigma\sqrt{2\pi}} \int_{-\infty}^{x} \exp\left[-\frac{1}{2}\left(\frac{x-m}{\sigma}\right)^2\right] dx \tag{1.144}$$

which has as an approximate value:

$$F(x) \approx \frac{\exp\left[-\frac{1}{2}\left(\frac{x-m}{\sigma}\right)^2\right]}{\sqrt{2\pi}\left[0.661\left|\frac{x-m}{\sigma}\right| + 0.339\sqrt{\left(\frac{x-m}{\sigma}\right)^2 + 5.51}\right]}$$

Figure 1.95 and Figure 1.96 represent the probability density function and the cumulative distribution function, expressed in percentage, of a variable reduced to its standard deviation and of null mean value.

The Gaussian distribution meets especially when the considered variable results from the additive effect of many random causes of comparable importance (central limit theorem).

When this distribution intervenes to represent the logarithm of a variable, a log-normal distribution is obtained.

This distribution is characterized by the following Gaussian coefficients:

0.0001%	0.001%	0.01%	0.1%	1%	10%	15.857%	50%
99.9999%	99.999%	99.99%	99.9%	99%	90%	84.143%	
±4.753	±4.265	±3.719	±3.091	±2.327	±1.282	±1	0

and by the following remarkable values:

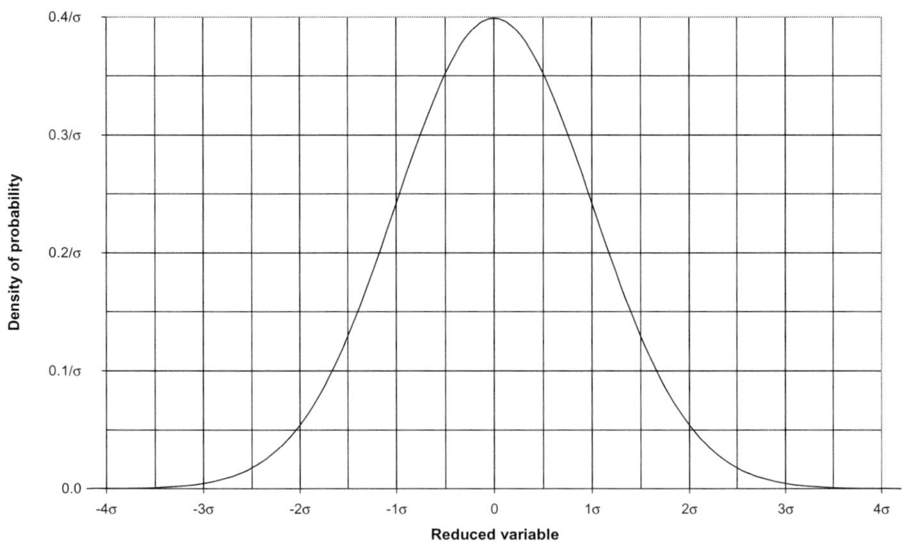

Figure 1.95 Probability density function of Gauss's law.

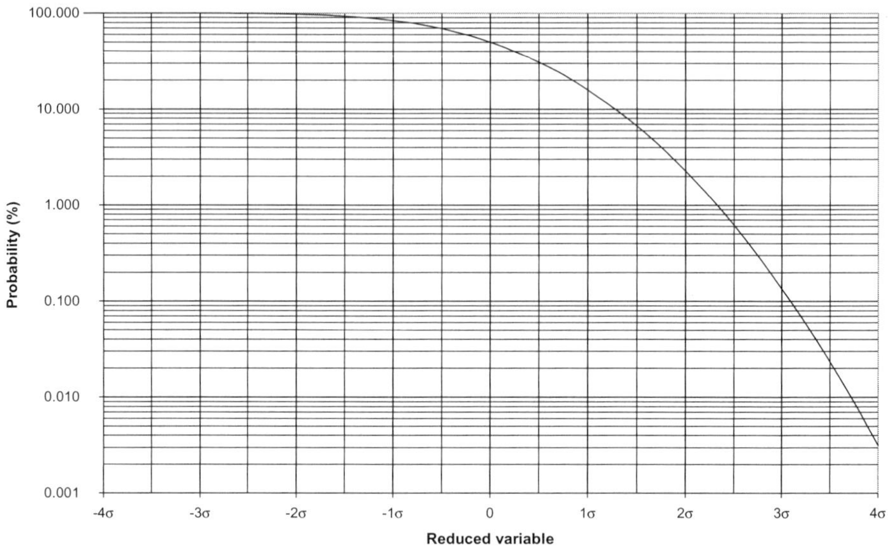

Figure 1.96 Cumulative distribution function of Gauss's law.

— Most probable value	$\exp(m - \sigma^2)$
— Median value	$\exp(m)$
— Variance	σ^2
— Mean value	$\exp(m + \sigma^2/2)$
— Mean square value	$\exp(m + \sigma^2)$

1.14.3 Rayleigh's Law of Distribution

This distribution applies to a positive continuous variable; the density of probability and the function of distribution are given by the relations:

$$p(x) = \frac{x}{q^2}\exp\left(-\frac{x^2}{2q^2}\right) \tag{1.145}$$

$$F(x) = 1 - \exp\left(-\frac{x^2}{2q^2}\right) \tag{1.146}$$

Figure 1.97 and Figure 1.98 represent the density of probability and the function of distribution, expressed in percentage, of a variable reduced to its most probable value q.

The Rayleigh's law presents the following characteristics:

- the probability that a variable X has a value lower than x is roughly proportional to the square of this value; for example, if the variable

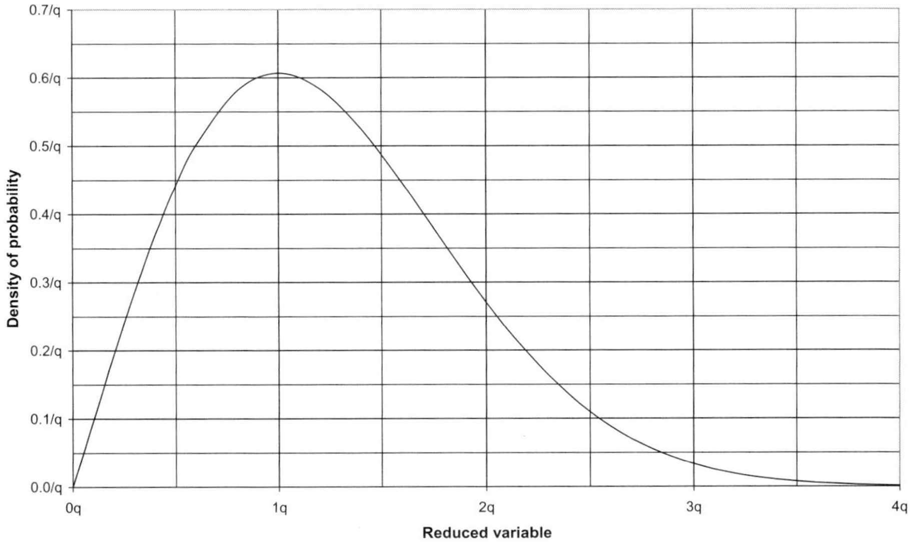

Figure 1.97 Probability density function of Rayleigh's law.

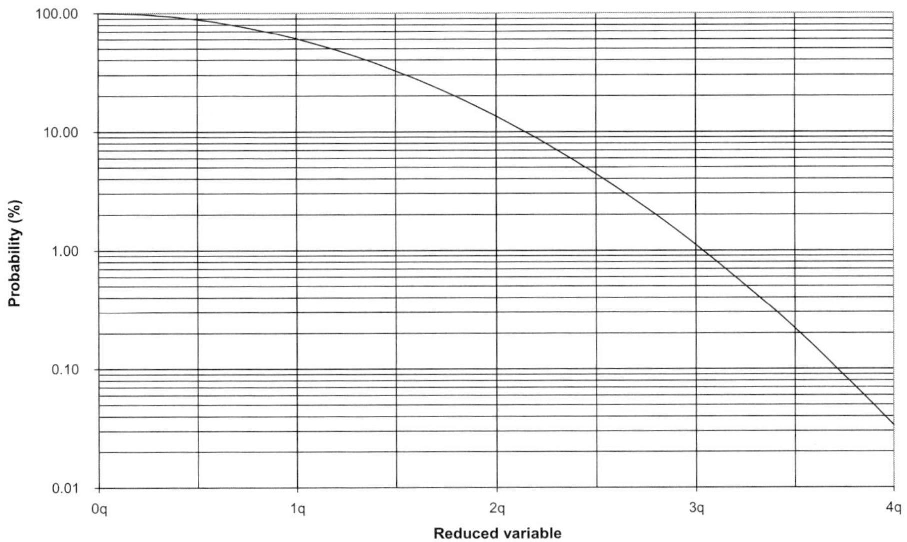

Figure 1.98 Cumulative distribution function of Rayleigh's law.

considered is a tension, the power decrease of 10 dB for each decade of probability,

- if one considers two independent variables x and y of null mean value and of same standard deviation σ, the random variable $r = \sqrt{x^2 + y^2}$ obeys a Rayleigh's distribution and the most probable value of r is equal to σ.

The distribution coefficients of the variable are:

0.001%	0.01%	0.1%	1%	10%	50%	90%	99%	99.9%	99.99%	99.999%
4.7985	4.2919	3.7169	3.0349	2.1460	1.1774	0.4590	0.1418	0.0447	0.0141	0.0045

and its remarkable values are:

— Most probable value q
— Median value $q\sqrt{2\ln 2} = 1.18q$
— Mean value $q\sqrt{\pi/2} = 1.25q$
— Mean square value $q\sqrt{2} = 1.41q$
— Standard deviation $q\sqrt{2-\pi/2} = 0.655q$

The Rayleigh's distribution thus intervenes particularly in the phenomena of diffusion where the variable results from the sum of a great number of vectors equipollent which amplitudes are of the same order of magnitude and which phases would have a uniform distribution.

1.14.4 Other Laws of Distribution

There are some other useful probability distributions such as:

1. Combined Gauss-Rayleigh distribution, where the random variable varies according to the GAUSS's law for the long-term and to the Rayleigh's law for the short-term with the following relations:

$$F_z(x) = 1 - \exp\left(-\frac{x^2}{z^2}\right)$$

$$p(z) = \frac{\exp\left[-\frac{1}{2}\left(\frac{\ln z - m}{\sigma}\right)^2\right]}{\sigma\sqrt{2\pi}z}$$

   where: z is the mean square value of the Rayleigh's variable which is distributed according to a log-normal law

2. Rice's distribution, where the random variable is the sum of two statistically independent Gaussian variables,

3. Nakagami *m* distribution, which fits various types of fade and includes as particular cases the normal distribution and the Rayleigh's distribution,
4. Nakagami-Rice *n* distribution, about the modulus of the vector resulting from the sum of a fixed vector and a vector varying according to the Rayleigh's distribution,
5. CHI-SQUARE distribution, where the random variable is the square of a Gaussian variable,
6. GAMMA distribution, which can be used to represent the rainfall rate distribution for percentages of time up to 10%,
7. Gumbel distribution, called law of extremes, which can be used to represent the high rainfall rate distribution for small percentages of time,
8. Pearson distribution . . .

CHAPTER 2

Principles of Digital Communication Systems

2.1 SIGNAL PROCESSING

The main objective of communication systems is to transmit high-quality voice, video, and data between users separated by a few meters or thousands of kilometers. The distance between the sites to be connected determines the type of transmission equipment, which can be metallic wires, optical fiber, microwave radio links, or satellite. At various stages along the route, numerous causes of distortion and additive noise under adverse conditions of propagation corrupt the signal and degrade the overall performance.

The purpose of this chapter is to describe the different steps of signal processing and to trade off basic parameters such as signal-to-noise ratio (SNR), signal distortion in amplitude and phase, group delay, coherence bandwidth, intermodulation, and interference in order to reach the system goals by optimizing the resources in terms of probability of error and bandwidth expenditure.

In an integrated services digital network (ISDN), the information to be transmitted passes through the following stages from one end to the other end:

1. Formatting, which transforms the source information into digital symbols
2. Source encoding, which produces analog-to-digital conversion
3. Channel coding, which reduces the probability of error or the SNR requirement
4. Encryption, which prevents unauthorized users from understanding messages and injecting false messages into the system
5. Routing, which is done by a switch to direct the message to its destination through the network
6. Multiplexing and multiple-access procedures, which combine signals issued from different sources so that they can share the same channel

Microwave Engineering: Land & Space Radiocommunications, by Gérard Barué

7. Synchronization, which allows the reconstruction of the message at the frame level or coordination with other users at the network level in order to use the resource efficiently
8. Modulation, which converts the symbols to waveforms that are compatible with the transmission channel and as bandwidth efficient as possible
9. The reverse processes at the reception side

2.1.1 Digital Multiplexing

2.1.1.1 Pulse Code Modulation The source of information is sampled and quantized to one of L levels and each quantized sample is digitally encoded into an l-bit codeword ($L = 2^l$ and reciprocally $l = \log_2 L$) according to the following:

- The sampling rate is fixed by Shannon's criterion, which states that the sampling frequency must be at least double the highest frequency to be transmitted.
- The quantization is done with a certain number of bits per code.

About telephony, for example, the sampling rate will be 8 kHz due to the fact that the bandwidth of a voice channel can be limited to the frequency band 300–3400 Hz; using 8 bits per code, this becomes to the usual bit rate of 64 kbits s^{-1}.

At this stage, quantization noise occurs corresponding to a quantization SNR that is proportional to the square of the number of coding levels, that is, $10 \log(L^2)$ in decibels as shown in Figure 2.1. Uniform quantization disadvantages the low signal levels whose probability of presence is much greater than that of high levels in telephony. To achieve a reasonably constant SNR, low signals must therefore be more amplified than stronger signals; this process is done by a compander, which has an approximate logarithmic characteristic (μ-law in North America and A-law in Europe) and the inverse operation is done by an expander at reception. Then, each quantized sample or PAM (pulse amplitude modulation) signal is encoded into a digital word called a PCM (pulse code modulation) signal.

2.1.1.2 Line Codes The PCM signal is made up of bit sequences of 1's and 0's which have to be transmitted in the correct form to be detected correctly; there are several types of pulse transmission which are called line codes. The choice of transmission code is guided by several important features:

- There must be no significant direct current for use by the transformer to avoid ground loops.
- The energy at low frequencies must be small to reduce the size of components.

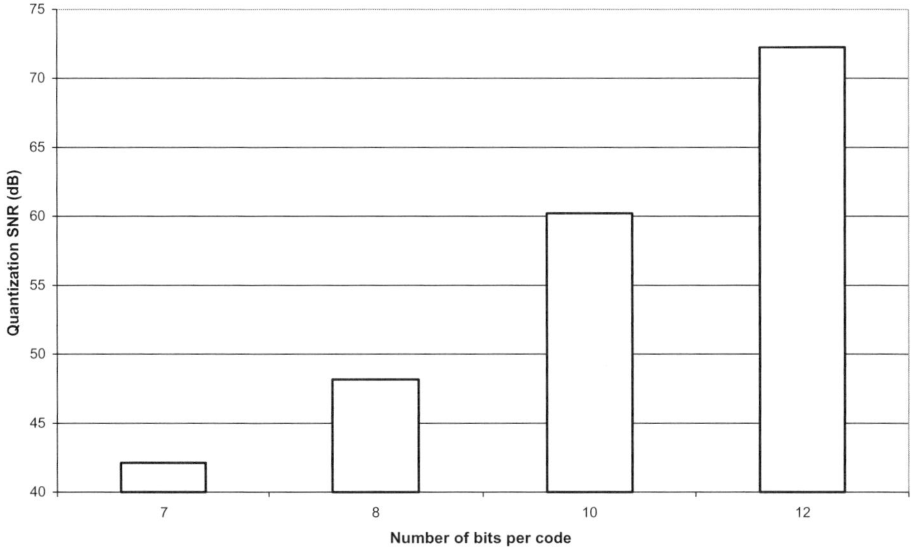

Figure 2.1 Quantization SNR versus number of bits per code.

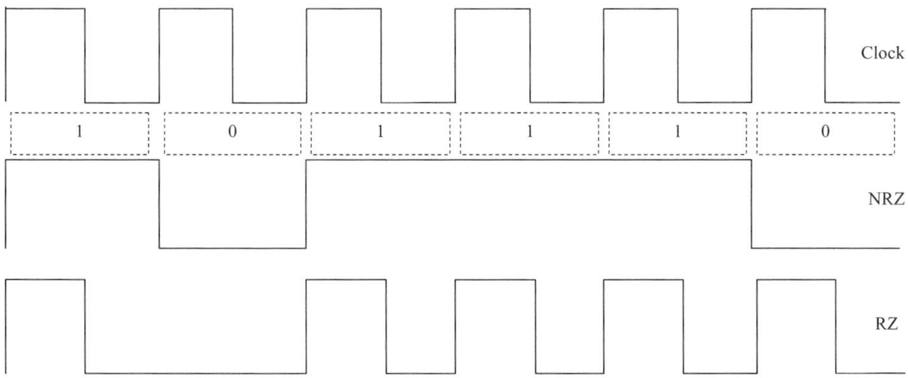

Figure 2.2 Comparison between NRZ and RZ codes.

- A significant number of 0 crossings must be available for clock frequency recovery.
- Good coding efficiency is necessary to minimize the required bandwidth.

The two main categories of line codes are the unipolar and the bipolar, and the pulses of line code are categorized as either no return to zero (NRZ) or return to zero (RZ). Figure 2.2 gives a representation of both line codes where we can see that NRZ code generates a higher direct current than RZ code. The most usual codes, shown on Figure 2.3, derive from the preceding basic ones:

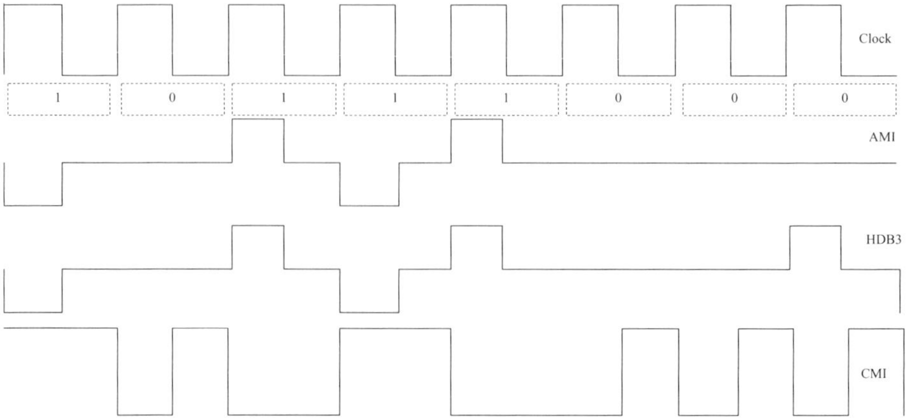

Figure 2.3 Comparison between AMI, HDB3, and CMI codes.

- Alternate mark inversion (AMI) or bipolar code, where the 1's are alternately positive and negative so there is no direct-current (DC) component in the spectrum; another advantage is that it allows the correction of errors as it is possible to detect a violation of the code by noise peaks.
- High-density bipolar three-zeros (HDB3) code, whose purpose is to limit to three the number of 0's in a long bit stream according to specified rules; this assures clock extraction in the regenerator of the receiver.
- Coded mark inversion (CMI), where the 1's are represented alternately by a positive and a negative state while the 0's are represented by a negative state at the first half of the bit interval and a positive state in the second interval.

Throughout the network as well as inside the same equipment or between two different equipments, signals may be changed from one code to another one; this conversion is done by a code converter.

2.1.1.3 High-Order Digital Multiplexing The digital baseband signal transmitted by microwave radio, satellite, or optical fiber systems is formed by time-division multiplexing (TDM), which involves periodically sampling numerous channels and interleaving them on a time basis. The digital primary multiplexer packages channels in groups of 24 in North America and 30 in Europe. For higher capacity links, several 24- or 30-channel bit streams are multiplexed to a higher order bit rate as 45 Mbits s^{-1}, which contains 672 channels in North America, or 140 Mbits s^{-1}, which provides 1920 channels in Europe. This higher order multiplexing can be done on either a nonsynchronous basis since the inception of PCM in the 1970s or a synchronous basis derived from the U.S. synchronous optical network (SONET) standard since 1988 and is also applicable to microwave radio systems; the term *synchronous*

TABLE 2.1 PDH and SDH Levels

Levels	Number of Channels	Bit Rate	Line Codes
	PDH for North American and Japanese Systems		
DS1	24	1.544 Mbits s^{-1}	Bipolar
DS2	96	6.312 Mbits s^{-1}	Bipolar (B6ZS)
DS3	672	44.736 Mbits s^{-1}	Bipolar (B3ZS)
DS4	4032	274.176 Mbits s^{-1}	Polar bipolar
	PDH for European Systems and Elsewhere		
1	30	2.048 Mbits s^{-1}	HDB3
2	120	6.448 Mbits s^{-1}	HDB3
3	480	34.368 Mbits s^{-1}	HDB3
4	1920	139.264 Mbits s^{-1}	CMI
5	7680	565.992 Mbits s^{-1}	CMI
	SONET/SDH Systems		
STS 1/STM 0		51.84 Mbits s^{-1}	CMI
STS 3/STM 1		155.52 Mbits s^{-1}	CMI
STS 12/STM 4		622.08 Mbits s^{-1}	CMI
STS 24/STM 8		1244.16 Mbits s^{-1}	CMI
STS 48/STM 16		2488.32 Mbits s^{-1}	CMI
STS 96/STM 32		4976.64 Mbits s^{-1}	CMI
STS 192/STM 64		9953.28 Mbits s^{-1}	CMI
STS 768/STM 256		39813.12 Mbits s^{-1}	CMI

Note: DS, digital signal; STS, synchronous transport signal; STM, synchronous transfer mode; B3ZS, B6ZS, bipolar with three- and six-zeros separation.

relates to the clocking system and the number of customer information pulses passing through the multiplexer per second. This number can vary plus or minus a few pulses per second in nonsynchronous systems but it is constant in synchronous systems; two processes have therefore evolved, the plesiochronous digital hierarchy (PDH) and the synchronous digital hierarchy (SDH). The main advantage of SDH is to allow the add-and-drop capability without a complete demultiplexing procedure and also multiplexing of tributaries that have different bit rates. Table 2.1 gives the levels for PDH and SDH.

2.1.2 Digital Modulation

The message to be transmitted is characterized by a sequence of digital symbols and is called a baseband signal or a bit stream.

2.1.2.1 Usual Modulation Types For radio transmission, the baseband signal has to modulate a sinusoid called a carrier wave in order to be converted to an electromagnetic field for propagation to the desired destination. The main objectives of the modulation process is to optimize the bandwidth efficiency in terms of transmitted bits per second per hertz in order to spare as much as possible the frequency spectrum resource.

1. *Frequency Shift Keying (FSK)* In FSK modulation, a fixed frequency tone is dedicated to the pulses 0 and another tone to the pulses 1; we can define the following modulation index:

$$m = \frac{f_0 - f_1}{R}$$

where f_0 = pulse 0 frequency
f_1 = pulse 1 frequency
R = bit rate

2. *Minimum Shift Keying (MSK)* MSK is a binary FSK where the two frequencies are selected to ensure that there is a 180° phase shift difference between the two frequencies in a single bit interval. Such a modulation type presents a relatively compact spectrum with out-of-band characteristics that are better than FSK, which reduces adjacent-channel interference.

3. *Phase Shift Keying (PSK)* There are several levels of PSK, from the simplest two phases, where the incoming bit stream is given a phase reversal of 180° every time a 1 changes to a 0 and vice versa, to quaternary (QPSK, or 4-PSK) and 8-PSK; by lumping together N bits of duration τ extending over the time $N\tau$ in order to get an N-bit symbol, there are $2N = M$ possible symbols. In BPSK, QPSK, and M-PSK, one signal is distinguished from another in phase but all are of the same amplitude.

4. *Quadrature Amplitude Modulation (QAM)* This modulation type, which combines both phase and amplitude modulation as an extension of PSK, offers an improved error performance; in the case of a 4-QAM, where two amplitude levels are used as inputs to a 2-PSK modulator, the result is identical to a 4-PSK. However, higher level QAM systems are distinctly different from the higher level FSK systems, as shown in Figure 2.4, which presents examples

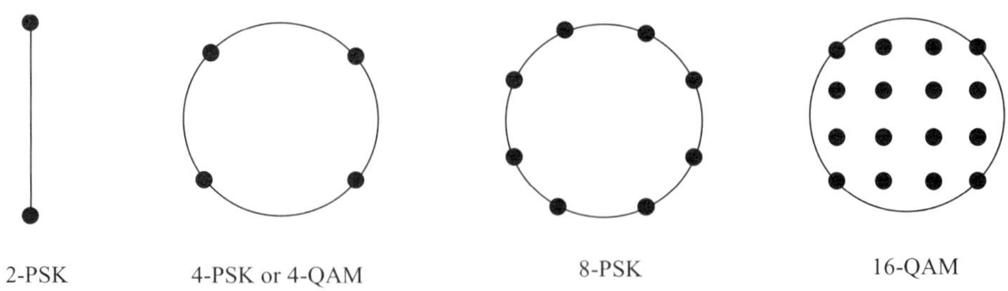

Figure 2.4 Examples of modulation types.

of modulation types. The various signal states in FSK are evenly distributed on a circle when they constitute a constellation diagram in QAM; therefore, the FSK signal has a constant envelope, unlike the QAM one. To keep a constant envelope as much as possible, the instantaneous carrier amplitude does not cross the zero point when the signal jumps from one state to another; this makes the signal less affected by nonlinearities.

2.1.2.2 Bandwidth Efficiency The main reason for moving to higher levels of QAM is the improvement in spectrum utilization efficiency; for example, a 64-QAM digital microwave radio system presents about the same spectrum efficiency as an analog system of similar capacity and it evolved QAM configurations 128, 256, and even up to 1024.

The spectral density η is simply the ratio of the bit rate R to the bandwidth B:

$$\eta = \frac{R}{B} \tag{2.1}$$

For M-PSK and M-QAM we can write

$$\eta = \frac{S}{B} \log_2 M \tag{2.2}$$

where $S = R/\log_2 M$ is the symbol rate in bauds. The effective bandwidth depends on the low-pass or bandpass filters in the modulation processes. The minimum filtering requirement for pulse transmission without performance degradation, that is, no intersymbol interference, is given by Nyquist's theorem, which states that if pulses are transmitted at a rate R, they will attain the full amplitude value if passed through a low-pass filter having a bandwith $R/2$ hertz.

The ideal Nyquist filter does not exist, but the filter characteristics commonly used present a coefficient on the order of 1.33, that is, the theoretical bandwidth is multiplied by this value; so, the effective spectrum efficiency for the usual modulations reaches approximately the following values:

M-QAM	4-QAM	16-QAM	64-QAM	128-QAM	256-QAM
$\log_2 M$	2	4	6	7	8
Spectrum efficiency, bits s^{-1} Hz^{-1}	1.5	3	4.5	5.25	6

2.1.2.3 Power Spectral Density and Transmitted Spectrum The distribution of the energy between the different states of the modulated signal as well as the occupied bandwidth depends on the type of modulation;

consequently, the power spectral density becomes larger as the occupied bandwidth decreases with equal bit rate. The normalized power spectral density of the signal for M-PSK and for M-QAM modulations can be expressed by the relation

$$P(f) = \tau_s \left(\frac{\sin(\pi f \tau_s)}{\pi f \tau_s} \right)^2 \tag{2.3}$$

where τ_s = symbol duration
f = frequency

Figure 2.5 represents the normalized power spectral density around the carrier central frequency for m values between 1 and 7, corresponding to $M = 2^m$ ranging between 2 and 128 states of the carrier, for parameter τ_s/T where the frequency f is replaced by the period $T = 1/f$. Figure 2.6 presents the normalized power spectral density for $m = 1, 2, 3$ to which correspond $M = 2, 4, 8$ states of the carrier.

Figures 2.7 and 2.8 show examples of typical transmitted spectra corresponding to various types of M-QAM modulation that are commonly encountered for PDH and SDH microwave links. We can compare the bandwidths of the corresponding signals to the bandwidth efficiency given in Section 2.1.2.2 and the normalized power spectral density indicated above.

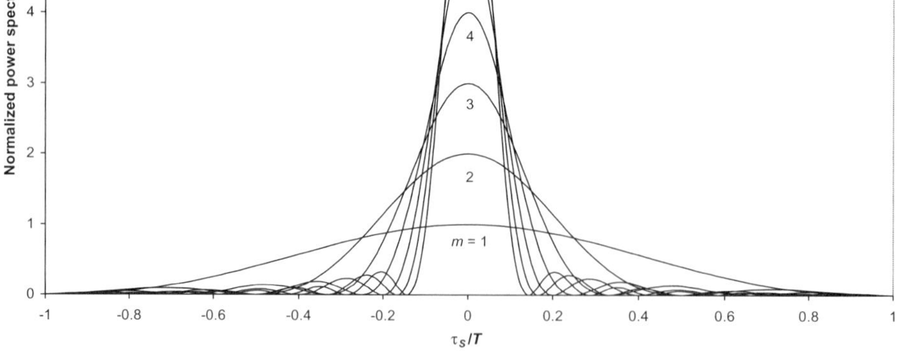

Figure 2.5 Normalized power spectral density.

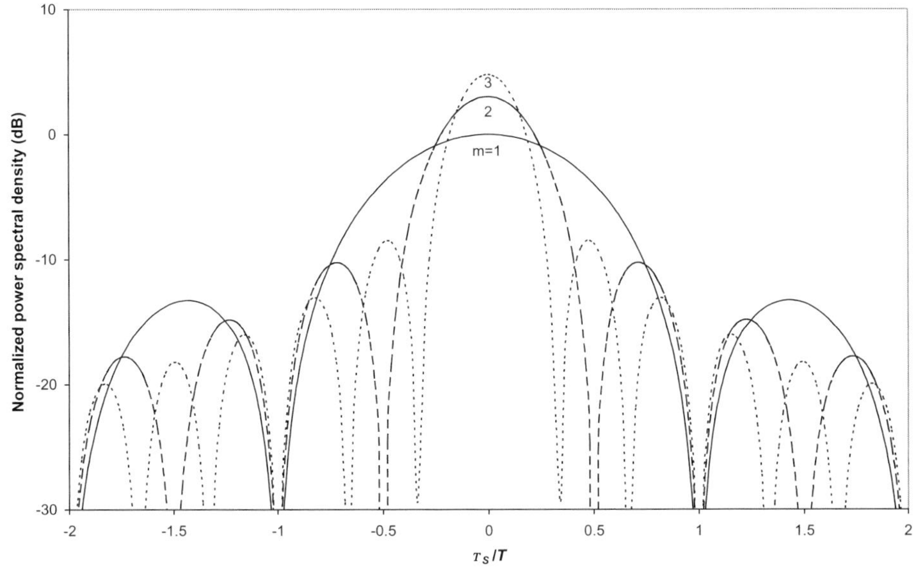

Figure 2.6 Normalised power spectral density expressed in decibels.

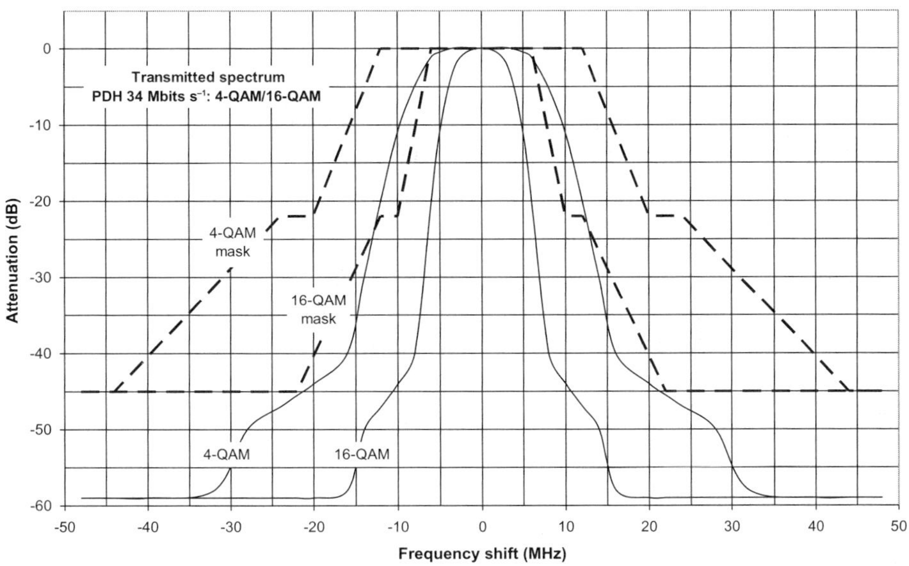

Figure 2.7 Transmitted spectra of typical PDH modulations.

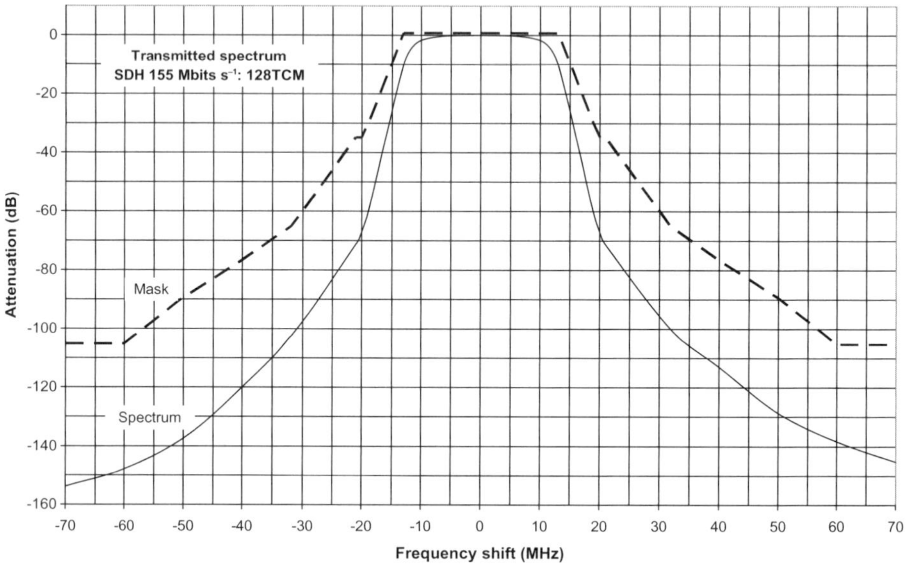

Figure 2.8 Transmitted spectrum of SDH-STM1 modulation.

2.2 THERMAL NOISE

2.2.1 Origin of the Thermal Noise

The origin of the thermal noise which develops in a conductor having a certain resistivity is the same as that of the thermal radiation of hot bodies; it is produced by the agitation of electrons and their shocks against the molecules of the conducting material. This motion causes the conductor to overheat when we pass an electric current, that is a flow of electrons, through it.

In particular, thermal noise is the source of the background noise observed at the output of the amplifiers of the signals we want to analyze; the level of the background noise is fundamental in radioastronomy because the electromagnetic radiation coming from the celestial bodies is of the same nature as the background noise and it is thus difficult to separate the signals we want to study from the thermal noise generated by the receiver.

2.2.2 Thermal Noise Voltage

The thermal noise of resistors constitutes an absolute limit and the electric voltage which develops at the terminals of a resistor in an interval of frequencies $\nu \pm B/2$ can be calculated using the Nyquist's formula:

$$\overline{e}^2 = 4kTRB \tag{2.4}$$

where $\bar{e}^2$ = mean-square value of tension of noise (V^2)
k = Boltzmann's constant (1.380664×10^{-23} J K^{-1})
T = absolute temperature of resistor (K)
R = resistance (Ω)
B = considered frequency bandwidth (Hz)

For instance, the tension of noise which would develop at the terminals of a resistor of 1 MΩ at the temperature of 300 K and in a frequency bandwidth of 1 MHz would have as a value

$$e = \sqrt{\bar{e}^2} = \sqrt{(4 \times 1.3807 \times 10^{-23}\ \text{J K}^{-1}) \times (300\ \text{K})(10^6\ \Omega)(10^6\ \text{Hz})} = 129\ \mu\text{V}$$

A resistor R is thus equivalent to a generator of electromotive force e in series with the resistor R supposed noiseless. Figure 2.9 presents the equivalent circuit of a resistor R affected by its thermal noise in the shape of a generator of noise associated with a noiseless resistor R. Such a generator connected to an external resistor R_{ext} will provide a maximum power when the latter has the same value as R, that is,

$$\frac{\bar{e}^2}{4R} = kTB \tag{2.5}$$

As the frequency ν does not intervene, we see that the thermal noise of resistors contains all the frequencies with the same intensity and consequently its spectrum is white (white noise).

2.2.3 Analogy between Thermal Radiation and Thermal Noise

In the preceding chapter we saw that an antenna transforms a radiated energy into an electric signal which is propagated in a circuit or on a line and reciprocally; in this last layout, the energy radiated disappeared from the line and all occurs as if this had been absorbed by a fictitious resistor replacing the antenna that we call the resistance of radiation.

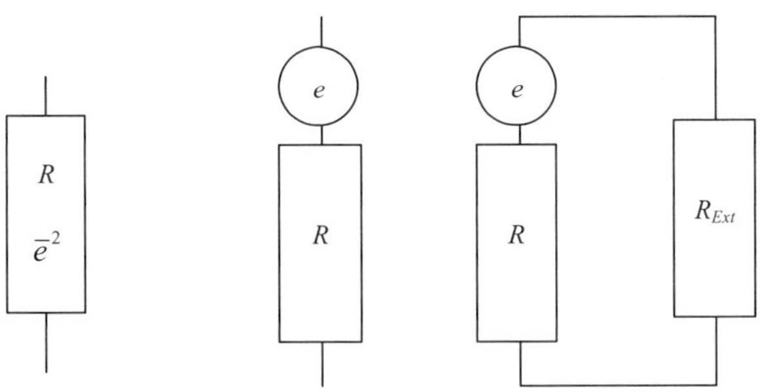

Figure 2.9 Equivalent circuit of resistor.

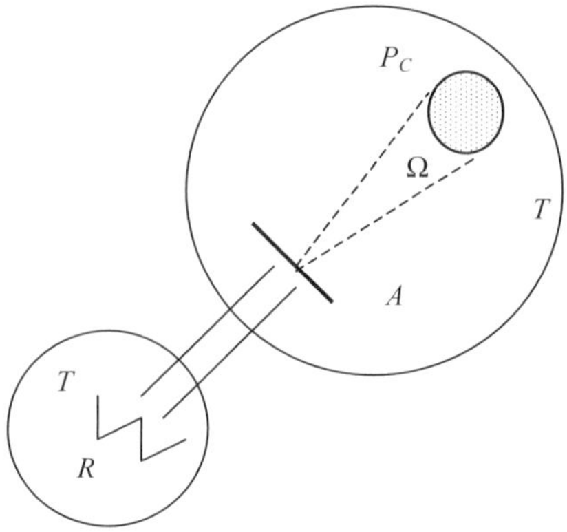

Figure 2.10 Thermal equilibrium between antenna and blackbody.

Here, we consider an antenna placed inside a blackbody risen at the absolute temperature T whose walls are perfectly absorbing at the considered frequencies and coupled by a lossless line to an external resistor R of the same value than its own resistance of radiation R_r at the same temperature T. In such a system, illustrated in Figure 2.10, there is an exchange of energy between the blackbody and the resistor; part of the energy radiated by the blackbody is collected by the antenna and absorbed by the resistor and, inversely, the tension of noise available at the terminals of the latter is radiated by the antenna and then absorbed by the walls of the blackbody. As the blackbody and the resistor are at the same temperature, the result of the exchange must be null, according to the second principle of thermodynamics, and consequently two energies that are propagated in opposite direction on the line are equal. Consider a portion P_C of the wall of the blackbody that is seen under a solid angle Ω and assign to the antenna a collecting area A and a gain G in this direction. The brightness of the element P_C in a frequency bandwidth B can be given by Rayleigh's law, valid for the usual frequencies in radio waves, starting from relation (1.137), that is, $2kTB/\lambda^2$.

The power flux density received under the solid angle Ω is then equal to $\Omega(2kTB/\lambda^2)$ and the power collected by the antenna, $A\Omega(2kTB/\lambda^2)$ On the other hand, the resistor produces a thermal noise power at the terminals of the resistance of radiation of the same value that is equal to kTB.

According to the definition of gain, the thermal noise power radiated by the antenna toward the element P_C is worth $G\Omega[kTB/(4\pi)]$. By equalizing the received power and the radiated power, we can write

$$G\Omega\frac{kTB}{4\pi} = A\Omega\frac{2kTB}{\lambda^2}$$

from which

$$G = 2\frac{4\pi A}{\lambda^2}$$

which is the gain of the antenna already defined by relation (1.24), at a coefficient of 2 more or less due to the fact that in this case the antenna receives not only one polarization, defined by a given direction of the electric field, but also all polarizations.

2.2.4 Apparent Noise Temperature

By measuring the noise tension at the terminals of a known resistor R placed in a medium of unknown temperature T, we can deduce from it the value of T. Let us now measure a noise tension produced by an unspecified process, for instance, by some energy of external origin; we can then define its apparent temperature T' using Nyquist's formula by considering that it is the temperature to which it would be necessary to rise the resistor in order to get the same noise tension without the external contribution.

2.2.5 Noise Figure

Consider a two-port circuit, having input and output terminals, of available gain G placed in a transmission line, as illustrated in Figure 2.11, the useful signals at the input and the output, P_{in} and P_{out}, as well as the noise power at the input, $P_{N,\text{in}}$, and output, $P_{N,\text{out}}$. The two-port circuit adds a certain contribution to the input noise and for that reason degrades the SNR at the output compared to that at the input, which yields

$$\frac{P_{\text{out}}}{P_{N,\text{out}}} < \frac{P_{\text{in}}}{P_{N,\text{in}}}$$

The noise figure F of the two-port circuit is given

$$F = \frac{[P_{\text{in}}/P_{N,\text{in}}]}{[P_{\text{out}}/P_{N,\text{out}}]} \tag{2.6}$$

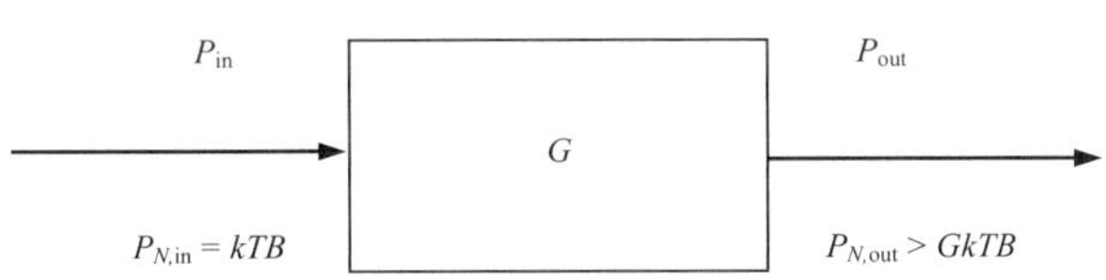

Figure 2.11 Two-port circuit.

Then,

$$P_{N,\mathrm{out}} = G(P_{N,\mathrm{in}} + P'_{N,\mathrm{in}}) = G(kTB + P'_{N,\mathrm{in}})$$

By regarding $P'_{N,\mathrm{in}}$ as the noise contribution of the two-port circuit brought back to the input and thus replacing this one by a perfect noiseless two-port circuit with an effective noise $P'_{N,\mathrm{in}}$, at the input, we can write

$$F = \frac{P_{\mathrm{in}}/(kTB)}{P_{\mathrm{in}}/(kTB + P'_{N,\mathrm{in}})} = 1 + \frac{P'_{N,\mathrm{in}}}{kTB}$$

Then,

$$P'_{N,\mathrm{in}} = (F-1)kTB \tag{2.7}$$

from which we deduce

$$P_{N,\mathrm{in}} + P'_{N,\mathrm{in}} = kTB + (F-1)kTB = FkTB$$

and finally,

$$P_{N,\mathrm{out}} = GFkTB \tag{2.8}$$

$$F = \frac{P_{N,\mathrm{out}}}{GkTB} \tag{2.9}$$

In the case of a chain made up of a number n of two-port circuits in cascade, as presented in Figure 2.12, we write

$$P'_{N,\mathrm{in}} = (F_1 - 1)kTB + \frac{(F_2-1)kTB}{G_1} + \cdots + \frac{(F_n-1)kTB}{G_1 \ldots G_{n-1}} = (F_{\mathrm{eq}} - 1)kTB$$

which gives the equivalent noise figure

$$F_{\mathrm{eq}} = F_1 + \frac{F_2 - 1}{G_1} + \cdots + \frac{F_n - 1}{G_1 \ldots G_{n-1}} \tag{2.10}$$

We see that if the first stage has a high gain, the noise contribution of the following stages is negligible, and we will consequently seek to place at the input

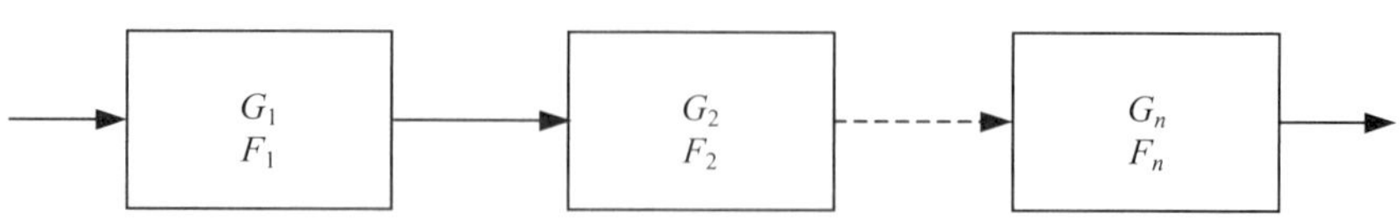

Figure 2.12 Chain of two-port circuits in cascade.

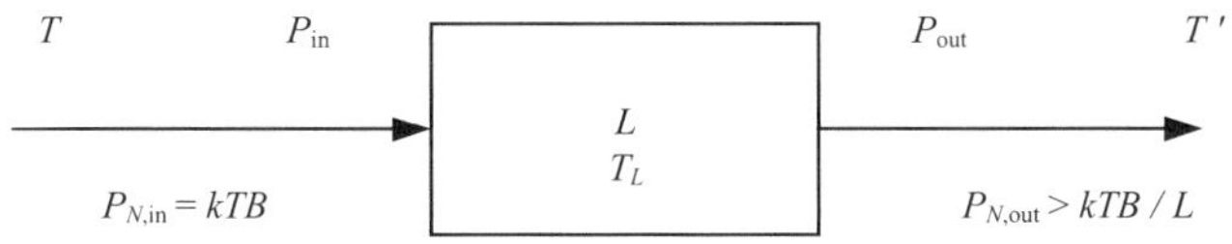

Figure 2.13 Two-port attenuator.

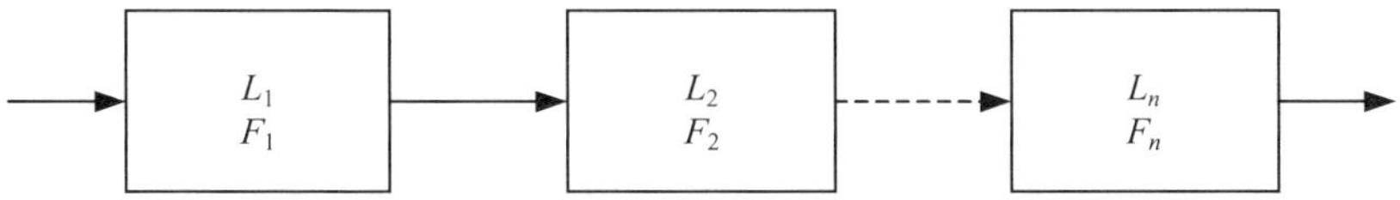

Figure 2.14 Chain of two-port attenuators.

a two-port circuit presenting the weakest noise figure as well as the highest possible gain.

Consider now a two-port circuit risen at the temperature T_L that brings an attenuation L, for instance, an attenuator or a transmission feeder line, as presented in Figure 2.13. Using relation (1.140) yields

$$T' = \frac{T}{L} + T_L\left(1 - \frac{1}{L}\right)$$

Attenuation being the opposite operation of gain, relation (2.9) becomes

$$F = L\frac{P_{N,\mathrm{out}}}{kTB}$$

As $P_{N,\mathrm{out}} = kT'B$, we obtain

$$F = L\frac{T'}{T} = 1 + (L-1)\frac{T_L}{T} \tag{2.11}$$

In the case of a chain of n two-port attenuators, presented in Figure 2.14, we have

$$P'_{bE} = (F_1 - 1)kTB + L_1(F_2 - 1)kTB + \cdots + L_1 L_{n-1}(F_n - 1)kTB = (F_{\mathrm{eq}} - 1)kTB$$

from which

$$F_{\mathrm{eq}} = F_1 + L_1(F_2 - 1) + \cdots + L_1 L_{n-1}(F_n - 1) \tag{2.12}$$

2.2.6 Equivalent Noise Temperature

The noise due to a two-port circuit can also be expressed in the form of an equivalent noise temperature T_{eq}, defined as the temperature of a noise source which, connected to the input of a perfect and noiseless two-port circuit, would give at the output the same noise as the considered two-port circuit. We can thus establish the relation between the equivalent noise temperature T_{eq} and the noise figure F of the receiver compared to the ambient temperature T_0 using relation (2.7):

$$P'_{N,\text{in}} = (F-1)kT_0B = kT_{eq}B$$

from which

$$T_{eq} = (F-1)T_0 \tag{2.13}$$

$$F = 1 + \frac{T_{eq}}{T_0} \tag{2.14}$$

Figure 2.15 shows the relation which exists between the noise figure, expressed in decibels, and the equivalent noise temperature compared to the ambient temperature $T_0 = 290\,\text{K}$. This equivalence makes it possible to measure the noise figure of a two-port circuit by means of a noise generator, which is in general a noise diode calibrated in the considered frequency band; Figure 2.16 illustrates the assembly employed.

We start by measuring the proper noise of the two-port circuit on the detector, the noise diode not being fed behaving like a resistor at the ambient temperature T_0, that is, FkT_0BG_0 with the nominal gain of the two-port circuit, G_0.

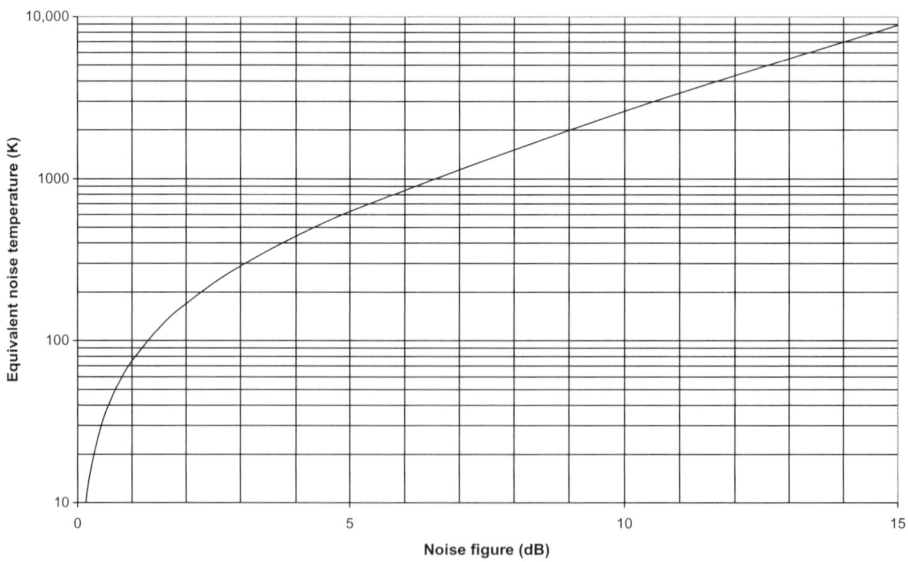

Figure 2.15 Equivalent noise temperature versus noise figure.

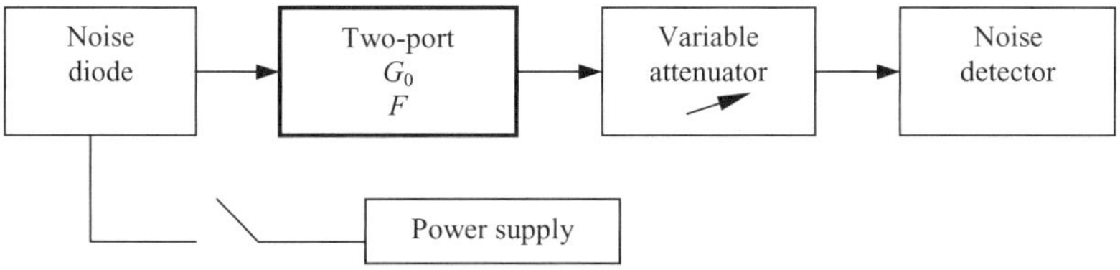

Figure 2.16 Measurement of noise figure of two-port circuit.

Then the noise diode, whose calibrated noise figure is F_D, is fed with a power supply and the variable attenuator is adjusted by a value A so that to obtain the same level as previously on the noise detector, that is, $(F_DkT_0B + FkT_0B)(G_0/A)$. As the noise level is the same during both measurements, we write

$$FkT_0BG_0 = (F_DkT_0B + FkT_0B)\frac{G_0}{A}$$

from which $F = F_D/(A - 1)$ or, in decibels, $F = F_D - 10\log(10^{0.1A} - 1)$. The noise generators are generally calibrated in equivalent noise temperature, so we use relation (2.14) to determine their noise figure.

In the case of a chain of n two-port circuits in cascade, formula (2.10) becomes, by replacing the noise figure of each circuit by their equivalent noise temperature:

$$T_{\text{eq}} = T_1 + \frac{T_2}{G_1} + \cdots + \frac{T_n}{G_1 \ldots G_{n-1}} \tag{2.15}$$

With a chain of n two-port attenuators, we write, from relations (2.12) and (2.14),

$$F_{\text{eq}} = 1 + \frac{T_{\text{eq}}}{T_0} = 1 + \frac{T_1}{T_0} + L_1\frac{T_2}{T_0} + \cdots L_1 \cdots L_{n-1}\frac{T_n}{T_0}$$

from which

$$T_{\text{eq}} = T_1 + L_1T_2 + \cdots + L_1 \ldots L_{n-1}T_n \tag{2.16}$$

2.2.7 Radiated Noise

Consider an amplifier coupled to an antenna, as illustrated in Figure 2.17. Starting with relation (2.8), we write

$$P_{N,\text{out}} = GFkTB \qquad P_{NR} = G_{ant}GFkTB \tag{2.17}$$

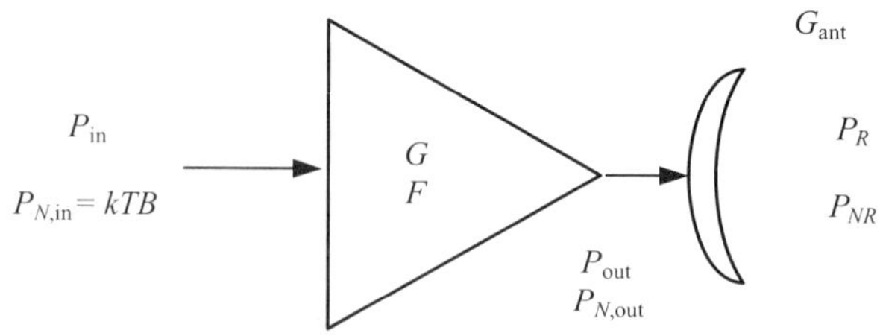

Figure 2.17 Radiated noise.

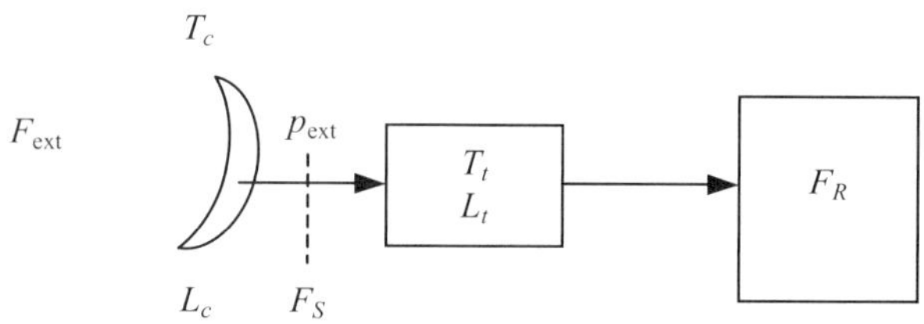

Figure 2.18 System of reception disturbed by external source of noise.

2.2.8 External Noise

2.2.8.1 Effect of External Noise The radioelectric noise emitted by matter and human activities, which constitutes a disturbing element for the telecommunications services, comes from various sources, and it is advisable to study their space and temporal distribution as well as their characteristics. The total power of noise received by the antenna is the sum of the contributions emitted by the various noise sources in the direction of the antenna and that are collected by the main lobe, such as by the side lobes.

The noise figure F_S of the whole reception system presented in Figure 2.18 can thus be given starting from relation (2.12), which leads to the expression

$$F_S = F_{ext} + (F_c - 1) + L_c(F_t - 1) + L_c L_t (F_R - 1) \tag{2.18}$$

where the noise figures are defined as follows:

- External noise figure F_{ext} is given by the relation

$$F_{ext} = \frac{p_{ext}}{kT_0B} = \frac{T_{ext}}{T_0}$$

where p_{ext} = noise power at lossless antenna port (W)
T_0 = reference ambient temperature (290 K)
B = bandwidth of receiver (Hz)
T_{ext} = equivalent external noise temperature (K)

- The noise figure associated with antenna losses, F_c, is given by relation (2.11):

$$F_c = 1 + (L_c - 1)\frac{T_c}{T_0}$$

where L_c = antenna losses
T_c = physical temperature of antenna (K)

- The noise figure associated with losses in the transmission line to the receiver, F_t, is given as

$$F_t = 1 + (L_t - 1)\frac{T_t}{T_0}$$

where L_t = transmission line losses
T_t = physical temperature of transmission line (K)

- The noise figure F_R is that of the receiver.

If $T_c = T_t = T_0$, which corresponds to the majority of the cases, expression (2.18) is simplified since $F_c = L_c$ and $F_t = L_t$ and becomes

$$F_S = F_{\text{ext}} - 1 + F_c F_t F_R = F_{\text{ext}} - 1 + L_c L_t F_R \tag{2.19}$$

Expressed in decibels, the noise figure and the external noise power become, while utilizing Boltzmann's noise at the temperature of reference T_0,

$$F_{\text{ext}} = 10\log F_{\text{ext}} \qquad \text{(dB)}$$
$$P_{\text{ext}} = 10\log p_{\text{ext}} \qquad \text{(dBW)}$$
$$B = 10\log B \qquad \text{(dBHz)}$$
$$P_{\text{ext}} = F_{\text{ext}} + B - 228.6\ \text{dB}\ (\text{W K}^{-1}\ \text{Hz}^{-1}) + 24.6\ \text{dB K} \qquad \text{(dBW)}$$

with Boltzmann's noise given as

$$10\log(1.380664 \times 10^{-23}\ \text{J K}^{-1}) = -228.6\ \text{dB}\ (\text{W K}^{-1}\ \text{Hz}^{-1})$$
$$10\log(290\ \text{K}) = 24.6\ \text{dBK}$$

For a very short vertical monopole antenna ($h << \lambda$) placed above a perfect ground, the mean-square value of the vertical component of the electric field corresponding to the external noise E_{ext} is given by the relation, in decibels (microvolts per meter),

$$E_{\text{ext}} = F_{\text{ext}} + 20\log f_{\text{MHz}} + B - 95.5 \tag{2.20}$$

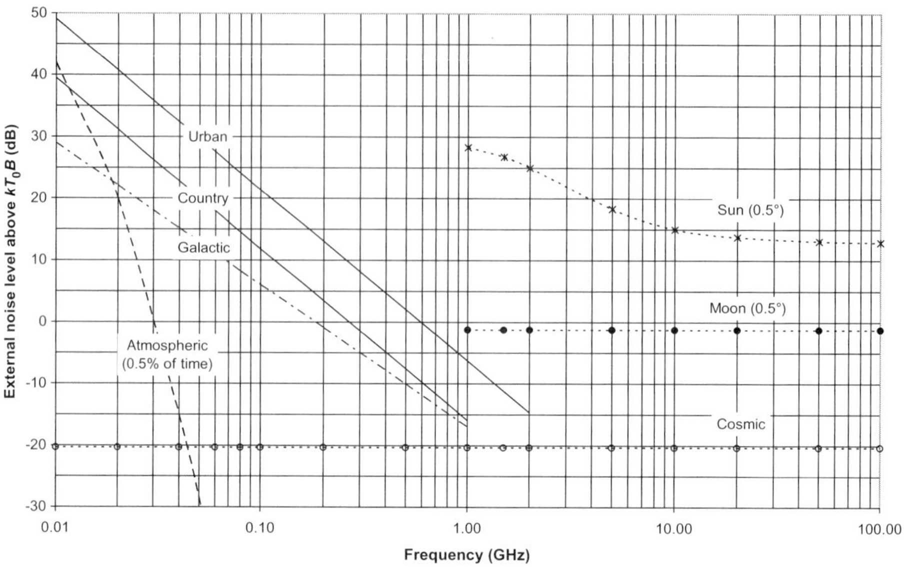

Figure 2.19 Mean values of external noise level received by omnidirectional aerial close to ground.

In the case of a half-wave antenna in free space, this expression becomes

$$E_{ext} = F_{ext} + 20 \log f_{MHz} + B - 99 \quad (2.21)$$

2.2.8.2 Sources of Noise Temperature

Principal Natural and Artificial External Sources of Noise Temperature Figure 2.19 (ITU-R PI.372) presents the mean value of the noise level above the background thermal noise kT_0B received by an omnidirectional aerial close to the ground. The urban and country sources correspond to human activity. The atmospheric source corresponds to the lightning activity and atmospheric electric discharges; the curve represents the value exceeded during 0.5% of time. It is also advisable to take account of the following:

- Sun and moon, whose influence is limited to their apparent solid angle of ~0.5°
- Milky Way, which arises as a broad belt of emission
- Fossil cosmic background temperature of 2.7 K

Figure 2.20 presents the mean equivalent temperature of the external noise sources according to the frequency by operating the transformation

$$T_{ext} = T_0 \times 10^{F_{ext}/10} \quad (2.22)$$

where F_{ext} is the external noise figure expressed in decibels.

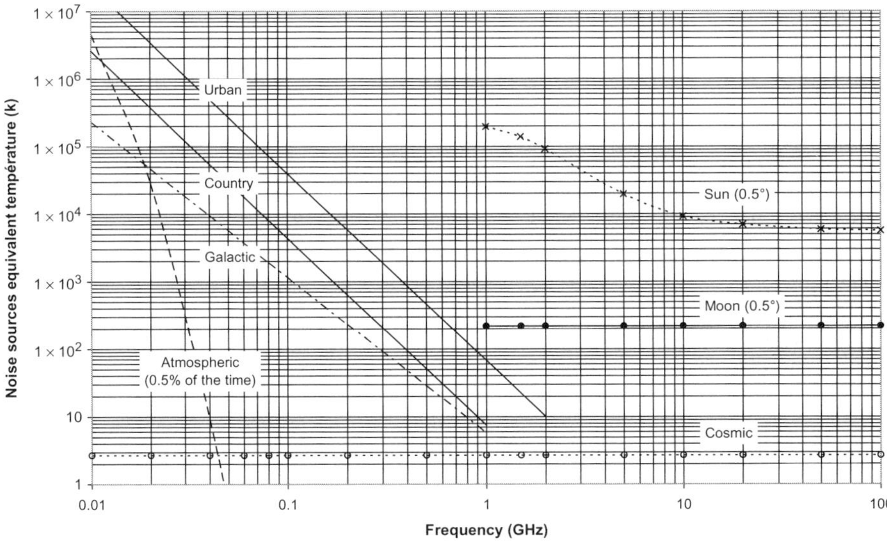

Figure 2.20 Mean values of external equivalent noise temperature received by omnidirectional antenna close to ground.

Radioelectric Emission Due to Gases of Atmosphere The atmosphere absorbs a part of the signal transmitted by the source, as we saw in Section 1.10, and thus it is possible to calculate the brightness temperature of the atmosphere starting from relation (1.140) for the gray body:

$$T' = \frac{T_S}{A_a} + T_a\left(1 - \frac{1}{A_a}\right)$$

by posing $T_s = 0$, since we are interested only in the contribution of the atmosphere, that is,

$$T' = T_a\left(1 - \frac{1}{A_a}\right) \tag{2.23}$$

where $T_a = T_0 = 290\,\text{K}$. Figure 2.21 presents the brightness temperature of the atmosphere for a vapor concentration of $7.5\,\text{g}\,\text{m}^{-3}$ that is to be related to Figure 1.74.

Radioelectric Emission Due to Hydrometeors As hydrometeors have a certain temperature, formula 2.23 can be applied in the case of attenuation due to the rain during the considered percentages of time which produces two cumulative effects:

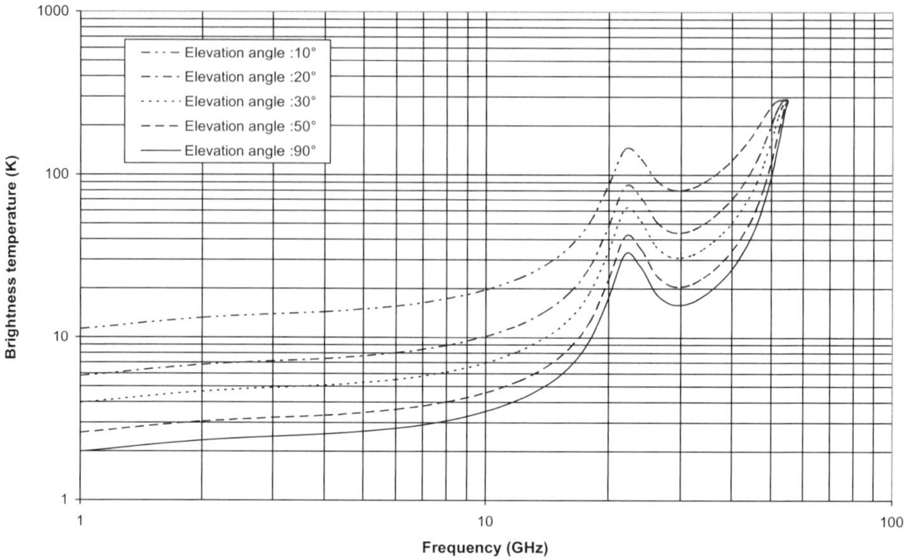

Figure 2.21 Brightness temperature of clear atmosphere (7.5 g m^{-3} H_2O).

- Additional loss of signal at time of crossing of zone of rain
- Increase in temperature of reception system by quantity ΔT, which can be determined from relation (1.140) as

$$\Delta T = T_R\left[1-\frac{1}{A_R}\right]$$

where A_R = attenuation calculated using relations (1.109) and (1.113)
T_R = temperature of rain, which is about 290 K

2.3 DIGITAL COMMUNICATION SYSTEMS DESIGN

2.3.1 Noise in Digital Communication Systems

The noise corresponds to the unwanted electric signals that arise from a variety of sources and are always present in electrical systems, as follows:

- Man-made noise due to human activity, such as electric transients and other radiating electromagnetic signals, that can be reduced or eliminated through filtering, shielding, and selection of appropriate frequency band,
- Natural noise, which includes galactic sources, atmospheric disturbances, and especially the proper thermal noise of the electronic circuits, that cannot be eliminated

2.3.1.1 Gaussian Random Noise We have seen that thermal noise is caused by the thermal motion of electrons in all dissipative components; according to the central limit theorem, the probability distribution of the sum of such statistical independent random variables approaches the Gaussian distribution whatever the individual distribution functions may be. The probability density function of the noise random function $n(t)$ of zero mean at any time t is thus given by the relation

$$p(n) = \frac{1}{\sigma\sqrt{2\pi}} \exp\left[-\frac{1}{2}\left(\frac{n}{\sigma}\right)^2\right] \tag{2.24}$$

which becomes, by considering the sum z of a DC component a and a Gaussian random variable n such as $z = a + n$,

$$p(x) = \frac{1}{\sigma\sqrt{2\pi}} \exp\left[-\frac{1}{2}\left(\frac{z-a}{\sigma}\right)^2\right] \tag{2.25}$$

where the variance σ^2 is equal to 1 in the normalized Gaussian density function.

2.3.1.2 Additive White Gaussian Noise The power spectral density of thermal noise is the same for all frequencies of interest in most communication systems; it emanates an equal amount of noise power per unit bandwidth at all frequencies from DC to about 10^{12} Hz. When the noise power has such uniform spectral density, it is called white noise, by analogy with the white light that contains all frequencies within the visible band of electromagnetic radiation. Other characteristics of white noise are that two different samples can be considered as uncorrelated and independent and that it is simply superimposed or added to the signal; for these reasons it is generally called additive white Gaussian noise (AWGN). In the modulation process, the two-sided power spectral density $N(f)$ is expressed in Watts per hertz by

$$N(f) = \frac{1}{2} N_0 \tag{2.26}$$

2.3.1.3 Detection of Binary Signals in Presence of AWGN During a given signaling interval T, a binary system will transmit one of two waveforms: $S_0(t)$ for a binary 0 and $S_1(t)$ for a binary 1; the received signal $R(t)$ is then composed of one of the two preceding signals $S(t)$ added to the noise $n(t)$ as presented in Figure 2.22, which shows the main elements of a digital signal detector. At the sampling stage, the signal S will be the result of the composition of a mean $\overline{S}_0$ or $\overline{S}_1$, depending on whether a 0 or a 1 was sent, and a zero-mean Gaussian random noise component, and we can write the conditional probability density functions as

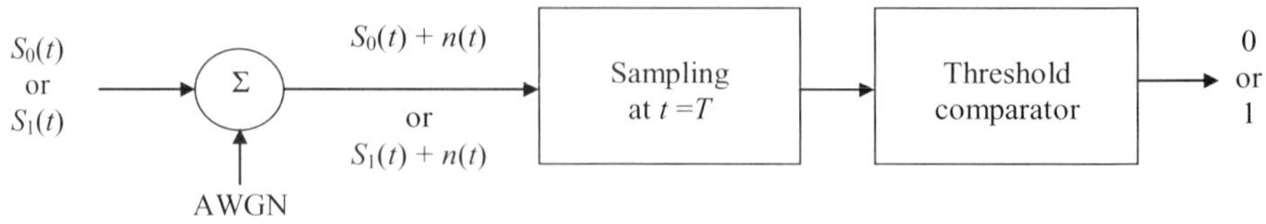

Figure 2.22 Digital signal detection.

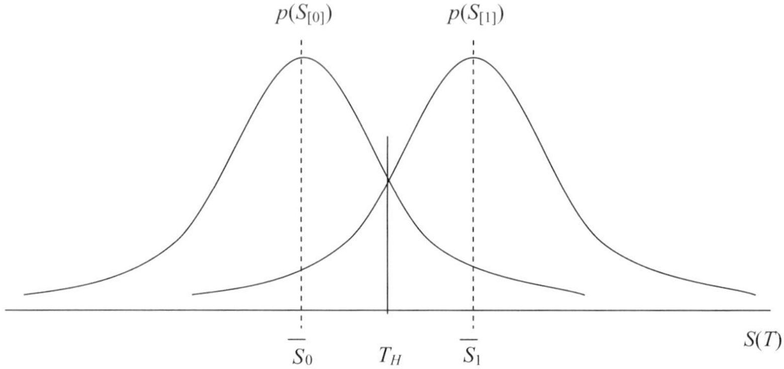

Figure 2.23 Conditional probability density functions.

$$
\begin{aligned}
p(S[_0]) &= \frac{1}{\sigma\sqrt{2\pi}} \exp\left[-\frac{1}{2}\left(\frac{S-\bar{S}_0}{\sigma}\right)^2\right] \\
p(S[_1]) &= \frac{1}{\sigma\sqrt{2\pi}} \exp\left[-\frac{1}{2}\left(\frac{S-\bar{S}_1}{\sigma}\right)^2\right]
\end{aligned}
\tag{2.27}
$$

The conditional probability density functions are presented on Figure 2.23 as well as the value T_H of the threshold of the comparator. The threshold level is generally equal to $(\bar{S}_0 + \bar{S}_1)/2$ so as to minimize the probability of making an incorrect decision; this strategy is known as minimum-error criterion.

2.3.1.4 Error Probability of Binary Detection The probability of an error is the sum of the probabilities of all the ways that an error can occur even on binary 0 and binary 1. Assuming that the a priori probabilities of error are equal for the 0's and 1's, we can write that the probability of a bit error P_B is numerically equal to the area under the tail of the probability density functions, that is,

$$
P_B = \frac{1}{\sigma\sqrt{2\pi}} \int_{TH}^{\infty} \exp\left[-\frac{1}{2}\left(\frac{S-\bar{S}_0}{\sigma}\right)^2\right] dS = \frac{1}{\sigma\sqrt{2\pi}} \int_{TH}^{\infty} \exp\left[-\frac{1}{2}\left(\frac{S-\bar{S}_1}{\sigma}\right)^2\right] dS
$$

Considering the new variable $x = (S - \bar{S}_1)/\sigma$, we can write

$$P_B = \frac{1}{\sqrt{2\pi}} \int_{TH}^{\infty} \exp\left[-\frac{1}{2}\left(\frac{x}{\sigma}\right)^2\right] dx = Q\left(\frac{\bar{S}_1 - \bar{S}_0}{2\sigma}\right) \tag{2.28}$$

where $Q(x)$ is the coerror function, commonly used for the probability under the tail of the Gaussian distribution and defined as

$$Q(x) = \frac{1}{\sqrt{2\pi}} \int_{x}^{\infty} \exp\left(-\frac{x^2}{2}\right) dx \tag{2.29a}$$

A good approximation for $x > 3$ is given by the relation

$$Q(x) = \frac{1}{x\sqrt{2\pi}} \exp\left(-\frac{x^2}{2}\right)$$

which can also be expressed using the error function erfc(x),

$$Q(x) = \frac{1}{2} \operatorname{erfc}\left(\frac{x}{\sqrt{2}}\right) \tag{2.29b}$$

as

$$\operatorname{erfc}(x) = \frac{2}{\sqrt{\pi}} \int_{x}^{\infty} \exp(-x^2)\, dx \approx 2Q(x\sqrt{2})$$

2.3.1.5 Signal-to-Noise Ratio The useful signal $S(t)$ affected by the AWGN signal $n(t)$ passes through a matched linear filter which is designed to provide the maximum SNR at its output for a given transmitted symbol waveform. At time interval T, the receiver delivers a signal component $\bar{S}_0$ or $\bar{S}_1$ and a noise component $\bar{n}$ which has average power equal to the variance σ^2. Therefore, the ratio of instantaneous signal power to average noise power (S/N) at time interval T can be expressed as

$$\left(\frac{S}{N}\right)_{T(0)} = \frac{\bar{S}_0^2}{\sigma^2}$$
$$\left(\frac{S}{N}\right)_{T(1)} = \frac{\bar{S}_1^2}{\sigma^2} \tag{2.30}$$

We demonstrate then that the maximum SNR has the value

$$\max\left(\frac{S}{N}\right)_T = \frac{2E}{N_0} \tag{2.31}$$

where the energy E of the input signal is

$$E = \int_{-\infty}^{+\infty} |S(f)|^2 \, df$$

Thus, the maximum SNR depends on the input signal energy and the noise power spectral density whatever the shape of the waveform of the signal could be. As the useful signal takes only two binary values, the detection process has, in fact, to minimize the argument of the relation (2.28) and then

$$\left(\frac{S}{N}\right)_T = \frac{(\bar{S}_1 - \bar{S}_0)^2}{\sigma^2} = \frac{2E_d}{N_0} \tag{2.32}$$

where E_d is the energy of the difference signal.

2.3.1.6 Error Probability of Binary Signal In a binary signal, the probability of error according to the SNR becomes

$$P_B = Q\left(\sqrt{\frac{E_d}{2N_0}}\right) \tag{2.33a}$$

If the amplitude of the signal is zero for binary 0 and A for binary 1, the energy E_d of the difference signal is $A^2\,T$, and the average energy per bit E_b is $A^2\,T/2$, expression (2.33a) becomes

$$P_B = Q\left(\sqrt{\frac{A^2T}{N_0}}\right) = Q\left(\sqrt{\frac{E_b}{N_0}}\right) \tag{2.33b}$$

Figure 2.24 presents the theoretical bit error probability P_B.

2.3.1.7 Error Probability for Modulated Signal In Figure 2.4 we presented some usual modulation types where the binary signal occupies different states of the modulated carrier. The noise jams the state points of the modulation and causes a random broadening of them, as shown in Figure 2.25; such conditions provoke a closing of the demodulator eye pattern and the degradation of the bit error ratio (BER). However, we can see that the distance between the points in the constellation diagram decreases when the number of states of the carrier increases; consequently, the noise that causes them to occupy a broader area around the required state points, as shown in Figure 2.25, gives a higher probability of states overlapping.

It is then necessary to increase the SNR in order to maintain a good BER; that is, the carrier level has to be increased by the same amount. We can see implicitly that the distance between the state points is reduced by 2 when the

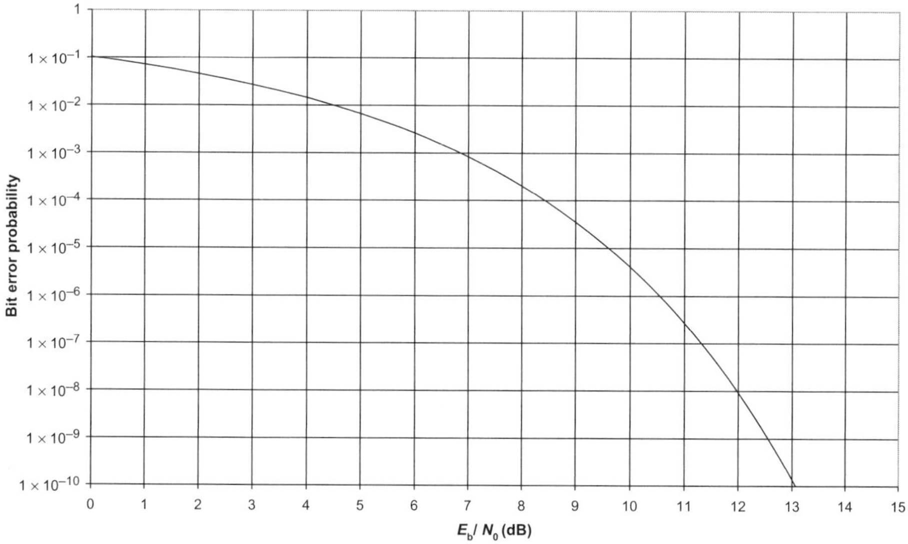

Figure 2.24 Bit error probability P_B versus E_b/N_0.

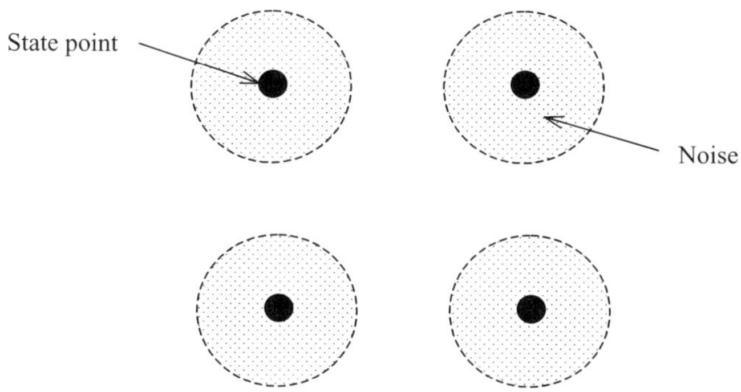

Figure 2.25 Constellation diagram in presence of noise.

number of states is multiplied by 4 and thus that the SNR has to be increased by 6 dB to keep the same BER.

When the receiver exploits the knowledge of the carrier's phase to detect the signals, the process is called coherent detection, whereas when the receiver does not utilize such phase reference information, the process is called non-coherent or differentially coherent detection. Figure 2.26 presents the BER distribution for some selected binary modulation schemes for which the formulas are

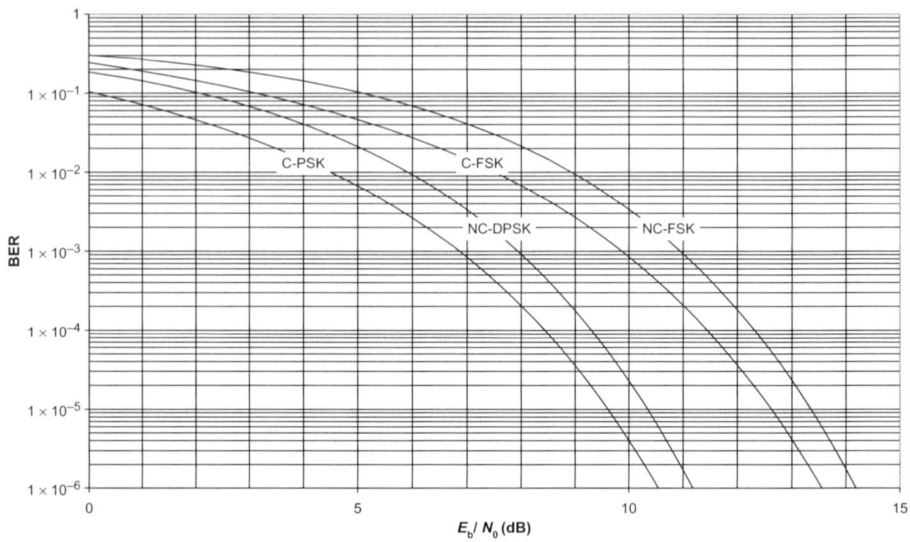

Figure 2.26 Probability of error for coherent and noncoherent modulations.

$$
\begin{aligned}
\mathrm{BER} &= \frac{1}{2\sqrt{\pi(E_b/N_0)}}\exp\left(-\frac{E_b}{N_0}\right) \quad \text{(coherent PSK)}\\
\mathrm{BER} &= \frac{1}{2}\exp\left(-\frac{E_b}{N_0}\right) \quad \text{(noncoherent DPSK)}\\
\mathrm{BER} &= \frac{1}{\sqrt{2\pi(E_b/N_0)}}\exp\left(-\frac{E_b}{2N_0}\right) \quad \text{(coherent FSK)}\\
\mathrm{BER} &= \frac{1}{2}\exp\left(-\frac{E_b}{2N_0}\right) \quad \text{(noncoherent FSK)}
\end{aligned}
\tag{2.34}
$$

For M-QAM modulations, the probability of error is given by the formula

$$
P_M = 2\,\mathrm{erfc}\left(\sqrt{\frac{3}{2(M-1)}\log_2 M\frac{E_b}{N_0}}\right) \tag{2.35}
$$

Figure 2.27 presents the respective distribution functions for various values of M by including a coding gain of 3 dB. The factor $10\log[2(M-1)/3]$ represents the increase in average power which is required to maintain a given level of performance as the number of signal state points increases; for instance, an increase by a factor 2 requires a 3-dB increase in power.

2.3.1.8 General Characteristics of Signal The following relations are enough, in a first approach, to determine the general characteristics of the transmitted signal:

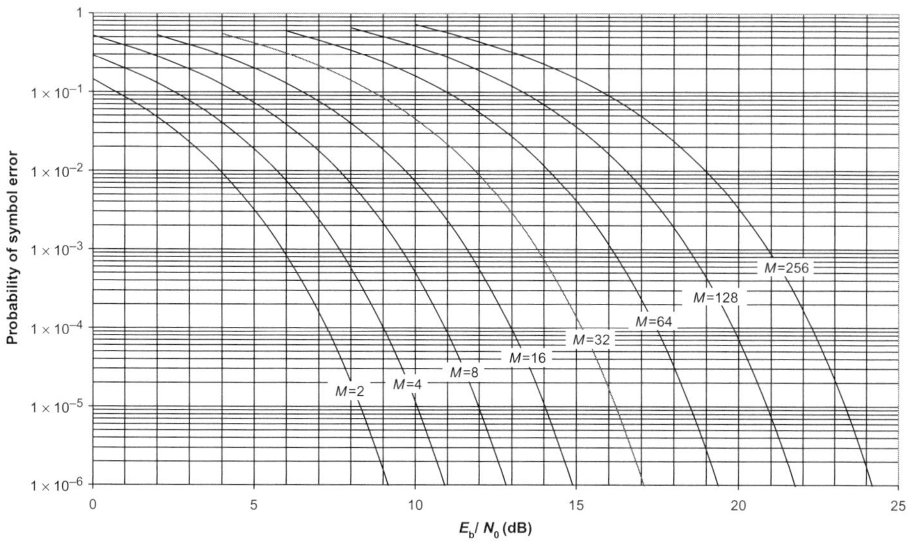

Figure 2.27 Probability of symbol error for *M*-QAM modulations.

- Bit time duration:

$$\tau = \frac{1}{R} \tag{2.36}$$

  where R is the bit rate.
- Symbol rate:

$$S = \frac{R}{mc}$$
$$m = \log_2 M \tag{2.37}$$

  where M is the number of states of the carrier and c is the coding ratio.
- Occupied bandwidth:

$$B = \frac{RN_y}{mc} = SN_y \tag{2.38}$$

  where N_y is the filtering Nyquist coefficient.
- Energy per bit:

$$E_b = \frac{C}{R} \tag{2.39}$$

  where C is the carrier power level.
- Noise spectral density:

$$N_0 = kT_e \tag{2.40}$$

  where k is Boltzmann's constant and T_e is the equivalent noise temperature of the system.

- Total thermal noise:

$$N = N_0 B \tag{2.41}$$

We can then write

$$\frac{E_b}{N_0} = \frac{C\tau}{N_0} = \frac{C}{N_0}\frac{1}{R} = \frac{C}{R}\frac{B}{N} = \left(\frac{C}{N}\right)\frac{B}{R} \tag{2.42}$$

2.3.1.9 Choice of Modulation The choice of a method of modulation rests on considerations that concern mainly the required power for the signal and the efficiency in the use of the frequency band. In general:

- PAM is simple to implement but the error ratio is high.
- PSK shows a better error characteristic but requires a great bandwidth.
- FSK is excellent for what precedes and offers the advantage of lending itself to multiphase modulation.
- QPSK, MSK, GMSK are variations of phase and minimum shift keying modulations.
- M-QAM makes it possible to still increase the number of states of the carrier, and therefore the spectral efficiency in terms of bits per second per hertz, but requires more energy as well as a greater linearity of the amplifiers.

The choice of modulation scheme thus appears as a tradeoff between spectrum efficiency and power efficiency, as shown by Figure 2.28.

2.3.1.10 Coding for Error Detection and Correction Channel coding refers to the class of signal transformations that are designed in order to improve communication performance by enabling the transmitted signals to better withstand the effects of various factors of degradation such as noise, fading, and jamming. The goal of channel coding is to reduce the probability of bit error P_B or the required E_b/N_0 at the cost of expending more bandwidth than would otherwise be necessary. However, the combined modulation and coding techniques for bandlimited channels allows to conciliate simultaneously performance goals that are normally contradictory or in conflict.

Forward Error Corrector The objectives of error correctors are to reduce the residual BER by several orders of magnitude and to increase system gain. This is achieved by encoding the bit stream prior to modulation by adding extra bits to the bit stream into the time domain according to specific rules; the combined set of information and parity bits is then modulated and transmitted. Such improvement, through the introduction of a redundancy into the message, is thus translated into an increase of the bit rate. Nevertheless, the additional bits do not contribute information to the transmitted message but serve to

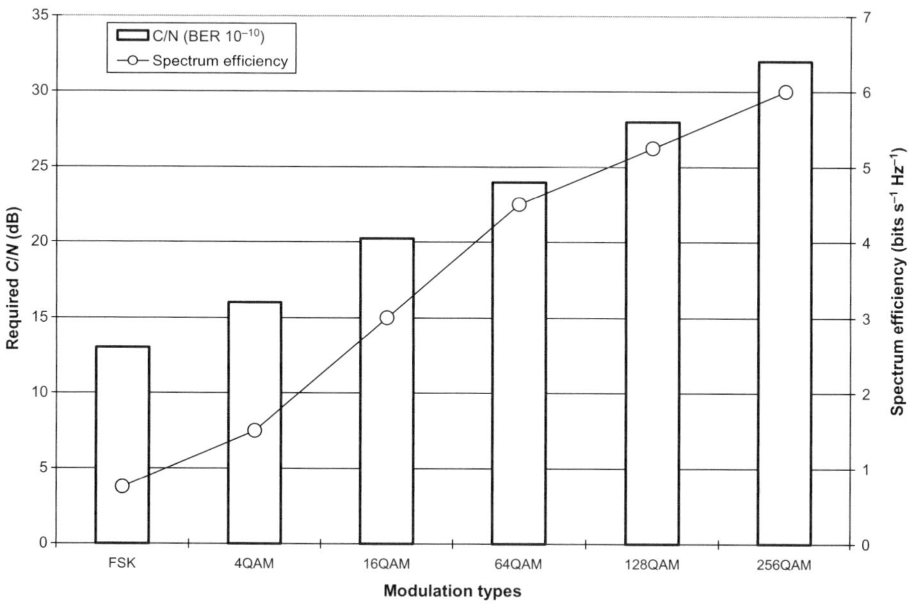

Figure 2.28 Comparative spectrum efficiency; bit rates are 34 Mbits s^{-1} for FSK, 4QAM, and 16QAM and 155 Mbits s^{-1} for 64TCM, 128TCM, and 256TCM.

allow the decoder in the receiver to recognize and therefore correct the errors that might arise during the transmission process.

Error correction codes can be broadly divided into three categories:

- Block codes (Hamming, BCH, Golay, Reed–Salomon, etc.), for which the coder associates X coded bits to each block of Y bits of information, the coding of one block being done independently of that of the blocks which precede it
- Convolutional codes (e.g., Viterbi), for which the coder associates X_0 coded bits to Y_0 bits of information, the coding being done on Z consecutive blocks; the code is then characterized by a length $X = ZY_0$
- Concatenated codes, which are combinations of both

There are also other codes, such as interleaving codes as a countermeasure against short error bursts; cyclic redundancy check, which is recommended in high-speed modems; and automatic request for repeat, which requires a return path.

In general, the bits of information are grouped in blocks of K bits and the coder then associates N bits to K bits ($N > K$), which constitutes a codeword; there are thus 2^K binary words of N bits. The ratio $Q = K/N$ is called the code efficiency and constitutes the first parameter defining a code. The transmitted digital rate R_C, or coded rate, is thus higher than the useful rate R_U and becomes

$$R_C = \frac{R_U}{Q}$$

The table below indicates the minimal necessary bandwidth, according to Nyquist's first criterion, to transmit an information rate $R_u = 1/\tau_u$ using a M-PSK modulation. If E_u is the average energy by bit of information in the absence of coding and E_c the average energy by transmitted coded bit, we have

$$E_c = QE_u$$

Calling B the noise bandwidth of the reception filter and considering the nominal power C of the carrier, we can write

$$\frac{C}{N} = \frac{E_c R_c}{N_0 B} = \frac{E_u R_u}{N_0 B} \tag{2.43}$$

that is, in decibels,

$$\left(\frac{E_u}{N_0}\right) = \left(\frac{E_c}{N_0}\right) + 10 \log \frac{1}{Q} \tag{2.44}$$

The $N - K$ additional bits of control brought by the coder take a certain energy which is used in return to increase the energy of the bits of information.

Number of States, M	Code Efficiency, Q	Minimum Bandwidth[a]
2	1	$1/\tau_u$
	$\frac{1}{2}$	$2/\tau_u$
	$\frac{2}{3}$	$3/(2\tau_u)$
4	1	$1/(2\tau_u)$
	$\frac{1}{2}$	$1/\tau_u$
	$\frac{2}{3}$	$3/(4\tau_u)$
	$\frac{3}{4}$	$2/(3\tau_u)$
8	1	$1/(3\tau_u)$
	$\frac{1}{2}$	$2/(3\tau_u)$
	$\frac{2}{3}$	$1/(2\tau_u)$
	$\frac{3}{4}$	$4/(9\tau_u)$
16	1	$1/(4\tau_u)$
	$\frac{1}{2}$	$1/(2\tau_u)$
	$\frac{2}{3}$	$3/(8\tau_u)$
	$\frac{3}{4}$	$1/(3\tau_u)$

[a]The really occupied bandwidth is in general 1.05–1.3 times the minimal Nyquist bandwidth.

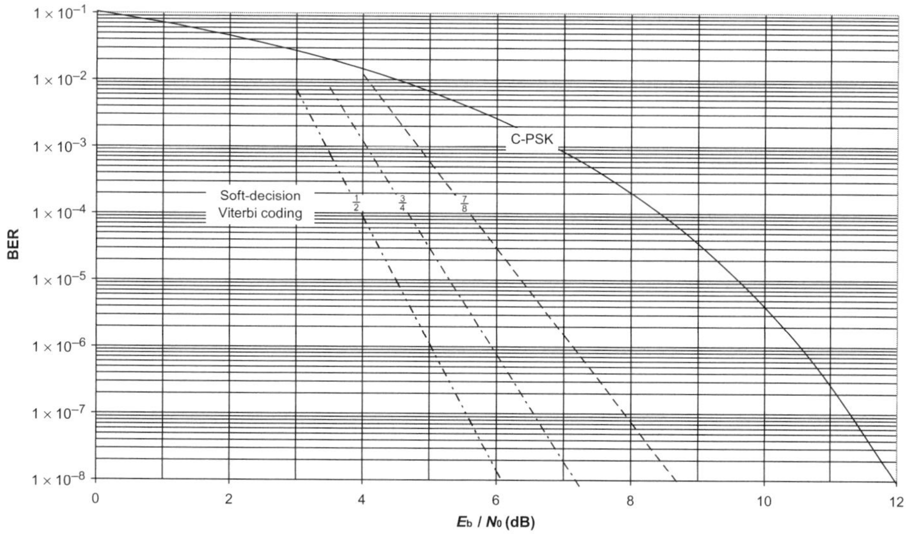

Figure 2.29 BER distribution versus coding efficiency.

Coding thus provides an additional gain, as shown in Figure 2.29, where the characteristics of C-PSK are compared with those obtained for various coding efficiencies of Viterbi type with soft decision. Thus coding allows us to:

- Improve the quality of the transmission for the same carrier power or to reduce this latter by preserving the same quality.
- Reduce, if necessary, the spectral density of the signal by using a spread-spectrum code.

Considering a coding gain G_c, relation (2.43) can then be written, in decibels, as

$$\left(\frac{C}{N}\right)=\left(\frac{E_b}{N_0}\right)+\left(\frac{R}{B}\right)-G_c \tag{2.45}$$

Treillis-Coded Modulation Trellis-coded modulation (TCM) was developed in order to improve the BER without increasing the bit rate; this process can achieve a coding gain of up to 6 dB in E_b/N_0.

The susceptibility of a modulation scheme to intersymbol interference depends on the distance between states in the signal state-space diagram. The trellis-coding technique obtains its advantage by effectively maximizing the physical Euclidean distance between the states of the constellation by expanding the signal set while avoiding bandwidth expansion. For M-QAM, the technique is called set partitioning; for m numbers of bits per symbol, each subset contains 2^{m-2} states and the number of states in the new constellation is 2^{m+1}. In such conditions, transitions between states other than adjacent states can occur and the free distance can be increased, with respect to the distance between adjacent states, by a factor $\sqrt{5}$ to which correspond a coding gain of about 3.5 dB in relation to the distributions of Figure 2.27.

2.3.2 Other Electrical Sources of Corruption

There are many others sources of degradation of the transmitted signals whose effects can be related to those of white noise, such as intermodulation products due to the nonlinearities of the circuits and interference by other systems operating in the same frequency bands.

2.3.2.1 Intermodulation Among the nonlinear circuits which generate intermodulation lines from the signals going through it, the power amplifier is the most important element of the transmission chain and generally is of two types:

- Traveling-wave tubes (TWT) or other types, such as klystron
- Solid-state transistors

The main characteristics of power amplifiers relate to the following

- Amplitude–frequency response
- Output–input power response
- Amplitude–phase modulation (AM–PM) conversion
- Intermodulation product level
- Group delay distortion
- Outband emission
- Residual amplitude modulation
- Noise figure
- Nonessential radiation

Generally, we can distinguish between several types of intermodulation which are described in recommendations ITU-R SM.1446 and ITU-R SM.2021:

Type 1: monocanal, where the useful signal is distorted by nonlinearities of the amplifier

Type 2: multichannel, where the various useful signals are distorted by nonlinearities of the same amplifier

Type 3: between transmitters, where several share the same nonlinear component

Type 4: by an active antenna that is shared by several channels

Type 5: by a nonlinear passive circuit, such as a circulator

The transfer function of a nonlinear amplifier in the intermodulation of two discrete frequencies can be represented by a Taylor series:

$$i_0 + k_1 e_{in} + k_2 e_{in}^2 + k_3 e_{in}^3 + k_4 e_{in}^4 + k_5 e_{in}^5 + \cdots$$

In a wave modulated in frequency with a continuous spectrum, the distortion in the proper band of the signal is due to third-order nonlinearities with an input–output response of the form

$$z(t) = y(t) + ay^3$$

When it is about a single carrier borrowing a nonlinear circuit, the discrete frequencies of the spectrum of modulation interfere between them by generating products of intermodulation which are comparable to a noise. When two carriers or more are simultaneously amplified by the same circuit presenting a certain compression of amplitude and a certain AM–PM conversion, crosstalk occurs between these carriers that is proportional to the product of the gain slope and the conversion factor; generally, the gain slope is approximately 0.05 dB MHz^{-1} and the AM–PM conversion is some degrees per decibel. The response in amplitude and the levels of intermodulation products which result from this, just as the noise figure, depend on the type of amplifier.

The frequency of the multiple lines constituting the intermodulation products that are generated by the combination of different carriers ($f_1, f_2, \ldots, f_n$) is given by the relation

$$f = m_1 f_1 + m_2 f_2 + \cdots + m_n f_n$$

where $m_1, m_2, \ldots, m_n$ are positive or negative integers. The integers $|m_1| + |m_2| + \cdots + |m_n|$ refer to the order of the intermodulation product.

In the usual case where the bandwidth is small compared to the radioelectric frequency, only the products of an odd nature enter the bandwidth of the amplifier; the most significant products which must be taken into account are, consequently, third order. When n carriers are amplified simultaneously, all the possible products of this type appear at the frequencies

$$f_{i,j} = 2f_i - f_j \qquad i, j = 1, 2, 3, \ldots$$

Third-order intermodulation products of the second type also appear at frequencies

$$f_{i,j,k} = \pm f_i \pm f_j \pm f_k$$

where only one minus is possible.

The level of the intermodulation products decreases according to the power at the input and the product order; this is why they are limited to third and

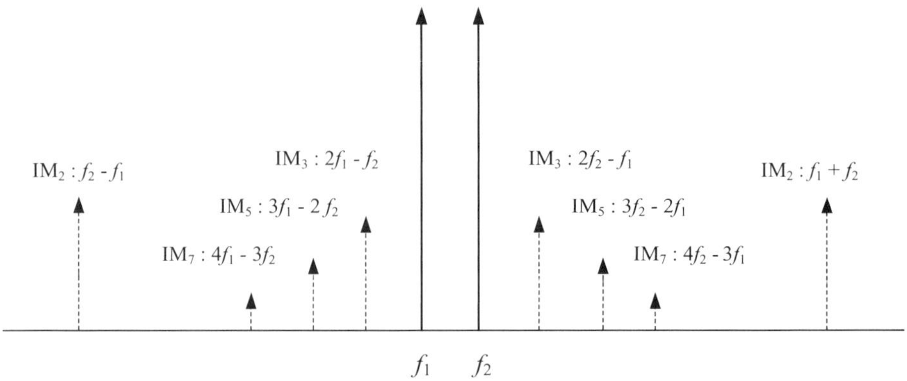

Figure 2.30 Relative position of intermodulation (IM) lines produced by two signals of frequency f_1, f_2.

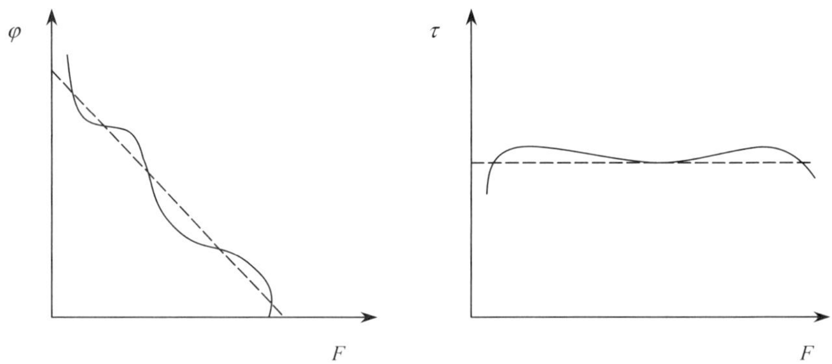

Figure 2.31 Variation of phase and group delay.

sometimes fifth order. Figure 2.30 presents intermodulation products of orders 2–5 that are generated by two signals of frequency f_1 and f_2 of equal intensity; the lines produced show their relative positions in the concerned frequency band.

The distortion of the propagation delay refers to the phase variation of the signal versus frequency; we deduce the group delay as shown in Figure 2.31. The group delay corresponds to

$$\tau = \frac{d\varphi}{d\omega} \qquad \omega = 2\pi F \tag{2.46}$$

Then

$$\tau = \frac{d\varphi}{2\pi dF}$$

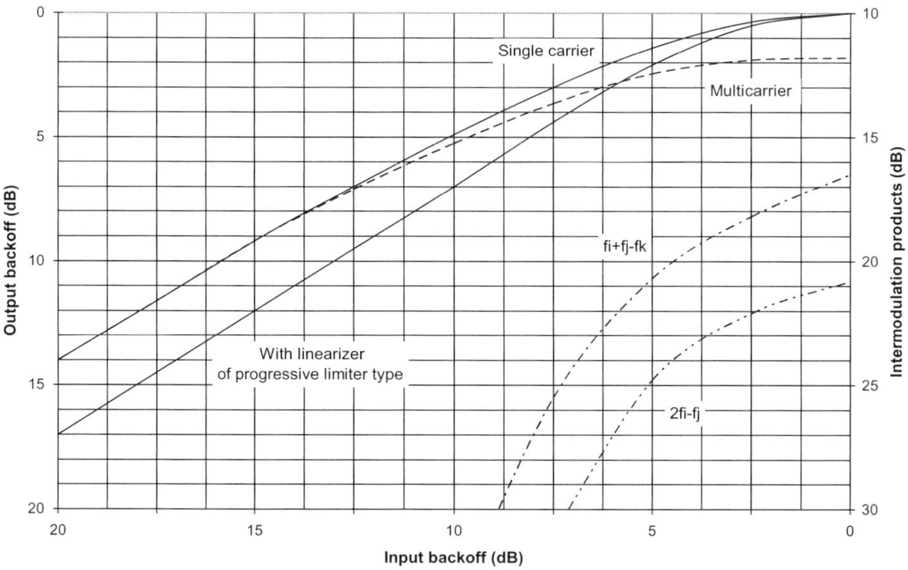

Figure 2.32 Input–output characteristics of TWT amplifier with and without linearizer.

The group delay τ, which is the ratio of the variation of the phase, expressed in radians, to the angular velocity, expressed in radians per second, has the dimension of time. The residual amplitude modulation is produced by the power supply while the nonessential radiations are mainly made up by the harmonics. The characteristics of the amplifier as well as the tolerance limits for all these phenomena are generally provided by the supplier or the operator; the power amplifiers, in particular those using TWTs, can be corrected by circuits of linearization.

Traveling-Wave Tube Amplifier The TWT constitutes an amplifier with a broad bandwidth that covers all the usable bandwidth of a satellite system, for example, with necessary qualities of uniformity of gain and group delay that make it possible to transmit simultaneously several carriers at various frequencies. However, the response in amplitude is not linear, which results in the creation of components of intermodulation comparable to noise when the number of carriers is important; these ones are all the more higher since the operation point of the tube approaches saturation.

Figure 2.32 presents the characteristics input-output of:

- An amplifier with TWT alone used in single- or multiple-carrier configurations
- An amplifier with TWT equipped with linearizer of progressive limiter type as well as the 3^{d} order intermodulation products of the two types of

the amplifier alone. The input backoff IB and the output backoff OB refer to the input and output levels compared to the operation point at saturation of the amplifier.

Figure 2.33 presents the ratios of the carrier to the first type of intermodulation product according to the output backoff which are obtained with two carriers in the case of a TWT alone and in a TWT with a linearizer of progressive limiter type. It is also advisable to take into account the noise figure of the TWT, which is usually in the range of 25–36 dB.

For greater convenience, we consider that all the carriers have the same power spectral density PSD defined by the ratio between the EIRP and the bandwidth B of the amplifier:

$$\mathrm{PSD} = \frac{\mathrm{EIRP}}{B} \tag{2.47}$$

In this case, the noise and the intermodulation products can be expressed in terms of output backoff, as well as the useful signal, and either in terms of carrier-to-noise ratio or carrier-to-intermodulation ratio as before.

Figure 2.34 presents the standard input–output characteristic of a TWT amplifier not linearized in the Ku band as well as the output backoff of the total noise and the intermodulation products. A relation that estimates the output backoff of total noise $\mathrm{OB}(N_{\mathrm{total}})$ (thermal noise + intermodulation products) of a nonlinearized TWT amplifier is, in decibels,

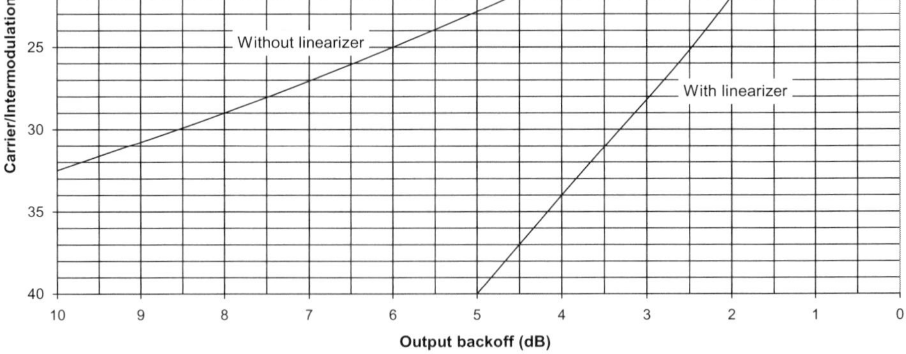

Figure 2.33 Carrier-to-intermodulation ratio of TWT amplifier with and without linearizer.

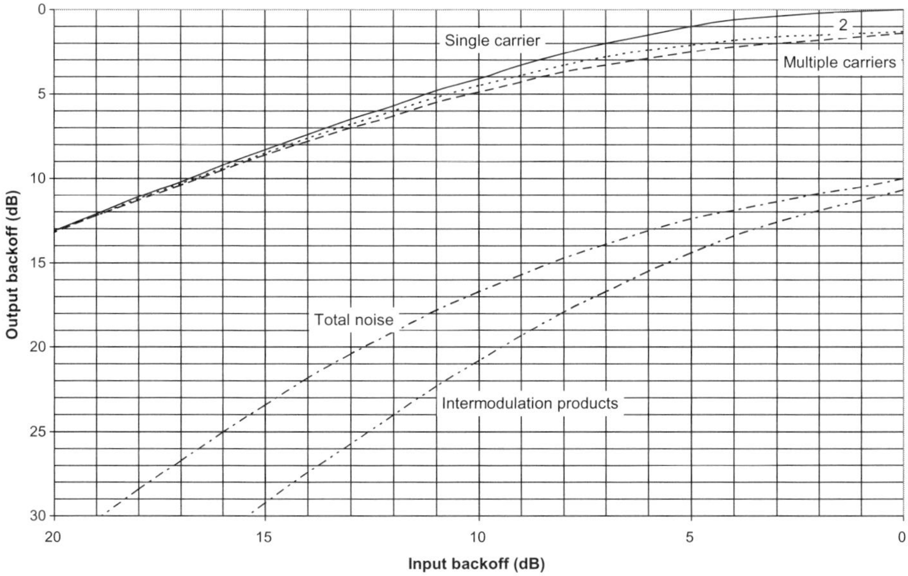

Figure 2.34 Input–output characteristics of nonlinearized TWT amplifier in Ku band.

$$\mathrm{OB}(N_{\mathrm{total}}) = 10 + 2\,\mathrm{OB}(C_{\mathrm{total}})^{0.93} \tag{2.48}$$

where $\mathrm{OB}(C_{\mathrm{total}})$ is the output backoff of the total power corresponding to all the carriers.

Solid-State Amplifier This type of amplifier is characterized by a greater linearity, such as by a smaller noise figure, than those of the TWT amplifier. Its principal characteristics are as follows:

- The compression point at 1 dB corresponds to the point of operation at which the gain varies by 1 dB relative to the linear characteristic of slope, 10 dB/10 dB.
- The third-order interception point makes it possible to plot the straight line of slope 30 dB/10 dB, giving the level of third-order intermodulation products.

Figure 2.35 gives the significance of these two characteristic points; it is seen that to a backoff of value Δ of the signal compared to the third-order interception point corresponds a backoff of 2Δ of the third-order intermodulation products.

Consequently, the compression point at 1 dB determines the point of operation nearest to the power at saturation P_{SAT} of the amplifier, and the higher is

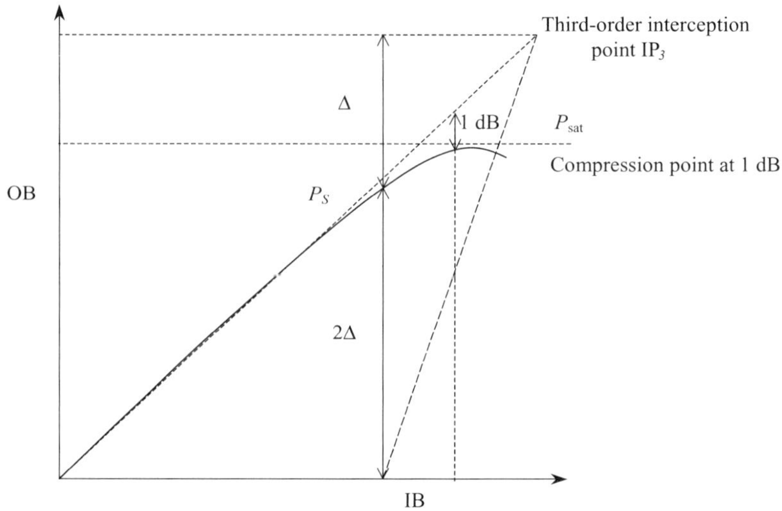

Figure 2.35 Typical operating points of solid-state amplifier.

the third-order interception point, the weaker are the intermodulation products. The compression point is in general sufficiently close to the power at saturation so that input and output backoff can be reduced independent of one or the other of these two characteristics. One can then write the relation between the third-order interception point IP_3, the level of the signal P_S, and that of the intermodulation products IM_3:

$$\mathrm{PI}_3 = P_S + \Delta = P_S + \frac{1}{2}\mathrm{IM}_3 \tag{2.49}$$

The level of the intermodulation products IM_3 can easily be measured by the method known as "two-tones," which consists in using two lines of the same level P_S shifted a few megahertz and to measure the level IM_3 of the intermodulation lines which appear on both sides, as presented in Figure 2.30.

It is then easy to define, as a first approximation, the output backoff OB which is necessary for a single carrier according to the minimal *C/N* for a given BER. Indeed, by supposing that a given *C/N* is required, we can write

$$\frac{C}{N} = \mathrm{IM}_3 \qquad \Delta = \frac{1}{2}\mathrm{IM}_3 \qquad \mathrm{OB} = \Delta - (\mathrm{IP}_3 - P_{\mathrm{sat}})$$

The table below provides, for instance, the characteristics of a 30-W amplifier in the frequency band 4.4–5 GHz.

Characteristics	Specifications	Measures
Minimum gain	16 dB	18.8 dB
Linearity	1 dB peak to peak (p-p) maximum	0.4 dB p-p
Compression point at 1 dB	+45.5 dBm typical	+45.6 dBm
Power at saturation	+46 dBm typical	+46 dBm
Third-order IM (two tones) (10 MHz)	–40 dBc minimum P_s = +34 dBm	–41 dBc

We then deduce the value of IP_3:

$$IP_3 = +34\text{ dBm} + \frac{1}{2}(41\text{ dBc}) = 54.5\text{ dBm}$$

For example, in a modulation requiring a *C/N* of 20 dB to obtain a BER < 10^{-12}, the minimal output backoff would have the value

$$OB = \frac{1}{2}(20\text{ dB}) - (54.5\text{ dBm} - 46\text{ dBm}) = 1.5\text{ dB}$$

If several carriers share the same amplifier, the calculation becomes more complex since it is necessary to add to the intermodulation products proper to each carrier the components resulting from the beating between carriers by taking into account their relative levels.

In any application, it is recommended to measure the threshold degradation of reception for a given BER by means of the assembly represented by Figure 2.36, where one varies the input backoff of the carrier by modifying the reception level so that the BER is maintained constant. Figure 2.37 presents the standard input–output characteristic of a solid-state amplifier in band C as well as the output backoff of the intermodulation products and the total noise.

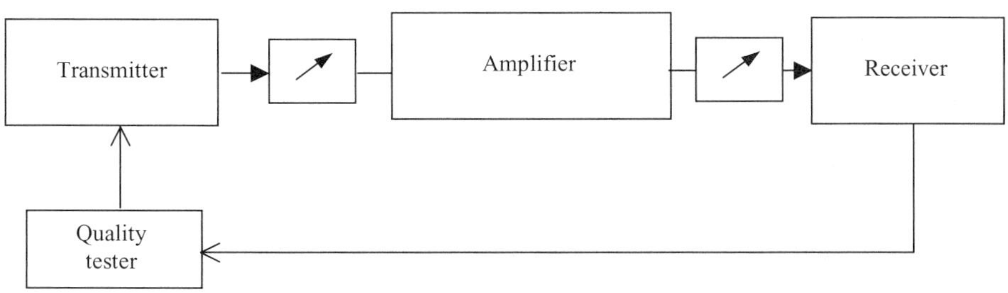

Figure 2.36 Measurement of degradation of reception threshold by intermodulation products.

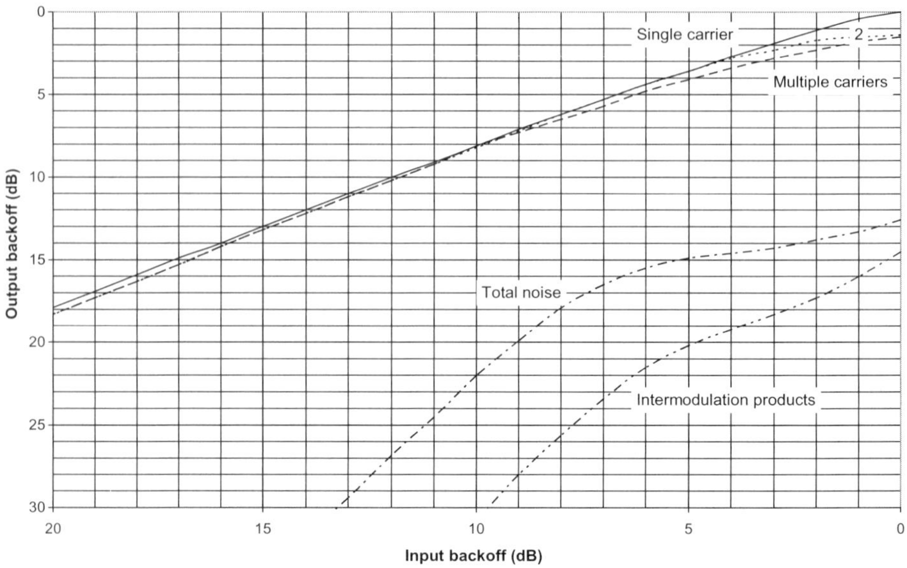

Figure 2.37 Input–output characteristic of solid-state amplifier in band C.

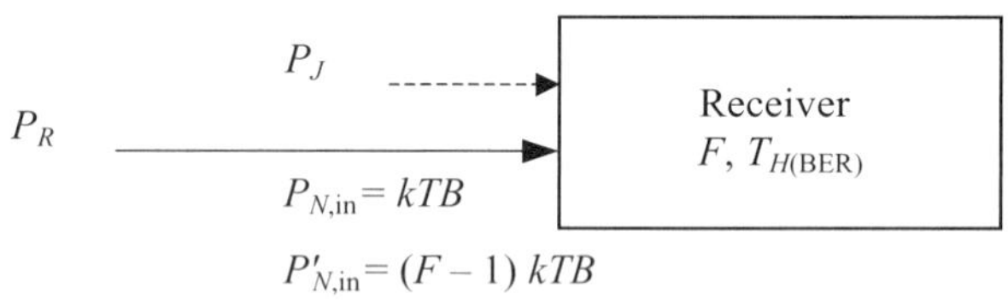

Figure 2.38 Receiver in the presence of a jammer.

2.3.3 Interference

2.3.3.1 Effect Due to Jamming The presence of a jammer at the reception can produce various effects, owing to its nature, which may be continuous or impulsive, such as degrading the threshold of the receiver for a given BER and consequently the uniform margin of the radio link.

We will consider here a permanent disturbing signal of comparable nature to the useful signal, which is generally the case in a network where the same frequency and polarization plan must be shared between the various radio links. As we saw previously, the threshold of the receiver, $T_{H(\mathrm{BER})}$, is an essential characteristic of the receiver which represents the minimal acceptable level of the useful signal for a given BER in the presence of its own white noise.

Figure 2.38 presents a receiver operating with the useful signal of level P_R which is affected by its own noise brought back to the input $P/_{N,\mathrm{in}}$ and by a disturbing signal P_J. We saw in Section 2.2.5 that the noise $P/_{N,\mathrm{in}}$ produced by a two-port circuit of bandwidth B and noise figure F at temperature T and brought back to the input is given by the relation (2.7):

$$P'_{N,\mathrm{in}} = (F-1)kTB$$

This noise power $P'_{N,\mathrm{in}}$ is added to that, $P_{N,\mathrm{in}}$, which exists at the input of the receiver because of ambient temperature and whose level in the same frequency bandwidth is

$$P_{N,\mathrm{in}} = kTB$$

which yields the total noise power at the input of the receiver,

$$N_R = P_{N,\mathrm{in}} + P'_{N,\mathrm{in}} = FkTB \tag{2.50}$$

called the reference background noise.

At the ambient temperature T_0 = 290 K of the reception equipment, the thermal noise calculated starting from Boltzmann's constant, which is given in Section 1.13.3 for the blackbody, has as an expression

$$kT_0 = (1.380664 \times 10^{-23}\ \mathrm{J\,K^{-1}}) \times 290\ \mathrm{K}$$

thus, in watts by hertz and decibels,

$$10\log(kT_0) = -204\ \mathrm{dB\ W\ Hz^{-1}} \tag{2.51}$$

The reference background noise can then be calculated, in dBW, using the general formula

$$N_R = F + 10\log(kT_0) + 10\log\left(\frac{R}{\log_2(M)}\right) + 10\log N_y \tag{2.52}$$

where F = noise figure (dB)
R = bit rate (bits s^{-1})
M = number of modulation levels
N_y = filtering Nyquist coefficient

We will consider the level of the useful signal P_R and that of the jammer P_J while posing

$$\frac{C}{N} = \frac{P_R}{N_R} \quad \text{(carrier-to-noise ratio)} \tag{2.53}$$

$$\frac{C}{I} = \frac{P_R}{P_J} \quad \text{(carrier-to-interference ratio)} \tag{2.54}$$

The disturbing signal of power P_J, which is supposed to occupy all the frequency bandwidth of the receiver and to behave like a white noise of Gaussian distribution, is added to the reference background noise N_R, which takes then the value N'_R, expressed in decibels:

$$N'_R = 10\log(10^{0.1N_R} + 10^{0.1P_J}) \tag{2.55}$$

Then we consider the threshold of the receiver, $T_{H(\mathrm{BER})}$, which is characterized by a SNR $(C/N)_{(\mathrm{BER})}$ for a given BER, expressed in decibels:

$$T_{H(\mathrm{BER})} = N_R + \left(\frac{C}{N}\right)_{(\mathrm{BER})} \tag{2.56}$$

N'_R being of comparable nature as N_R, the characteristic $(C/N)_{(\mathrm{BER})}$ for the same BER remains unchanged and the threshold of the receiver in the presence of the jammer, expressed in decibels, becomes

$$T'_{H(\mathrm{BER})} = N'_R + \left(\frac{C}{N}\right)_{(\mathrm{BER})} \tag{2.57}$$

We can also write, in real values,

$$\begin{aligned}\frac{T'_{H(\mathrm{BER})}}{T_{H(\mathrm{BER})}} &= \frac{N'_R}{N_R} = \frac{P_J + N_R}{N_R} = \frac{P_J}{N_R} + 1 = \frac{P_J}{T_{H(\mathrm{BER})}}\frac{T_{H(\mathrm{BER})}}{N_R} + 1 \\ &= \frac{P_J}{T_{H(\mathrm{BER})}}\left(\frac{C}{N}\right)_{(\mathrm{BER})} + 1\end{aligned}$$

from which the general expression of threshold degradation expressed in decibels is

$$T'_{H(\mathrm{BER})} - T_{H(\mathrm{BER})} = 10\log[1 + 10^{0.1[P_J - T_{H(\mathrm{BER})} + (C/N)](\mathrm{BER})}] \tag{2.58}$$

which depends only on the level of the jammer and the characteristics provided by the manufacturer for the nominal threshold $T_{H(\mathrm{BER})}$ and the $(C/N)_{(\mathrm{BER})}$ for a given BER.

Figure 2.39 presents the level of thresholds for BER 10^{-3} and 10^{-6} according to the level of the jammer by considering a $(C/N)_{(\mathrm{BER})}$ of 25 dB for BER 10^{-3}, which is a current value in the case of a modulation TCM128 at the bit rate of 155 Mbits s^{-1}.

Figure 2.40 shows the degradation of the threshold level according to the ratio of the nominal threshold level to the jammer level for various typical values of $(C/N)_{(\mathrm{BER})}$ corresponding to various modulations. The equipment to

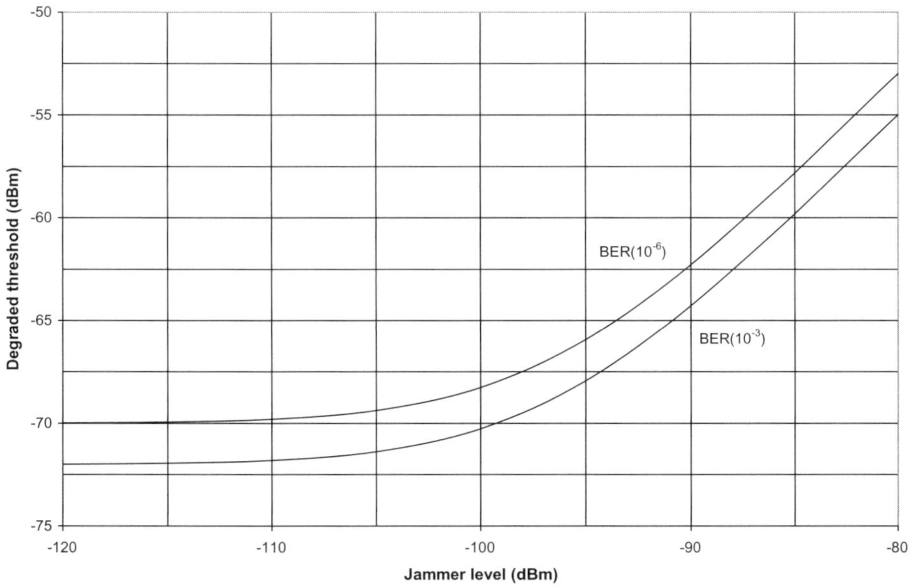

Figure 2.39 Degraded thresholds according to jammer level for BER 10^{-3} and 10^{-6} for typical $(C/N)_{(\mathrm{BER})}$ of 25 dB at BER 10^{-3}.

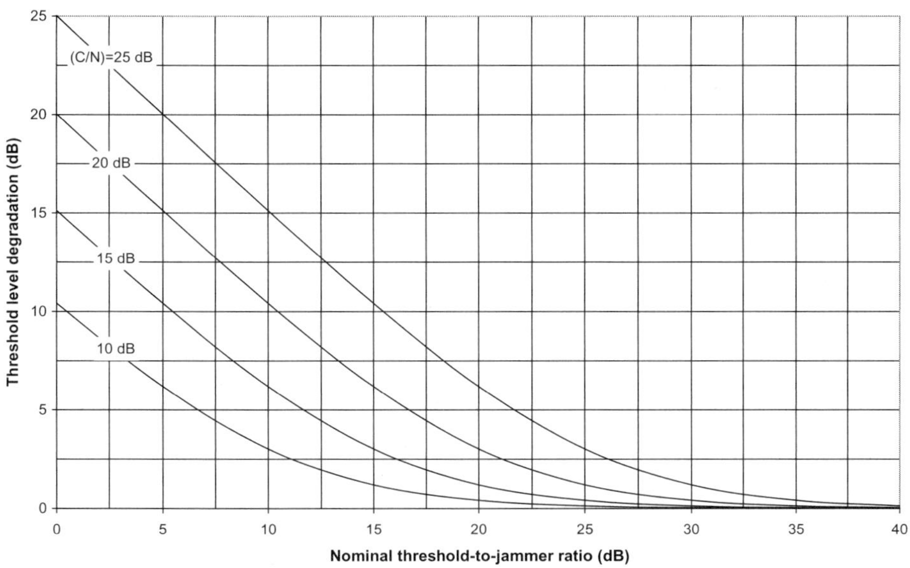

Figure 2.40 Degradation of threshold level according to nominal threshold-to-jammer ratio for various values of $(C/N)_{(\mathrm{BER})}$.

which Figure 2.39 refers belongs to the SDH-TCM128 type operating at the bit rate of 155 Mbits s^{-1} and at the frequency of 8 GHz with the following receiver characteristics:

Noise figure: 3 dB
Threshold for BER 10^{-3}: –72 dBm
Threshold for BER 10^{-6}: –70 dBm
Number of modulation levels: 128

Using the formulas given above, we find

$$N_R = -97\,\text{dBm}$$

$$\left(\frac{C}{N}\right)_{(\text{BER})} = 25\,\text{dB for BER } 10^{-3}$$

$$\left(\frac{C}{N}\right)_{(\text{BER})} = 27\,\text{dB for BER } 10^{-6}$$

2.3.3.2 Calculation of Level of Jamming Jammers are characterized by their frequency, which is in the vicinity of that of the receiver, their power level, and their bandwidth. We will consider jammers whose spectral characteristics are close to those of the useful signals, as is the case when studying jamming inside a network in the process of definition; the effects of jammers of a different nature will have to be the subject of another study.

The curve presented in Figure 2.41 makes it possible to calculate the protection brought about by the selectivity of the receiver according to the frequency difference between the useful carrier and the jammer in SDH modulation. It is thus very useful to ask the manufacturer to provide this type of curve, particularly where it is necessary to consider crossed interferences between modulations of various types that occupy different bandwidths.

Figure 2.42 presents a measuring bench of the interference reduction factor (IRF), also called net filter discrimination (NFD), comprising a transmitter that is used for jamming by means of a coupler, a radio link of which one measures the BER in the vicinity of the threshold. The method consists in measuring the level of the jammer P_J that degrades the threshold $T_{H(\text{BER})}$ of a value X decibels for various frequency deviations in the vicinity of the carrier. For example, the values indicated in Figure 2.42 and the table below refer to an equipment SDH at 155 Mbits s^{-1} for a degradation $X < 1$ dB of the threshold for BER 10^{-3} corresponding to frequency deviations ΔF of 0, 28, and 56 MHz.

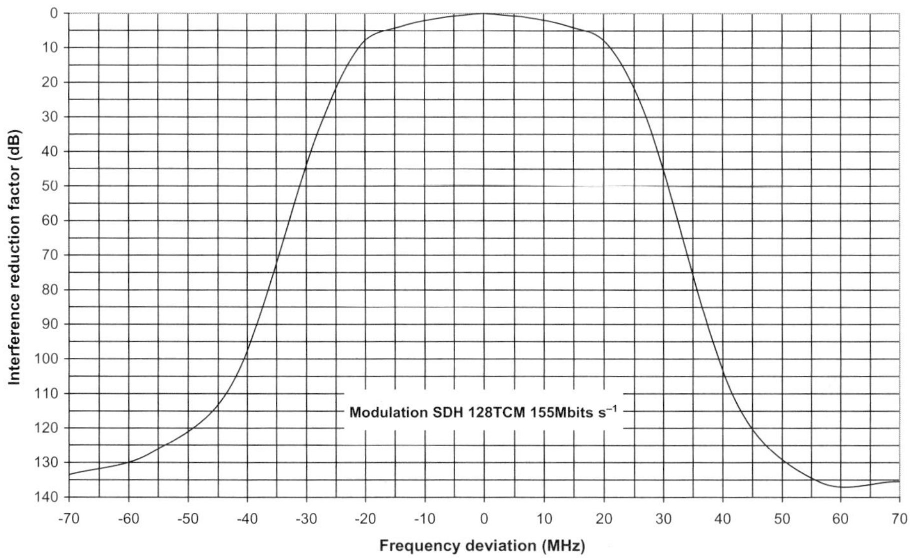

Figure 2.41 Interference reduction factor IRF (NFD).

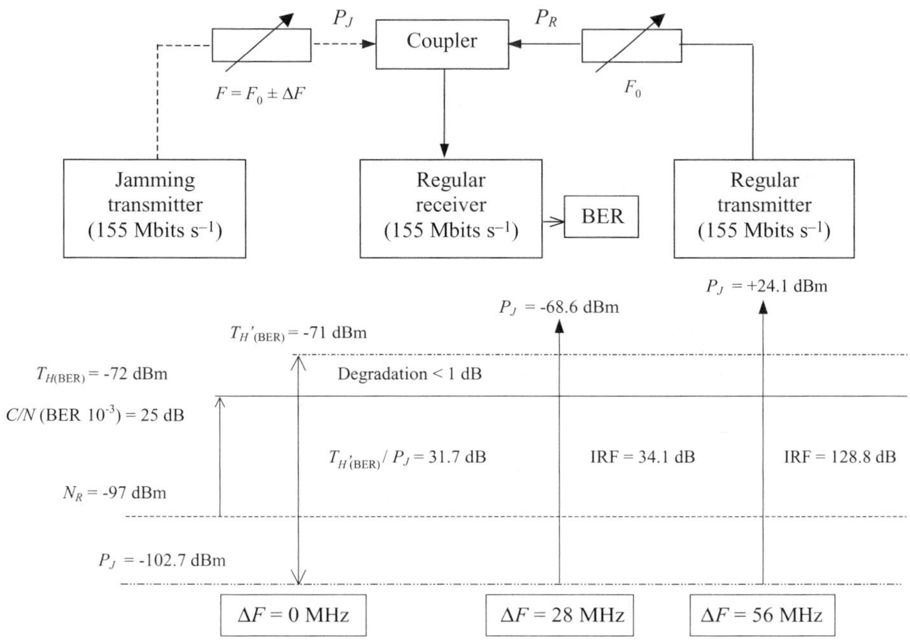

Figure 2.42 Measuring bench of characteristic IRF (NFD).

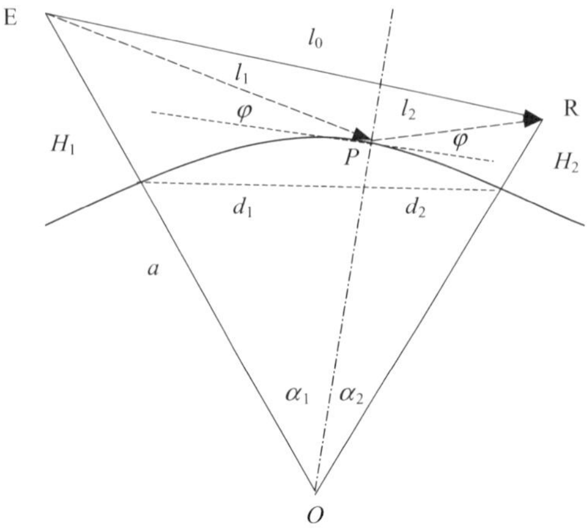

Figure 2.43 Reflection on ground.

ΔF (MHz)	P_J (dBm)	$T'_{H(\mathrm{BER})}/P_J$ ($X < 1\,\mathrm{dB}$) (dB)	IRF (dB)
0	–102.7	31.7	0
28	–68.6	–2.4	34.1
56	+24.1	–95.1	126.8

The IRF is calculated for a given ΔF as

$$\mathrm{IRF} = \left[\frac{P'_{S(\mathrm{BER})}}{P_P}\right]_{0\,\mathrm{MHz}} - \left[\frac{P'_{S(\mathrm{BER})}}{P_P}\right]_{\Delta F} \tag{2.59}$$

2.3.4 Effects of Reflection

2.3.4.1 Reflection on Ground Figure 2.43 presents a microwave link affected by reflection on the ground. The geometric elements to take into account are the heights of the antennas, H_1 and H_2, which are in general very small compared to the distance, which is small even with respect to the effective Earth radius a; the direct and reflected paths l_0, l_1, and l_2; and the angle of reflection φ. Consider the pathlength difference Δl between the direct ray and the reflected ray:

$$\Delta l = l_1 + l_2 - l_0 \tag{2.60}$$

For any triangle applied to EPR we have

$$\cos\left(\frac{\widehat{\mathrm{EPR}}}{2}\right) = \sqrt{\frac{p(p - l_0)}{l_1 l_2}} \qquad p = \frac{l_0 + l_1 + l_2}{2}$$

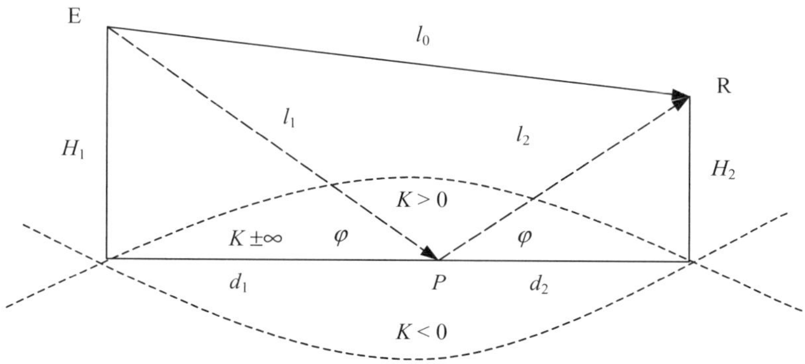

Figure 2.44 Reflection on effective Earth.

which yields

$$\cos\left(\frac{\pi - 2\varphi}{2}\right) = \sin\varphi = \sqrt{\frac{p(p - l_0)}{l_1 l_2}}$$

We can write

$$\Delta l = \frac{4 l_1 l_2 \sin^2\varphi}{l_0 + l_1 + l_2} \tag{2.61}$$

As, in the general case, the angle φ is small, this expression becomes

$$\Delta l \approx \frac{2\varphi^2 d_1 d_2}{d} \tag{2.62}$$

A more convenient representation is given in Figure 2.44, which takes into account all possible variations of the effective Earth radius factor K by introducing only second-order error on the path difference and the location of the reflection point P. In such a representation, the geometric elements of the reflection become, for $K = \pm\infty$ and angles expressed in radians,

$$\frac{d_1}{d} = \frac{H_1}{H_1 + H_2} \qquad \frac{d_2}{d} = \frac{H_2}{H_1 + H_2}$$

where

$$\varphi \approx \frac{H_1}{d_1} = \frac{H_2}{d_2} = \frac{H_1 + H_2}{d_1 + d_2}$$

which yields

$$\Delta l \approx \frac{2H_1H_2}{d} \tag{2.63}$$

The reflection angle φ according to the effective Earth radius is given by the following approximate formula, starting from relation (1.84):

$$\varphi \approx \frac{H_1 - y}{d_1} = \frac{H_2 - y}{d_2} \qquad y = \frac{d_1 d_2}{2a}$$

Incidence on Received Field The level of the received field results from vector addition of the direct wave E_0 and the reflected wave E_R, of which the ratio between the modules, $|E_R|/|E_0|$, depends on the reflection coefficient ρ and the phase difference θ is a function of the ratio of their path difference Δl to the wavelength λ. The direct and reflected waves are composed as represented in the Figure 2.45. The phase difference θ, expressed in radians and to which it is advisable to add the phase due to the reflection presented in Figure 1.25, is given by the relation

$$\theta = \frac{2\pi\Delta l}{\lambda} \quad (+\sim\pi \text{ for small incidence angles}) \tag{2.64}$$

The ratio between the resulting wave and the direct wave is written as

$$\frac{E}{E_0} = 1 + \rho e^{j\theta}$$

with $E_R = \rho E_0$, where ρ is the reflection coefficient. The modulus of the resulting wave is deduced from the following relation for any triangle:

$$E^2 = E_0^2 + E_R^2 - 2E_0E_R\cos\theta$$

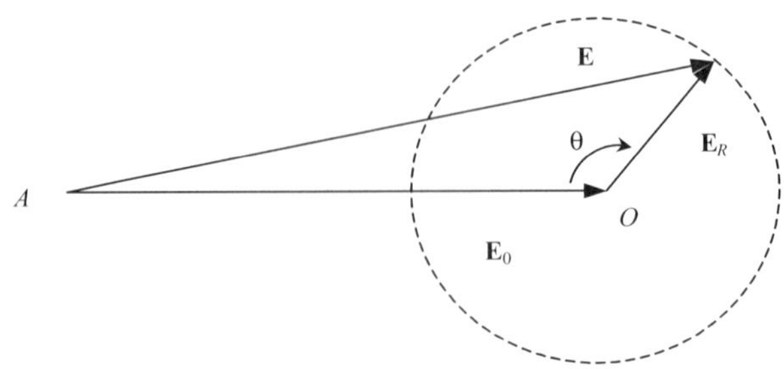

Figure 2.45 Vectorial composition of direct and reflected waves.

where

$$\frac{E}{E_0} = \sqrt{1 + \rho^2 + 2\rho\cos\theta} \tag{2.65}$$

Differential Time Delay The differential time delay, expressed in nanoseconds, between the direct wave and the reflected wave is given by the relation

$$\tau = \frac{\Delta l}{c} \times 10^9 \approx \frac{\Delta l}{0.3} \tag{2.66}$$

where c is in meters per second and Δl in meters.

Interference Fringes The path difference between the direct wave and the reflected wave produces interference fringes in the vertical plane of reception, as represented in Figure 2.46, and the distance between two minima or two maxima can be given by the relation

$$\Delta l = \frac{2H_1(H_2 + \delta h)}{d} - \frac{2H_1 H_2}{d} = \lambda$$

from which

$$\delta h = \frac{\lambda d}{2H_1} \tag{2.67}$$

The pathlength difference can also be expressed in terms of wavelengths, as there is a minimum for each half wavelength, by the formula

$$N_W = \frac{\Delta l}{\lambda}$$

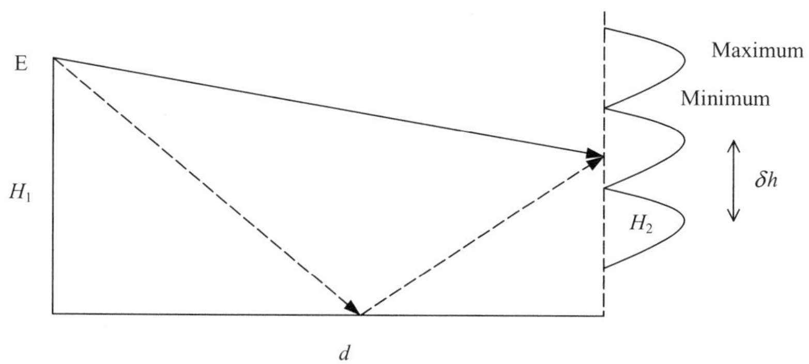

Figure 2.46 Interference fringes.

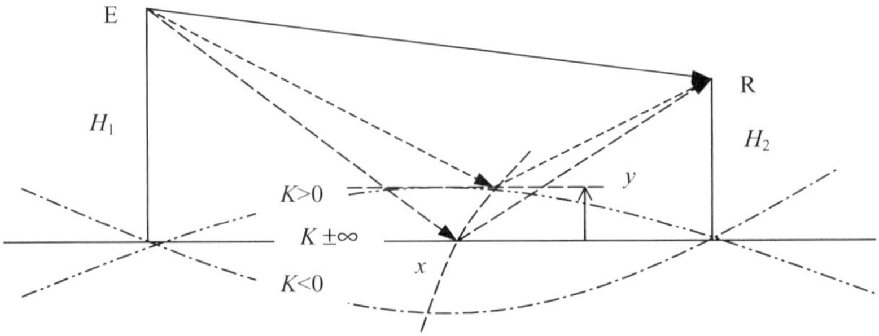

Figure 2.47 Reflection in case of variation of refractivity.

Influence of Variation of Refractivity of Atmosphere It is now advisable to define the consequences of the reflection on the ground for all possible values of the factor K while varying the effective Earth radius a as shown in Figure 2.47. The difference in altitude y relative to the path profile corresponding to infinite Earth radius according to the factor K is given by relation (1.84). For simplicity, we will be interested in horizontal reflections for all values of the factor K without taking account of the rotundity of the effective Earth, which results in second-order error about the locus of the reflection point. Indicating by x the distance from the reflection point to the transmitter and using the following relations for any triangle, we write

$$\frac{H_1 - y}{x} = \frac{H_2 - y}{d - x} = \frac{H_2 - H_1}{d - 2x} = \frac{H_1 + H_2 - 2y}{d}$$

which yields

$$(H_1 - y)(d - x) = (H_2 - y)x$$

Thus

$$x = \frac{d(H_1 - y)}{H_1 + H_2 - 2y} \tag{2.68}$$

and

$$y = \frac{H_1(d - x) - H_2 x}{d - 2x} \tag{2.69}$$

The locus of the reflection point is a hyperbole whose branches pass respectively by point E and point R, the upper one corresponding to the highest point and the lower one to the lowest point. In the case of reflection on the ground or in the low layers of the atmosphere, we will be interested in the

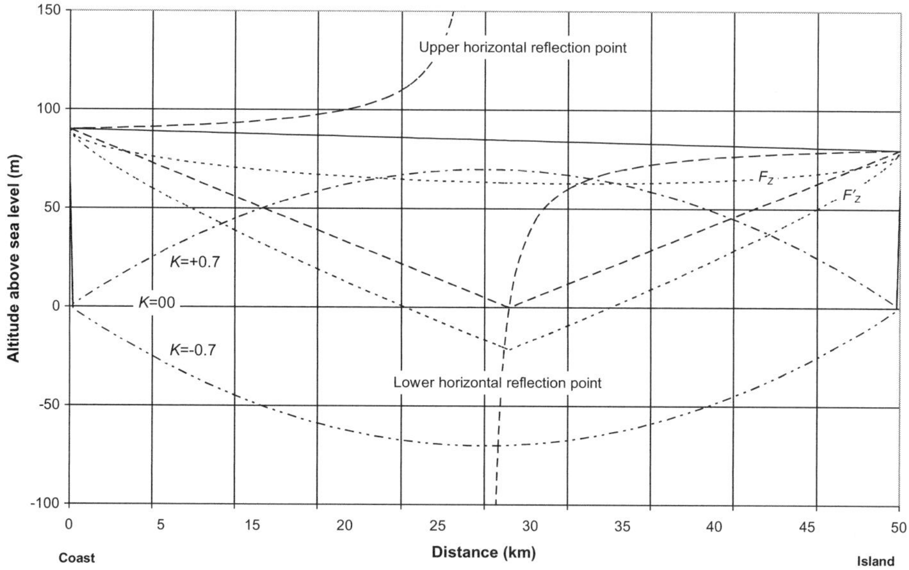

Figure 2.48 Radio link with reflection on sea.

branch which passes by the point whose altitude is the lowest, as shown in the Figure 2.48, that presents a 50-km radio link realized above the sea at the frequency of 8 GHz or by the upper branch in the case of reflection on an elevated layer.

Figures 2.49 and 2.50 present the distributions of the resulting field and the differential time delay that were obtained while varying continuously the factor K between the arbitrary values of −0.7 to +0.7 (either $1/K \in -1.5, +1.5$). The number of fades, and consequently the fading speed in the whole range of variation of the vertical gradient of the refractivity, increases in the inverse ratio than the size of the interference fringes.

Figure 2.51 presents the differential attenuation in the bandwidth of the receiver (28 MHz), that is, the amplitude of the deformation of the envelope of the transmitted signal compared to its level at the central frequency. Figure 2.52 shows the distortion undergone by the envelope of the signal at the reception due to the selective effects of reflection versus differential attenuation.

Influence of Ground on Antenna Radiation Pattern When the size of the antenna is small with respect to the interference fringes, the resulting received field can be regarded as an equiphase plane wave; the transmitted signal undergoes only a fluctuation of the amplitude as well as a distortion that is a function of the differential delay between the direct wave and the reflected wave. However, when the vertical dimension of the antenna is of the same order of magnitude or higher than that of the interference fringes, the received field cannot be regarded any more as equiphase on the whole of its aperture;

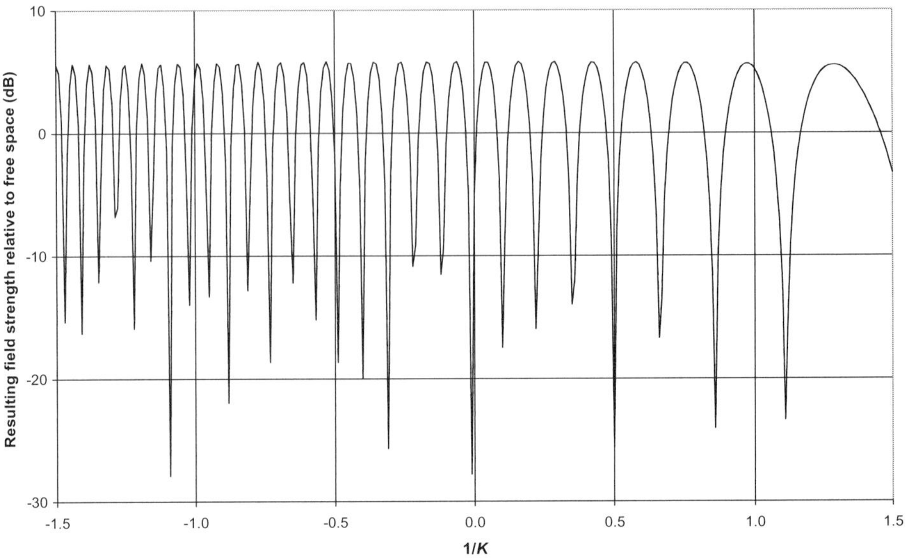

Figure 2.49 Field strength resulting from reflection on sea versus factor K.

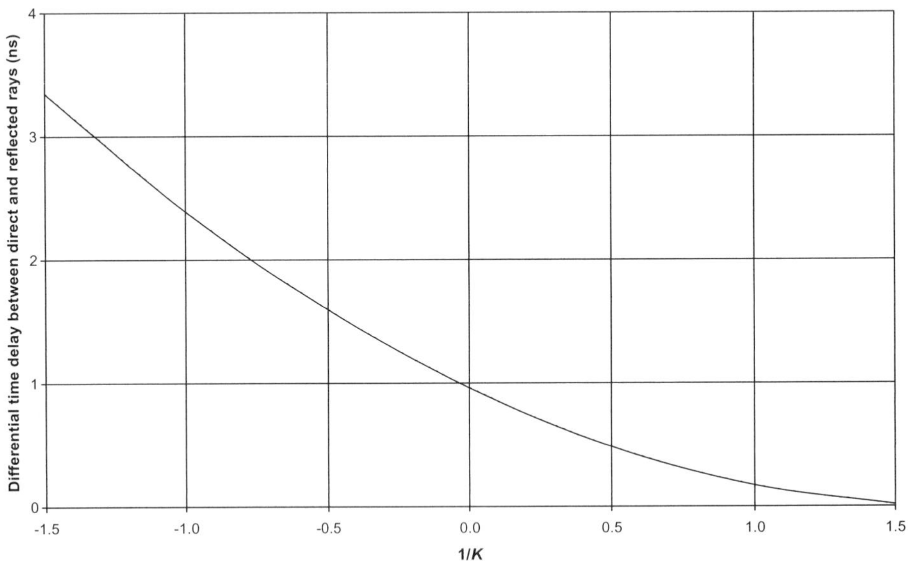

Figure 2.50 Differential time delay versus factor K.

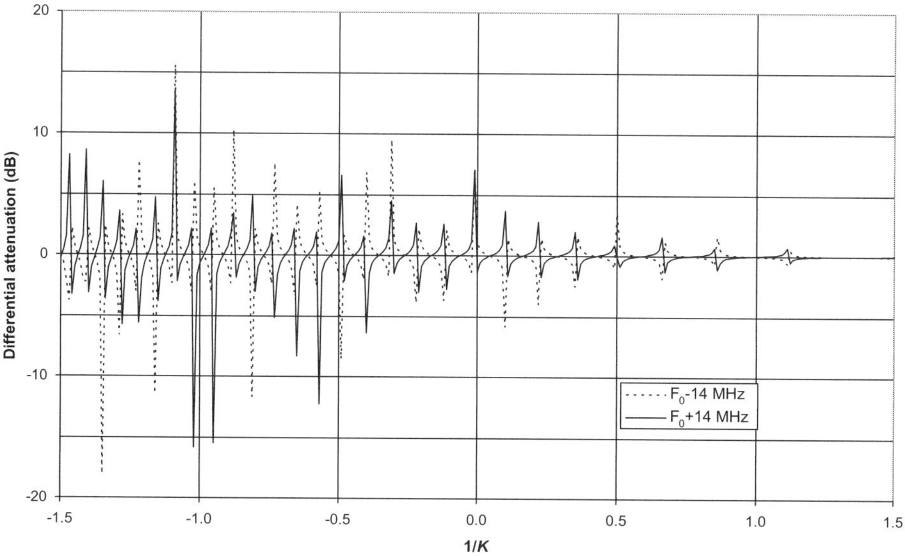

Figure 2.51 Differential attenuation versus factor K.

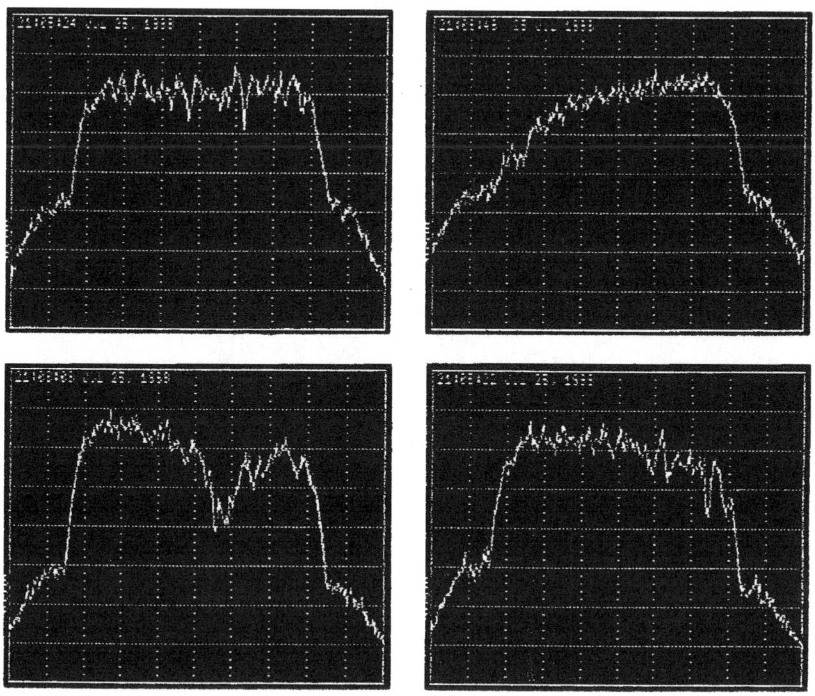

Figure 2.52 Distortion of received signal envelope due to reflection.

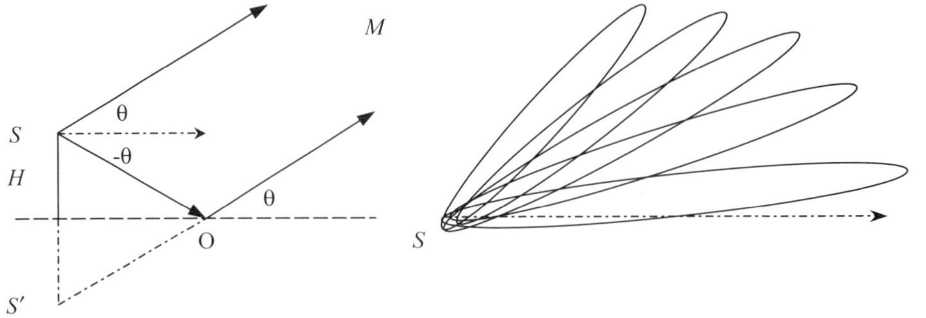

Figure 2.53 Foliation of antenna radiation pattern.

in such conditions, the radiation pattern can be modified to a significant degree.

Figure 2.53 presents the distorted radiation pattern that results from the presence of a reflecting plane in the vicinity of a source placed at a height H; note that there is always a minimum in the direction of the reflecting plane or the horizon. The magnitude of the electric field at the point M at long distance has as an expression

$$E_M = E_0 \sin\left[\frac{2\pi H}{\lambda} \sin\theta\right]$$

A foliation of the radiation pattern occurs which presents maxima and minima for the values θ_M and θ_m of the elevation angle, given by the relations

$$\sin\theta_M = \frac{(2k+1)\lambda}{4H} \qquad \sin\theta_m = \frac{k\lambda}{2H}$$

It is necessary thus to reduce as much as possible the level of the reflected wave while taking advantage of the possible masks along the path profile by reducing the height of the antennas or by using sufficiently directive antennas toward the zone of reflection.

2.3.4.2 Multiple Reflections In the case of guided propagation inside a layer of the atmosphere, as represented in Figure 2.54 by rectilinear trajectories, multiple reflections occur and the resulting field varies quickly according to the distance, revealing a succession of reinforcements and depressions called "radioelectric holes." Figure 2.55 illustrates the vectorial method that makes it possible to gradually calculate the value of the resulting field by employing the following relations:

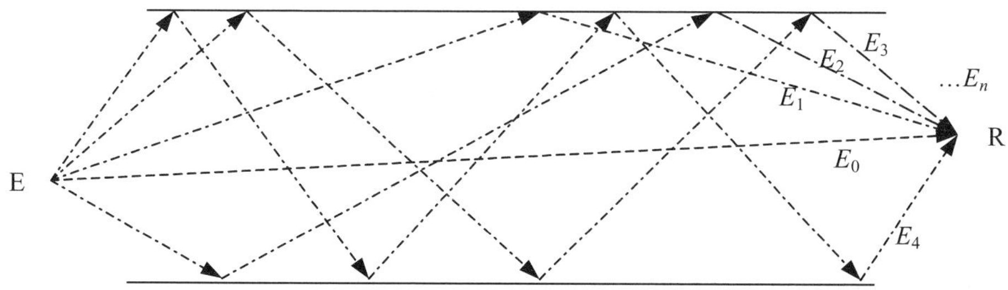

Figure 2.54 Guided propagation inside atmospheric layer.

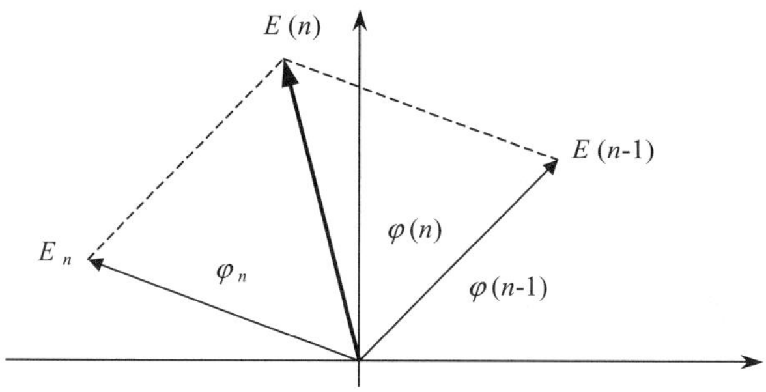

Figure 2.55 Vectorial composition of resulting field.

$$E(n) = \sqrt{(x(n-1) + x_n)^2 + (y(n-1) + y_n)^2} \tag{2.70}$$

$$\varphi(n) = \begin{cases} \arccos\left(\dfrac{x(n-1) + x_n}{E_n}\right) & \text{if } y(n-1) + y_n > 0 \\ \pi - \arccos\left(\dfrac{x(n-1) + x_n}{E_n}\right) & \text{if } y(n-1) + y_n < 0 \end{cases}$$

where

$$x_n = \frac{E_n}{E_0} \cos \varphi_n \qquad y_n = \frac{E_n}{E_0} \sin \varphi_n \qquad \frac{E_n}{E_0} = \rho^n$$

Figure 2.56 shows, for instance, the computed field strength at frequency 8 GHz resulting from multiple reflection of order $n = 5$ according to the distance starting from the transmitter by considering a 30-m-thick ducting layer. We can see that weak variations of radio-meteorological conditions as well as of the height of the layer may involve important and very fast fluctuations of the resulting field strength and differential delay; certain zones are prone to field reinforcements while others are in depressions which can sometimes last several hours.

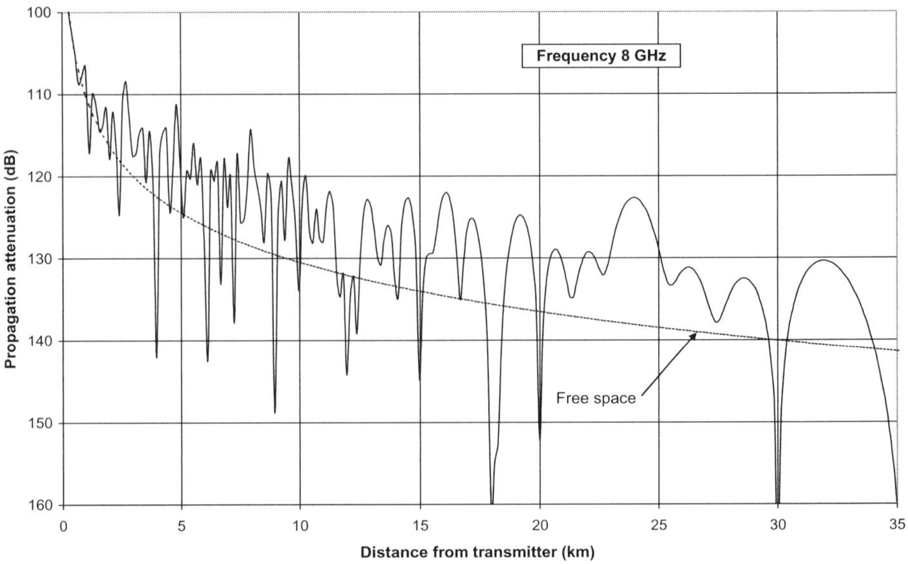

Figure 2.56 Resulting field strength versus distance in ducting layer.

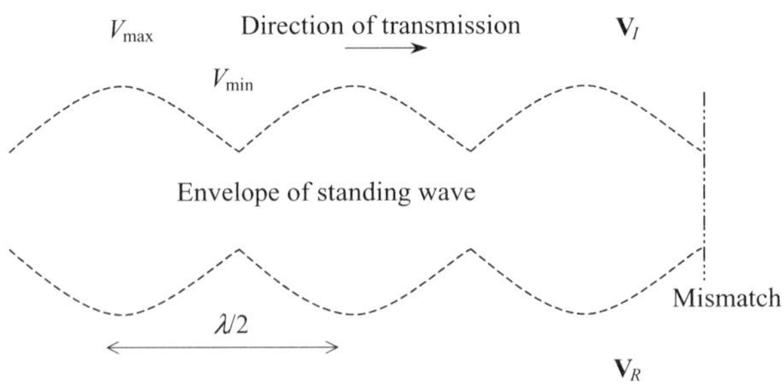

Figure 2.57 Standing wave.

2.3.4.3 Standing Wave

Properties of Standing Wave Mismatch of the end of transmission lines, such as coaxial cables and waveguides that are used to connect radio equipment to antennas, causes a reflection of the transmitted wave; the wave thus reflected interferes with the incident wave as illustrated in Figure 2.57 and the line becomes the seat of a standing wave. Figure 2.57 shows the variations of the level of the standing wave, whose envelope has the form of a cycloid, which one would observe by using a split line. The amplitude of the standing wave varies throughout the line by presenting bellies of tension V_{max} and nodes of tension V_{min}. The bellies and the nodes of tension are fixed and their spacing corresponds to a half wavelength; by definition, considering the tensions V_I and V_R of the incident and reflected waves, we have the following:

- the reflection coefficient:

$$\rho = \frac{V_R}{V_I} \tag{2.71}$$

- the voltage standing-wave ratio (VSWR):

$$S = \frac{V_{max}}{V_{min}} \tag{2.72}$$

Because of the vectorial composition of the direct reflected waves, we can write

$$V_{max} = V_I + V_R \qquad V_{min} = V_I - V_R$$

Thus

$$S = \frac{V_I + V_R}{V_I - V_R} = \frac{1+\rho}{1-\rho} \tag{2.73}$$

$$\rho = \frac{S-1}{S+1} \tag{2.74}$$

If we consider the incident power P_I and the reflected power P_R, we can also write

$$\frac{P_R}{P_I} = \left(\frac{V_R}{V_I}\right)^2 = \rho^2$$

Thus

$$\rho = \sqrt{\frac{P_R}{P_I}} \tag{2.75}$$

$$S = \frac{1+\sqrt{P_R/P_I}}{1-\sqrt{P_R/P_I}} \tag{2.76}$$

The VSWR can be measured in various ways using a reflectometer, a split line, or a directive coupler. Figure 2.58 presents the values of the reflection coefficient ρ and the VSWR S versus the ratio between reflected and incident powers.

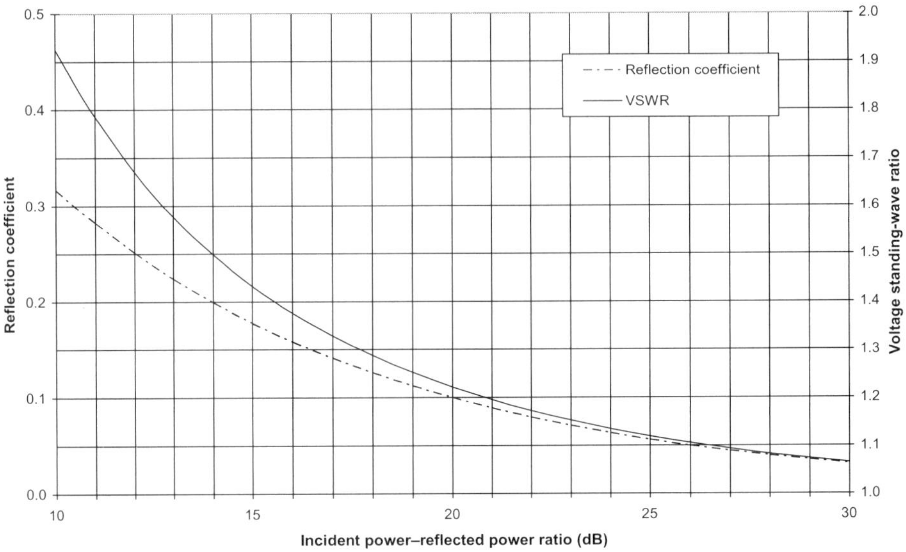

Figure 2.58 Reflection coefficient ρ and VSWR versus ratio of incident power to reflected power.

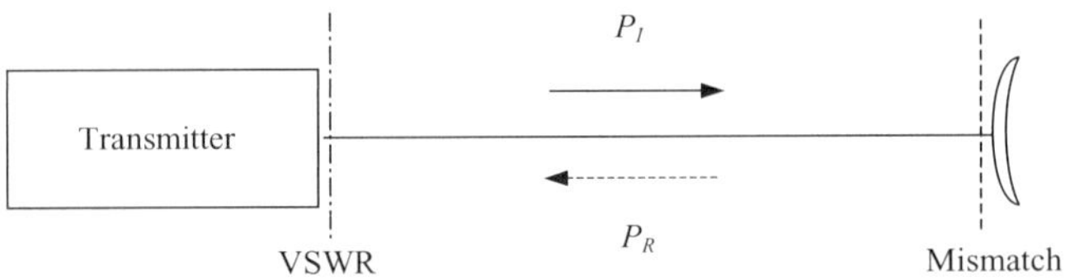

Figure 2.59 Standing wave in transmission feeder line.

Effect of Standing Wave on Transmission Assuming the specific transmission delay to be about 3.5 ns m^{-1} in a coaxial cable and 4 ns m^{-1} in a waveguide, the ratio between the reflected and incident powers must be reduced as the differential delay brought by the transmission line is increased, because of its length, in order to reduce as much as possible the level of the intermodulation products and to avoid degradation of reception conditions such as the total signature of the receiver, as we will see it in the next chapter. For example, Figure 2.59 shows a transmitter connected to an antenna through a feeder line where occurs a standing wave due to mismatch at its end. Reflected power P_R interferes with incident power P_I all along the feeder line and the complete calculation consists in solving a convolution integral using the specific attenuation of the feeder line as well as the differential delay between both and the symbol time duration of the modulation.

For greater convenience, we will consider that the length of the feeder line is great enough so that the differential delay is great compared with the symbol time duration. In such a case, the reflected signal is totally decorrelated from the incident one, and thus similar to a white noise, and the incident power–reflected power ratio at midpath of the feeder line will be empirically considered as a carrier-to-noise ratio. For example, Figure 2.60 shows the computed BER versus VSWR for various commonly used modulations in a 3-dB-loss transmission feeder line.

There is another way to compute the effect of VSWR on the modulation diagram that consists of considering the distortion at the level of the state points of the constellation due to the amplitude modulation in the transmission line. Consider the amplitude ripple α shown on Figure 2.61. The carrier-to-noise ratio is calculated from the amplitude ripple according to the formula

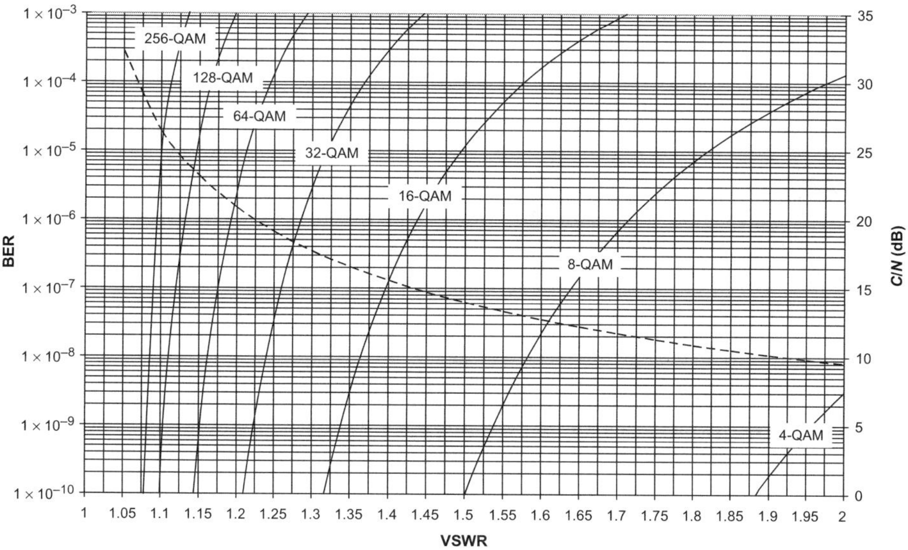

Figure 2.60 BER versus VSWR of 3-dB-loss feeder line.

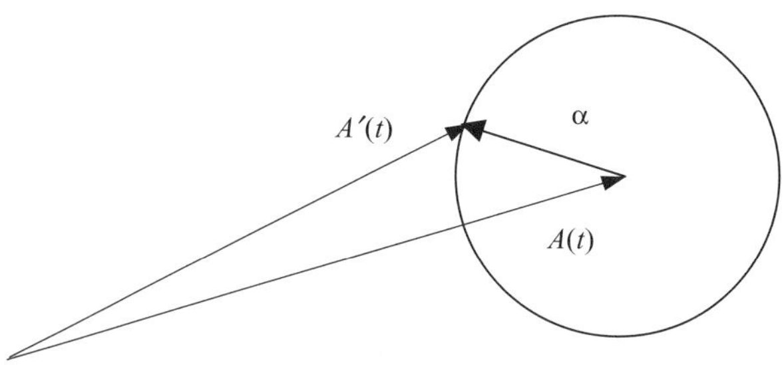

Figure 2.61 Distortion due to amplitude modulation in feeder line.

$$\frac{C}{N} = -20 \log \frac{10^{\alpha/20} - 1}{10^{\alpha/20} + 1}$$

The permissible amplitude ripple, expressed in decibels, is given by the following table:

4-QAM	16-QAM	64-QAM	128-QAM	256-QAM
$\alpha = 3\,\text{dB}$	$\alpha = 1.41\,\text{dB}$	$\alpha = 0.62\,\text{dB}$	$\alpha = 0.3\,\text{dB}$	$\alpha = 0.14\,\text{dB}$

Figure 2.62 shows the corresponding theoretical values of admissible VSWR for the relevant modulation types that are close to the preceding values.

2.3.5 Jitter, Wander, Phase Transients and Network Synchronization

Most of telecommunication networks include digital switching and transmission systems that require synchronization. Synchronization refers to an arrangement for operating digital switching and transmission systems at a common clock rate. ISDN and SONET transmission equipment require highly reliable and accurate synchronization networks. Jitter, wander, and phase transients are impairments of the digital signals caused by the clocks of the multiple systems that transport these signals. *Jitter* is defined as the short-term

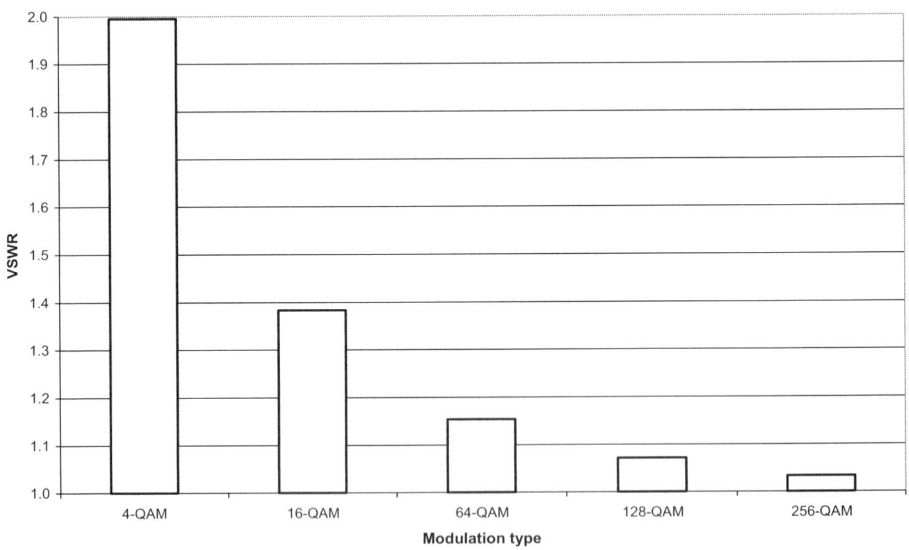

Figure 2.62 Admissible VSWR for commonly used modulation types.

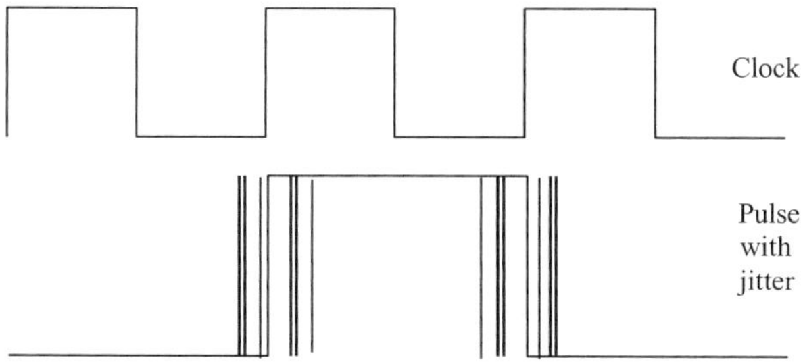

Figure 2.63 Example of jitter.

variations of the significant instants, that is, the pulse rise or fall points, of a digital signal from their ideal positions in time and *wander* as their long-term variations; the short term implies phase oscillations with spectral components greater than or equal to 10 Hz and the long-term spectral components less than 10 Hz. *Phase transients* are large excursions in phase of limited duration.

2.3.5.1 Jitter Jitter can affect digital signals in two ways; high-speed jitter may lead to errors due to inability of equipment to sample the incoming bit stream correctly and may lead to overflow or underflow of synchronizer buffers. Figure 2.63 shows an example of the jitter of a pulse by comparison with the clock. The main causes of jitter are regenerators, due to timing extraction circuits, and multiplexers, owing to the justification process and decision threshold crossing. In SDH multiplexers, jitter is generated by the pointer processing technique into the payload they carry; as the transmission link may have many regenerators and various clocks, jitter accumulates as the signal progresses along the link.

2.3.5.2 Wander The main source of wander is clocks; synchronized clocks use some control mechanism to maintain a lock with their reference that leads to wander. Wander can affect service in the following two ways: frame slip and clock holdover if it loses its references.

2.3.5.3 Phase Transients A phase transient refers to an unusual large and sudden change in phase of a signal with respect to the surrounding phase variations of that signal which is caused by rearrangement activities in clocks and by pointer adjustments for payload signals carried on SONET.

2.3.5.4 Network Synchronization Engineering Recommendations ITU-T G.822 to 6.825 have been established to control frame slip rate, jitter, and wander within digital PDH and SDH networks. Higher level QAM signals are extremely susceptible to degradation in the carrier-to-noise ratio caused by carrier and timing phase jitter and require very tight jitter control. Phase jitter on a carrier, local oscillator, or recovered carrier causes circular spreading of the constellation states and eye pattern closure which is similar to high noise level. Slip buffers are used in PDH systems to separate writing and reading data mechanisms; any variations in phase or frequency less than the length of the buffer are then absorbed with no effect on the carried traffic. In SDH systems, the hierarchical method of synchronization establishes a master–slave relationship where the slave receives timing from the upstream message. Synchronization system units (SSUs), equipped with high-stability clocks and placed at various places of the network under the supervision of a master clock, absorb the jitter and wander.

CHAPTER 3

Microwave Line-of-Sight Systems

3.1 ENGINEERING OF LINE-OF-SIGHT SYSTEMS

3.1.1 Introduction

Microwave radio systems are regarded as line of sight when the waves transmitted between the two stations do not meet any obstacle most of the time. In the engineering phase of this type of microwave radio link, it is advisable to take account of the following factors and dominant effects of propagation which have been treated in preceding chapters:

- Attenuation by diffraction due to obstacles on path which block the way under conditions of subrefraction (K_{min})
- Absorption by gases of atmosphere
- Absorption by solid particles in atmosphere
- Attenuation and depolarization by hydrometeors (e.g., rain, ice crystals)
- Fading imputable to multipath, reflection on ground, or ducting under conditions of superrefraction ($dN/dh < -100\,\mathrm{N\,km^{-1}}$)
- Amplitude–phase distortion of signal due to selective fading and propagation differential time delay in event of multipath, reflection, or ducting conditions
- Vertical variation of launch and arrival angles of rays at antennas due to changes in vertical refractivity gradient
- Jamming by other radio links of network which share same frequency plan and polarization or by other external sources of emission

3.1.2 Establishment of Radio Link Path Profile

This task can be carried out in various ways:

- By geological survey maps, preferably on scale 1/25,000 in order to distinguish as well as possible details which relate to main obstacles

Microwave Engineering: Land & Space Radiocommunications, by Gérard Barué

- By statements determined from air photographs
- By land surveys carried out on terrain
- By digital geographic databases

3.1.2.1 Determination of Azimuth and Distance It is advisable to define, first of all, the geographical coordinates of both stations with the greatest care so as to locate them on a map whose scale corresponds to the required precision for determination of distance and altitude of the relief and the height of the obstacles that can be drawn up on the path in the plane of the Earth great circle. We can employ the following formulas whose precision, about 1/100 of a degree for angles and 10 m for distances, is in general sufficient to study geological survey maps at the scale of 1/25,000, to carry out land surveys on the terrain, or to materialize the azimuths of the antennas by using a theodolite:

- Calculation of azimuth[1] from a station A toward station B:

$$A_z = A_0 + \frac{L}{2}\sin l + \arctan\left(L\cos l \frac{1-(\varepsilon \sin l)^2}{(l_A - l_B)(1-\varepsilon^2)} \right) \tag{3.1}$$

where

$$L = L_A - L_B \qquad l = \frac{1}{2}(l_A + l_B)$$

and

$$A_0 = \begin{cases} 180^\circ & \text{if } l_A - l_B > 0 \\ 360^\circ & \text{if } l_A - l_B < 0,\ L_A - L_B > 0 \\ 0^\circ & \text{if } l_A - l_B < 0,\ L_A - L_B < 0 \end{cases}$$

where L_A, l_A = longitude and latitude of station A (deg)
L_B, l_B = longitude and latitude of station B (deg)
ε = eccentricity of Clarke's ellipsoid (1866)

with

$$\varepsilon = \sqrt{\frac{a^2 - b^2}{a^2}} = 0.082271854$$

[1]The International Conventions adopt as a reference for longitude the meridian line passing by Greenwich, which is an old observatory located in a suburb of London in England.

where

$$a = 6378206.4\,\text{m} \qquad b = 6356583.8\,\text{m}$$

- Calculation of distance between stations A and B on arc of Earth great circle:

$$d = \frac{111.3177 L \cos l}{\sqrt{1-(\varepsilon \sin l)^2}\, \sin\left[\arctan\left(L \cos l \dfrac{1-(\varepsilon \sin l)^2}{(l_A - l_B)(1-\varepsilon^2)}\right)\right]} \tag{3.2}$$

3.1.2.2 Data Necessary for Establishment of Path Profile The path profile can be traced from a cartographic database using the appropriate software or from any topographic statement while taking particular care of the high points of the terrain and the possible obstacles. It is appropriate, however, to stress the importance of the grid of a digitized chart, that is, the size of the elements of the surface that were used to define the average altitude of each considered point. For example, a portion of terrain about 20 m × 30 m corresponds to 1″ of arc under a latitude of 45° and a portion of 100 m × 150 m to 5″; a chart digitized with the step of 5″ can thus generate an awkward uncertainty on altitudes in rough terrain. By comparison, an interval of 1 mm, generally readable on a chart, represents a distance of 25 m on the scale 1/25,000 and 100 m on the scale 1/100,000. Moreover, the large-scale maps provide a number of details, such as level lines or equal-altitude lines as well as the nature of the obstacles (e.g., villages, cities, stretches of water, meadows, woods, forests); at first approximation, we retain an on-ground height of 5–10 m for vegetation, 20 m for villages and wood, and 30 m or more for forests and cities. Figure 3.1 shows, as an example, the variance in altitude which has been observed between a geographic terrain database and precise geographical positioning system (GPS) readings that have been performed on about 600 plots throughout a path 900 km long in the Middle East. The distribution obeys a Gaussian law which defines a standard deviation of 4 m so that less than 1% of the plots may present an altitude error of 10 m.

To carry out a path profile directly from a geological survey map, we start by locating the two stations and connect them by a straight line. Then, we plot the distance and altitude of the ground starting from one of the stations, with a step ranging between 0.1 and 1 km, by carefully noting the location of the high points, the low points, and the zones which can constitute a reflective surface.

It is necessary, at this stage, to take account of the following elements:

- Modern geological survey maps are prepared so that the respective positions of the objects on the surface of Earth and their images on the paper are bound by a mathematical relation which preserves the angles but modifies the lengths and surfaces.

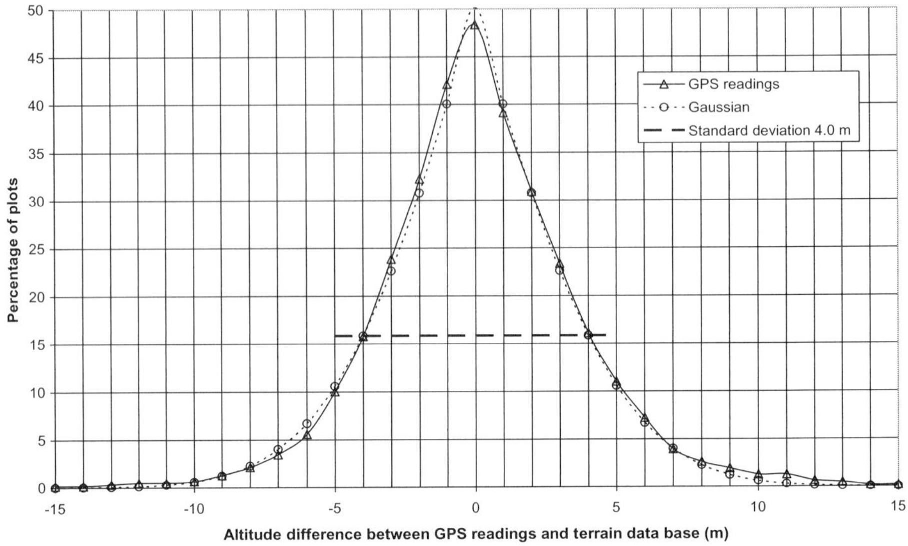

Figure 3.1 Distribution of error in altitude of terrain database.

- The precision of a chart strongly depends on its scale.
- The radius of the first Fresnel's ellipsoid, which corresponds to some extent to the "useful volume" occupied by most of the energy of the electromagnetic wave which propagates between two points and is given by relation (1.65), is in general about several tens to several hundreds of meters.

Consequently, we will record the highest altitude in a zone about a hundred meters or more on both sides of the path line and will make use of formulas (3.1) and (3.2) to calculate the exact position of a given obstacle with respect to the path line or to define the points of passage of a chart to another one without loss of precision.

3.2 DESIGN OF LINE-OF-SIGHT MICROWAVE RADIO LINK: INTERFEROMETRIC METHOD

3.2.1 Layout of Path Profile

Now, data collected previously on a diagram that gives the trajectory of the direct radioelectric ray are compared to the path profile of the link for notable values of the factor K. We saw in Section 1.8, in relation with the influence of the atmosphere, that the trajectory of an electromagnetic wave is curved and the radius of curvature depends on the value of the vertical refractivity gradient. For this reason, certain authors use a model of propagation where the terrestrial radius remains constant and the trajectory of the electromagnetic wave is curved. However, as we saw in Section 2.3.4, which treats the effects of reflection on propagation, it is more convenient to represent the trajectory of the rays as rectilinear and to vary the effective Earth radius.

The path profile, characterized by an altitude and a given height obstacle at a given distance, is thus corrected by relation (1.84), which determines the terrestrial curvature according to the factor K, of which the most significant values should be considered for the definition of the architecture of the stations. The characteristic K values to take into account are as follows:

- Median value K (50%)
- Minimal values of $K > 0$, or $K_{\min}$, given by Figures 1.53 and 1.54, which correspond to extreme conditions of subrefraction
- Minimal values of $K < 0$, which correspond to extreme conditions of superrefraction defined in Section 1.8 for geographical zone considered

The method suggested here consists in plotting several diagrams, by respecting the criteria of clearance defined below, in the following way:

- Path profile of link for median value of factor K (50%), which is close to $\frac{4}{3}$ in temperate continental climate and infinite in desert climate
- Path profile showing direct and reflected rays for broad range of variation of effective Earth radius according to factor K, which varies between $K_{\min}$ and minimal value $K < 0$, which corresponds to arbitrary value of vertical refractivity gradient of $-400\,\mathrm{N\,km^{-1}}$ or less in high superrefractive zones about horizontal reflection
- Average altitude and regression line of path profile
- Path profile representing direct and reflected rays after transformation according to regression line used as altitude of reference which concerns diffuse reflection

3.2.2 Clearance Criteria

Recommendation ITU-R P.530 recommends using the following clearance criteria for temperate and tropical climates:

- For the median value K (50%), $h/R_{F1} \le -1$ for the highest obstacle.
- For the minimum positive value $K_{\min}$ (99.9% of any month):
  (a) In temperate climate

$$\frac{h}{R_{F1}} \le \begin{cases} 0 & \text{for single source of obstruction} \\ -0.3 & \text{for obstruction extending on whole portion of path} \end{cases}$$

(b) In tropical climate

$$\frac{h}{R_{F1}} \leq -0.6 \quad \text{for paths of more than 30 km}$$

For the lower antennas, in vertical space diversity, the clearance can lay between 0 and –0.6 according to the nature of the obstruction for K (50%); it should be pointed out that space diversity takes effect principally during superrefraction when clearance is maximum.

3.2.3 Regression Line and Terrain Roughness

Here, we need to calculate the average slope of the terrain as well as its roughness, which can be used as criteria in the prediction of the performance of the radio link, and to determine the zone of reflection and its proper coefficient of reflection. The method consists of first calculating the average height:

$$\bar{y} = \frac{\sum_{i=1}^{n} y_i}{n} \tag{3.3}$$

Then the average quadratic value of the relief is given as

$$r = \sqrt{\sum_{i=1}^{n} \frac{(y_i - \bar{y})^2}{n-1}} \tag{3.4}$$

where n = number of points of profile
y_i = altitude corresponding to point i among n

The regression line is determined by the relation

$$y = ax + b \tag{3.5}$$

where x is the distance from one end with

$$a = \frac{\sum_{i=1}^{n} (x_i y_i) - \left(\sum_{i=1}^{n} x_i \sum_{i=1}^{n} y_i\right) \Big/ n}{\sum_{i=1}^{n} x_i^2 - \left[\sum_{i=1}^{n} x_i\right]^2 \Big/ n}$$

$$b = \frac{\sum_{i=1}^{n} y_i}{n} - a\frac{\sum_{i=1}^{n} x_i}{n}$$

by considering the point of altitude and distance (x_i, y_i). The roughness of the terrain is then given by the relation

$$R = \sqrt{rr^*} \tag{3.6}$$

where

$$r^* = \sqrt{\frac{\sum_{i=1}^{n}(y_i - y_i^*)}{n-1}}$$

with y_i^* the height of point (x_i, y_i) compared to the regression line, that is, $y_i^* = y_i - (ax_i + b)$.

3.2.4 Reflection Zone

We start by locating the horizontal reflection point, of which the effects are prevalent under the conditions of superrefraction, by using the method described in Section 2.3.4 and making it correspond to a conservatory reflection coefficient of 0.9–1. Other zones of reflection can also appear on the path profile, in particular along the regression line which represents the average slope of the terrain; corresponding to these zones there is a diffuse reflection coefficient that can be calculated using formulas in Section 1.7. As far as possible, the reflection zone has to be shielded by an obstacle or moved to a poorer reflecting surface such as rougher terrain at vegetation by adjusting the antenna height at one end or both ends of the path.

3.2.5 Differential Time Delay

The differential time delay between the direct ray and reflected ray can be calculated using relation (2.66). The differential time delay must be associated with the ratio between the level of the direct ray and that of the reflected ray in order to compare them with the strength property of the receiver facing interference, which is called the system signature.

3.2.6 Differential Attenuation

The in-band power difference, which is defined as the peak-to-peak difference in attenuation across the frequency bandwidth occupied by the useful signal, can be calculated from relation (2.65) comparing the values obtained for the extreme frequencies to that at the central frequency; this type of impairment of the transmitted signal, which is illustrated in Figures 2.51 and 2.52, relates to the capacity of adaptive equalization of the receiver.

3.2.7 Reception in Space and/or Frequency Diversity

Frequency-selective fading due to multipath propagation results in time-varying perturbations, and the application of diversity techniques can provide a high improvement of the performance as far as the correlation between the signals received from various paths is low. The most common methods are described as double space or frequency diversity and can be used either singly or in combination, such as quadruple space and frequency diversity. The interferometric method aims at obtaining the covering of the interference fringes which result from reflection and multipath in a two-ray model, as described in Section 2.3.4, so that when one receiver is under adverse conditions, the other profits from favorable conditions; it is thus a question of seeking a correlation in opposition of the fields received in diversity.

In the case of space diversity, which is carried out in the vertical plane in line-of-sight links, the spacing between the main and diversity antennas will thus have to be such that when one is located in a zone of minimal field, the other is in a zone of maximum field; this condition is obtained when the spacing between the antennas at station B, $(\Delta h)_B$, is equal to an odd multiple of the half-interference fringe given by relation (2.67), where H_A is the height of the emitting antenna of station A above the reflection plane:

$$(\Delta h)_B = n\frac{\delta h}{2} = n\frac{\lambda d}{4H_A} \qquad n = 1, 3, 5 \tag{3.7}$$

In the case of frequency diversity, the necessary spacing in frequency can be calculated starting from relations (2.63) and (2.64) in the following way:

- Path difference between direct wave and reflected wave:

$$\Delta l \approx \frac{2H_A H_B}{d}$$

- Phase difference between direct wave and reflected wave:

$$\theta \approx \frac{2\pi \Delta l}{\lambda} \quad (+\sim\pi \text{ for low grazing angles according to Figure 1.25})$$

We then calculate the wavelength λ' so that the phase difference between the direct wave and the reflected wave at the frequency diversity F' is equal to

$$\theta' \approx \frac{2\pi \Delta l}{\lambda'} = \theta \pm \pi$$

for the same path difference Δl since there is only one antenna at each end, which yields

$$\frac{2\pi \Delta l}{\lambda'} = \frac{2\pi \Delta l}{\lambda} \pm \pi$$

and thus

$$\frac{F'}{F} = \frac{\lambda}{\lambda'} = 1 \pm \frac{\lambda}{2\Delta l} \tag{3.8}$$

In the case of a quadruple space and frequency diversity, where we emit simultaneously on the main and space diversity antennas, it is necessary to determine the vertical spacing between the antennas for a given frequency spacing in order to get the best possible compromise for the covering of the interference fringes; such an operation is relatively easy when using a model which takes into account all the parameters of the radio link.

As a result of the usual constraints on the frequency plan, we prefer space diversity, which in addition offers much more flexibility than frequency diversity because of the required spacing between both channels inside the frequency plan.

3.2.8 Example of Line-of-Sight Radio Link

We take as an example a 40-km link established at the frequency 8 GHz between stations A and B located in a temperate continental climate and whose respective altitudes are 180 and 280 m. The table below provides the data necessary to establish the path profile by indicating the most representative points of the relief with a step of about 1 km:

Distance/Station A (km)	Altitude (m)	On Ground (m)	Nature
0	180		
1–3	180	25	Town
4	170	15	Suburb
5	167	10	Bush
6	162	0	Bush
7–19	160	0	Lake
20	165	0	Meadow
21	178	0	Meadow
23	190	20	Wood
25	194	20	Wood
28	183	20	Wood
29	190	30	Forest
31	203	30	Forest
33	195	30	Forest
35	180	30	Forest
37	185	20	Wood
38	189	20	Wood
39	240	5	Bush
40	280	0	

On the path profile illustrated by Figure 3.2, we see that the first Fresnel ellipsoid is totally cleared off for the median value K (50%) = $\frac{4}{3}$ with the heights for the antennas chosen to be 40 m at station A and 20 m at station B.

We notice in Figure 3.3, which represents the path profile for extreme values of K, that:

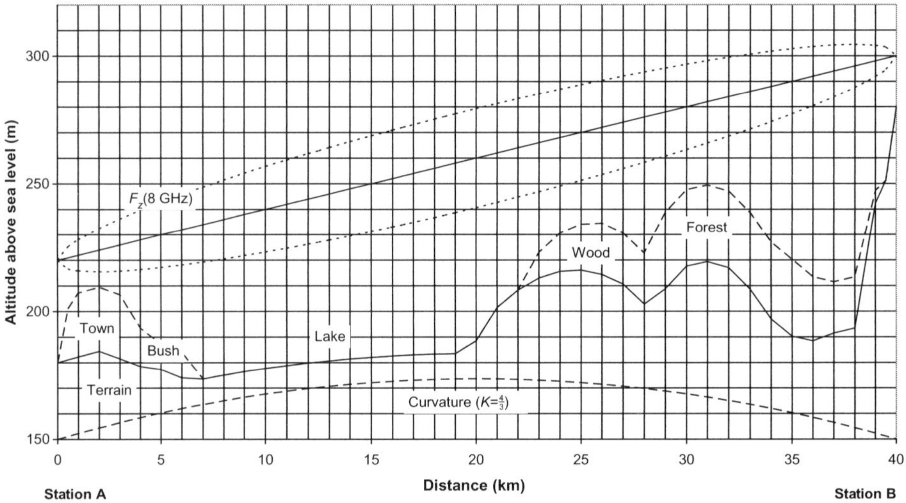

Figure 3.2 Representation of path profile for K (50%) = $\frac{4}{3}$.

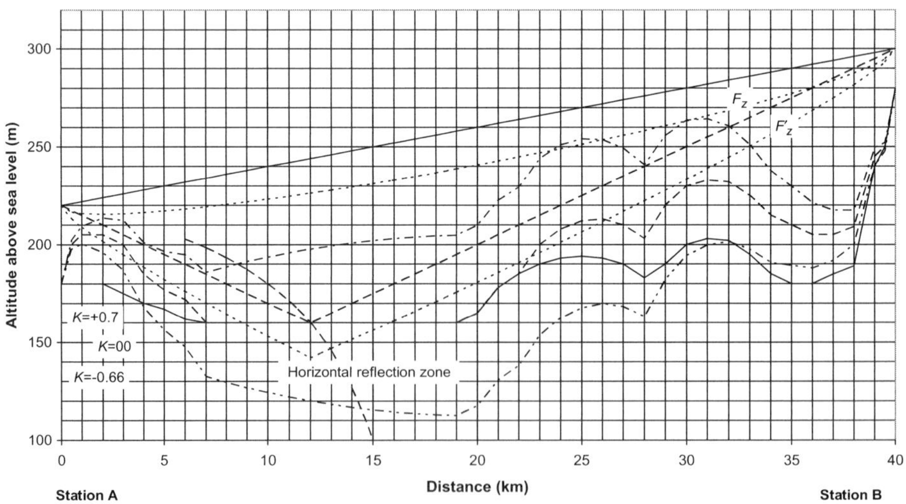

Figure 3.3 Representation of direct and reflected rays and effective Earth radius for extreme values of factor K.

- The clearance of the direct ray compared to the principal obstacle consisting of the wooded hill located 24 km from station A is –0.63 for $K_{min} = 0.7$ relative to 99.999% of the time.
- The horizontal reflection zone, delimited by the first Fresnel's ellipsoid of the reflected ray, corresponds to a large body of water which is cleared from the principal obstacle for all values of the factor K apart from K_{min}.

It is seen that the height chosen for the antennas at each end of the connection answers well the criteria of clearance recommended in Section 3.2.2 with a comfortable margin; the attenuation due to the main obstacle, given by Figure 1.70, is close to 0 dB for K_{min}.

Figure 3.4 indicates the average altitude of the path profile as well as the regression line; the regression line represents the average slope of the terrain, which here is 1.21 mrad, and the roughness is about 18 m. Figure 3.5 presents the path profile after transformation according to the regression line used as altitude of reference. In the case of this link, which is not symmetric, Figure 3.3 presents the horizontal reflection in the low layers of the atmosphere, in the vicinity of the ground where can occur an important vertical gradient of temperature and hygrometry, or a stretch of water, and Figure 3.5 presents the regressive reflection, or diffuse reflection on the ground. The two zones of reflection, horizontal and regressive, are confused when the link is established on a terrain of null slope, which is not the case here.

We see in Figure 3.5 that the zone of regressive reflection is less cleared from the principal obstacle than the zone of horizontal reflection that is dominating, particularly in conditions of superrefraction; therefore, only horizontal

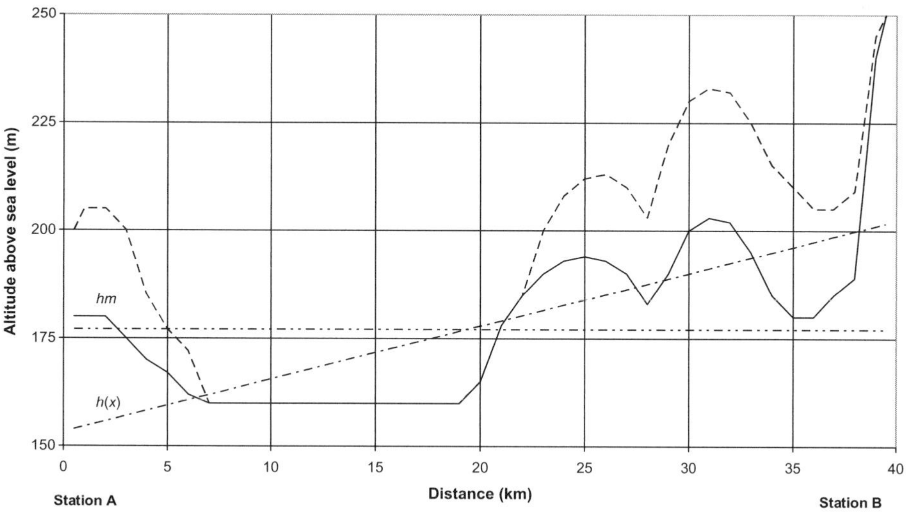

Figure 3.4 Average altitude and regression line.

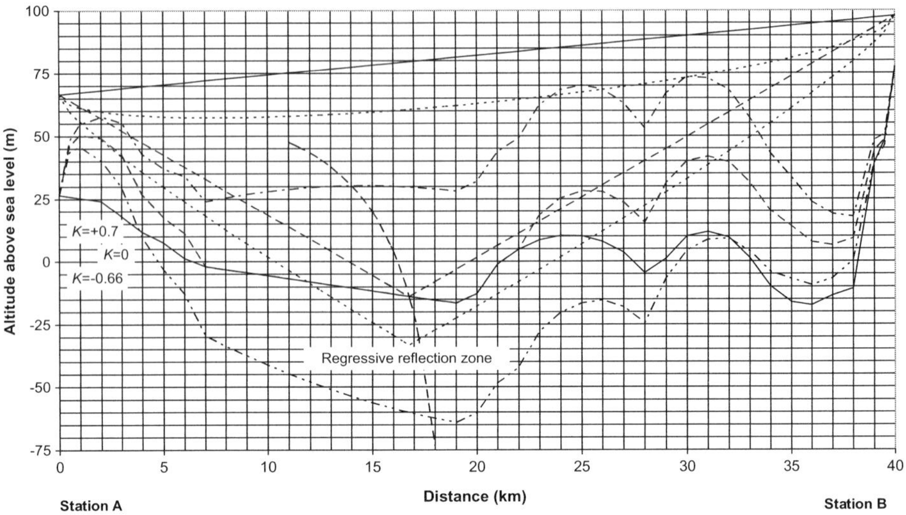

Figure 3.5 Regressive reflection.

reflection has to be considered. The average altitude of the zone of reflection being 160 m and the altitude of the antennas known at the two ends, we can calculate the spacing of the interference fringes using relation (2.67); for $K = \infty$ (intermediate between $K_{\min}$ and $K < 0$) and with the following values:

	Station A	Station B
Altitude at ends, m	180	280
Transmission antenna height, m	40	20
Average altitude of reflection zone, m	160	160

this gives

$$\delta h_A = \frac{\lambda d}{2H_B} = 5.3\,\text{m} \qquad H_B = 280\,\text{m} + 20\,\text{m} - 160\,\text{m} = 140\,\text{m}$$

$$\delta h_B = \frac{\lambda d}{2H_A} = 12.5\,\text{m} \qquad H_A = 180\,\text{m} + 40\,\text{m} - 160\,\text{m} = 60\,\text{m}$$

To optimize the space diversity, the necessary spacing between the antennas must be, according to relation (3.7), about 2.7 m (or 8 m for $n = 3$) at station A and about 6.5 m at station B.

Figure 3.6 thus illustrates the architecture that has been retained for the localization of the main and diversity antennas; it will be noticed that the

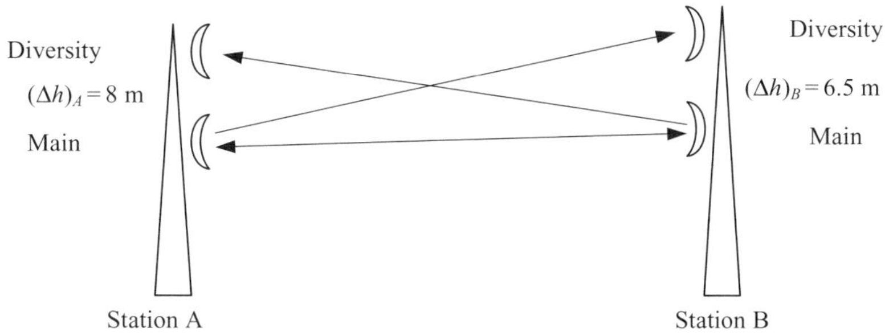

Figure 3.6 Architecture of stations.

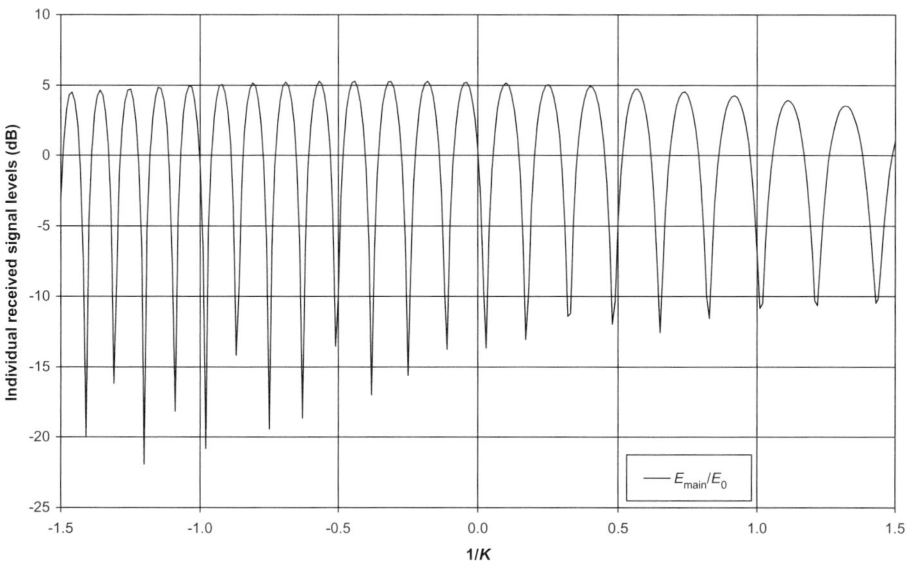

Figure 3.7 Received field on main antenna at station B resulting from horizontal reflection.

diversity antennas are placed above the main ones in order to reduce the branching losses to the transmitters without affecting the clearance of the link and to take advantage of more spread interference fringes.

Figure 3.7 shows the variation of the computed field on the main antenna at station B by using relation (2.65), which results from summation of the direct wave and reflected wave, for all possible values of the factor K and by supposing a reflection coefficient[2] of 1. We can see that the received field

[2]The coefficient of reflection can be calculated with precision using the method described in the Section 1.7; however, the experiment shows that the relief can be completely "gummed" by the conditions which reign in the low layers of the atmosphere in the vicinity of the ground, in particular in a period of superrefraction or ducting, and the coefficient of reflection can then be close to unity.

suffers important variations of amplitude; these variations of amplitude are of course accompanied by important variations of phase, as shown in Figure 2.45, which represents the vectorial composition of the direct and reflected rays.

Figure 3.8 presents, with respect to the beam slope, the apertures of the selected 2.4-m antennas (±0.6°), the reflection angle, and the launch and arrival angles of the direct and reflected beams, and Figure 3.9 shows the relative levels of the direct and reflected beams which result from the angular discrimination by the antennas:

- In Figure 3.8, direct and reflected rays are included for all the possible values of the factor K inside the total half-power radiation pattern of the antennas.
- In Figure 3.9, the levels of the direct and reflected rays are very close together and the reflected signal can even exceed the direct signal in conditions of ducting; in this case, it is advisable to tilt the antennas upward in order to increase the direct-to-reflected power ratio, as we will see below.

Figure 3.10 presents the computed differential time delay between the direct and reflected rays on the main and diversity antennas, and Figure 3.11 shows the differential attenuation of the signal spectrum which results from the reflection inside the whole bandwidth (±14 MHz). Figures 3.12 and 3.13 show

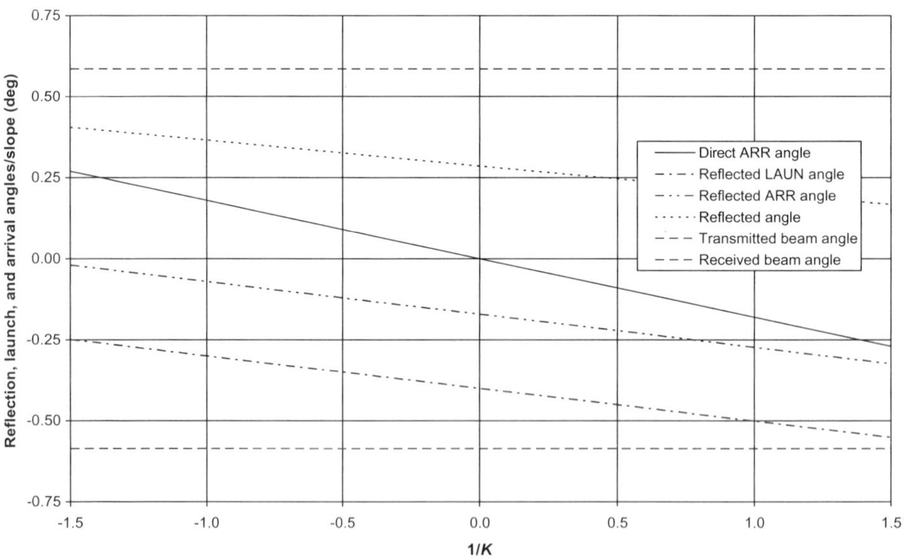

Figure 3.8 Comparative diagram of angles of reflection, aperture of antennas, and launch (LAUN) and arrival (ARR) angles of rays.

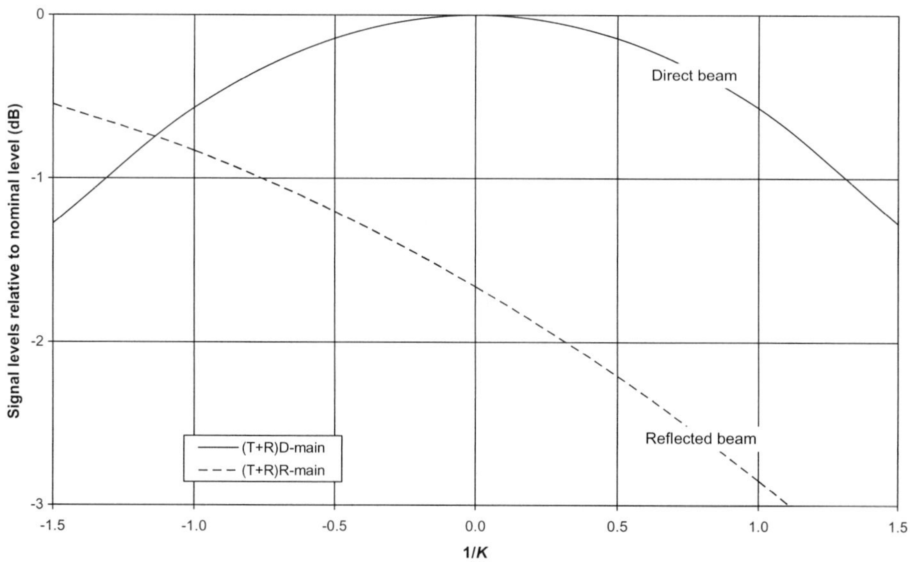

Figure 3.9 Cumulated discrimination of direct and reflected beams versus antenna emission and reception.

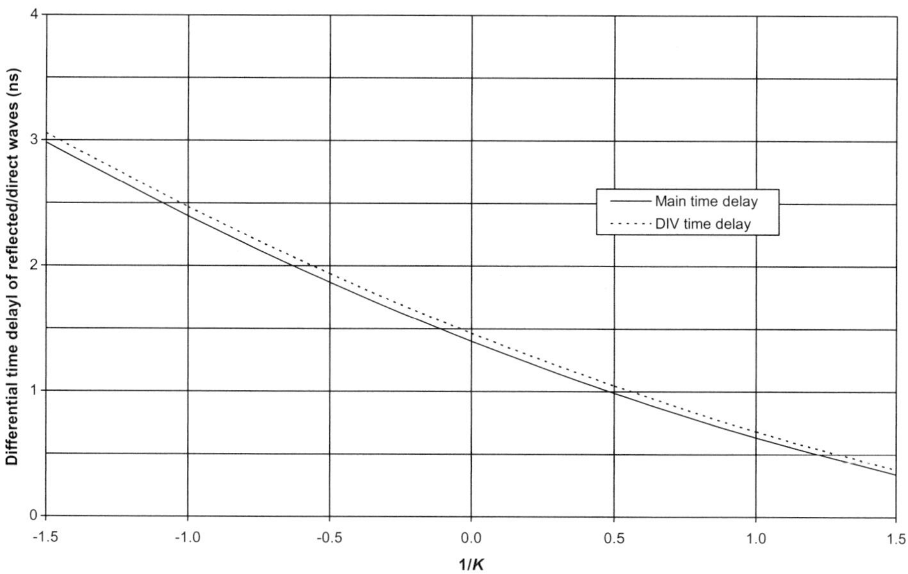

Figure 3.10 Differential time delay.

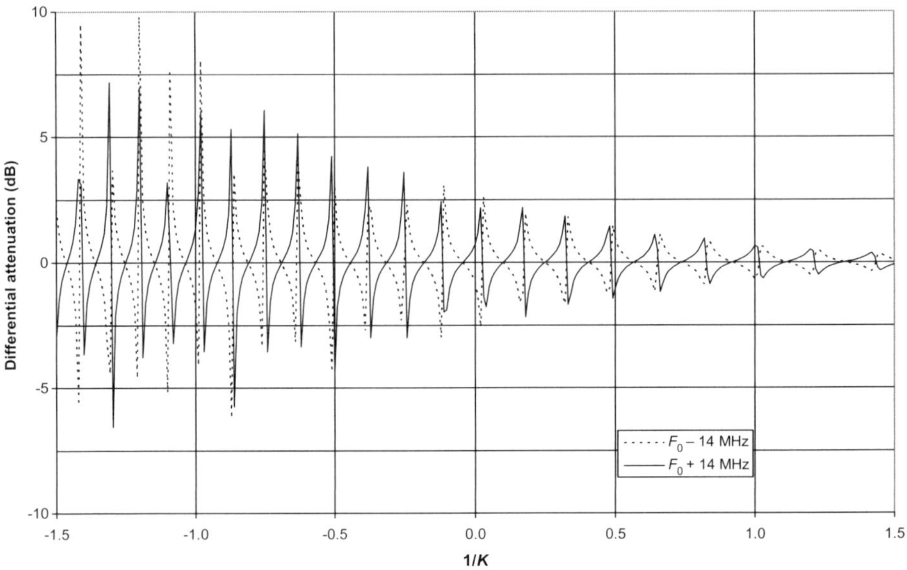

Figure 3.11 Differential attenuation in whole bandwidth.

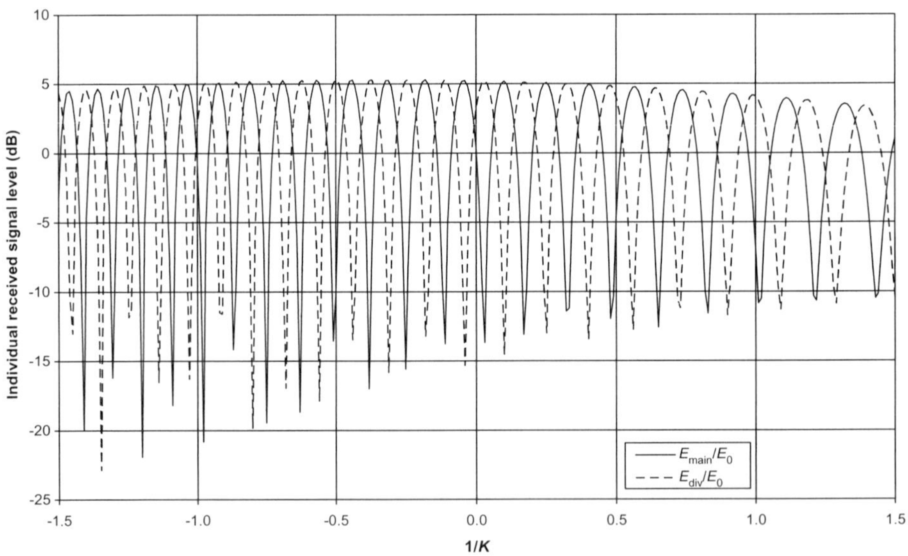

Figure 3.12 Individual received signals on main and diversity antennas with spacing of 6.5 m (A → B).

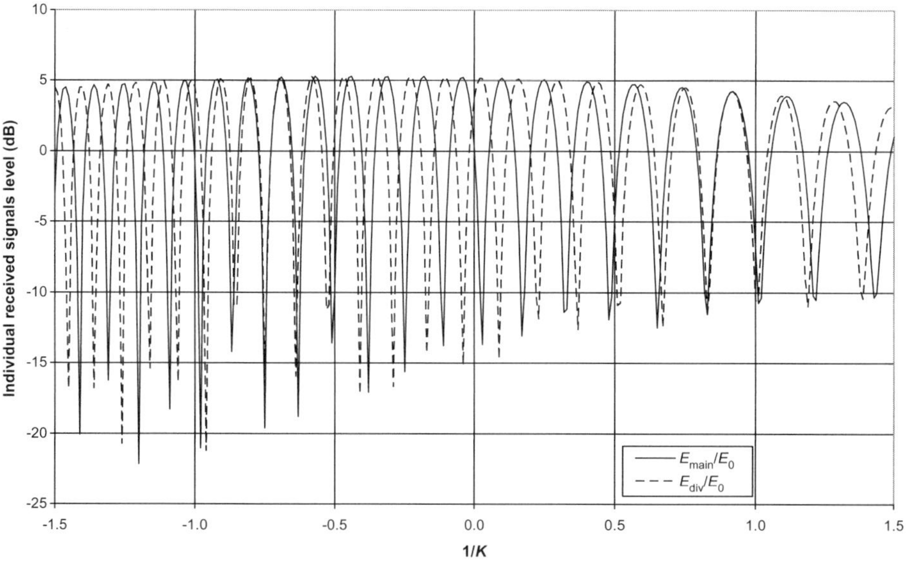

Figure 3.13 Individual received signals on main and diversity antennas with spacing of 8 m (B → A).

the respective computed levels of the individual signals that are received simultaneously on the main antenna and the diversity antenna.

Figures 3.14 and 3.15 present the computed combined signal in each direction, that is, the highest of both received signals in the case, for example, of switching by selection of the strongest one; we see that the combined signal is well smoothed at station B but the overlapping is not perfect for some values of factor K at station A, because the spacing between the antennas corresponds to three half-interference fringes instead of a single one, but the amplitude of variation is less than that of the individual signals. We see that, due to the sought correlation in opposition, the combined field is practically always higher than the value which corresponds to free space; moreover, the strongest signal is that whose distortion is the lowest since it corresponds to the most favorable configuration in terms of phase difference between the direct and reflected beams, as we saw in Section 2.3.4. To replace space diversity by frequency diversity, it would be necessary to use frequency separation ΔF, calculated as follows using formula (3.8):

$$\Delta l = 0.42\,\mathrm{m}$$

that is, ΔF = ±357 MHz for F_0 = 8 GHz. This frequency separation is much more important than that which may be used in the 8-GHz band where channeling is usually of 28 MHz. The recourse to frequency diversity in association with space diversity makes it possible to notably improve the conditions of reception. The best combination is obtained, in general, while emitting at high

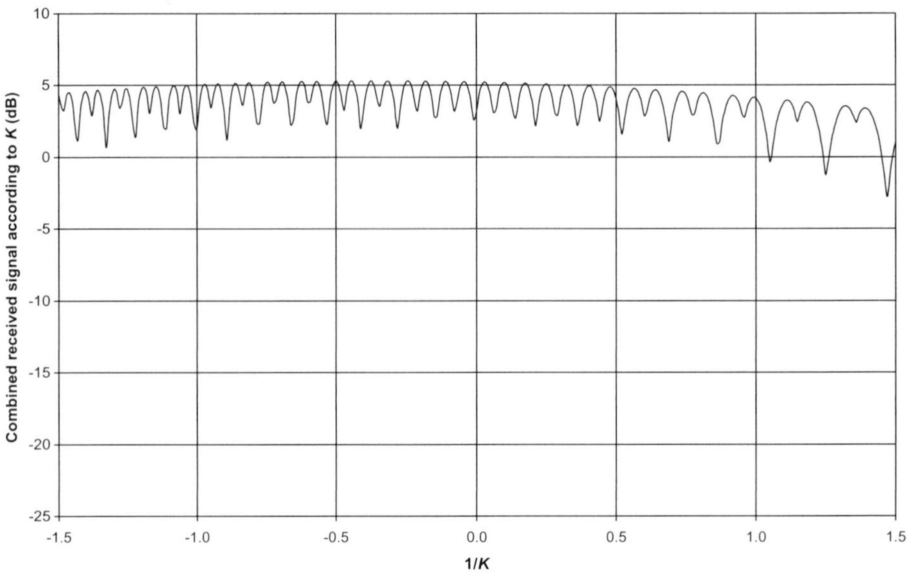

Figure 3.14 Combined received signal in space diversity (A → B).

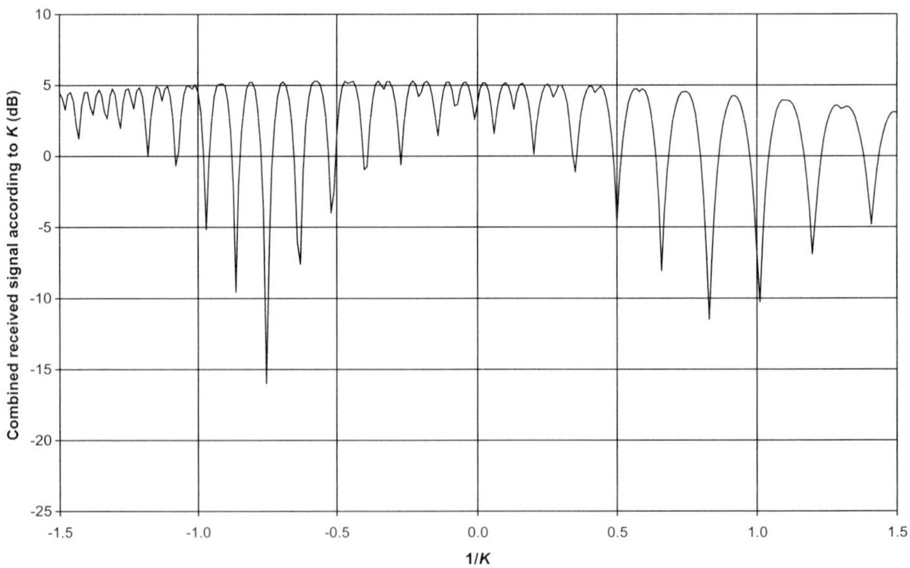

Figure 3.15 Combined received signal in space diversity (B → A).

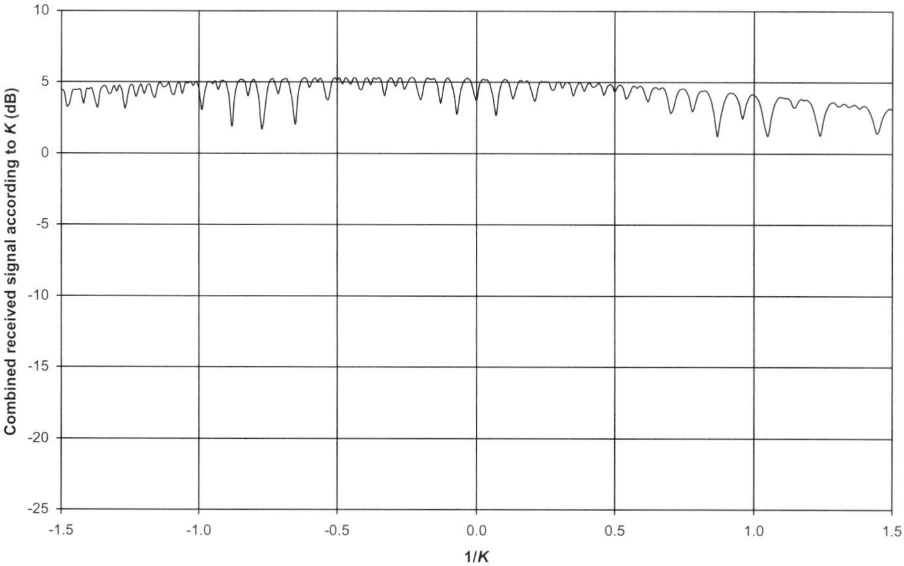

Figure 3.16 Combined received signal in space and frequency diversity (A → B).

frequency on the low antenna and, conversely, at low frequency on the high antenna. Although such a configuration is not justified here, Figures 3.16 and 3.17 show the computed result for the combined signal by employing quadruple diversity of space and frequency by preserving the same spacing between the antennas and using a frequency separation of 28 MHz.

3.2.9 Recordings in Ducting Conditions and Reflection on Sea

3.2.9.1 Case of Terrestrial Link in Desert Climate The effectiveness of the above-mentioned interferometric method has been checked on a test link (X–Y) located in an area of the Persian Gulf which meets the most marked conditions of ducting with a factor P_L of 90% during the summer months, as shown in Figure 1.38. Figure 3.18 presents the path profile of this 35-km link that was carried out in space diversity only at a bit rate of 155 Mbits s^{-1} in the 8-GHz frequency band. The following table presents the characteristics of both stations:

	Station X	Station Y
Main antenna height, m	53.3	84.4
Diversity antenna height, m	58.9	90.7
Spacing, m	5.6	6.3

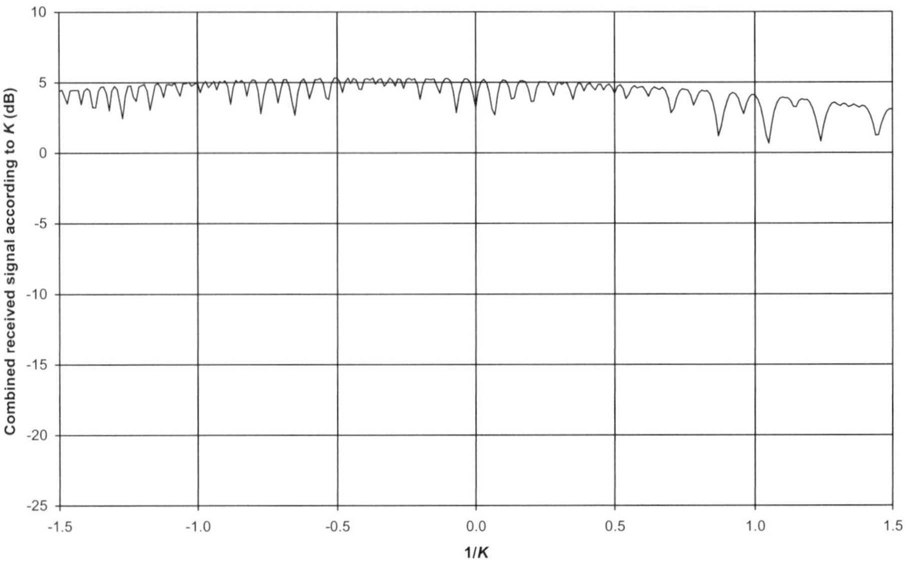

Figure 3.17 Combined received signal in space and frequency diversity (B → A).

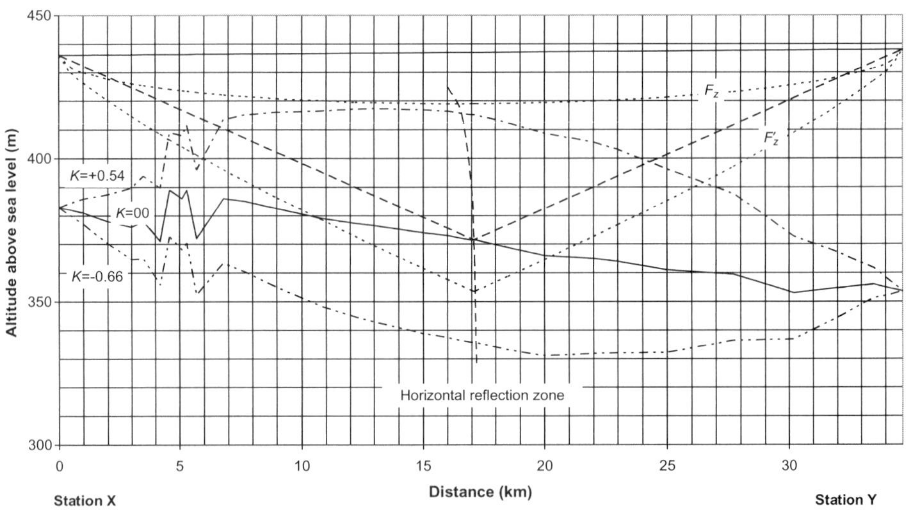

Figure 3.18 Path profile test link X–Y in desert climate.

The diversity antennas have been installed above the main ones in order to reduce the feeder losses at the transmitter part, and the spacing corresponds to the interference half fringes. The profile reveals a long sandy and desert strip of land, completely out of vegetation, which constitutes a zone of reflection with a very weak roughness factor. The points of horizontal and regressive reflection are well cleared; the first gives place at night, during the appearance of surface ducting layers, to very strong fluctuations and the second during the day, when the atmosphere is brewed by important movements of convection due to the sun heating, to slow variations of the received field.

The figures which follow show measurement results obtained during one night in July on the receiver connected to the main antenna of station X during the appearance of a radioelectric hole which lasted several hours by generating a median depression of some 20 dB.

Figures 3.19 and 3.20 show the level of the instantaneous field that was recorded over two 45-min successive periods by using a sampling rate of 10 Hz.

Figure 3.21 shows the levels of the instantaneous individual fields received simultaneously on the main antenna and the diversity antenna over a 1-min period.

Figure 3.22 presents the combined field resulting from the selection of the strongest signal.

Figure 3.23 shows the speed of variation of the instantaneous individual fields received over a period of 5 min.

Figure 3.24 shows the activity of the hitless switching function of the receivers.

Figure 3.25 presents the distribution of the individual and combined fields received over a period of 10 min (2:10–2:20).

Figure 3.26 presents the coefficient of correlation observed over the same period between the signals received simultaneously on the main and diversity antennas.

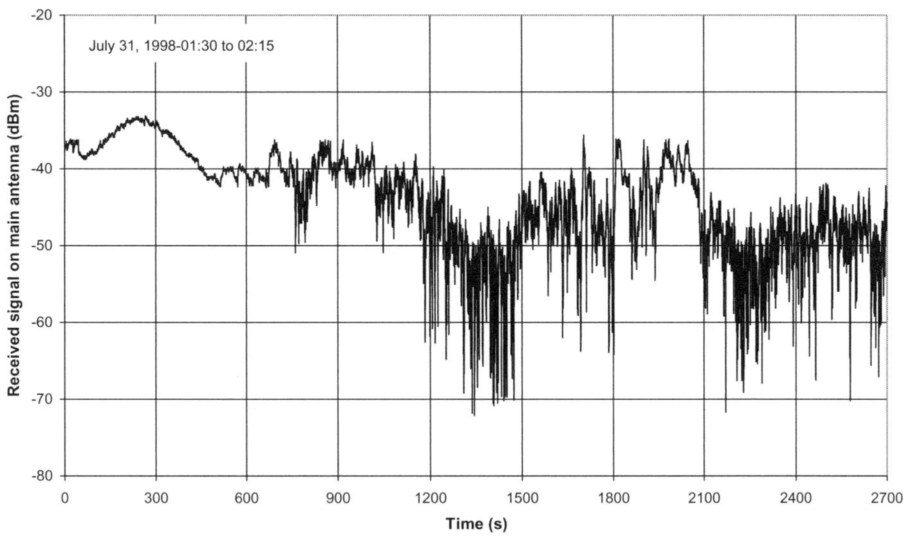

Figure 3.19 Field strength recording at station X (first period).

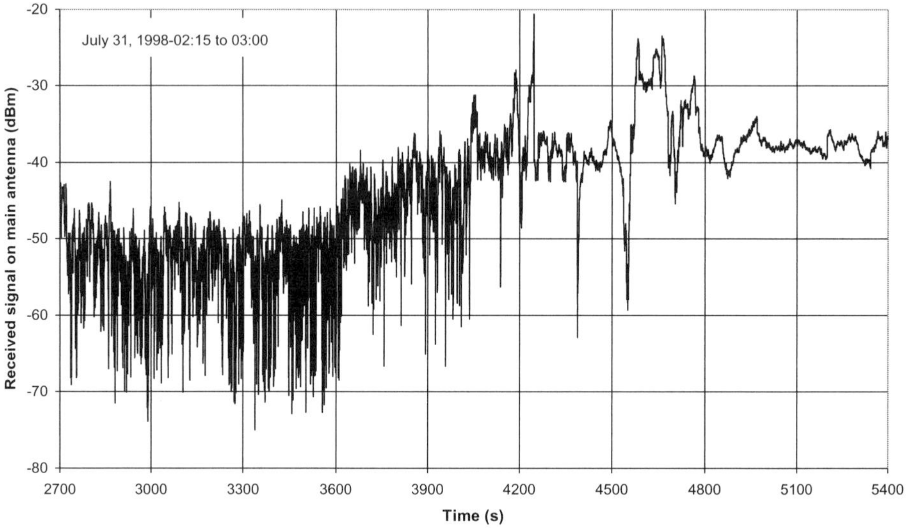

Figure 3.20 Field strength recording at station X (second period).

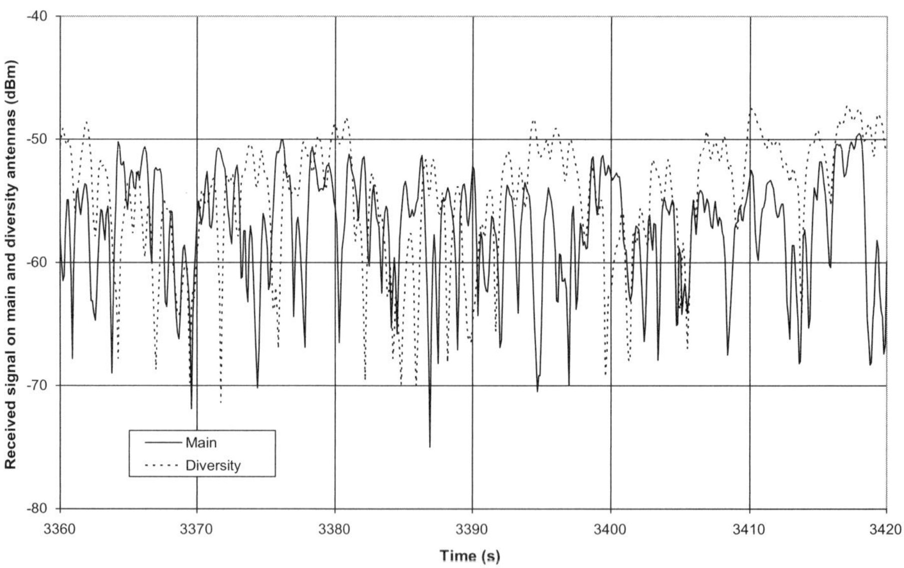

Figure 3.21 Simultaneous recording of instantaneous signals received on main and diversity antennas.

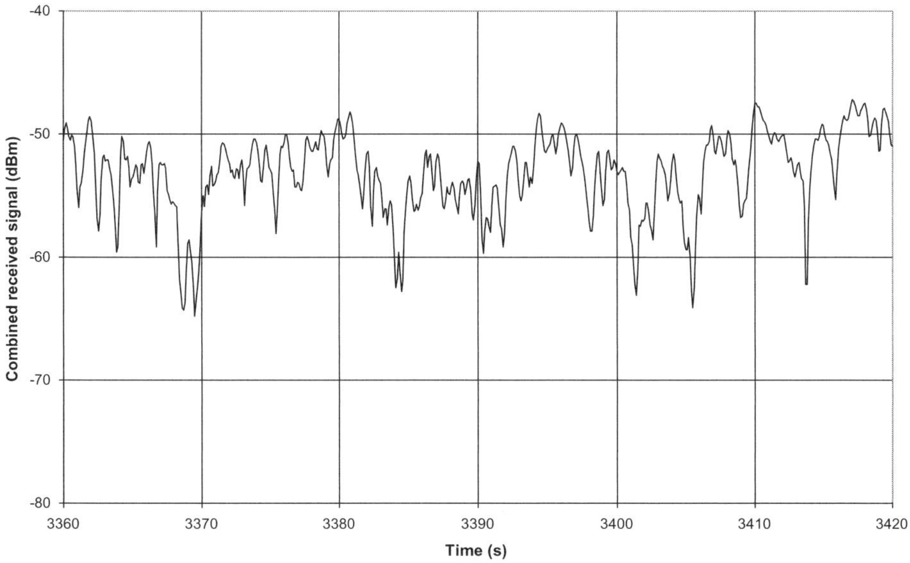

Figure 3.22 Combined received field by selection of strongest signal.

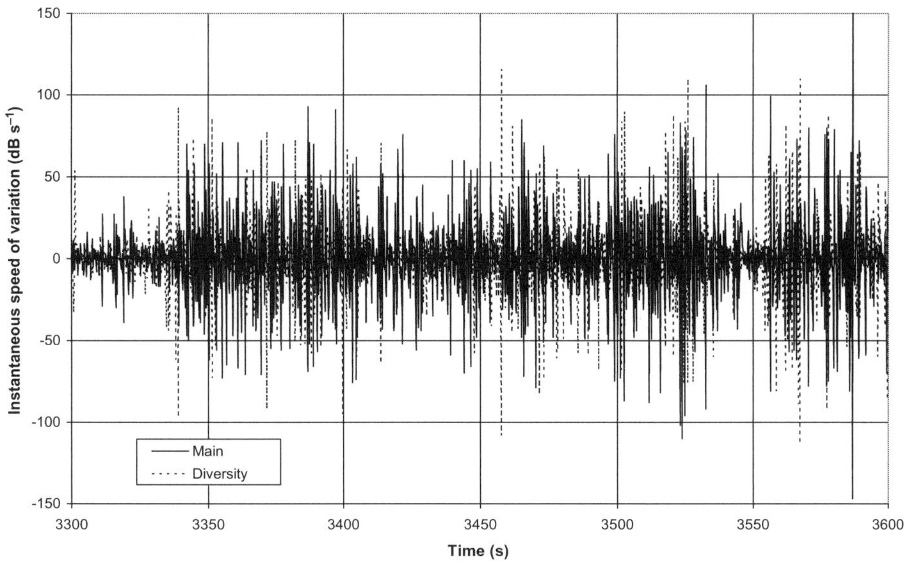

Figure 3.23 Speed of variation of both received individual fields.

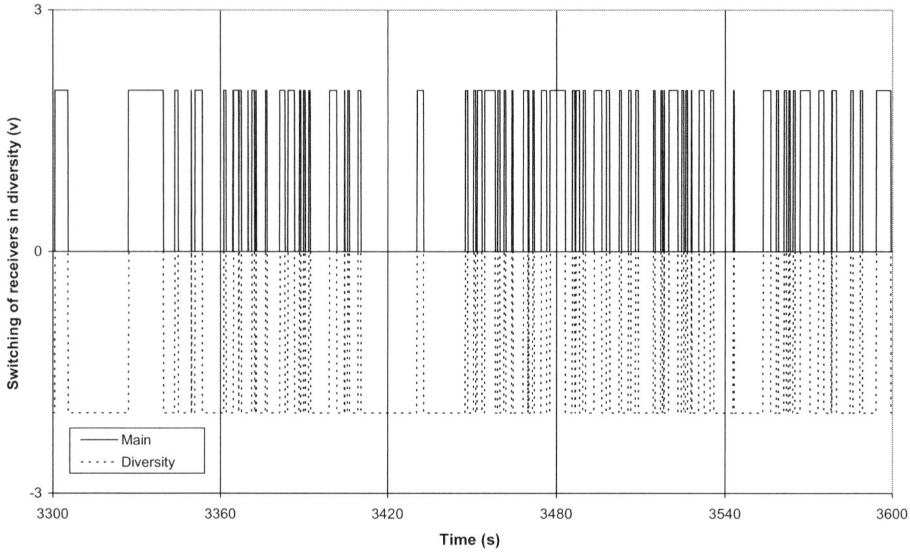

Figure 3.24 Switching activity of both receivers in diversity.

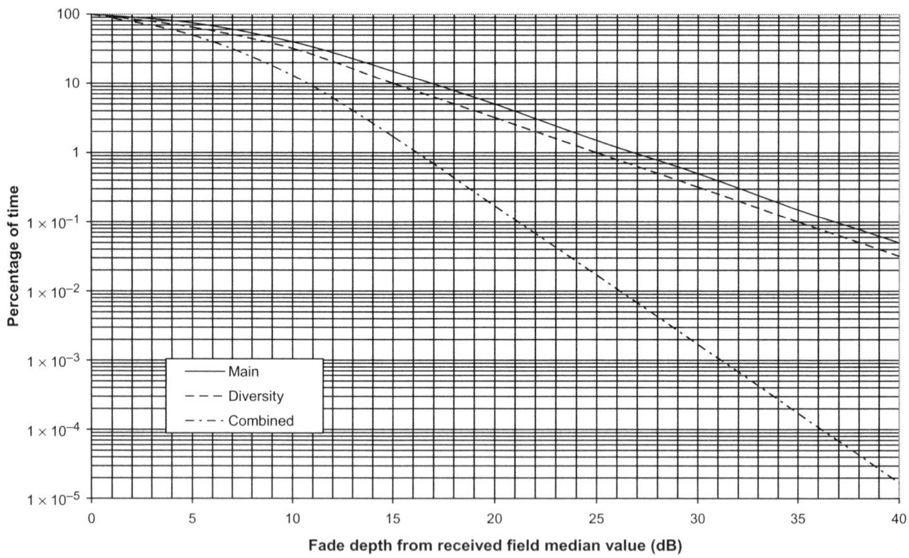

Figure 3.25 Distribution of individual and combined received fields (July 31, 1998–02:10 to 02:20).

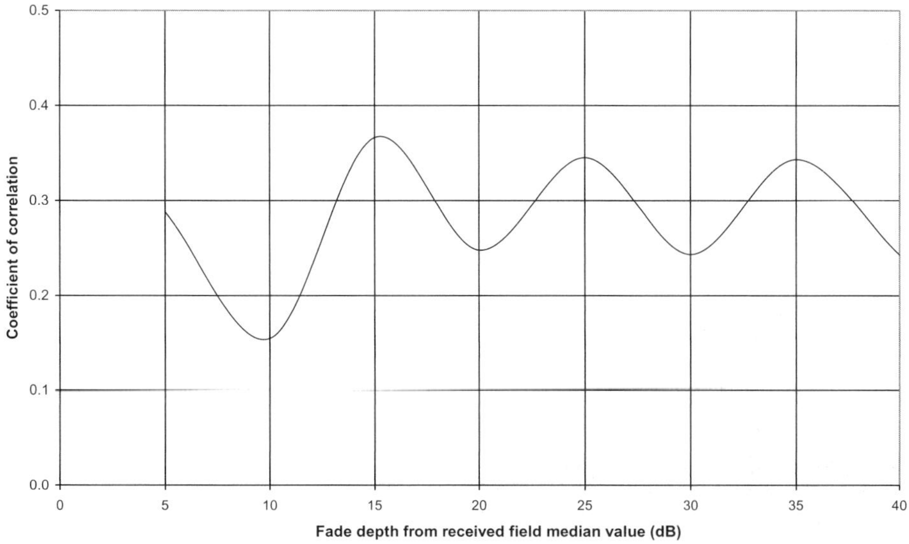

Figure 3.26 Coefficient of correlation between main and diversity signals (July 31, 1998–02:10 to 02:20).

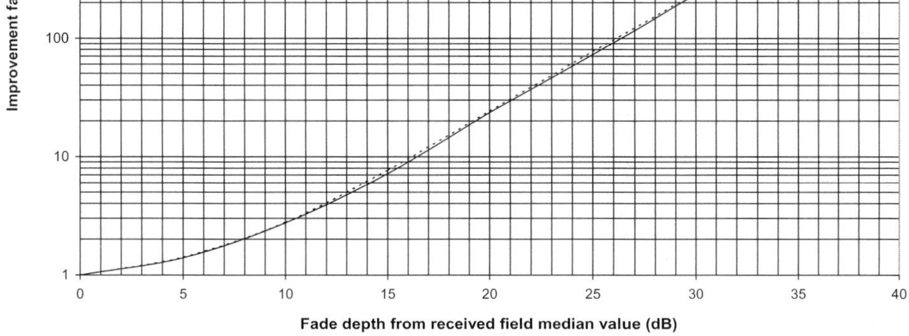

Figure 3.27 Improvement Factor (measured and theoretical limit) (July 31, 1998–02:10 to 02:20).

Figure 3.27 presents the improvement factor that has been observed over the same period by comparison with the theoretical limit for two Rayleigh distributed independent signals.

The improvement factor and the diversity gain, according to the coefficient of correlation, for two Rayleigh distributed random signals are given by formulas (1.115) and (1.116); Figure 3.25 shows that both individual received fields correspond well to a Rayleigh's distribution as the slope is 10 dB per decade beyond a fading depth of about 15 dB. This recording proves that the interferometric method of correlation in opposition of the fields which are received in space diversity makes it possible to reduce the coefficient of correlation to a very low value during ducting conditions and to thus obtain an improvement factor which is close to the theoretical limit, as shown in Figures 3.26 and 3.27. Figure 3.21 illustrates well the variation in opposition of the instantaneous fields which are received on the main and diversity antennas; indeed, when one of the receivers is affected by fading, the other records a field maximum which is also more stable.

As this method applies to a very broad range of variation of the air vertical refractivity gradient, the performances remain optimized until under the conditions of ducting, which are generally at the origin of the unavailability of the links, provided that the median depression of the field leaves a sufficient margin compared to the receiver threshold. Speeds of variation of the field up to 200 dB s^{-1} were also observed during this period of ducting; these very fast fluctuations relate to the capacity of equalization of the receiver. In spite of such extreme conditions of propagation, which occurred practically every night for continuous periods of several hours, the performance of the link remained satisfactory and complied with performance objectives. However, in order to face depressions of the median value of the field that could reach 30 dB, power amplifiers were installed to increase the margin to threshold.

3.2.9.2 Case of Long-Haul Link over Sea in Temperate Climate Figure 3.28 presents the profile of an 80-km link comprising a vast zone of reflection on the sea that was established in temperate climate ($P_L = 5\%$).

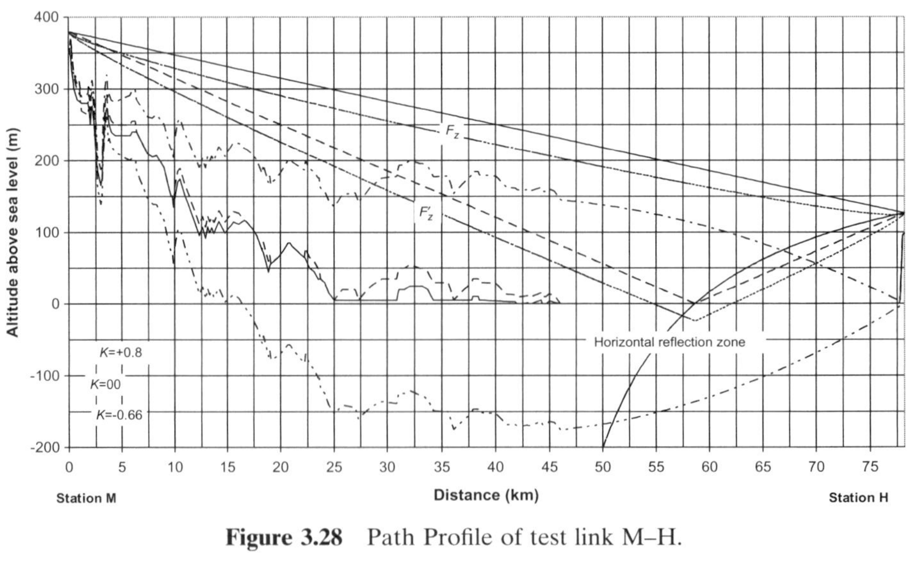

Figure 3.28 Path Profile of test link M–H.

Figure 3.29 shows the variations of the combined field that has been computed for station M using the interferometric method for a space diversity of 7 m, corresponding to one interference half fringe only, for all possible values of the vertical refractivity gradient.

Figure 3.30 shows the result of the recording of the instantaneous fields received on the main and diversity antennas during a period of 40 min; notice that the main and diversity fields are practically correlated in opposition.

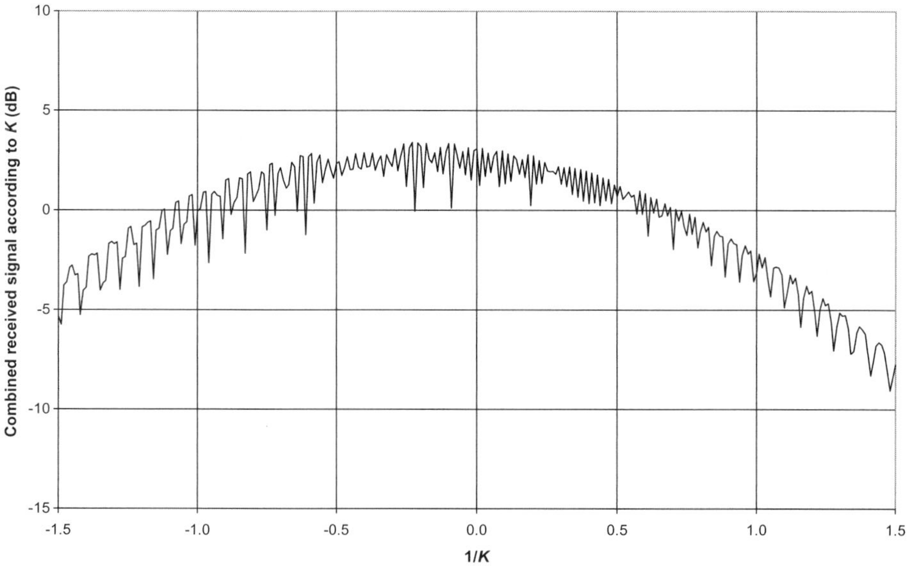

Figure 3.29 Computed combined received field (H → M).

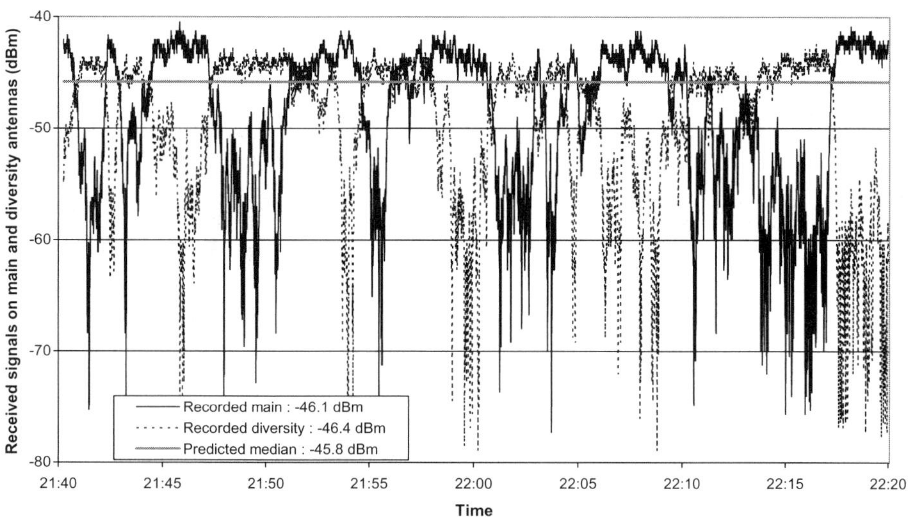

Figure 3.30 Individual received fields in space diversity (October 8, 2003).

Figure 3.31 shows the instantaneous combined field obtained by selection of the strongest signal from the recording; this result, which presents remarkable stability, fits in with the computed value by interferometry presented in Figure 3.29, and the difference of ~2 dB between the individual and combined medians is close to the combination gain by selection given later in Figure 3.57.

Figure 3.32 presents the variation speeds observed on both received fields that correspond to alternate periods; it is seen that when one of the fields

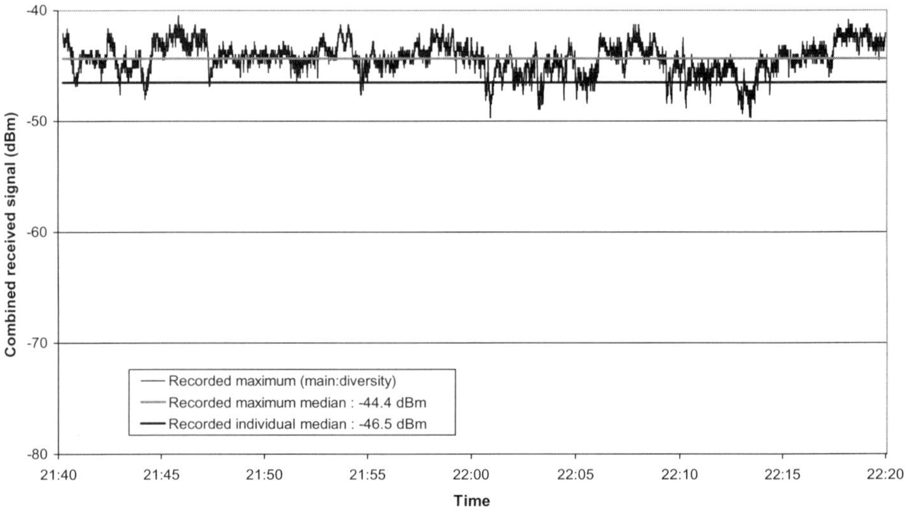

Figure 3.31 Combined received field in space diversity (October 8, 2003).

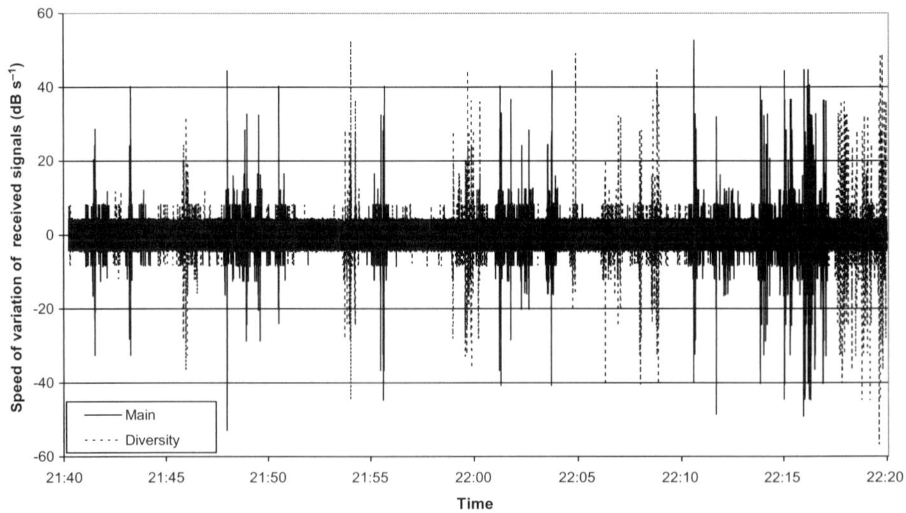

Figure 3.32 Speed of variation of received fields (October 8, 2003).

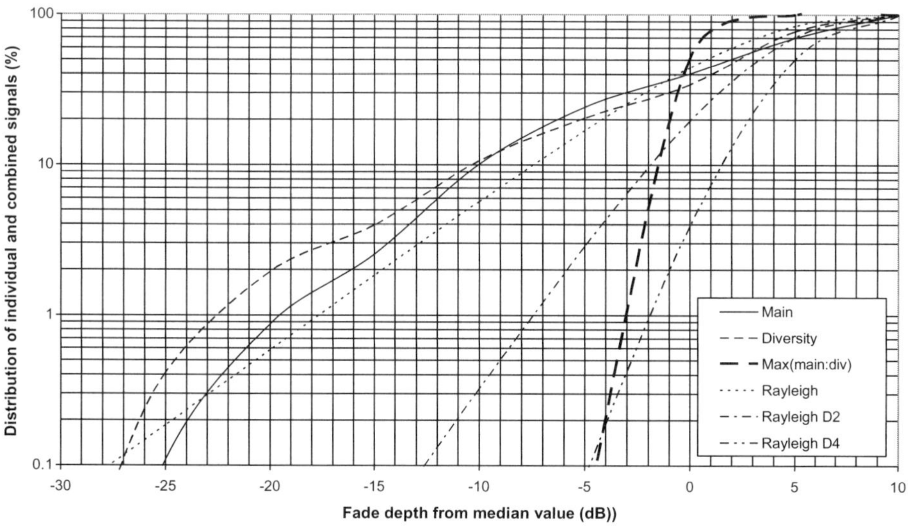

Figure 3.33 Distribution of received fields compared to theoretical limits (October 8, 2003–21:40 to 22:20).

strongly varies, the other remains stable, which thus makes it possible to obtain an optimal response of the equipment at reception.

Figure 3.33 presents the distribution according to the percentage of time of:

- Individual received fields measured on main and diversity antennas compared to theoretical Rayleigh's distribution
- Combined field obtained by selection of strongest signal [Max (main: div)]
- Combined fields calculated for diversity of order 2 [Rayleigh D2] and order 4 [Rayleigh D4] by considering a null coefficient of correlation

The combined field by selection of the strongest signal [Max (main:div)] offers a slope of 1.5 dB per decade, which is much more abrupt than that which corresponds to a double diversity (5 dB per decade) or even to a quadruple diversity (2.5 dB per decade). We can estimate, at first approximation, the diversity gain which corresponds to this distribution by comparing the minimum level recorded for the combined field to that of the individual fields; for example, for 0.1% of time,

$$\text{Diversity gain} \approx 25\,\text{dB} - 5\,\text{dB} = 20\,\text{dB}$$

which results in a very high improvement factor which cannot be measured but is evaluated by extrapolation as higher than 100,000.

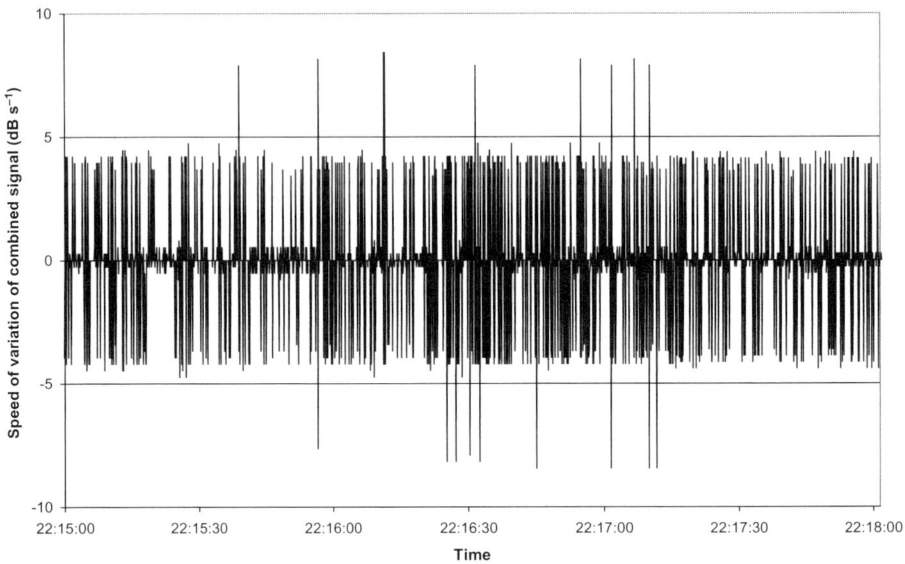

Figure 3.34 Speed of variation of combined received field.

Figure 3.34 shows the speed of variation of the combined field over a period of 3 min that is about five times less than that of the individual fields illustrated by Figure 3.32. By comparing these results with those obtained previously on the terrestrial link in Section 3.2.9.1, we deduces the following:

1. With a variation of about 10 dB per decade, the individual distribution of the received fields which results from either multipath or a reflection on the sea obeys Rayleigh's law very well
2. Using the interferometric method, the combined field presents the following:
   - In the case of multipath, a distribution approaching that which corresponds to a double diversity with a null coefficient of correlation
   - In the case of a reflection on the sea, a distribution approaching a quadruple diversity; in other words, the combination by correlation in opposition of two random signals coming from the same source and received in space diversity after a reflection would result in a signal quasi-determinist

3.2.10 Antenna Discrimination

We saw in Section 1.8.5, that, because of the variation of the vertical refractivity gradient of the propagation medium, the direct and reflected launch and arrival angles of the rays strongly vary. Launch and arrival angles of the direct and reflected rays are represented according to the factor K in Figure 3.8 in the example given in Section 3.2.8, and the associated field strengths, because of the cumulative discrimination by both emission and reception antennas, are presented in Figure 3.9.

We noticed that, in conditions of superrefraction, the level of the reflected ray can be close to or even higher than that of the direct ray; it may result in an interruption of the link if the differential delay is in addition higher than the acceptable value by the receiver which is determined by its characteristic signature, as we will show.

It appears that a positive tilt of both the main and diversity antennas increases to a significant degree the discrimination of the direct signal relative to the reflected one and that, in contrast, a negative tilt can only affect the link by increasing the fade depth and speed. In addition, it is often necessary to use high-gain antennas on long-haul links, which are consequently very directive, and the level of the direct ray, because of variation of launch and arrival angles, can be strongly affected; thus angle diversity can result in a significant improvement by offering an apparent broader combined beamwidth.

A solution to this problem consists of slightly tilting the antennas upward in order to reduce the level of the reflected signal; however, as the direct ray may be degraded at the same time, this solution is in general applied to diversity antennas with the result presented in Figure 3.35 for a tilt in elevation of 0.4° upward.

In such conditions, the direct ray is always stronger than the reflected one.

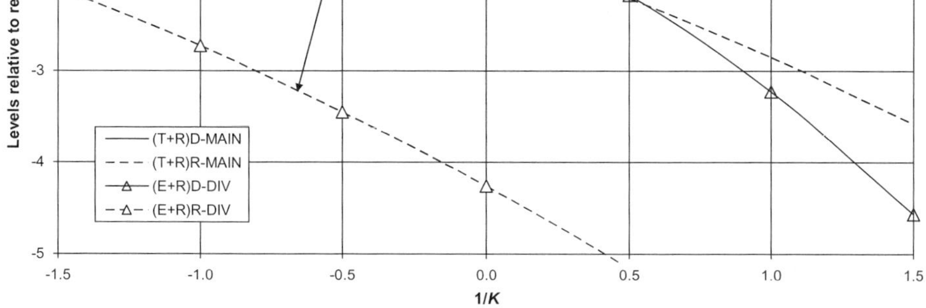

Figure 3.35 Improvement in discrimination of direct and reflected rays by positive tilt in elevation of antennas: horizontal reflection on link A towards B; cumulated discrimination of transmission and reception antennas (Diameter 2.4 m).

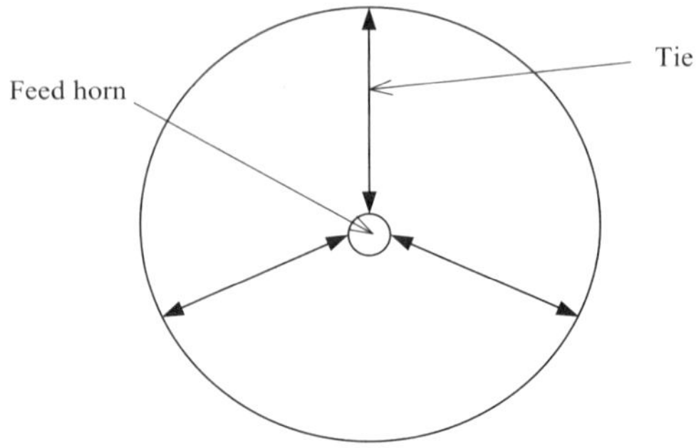

Figure 3.36 Adjustment of tie length.

3.2.11 Pointing of Antennas

Precision in pointing the antennas is very important in the final quality of the microwave links since their directivity is sufficient to significantly reduce the effects due to reflections on the path and to optimize the conditions of reception with regard to the refractivity changes of the atmosphere, multipath, and ducting.

3.2.11.1 Centering of Source of Illumination This operation must be carried out before hoisting the antenna when the reflector is in a vertical position and in the required polarization, as shown in Figure 3.36, by means of ties; tolerance of about one millimeter over the length of the ties makes it possible to reduce the difference between the radioelectric axis and the axis of revolution to less than 0.1°.

3.2.11.2 Prepointing in Azimuth First, after the antenna is installed in the tower, it can be prepointed by a compass or, to increase the precision, using a theodolite; the high directivity, presented in Figure 1.12, of antennas generally used in long-haul links of the infrastructure requires a certain precision for this operation in order to avoid important wastes of time during their implementation by quickly obtaining an exploitable radioelectric signal for the final pointing operation.

The operation consists in being placed, as shown in Figure 3.37, in such a way that the antenna is seen in a plane perpendicular to the azimuth at a certain distance from the tower. If a compass is used, the azimuth angles are measured with respect to magnetic north considering correction of the magnetic declination whose value is reproduced on geological survey maps; moreover, it is necessary to move away from metal masses such as the tower, electric power lines, or underground structures liable to create a magnetic deviation.[3]

[3]The magnetic declination and its annual variation can be raised on geological survey maps with a precision of about 0.1° or on a chart of equal declination curves with a resolution of 5′ and the local magnetic deviation can be measured by aiming at a reference mark whose layer is known; the measurement error made with a precision compass can thus, by taking all the precautions necessary, be reduced to less than 1°.

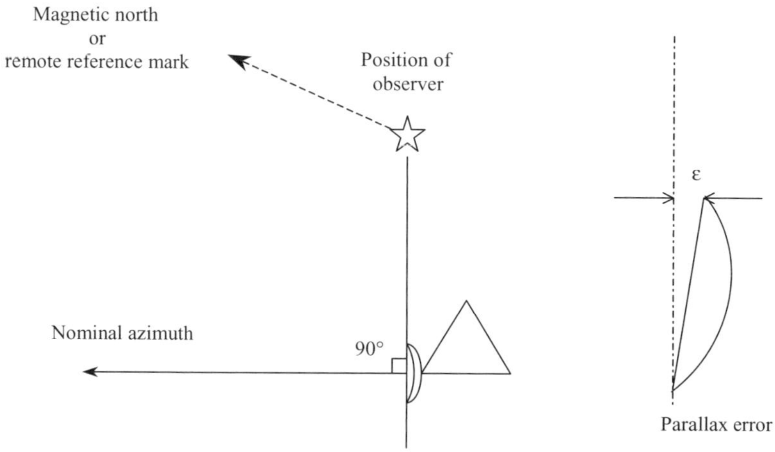

Figure 3.37 Prepointing using precision compass or theodolite.[4]

Second, it is necessary to install the theodolite at a site at the same time aiming the remote reference mark, whose direction is used as a reference compared to true north, and the antenna.

3.2.11.3 Final Pointing

The final pointing is carried out in three phases:

- Presetting in elevation
- Pointing in azimuth
- Pointing in elevation

Presetting in Elevation We proceed to this preliminary operation using an inclinometer, as shown in Figure 3.38, which is applied to the back ring that constitutes by construction the reference to the axis of revolution of the antenna; this device makes it possible to adjust the elevation angle of the antenna with a precision which can reach ±0.1°. The antenna can be thus prepointed according to the slope of the path, calculated using the relation

$$\varepsilon = \frac{H_r - H_e}{d} \quad \text{(mrad)} \tag{3.9}$$

[4]In the event a theodolite is used, the maximum error which can be made with, for example, an indetermination of 1″ of arc on the geographical coordinates, i.e., approximately 30 m in latitude and longitude, and a distance of about 10 km between the station and the reference mark is on the order of 0.25°. When an antenna is observed that way, it is important to see the two sides as shown in the figure; for instance, a parallax of 1 cm in a 3-m antenna would give an error of

$$e \approx \frac{\varepsilon}{D} = \frac{0.01\,\text{m}}{3\,\text{m}} = 3.33\,\text{mrad} \approx 0.2^\circ \quad (1^\circ \approx 17.45\,\text{mrad})$$

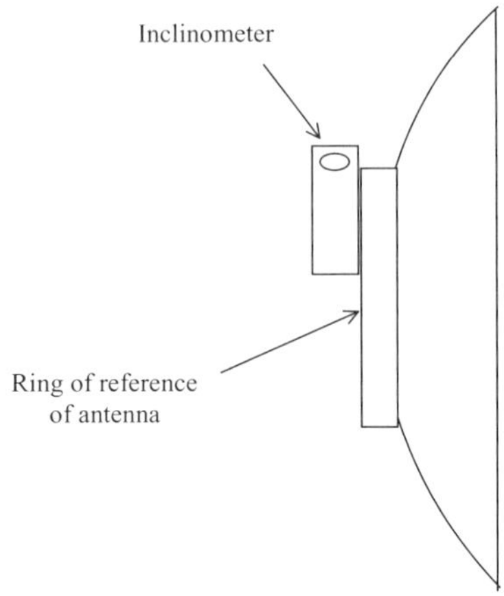

Figure 3.38 Presetting elevation angle of antenna.

where H_e = altitude of transmitting antenna (m)
H_r = altitude of receiving antenna (m)
d = pathlength (km)

Final Pointing in Azimuth When the connection is established, the best way for proceeding to the final adjustment consists in using the symmetry of the antenna pattern,[5] as shown in Figure 3.39. We proceed in the following way for each direction of transmission:

- With the transmission antenna remaining fixed, the reception antenna is depointed in a direction, called position 1, such that the field drops by a certain value (10 dB, e.g., in order to not fall on a secondary lobe).
- This position is located by means of a sufficiently precise mechanical indicator
- The antenna is swiveled in the other direction until the field level reaches the same value as previously, called position 2.
- The average position between positions 1 and 2 that corresponds to the optimal position which coincides with the nominal azimuth of the path.

Final Pointing in Elevation The method of final pointing in elevation is practically the same one as in azimuth; however, as the propagation medium is not

[5]The traditional method of pointing on the maximum of received field can induce an important error insofar as the top of the antenna radiation pattern is relatively flat and as the fluctuations of the field level due to the propagation may be of the same order of magnitude as the variations related to the depointed positions of the antenna. In the method suggested here, the variations of the signal related to the depointing of the antenna on the respective positions 1 and 2 are significantly higher and the fluctuations due to the propagation involve only a tiny error since they refer to the abrupt and symmetrical zones of the radiation pattern.

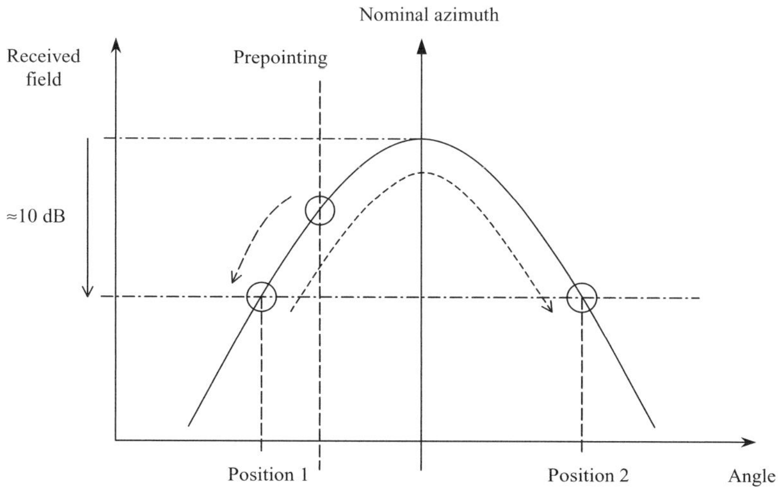

Figure 3.39 Method of pointing using symmetry of antenna radiation pattern.

symmetrical because of the reflection on the ground, we will take care that the final position of the antenna is coherent with the computed value. Moreover, the maximum received field may correspond to the sum of the direct and reflected rays, which can only harm the required discrimination of the reflected ray at the time of the operation of adjustment in elevation.

The antenna of diversity—in the case of space diversity or quadruple diversity of space and frequency—will be slightly tilted upward of the value defined for angle diversity, as we saw in Section 3.2.10, which aims to reduce the influence of the rays reflected by the ground or the low layers of the atmosphere and to take account of the extreme variation of the launch and arrival angles of the direct rays under the conditions of ducting propagation. In all cases, a positive tilt in elevation of both main and diversity antennas is preferable to a negative tilt toward the ground.

Checking Plane of Polarization The verticality or horizontality of the feed horn must be ensured with a precision of about 1° in order to not degrade the intrinsic values for isolation and decoupling of polarization.

3.2.11.4 Effect of Misalignment Figure 3.40 presents the radiating main lobe of an antenna where the transmission axis is off the antenna axis. We can see that, in addition to the reduction in gain of the antenna relative to its maximum value on the nominal axis, the apparent radiation pattern is also reduced; in such conditions of transmission, the variation of the received field due to the variation of the launch and arrival angles according to changes of the vertical refractivity gradient will be amplified and cause a certain instability of the link with faster and deeper fading.

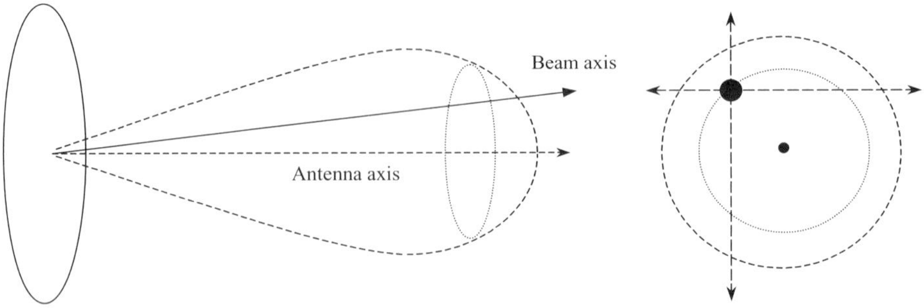

Figure 3.40 Misaligned antenna.

3.2.11.5 Behavior in Extreme Climatic Conditions The stability of the structure which receives the antennas is very important as it must be able to support not only their weight but also the overloads which can occur in the event of bad weather conditions (snow, ice, wind). As the microwave stations of telecommunications are generally located on exposed sites, it is necessary to take account of the conditions of most extreme snow and wind that occur in the area. Generally, the stresses of the structure in twist and sway should not generate deformations such as a loss of gain due to the misalignment of the antenna that is higher than 3 dB; a larger tolerance can however be allowed if the margin to the threshold is sufficiently important to compensate for the other causes of path loss which can intervene simultaneously. The total deflection angle, $\Delta\alpha$, of the antenna, which corresponds to the average quadratic value of the deflection angles in azimuth and elevation due to the structural deformations, should not thus exceed half of the total half-power beamwidth, whose value is given by relation (1.38), which results in the expression

$$\Delta\alpha = \pm\sqrt{(\Delta\alpha_{\text{Azimuth}})^2 + (\Delta\alpha_{\text{Elevation}})^2} \leq \pm\frac{1}{2}\alpha_T$$

3.3 LINK BUDGET

3.3.1 Introduction

As we saw in Sections 3.1 and 3.2, the following should be considered in order to optimize the performance for all conditions of propagation which can arise:

- Manner of carrying out the path profile of the link
- Rules of minimum clearance to be respected with regard to obstacles
- Method to apply in order to face problems of reflection

The object of this section is, on the one hand, to define the objectives of performance in terms of quality and availability and, on the other hand, to determine the characteristics of the equipment to implement in order to reach them taking into account the economic and technical constraints. However, the ITU recommendations refer to the hypothetical digital reference circuit (HDRC) and the hypothetical digital reference path (HDRP) with the intention, as a guide to designers and planners, of establishing equipment design performance objectives that are compatible with the requirements of the network: "It is not intended that it will be quoted in specifications of real systems, for acceptance tests, or for operational agreements" (ITU-R F.557 and ITU-R Rec.594).

3.3.2 Objectives of Performance

We will be interested here in an international digital connection forming part of an ISDN whose parameters and objectives of performances are mainly specified by ITU-T G.821 when operating below the primary rate (n 64 kbits s^{-1} with $1 \leq n \leq 24$ or 31) and ITU-T G.826 at or above the primary rate (n 64 kbits s^{-1} to 3500 Mbits s^{-1}).

3.3.2.1 Recommendations ITU-T G.821 and Associated ITU-R Recommendation ITU-T G.821 defines three distinct quality grades for a connection having an overall length of 27,500 km; Figure 3.41 shows the distribution of the various sections. The objectives of performances that are allocated to the various sections are summarized in the table below; the rule of proportionality applies in theory to distances ranging between 280 and 2500 km but can be extended to shorter lengths. The error performance refers to any month (also called in the past the most unfavorable month) and are based on the following events and parameters:

- *Errored second (ES)*: a 1-s period in which at least an error occurred or during which a loss of signal (LOS) or an alarm indication signal (AIS) is detected
- *Severely errored second (SES)*: a 1-s period which has BER $\geq 10^{-3}$ or during which a LOS or an AIS is detected

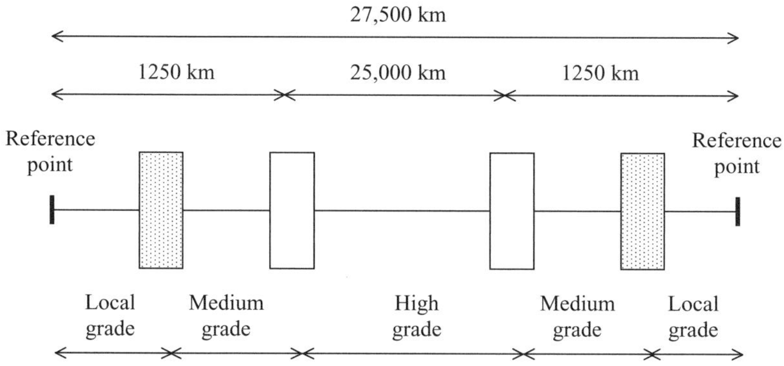

Figure 3.41 Distribution of sections according to ITU-T G.821.

- *Errored second ratio* (*ESR*): the ratio of ES to total seconds in available time during a fixed measurement interval
- *Severely errored second ratio* (*SESR*): the ratio of SES to total seconds in available time during a fixed measurement interval

		High-Quality Grade HDRC, $n \times 64\,\text{kbit s}^{-1}$ (40%)		Medium- and Local-Quality Grade $2 \times [(2 \times 15\%)$ $(2 \times 1250\,\text{km})]$	
Performance Objectives	ITU-T G.821 HDRC, $n \times 64\,\text{kbit s}^{-1}$ (27,500 km)	ITU-T G.821 (25,000 km)	ITU-R F.634, F.594, F.557, F.695 (2500 km)	ITU-T G.821	ITU-R F.696, F.697, F.557
Errored seconds	0.08 8%	0.032 3.2%	0.32% $0.32 \times (L/2500)\%$[a] (any month)	$2 \times (0.012 + 0.012)$ 4.8%	Medium 2.4%, local 2.4%
Severely errored seconds ($BER > 10^{-3}$)	0.002 0.2% (0.1% + 0.1%)[b]	0.0004 0.04% (+0.05%)[b]	0.054% $0.054 \times (L/2500)\%$[a] (any month)	$2 \times (0.00015 + 0.00015)$ 0.06% (+0.05%)[b]	Medium 0.08%, local 0.03%
Unavailability $BER > 10^{-3}$ (10 s consecutives)			$0.3 \times (L/2.500)\%$[a] (average year)		0.2–0.5%

[a]Length of section, L, ranges between 280 and 2500 km but shorter distances can be considered.
[b]Tolerance for unfavorable conditions of propagation on section of 2500 km HDRP for radio relay systems.

Figure 3.42 presents the main parameters values according to the high grade circuit length.

The total allocation of 0.002 SESR is subdivided into each circuit classification as follows:

(a) 0.001 SESR is divided between:
- High grade, 0.00015
- Medium grade, 0.00015
- Local grade, 0.0004

For a satellite HDRP operating in the high-grade portion there is a block allowance of 0.0002 SESR

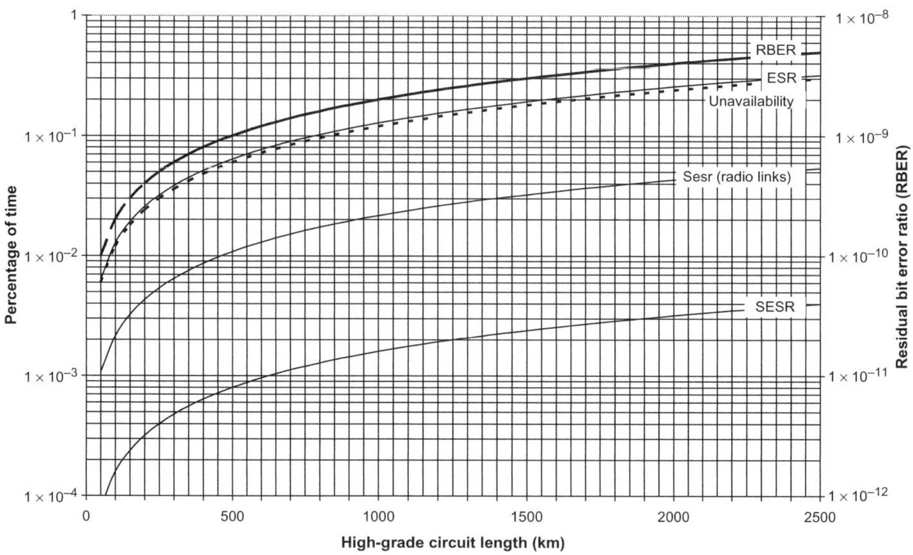

Figure 3.42 Performance objectives for high-grade circuit (2500 km).

(b) 0.001 SESR is a block allowance to both medium- and high-grade classifications to accommodate the occurrence of adverse network conditions occasionally experienced on transmission systems during the worst month of the year.

Because of the statistical nature of the occurrence of worst-month effects in a worldwide connection, it is considered that the following allowances are consistent with the total 0.001 SESR figure:

- 0.0005 SESR to a 2500-km HDRP for radio relay systems that can be used in high- and medium-grade portions of the connection
- 0.0001 SESR to a satellite HDRP

The availability objectives for radio relay systems refer to a period of observation much longer, usually one year or more, and have as a finality to take account of many factors, such as the reliability of the various equipments, the interferences, the propagation conditions, and the maintenance actions. They are summarized as follows for a bidirectional connection:

- A period of unavailable time begins at the onset of 10 consecutive SESs, in either one or both directions; this 10 s is considered to be part of unavailable time.
- A new period of available time begins at the onset of 10 consecutive non-SES events in either one or both directions; this 10 s is considered to be part of available time.

According to ITU-R F.557, the overall bidirectional availability objective appropriate to a 2500-km HDRP should be 99.7% of the year as a provisional value, but the objectives selected may fall in the range 99.5–99.9%. Recommendation ITU-R F.695 defines an availability objective for radio relay links forming part of a high-grade section which is given by the formula

$$A = 100 - \left(0.3 \times \frac{L}{2500}\right)\%$$

where L is between 280 and 2500 km. For medium and local grades, an unavailability objective is given respectively by ITU-R F.696 and ITU-R F.697 within the range of 0.2–0.5%. The unavailability allowance is usually equally shared between propagation, equipment and maintenance. A useful parameter for real digital radio relay links forming part of a high-grade circuit within an ISDN, called residual bit error ratio (RBER), was given by former ITU-R Rec.634 according to the formula

$$\text{RBER} \leq \frac{L \times 5 \times 10^{-9}}{2500}$$

3.3.2.2 Recommendations ITU-T G.826 and ITU-T G.827 Recommendation ITU-T G.826 comprises a national section at each end and an international portion belonging to four assumed intermediate countries as presented in Figure 3.43. The boundary between the national and international portions is defined to be an international gateway (IG), which may correspond to a cross-connect, a higher order multiplexer, or a switch. To each national portion is allocated a fixed block allowance of 17.5% of the end-to-end performance objectives by affecting moreover a multiplicative factor to it according to the air route distance between the path endpoint (PEP) and the IG as follows:

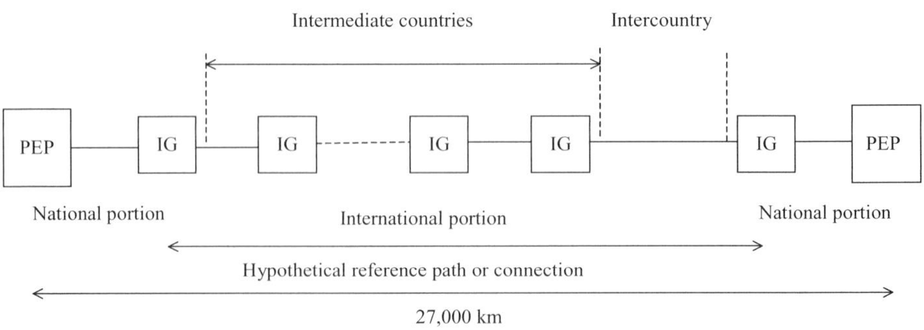

Figure 3.43 Hypothetical reference path or connection.

- If <1000 km, the routing factor is 1.5.
- If ≥1000 km and ≤1200 km, the route length is taken to be 1500 km.
- If ≥1200 km, the routing factor is 1.25.

The routing factor applies also to the international portion between consecutive IGs.

Recommendation ITU-T G.826, which can be related to ITU-R F.1092 for the international radio relay systems and ITU-R F.1189 for the national ones, specifies the end-to-end performance objectives for a 27,500-km HDRC or HDRP by using the following events and parameters:

- *Errored block* (*EB*): a block in which one or more bits are in error
- *Errored second* (*ES*): a 1-s period in which:
  (a) Occur one or more errored blocks or at least one defect for paths
  (b) One or more bits are in error or LOS or AIS is detected for connections
- *Severely errored second* (*SES*): a 1-s period which:
  (a) Contains ≥30% EBs or at least one defect for paths, SES being a subset of ES
  (b) Has BER $\geq 10^{-3}$ or during which LOS or AIS is detected for connections
- *Background block error* (*BBE*): an errored block not occurring as part of an *SES*
- *Errored second ratio* (*ESR*): the ratio of ES to total seconds in available time during a fixed measurement interval applicable to both paths and connections
- *Severely errored second ratio* (*SESR*): the ratio of SES to total seconds in available time during a fixed measurement interval applicable to both paths and connections
- *Background block error ratio* (*BBER*): the ratio of BBE to total blocks in available time during a fixed measurement interval, excluding all blocks during SES in the count of total blocks, applicable only to paths

Periods of consecutive SESs persisting for T seconds ($2 \leq T < 10$) may be considered as periods of unavailability if these events have a severe impact on service.

The table below specifies the end-to-end performance objectives for a 27,500-km HDRC or HDRP according to the data rates that are applicable to each direction over a suggested evaluation period of one month:

	$n \times 64\,\text{kbits s}^{-1}$	1.5–$5\,\text{Mbits s}^{-1}$	>5–$15\,\text{Mbits s}^{-1}$	>15–$55\,\text{Mbits s}^{-1}$	>55–$160\,\text{Mbits s}^{-1}$	>160–$3500\,\text{Mbits s}^{-1}$
Bits per block	NA	800–5000	2000–8000	4000–20,000	6000–20,000	15,000–30,000[a]
ESR	0.04	0.04	0.05	0.075	0.16	(—)[b]
SESR	0.002	0.002	0.002	0.002	0.002	0.002
BBER	NA	2×10^{-4}	2×10^{-4}	2×10^{-4}	2×10^{-4}	10^{-4}

[a]As currently defined, VC-4-4c is a 601-Mbits s^{-1} path with a block size of 75,168 bits per block and a BBER of 4×10^{-4}.
[b]The ESR objectives tend to lose significance for applications at high bit rates above 160 Mbits s^{-1}.

As a comparison, for the high-grade international portion of 25,000 km, the two recommendations allocate the following objective for the SESR:

- ITU-T G.821: 0.4×0.001, which is equivalent to 35 SES / 24 h averaged over one month
- ITU-T G.826: 0.6×0.002, which is equivalent to 101 SES / 24 h averaged over one month

The availability performance parameters and objectives for end-to-end international constant-bit-rate digital paths are given by ITU-T G.827, coupled with ITU-R F.1492 and ITU-R F.1493 for respectively international and national portions. Three performance level categories are defined:

(a) Standard-priority level
(b) High-priority level
(c) Pre-emptible priority level

The following formula specifies the length categories in 100-km intervals:

$$100(i-1) \le L < 100i \qquad i = 1, 2, \ldots$$

The routing factor is as follows:

- If <1000 km, the routing factor is 1.5.
- If ≥1000 km and ≤1200 km, the route length is 1500 km.
- if ≥1200 km, the routing factor is 1.25.

Each category is associated with an integer variable k which is used in the following formulas to determine the availability performance objectives for a path element (PE) of length L:

$$AR_{jS} = 1-(A_{jS} + k \times X_{jS}) \quad \text{for standard priority}$$
$$AR_{jH} = 1-(A_{jH} + k \times X_{jH}) \quad \text{for high priority}$$

where

$$j = \begin{cases} 1 & \text{PE length} < 2500\,\text{km} \\ 2 & \text{for } 2500\,\text{km} < \text{PE length} < 5000\,\text{km} \\ 3 & \text{for } 5000\,\text{km} < \text{PE length} < 7500\,\text{km} \\ 4 & \text{for PE length} > 7500\,\text{km} \end{cases}$$

and *k*, *A*, and *X* are given by the following tables:

km	0	100	200	300	400	500	600	700	800	900
0		1	2	3	4	5	6	7	8	9
1,000	10	11	12	13	14	15	16	17	18	19
2,000	20	21	22	23	24	25	1	2	3	4
3,000	5	6	7	8	9	10	11	12	13	14
4,000	15	16	17	18	19	20	21	22	23	24
5,000	25	1	2	3	4	5	6	7	8	9
6,000	10	11	12	13	14	15	16	17	18	19
7,000	20	21	22	23	24	25	1	2	3	4
8,000	5	6	7	8	9	10	11	12	13	14
9,000	15	16	17	18	19	20	21	22	23	24
10,000	25	26	27	28	29	30	31	32	33	34

		Length < 2500 km		2500–5000 km		5000–7500 km		≥ 7500 km	
Path Element[a]	Level	A_1	X_1	A_2	X_2	A_3	X_3	A_4	X_4
IPCE	Standard	0	3	75	4	150	5	250	5
	High	0	0.6	15	0.8	30	1	50	1
NPE	Standard	0	5	100	8	275	5	375	5
	High	0	0.8	20	1.6	55	1	75	1
ICPCE	Standard	0	25	100	40	275	25	375	25
	High	0	0.8	20	1.6	55	1	75	1

Note: Satellite links may be deployed in any single path element or any contiguous combination of them.

[a]IPCE, international path core element; NPE, national path element; ICPCE, intercountry path core element.

The duration of unavailable time must be greater than 10 s in order to meet the definition of unavailability and the observation period is recommended to be one year. The end-to-end performance objectives for a 27,500-km HDRP at or above the primary rate are as follows.

	Data Rate 1.5–40 Mbit s^{-1}	
	Availability Ratio (AR)	Outage Intensity (OI)
High priority	98%	70
Standard priority	91%	250
Preemptive priority	FFS	FFS

Note: Values for OI are based on MTTR of 4 h.

The PE outage intensity objectives are determined as

$$\mathrm{OI}_{jS} = B_{jS} + k \times Y_{jS} \quad \text{for standard priority}$$

$$\mathrm{OI}_{jH} = B_{jH} + k \times Y_{jH} \quad \text{for high priority}$$

where j and k are given above and B and Y are as follows:

		Length < 2500 km		Length 2500–5000 km		Length 5000–7500 km		Length ≥ 7500 km	
Path Element	Level	B_1	Y_1	B_2	Y_2	B_3	Y_3	B_4	Y_4
IPCE	Standard	4	0.6	14	1	35	1.4	65	2
	High	1	0.2	3	0.3	8	0.4	15	0.6
NPE	Standard	5	0.6	18	1	40	1.6	75	2
	High	1	0.4	4	0.4	10	0.4	16	0.6
ICPCE	Standard	5	0.6	18	1	40	1.6	75	2
	High	1	0.4	4	0.4	10	10.4	16	0.6

The OI objectives are based on a MTTR value of 4 h; for end-to-end calculations, it is more convenient to use the outage intensity than the mean time between service outage. It is recognized that some events due to anomalous working conditions, such as propagation impairments for radio applications, can give rise to self-healing unavailability events; such events, usually much shorter, are not taken into account.

3.3.2.3 Recommendation ITU-T G.828 Recommendation ITU-T G.828 defines end-to-end error performance parameters and objectives for international constant-bit-rate synchronous digital paths in accordance with SDH technology. The following gives the end-to-end objectives for an HDRP of 27,500 km over any month:

Bit Rate (kbit s^{-1})	Type of Connection	Block s^{-1}	ESR	SESR	BBER	SEPI
1,664	VC-11, TC-11	2,000	0.01	0.002	5×10^{-5}	
2,240	VC-12, TC-12	2,000	0.01	0.002	5×10^{-5}	
6,848	VC-2, TC-2	2,000	0.01	0.002	5×10^{-5}	
48,960	VC-3, TC-3	8,000	0.02	0.002	5×10^{-5}	
150,336	VC-4, TC-4	8,000	0.04	0.002	10^{-4}	
601,344	VC-4-4c, TC-4-4c	8,000	—	0.002	10^{-4}	
2,405,376	VC-4-16c, TC-4-16c	8,000	—	0.002	10^{-4}	
9,621,504	VC-4-64c, TC-4-64c	8,000	—	0.002	10^{-3}	

Note: ESR objectives tend to lose significance for applications at high bit rates above 160 Mbits s^{-1}. SEPI, severely errored period intensity (objectives require further study).

3.3.3 Transmission Link Budget

It is a question here of drawing up the budget of the losses and gains which are met throughout the transmission chain and which make it possible to calculate the median power received as well as the margin to threshold; the threshold of the receiver is characterized by a level of signal power $T_{H(\mathrm{BER})}$ which corresponds in general to a BER of 10^{-6} or 10^{-3} due to thermal noise.

3.3.3.1 Median Received Power and Thermal Noise Margin The median power received by the receiver, expressed in decibels, is given by the relation

$$P_R = P_E - p_{BE} - p_{FE} + G_E - A_{EL} - A_a - A_{OB} + G_R - p_{FR} - p_{BR} \tag{3.10}$$

where P_R = received median power (dBW)
P_E = emitted power (dBW)
p_{BE} = transmitter branching loss (dB)
p_{FE} = feeder loss to transmission antenna (dB)
G_E = transmission antenna gain (dBi)
A_{FS} = free-space loss (dB)
A_a = atmospheric attenuation (dB)
A_{OB} = obstacle median diffraction loss (dB)
G_R = reception antenna gain (dBi)
p_{FR} = feeder loss to reception antenna (dB)
p_{BR} = receiver branching loss (dB)

The margin to the thermal threshold or margin against uniform fading, expressed in decibels, is given by the expression

$$M = P_R - T_{H(\mathrm{BER})} \tag{3.11}$$

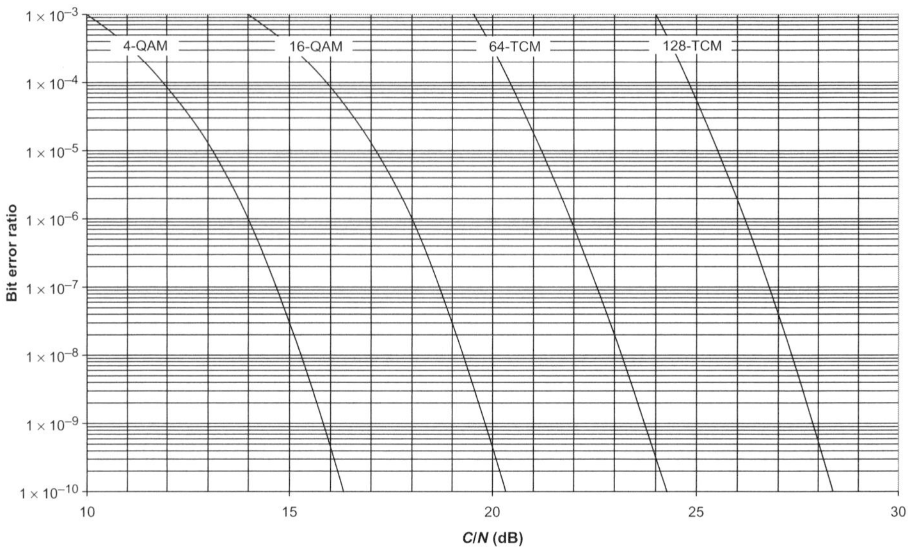

Figure 3.44 BER according to carrier-to-noise ratio for usual modulations: 4- and 16-QAM, 34 Mbits s^{-1}; 64- and 128-TCM, 155 Mbits s^{-1}.

In addition to the thermal threshold, the manufacturers also indicate another characteristic of the receiver, called selective margin, which refers to the capability of the equipment against the selective fading; this value is in general much higher than that of the calculated thermal threshold margin which one usually meets and does not intervene in this phase of calculation.

3.3.3.2 Receiver Thermal Threshold Figure 3.44 presents, as an example, the typical characteristics of the BER according to the carrier-to-noise ratio for four usual types of modulation which make it possible to define the corresponding thermal threshold levels for BER 10^{-3} and 10^{-6}:

- QAM differential coding with 4 and 16 amplitude–phase states at rate of 34 Mbits s^{-1} for PDH equipment of transmission
- TCM with 64 and 128 amplitude–phase trellis-coded states at bit rate of 155 Mbits s^{-1}, for SDH equipment of transmission

These characteristics can be given according to the carrier–to–noise ratio C/N or the bit energy–to–noise spectral density ratio E_b/N_0, which are bound by the following general relations depending on the modulator and demodulator efficiency:

$$\frac{E_b}{N_0} = \frac{C\theta}{N_0} = \frac{C}{DN_0} = \left(\frac{\mathrm{C}}{\mathrm{N}}\right)\frac{B}{D} \tag{3.12}$$

where E_b = energy per bit (J)
C = carrier received level (W)
N_0 = noise spectral power density (W Hz^{-1})
R = bit rate (bits s^{-1})
θ = bit duration, = $1/R$ (s)
B = receiver bandwidth (Hz)
N = total noise power, = $N_0 B$ (W)

The bandwidth occupied by the signal depends in general on the type of modulation and coding.

3.4 METHODS OF PREDICTION

The methods we will consider for the design of line-of-sight radio relay systems are given by the ITU-R P.530 series:

1. ITU-R P.530-8 (1999)
2. ITU-R P.530-9 (2001) to ITU-R P.530-12 (2007)

These methods have been developed on the basis of specific climate and topographical conditions in order to allow the prediction of some of the most important propagation parameters and mechanisms that can occur independently or in combination:

- Fading and enhancement due to atmospheric multipath or beam spreading (defocusing)
- Attenuation due to hydrometeors
- Reduction of cross-polarization discrimination in multipath or precipitation conditions

However, we should only apply them at the end of the study of the architecture of the link, discussed in preceding chapters, whose objective it is to determine the optimal height of the antennas in order to reduce as far as possible the effects due to diffraction, reflection, and ducting.

3.4.1 Recommendation ITU-R P.530-8

This recommendation (1999) is mainly based on the climatic variable P_L, which corresponds to the percentage of time that the refractivity gradient in the lowest 100 m of the atmosphere is more negative than $-100\,\text{N}\,\text{km}^{-1}$ in the estimated average worst month, generating multipath conditions.

3.4.1.1 Geoclimatic Factor In the case of inland links, in which either the entire path profile is above 100 m altitude relative to sea level or part or all of the path is below 100 m altitude for a link entirely within 50 km of the coastline, the geoclimatic factor K is estimated, if the measured value is not available for the path location in question, for the average worst month using the relation

$$K = 5\times 10^{-7} \times 10^{-0.1(C_0 - C_{\text{lat}} - C_{\text{lon}})} P_L^{1.5} \tag{3.13}$$

where:

- Coefficient C_0, expressed in decibels, is related to the lowest antenna and type of terrain above which the link passes entirely or partially according to the table below.
- Coefficient C_{lat}, expressed in decibels, is linked to the latitude ξ by the relation

$$C_{\text{lat}} = \begin{cases} 0\,\text{dB} & \text{for } \xi \le 53\,°\text{N or }°\text{S} \\ -53 + \xi\,(\text{dB}) & \text{for } 53\,°\text{N or }°\text{S} \le \xi \le 60\,°\text{N or }°\text{S} \\ 7\,\text{dB} & \text{for } \xi \ge 60\,°\text{N or }°\text{S} \end{cases}$$

- Coefficient C_{Lon}, expressed in decibels, is linked to the longitude by the relation

$$C_{\text{lon}} = \begin{cases} 3\,\text{dB} & \text{for longitudes of Europe and Africa} \\ -3\,\text{dB} & \text{for longitudes of North and South America} \\ 0\,\text{dB} & \text{for all other longitudes} \end{cases}$$

- Climatic variable P_L, which is given in Figures 1.36–1.39 for the months of February, May, August, and November, from which one will retain the highest value for the worst month

Altitude of Lowest Antenna and Type of Path Terrain	C_0 (dB)
Low-altitude antenna (0–400 m)—plains: overland or partially overland links, with lower antenna altitude less than 400 m above mean sea level, located in largely plains areas	0
Low-altitude antenna (0–400 m)—hills: overland or partially overland links, with lower antenna altitude less than 400 m above mean sea level, located in largely hilly areas	3.5
Medium-altitude antenna (400–700 m)—plains: overland or partially overland links, with lower antenna altitude in range 400–700 m above mean sea level, located in largely plains areas	2.5
Medium-altitude antenna (400–700 m)—hills: overland or partially overland links, with lower antenna altitude in range 400–700 m above mean sea level, located in largely hilly areas	6

Altitude of Lowest Antenna and Type of Path Terrain	C_0 (dB)
High-altitude antenna (>700 m)—plains: overland or partially overland links, with lower antenna altitude more than 700 m above mean sea level, located in largely plains areas	5.5
High-altitude antenna (>700 m)—hills: overland or partially overland links, with lower antenna altitude more than 700 m above mean sea level, located in largely hilly areas	8
High-altitude antenna (>700 m)—mountains: overland or partially overland links, with lower antenna altitude more than 700 m above mean sea level, located in largely mountainous areas	10.5

In the case of coastal links passing over or near large bodies of water, we estimate the geoclimatic factor using the relations

$$K = \begin{cases} 10^{(1-r_c)\log K_i + r_c \log K_{cl}} & \text{for } K_{cl} \geq K_i \\ k_i & \text{for } K_{cl} < K_i \end{cases} \tag{3.14}$$

where r_c = fraction of path profile below 100 m altitude above mean level of body of water in question and within 50 km from coastline but without intervening height of land above 100 m

K_i = terrestrial component given by expression for K in equation (3.13)

K_{cl} = maritime component given by relation

$$K_{cl} = 2.3 \times 10^{-4} \times 10^{-(0.1C_0 - 0.011\xi)}$$

In the case of coastal links over or near medium-size bodies of water, K can be estimated from

$$K_G = \begin{cases} 10^{(1-r_c)\log K_i + r_c \log K_{cm}} & \text{for } K_{cm} \geq K_i \\ K_i & \text{for } K_{cm} < K_i \end{cases}$$

where

$$K_{cm} = 10^{0.5(\log K_i + \log K_{cl})}.$$

3.4.1.2 Path Inclination and Elevation Angle From antenna heights above sea level or some other reference, the magnitude of the path inclination in milliradians is calculatd from

$$|\varepsilon_P| = \frac{|H_r - H_e|}{d} \tag{3.15}$$

where H_e = transmitting antenna height (m)
H_r = receiving antenna height (m)
d = pathlength (km)

The elevation angle for the transmitting and receiving antennas is equal to the path inclination with corresponding sign.

3.4.1.3 Percentage of Time of Fade Depth We calculate the percentage of time p_w that fade depth A is exceeded in the average worst month from the relation[6]

$$p_w = Kd^{3.6} f^{0.89} (1 + |\varepsilon_P|)^{-1.4} \times 10^{-A/10} \tag{3.16}$$

where A is the fade depth in decibels and f the frequency in gigahertz with a lower limit f_{min} = 15/d.

3.4.1.4 Method for Various Percentages of Time The method given below for predicting the percentage of time of the average worst month that any fade depth is exceeded combines the deep-fading distribution given in the preceding section and an empirical interpolation procedure for shallow-fading down to 0 dB below the median value. Initially, we calculate the multipath occurrence factor p_0, that is, the intercept of the deep-fading distribution with the percentage-of-time axis using relation (3.16) for A = 0 dB:

$$p_0 = Kd^{3.6} f^{0.89} (1 + |\varepsilon_p|)^{-1.4} \tag{3.17}$$

Then, provided that $p_0 < 2000$, we calculate the value of fade depth A_t at which the transition occurs between the deep-fading distribution and the shallow-fading distribution as predicted by the empirical interpolation procedure:

$$A_t = 25 + 1.2 \log p_0 \tag{3.18}$$

1. If the considered fade depth A is equal to or greater than A_t, the percentage of time that A is exceeded is given by

$$p_w = p_0 \times 10^{-A/10} \tag{3.19}$$

which is equivalent to relation (3.16)

[6]Equation (3.16) was derived from fading data on paths with lengths in the range 7–95 km, frequencies in the range 2–37 GHz, path inclinations in the range 0–24 mrad, and grazing angles between 1 and 12 mrad. Checks using several other sets of data for paths up to 237 km in length and frequencies as low as 500 MHz suggest, however, that it is valid for larger ranges of pathlength and frequency.

2. If the considered fade depth A is less than A_t, we calculate the percentage of time p_t that A_t is exceeded using the relation

$$p_t = p_0 \times 10^{-A_t/10} \tag{3.20}$$

which is equivalent to relation (3.16) for $A = A_t$.

Next we calculate the value of parameter q'_a from the transition fade A_t and the corresponding transition percentage of time p_t with the relation

$$q'_a = -\frac{20\log\{-\ln[(100 - p_t)/100]\}}{A_t} \tag{3.21}$$

Then

$$q_t = \frac{q'_a - 2}{(1 + 0.3 \times 10^{-A_t/20}) \times 10^{-0.016 A_t}} - 4.3\left(10^{-A_t/20} + \frac{A_t}{800}\right) \tag{3.22}$$

Lastly, the percentage of time p_w that the considered fade depth A is exceeded is calculated as

$$p_w = 100\left[1 - \frac{1}{\exp(10^{-q_\alpha A/20})}\right] \tag{3.23}$$

with

$$q_a = 2 + (1 + 0.3 \times 10^{-A/20})\left[q_t + 4.3\left(10^{-A/20} + \frac{A}{800}\right)\right] \times 10^{-0.016A} \tag{3.24}$$

Figure 3.45 presents the distribution of the exceeded fade depth A according to the multipath occurrence factor p_0 as parameter.

3.4.1.5 Prediction of Nonselective and Selective Outages Outage is defined as the probability that the BER is higher than a given threshold.

The probability of outage P_{ns} due to the nonselective component of fading is given by

$$P_{\text{ns}} = \frac{p_w}{100} \tag{3.25}$$

where p_w, calculated with relation (3.19) or (3.23) as the case may be, is the percentage of time that the flat fade margin M corresponding to the specified

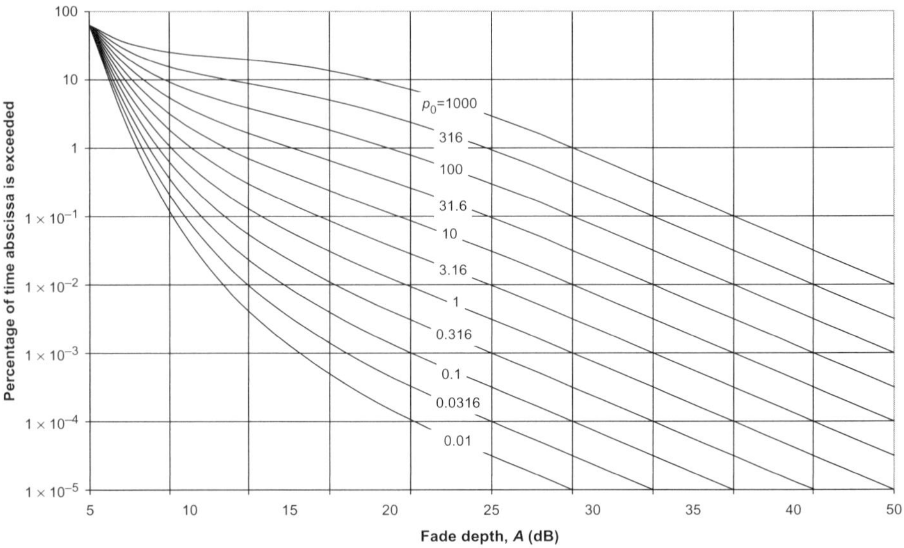

Figure 3.45 Percentage of time p_w that fade depth A is exceeded in an average worst month versus p_0.

BER, given by relation (3.11), is exceeded. The multipath activity parameter is given by the relation

$$\eta = 1 - \exp(-0.2P_0^{0.75}) \tag{3.26}$$

where $P_0 = p_0/100 = P_w \times 10^{A/10}/100$ is the multipath occurrence parameter.

3.4.1.6 Prediction Method for Enhancement It is necessary to predict the level that the enhancements can reach in multipath fading conditions in order to not exceed the zone of linearity of the receivers and to calculate the levels of interference that may affect the other links which share the same frequency and polarization plan; automatic transmitted power control (ATPC) is usually employed to reduce their effects.

The average worst-month enhancement above 10 dB is predicted using

$$p_w = 100 - 10^{(-1.7+0.2A_{0.01}-E)/3.5} \tag{3.27}$$

where E in decibels is the enhancement not exceeded for p percent of the time ($E < 10$ dB) and $A_{0.01}$ is the predicted deep fade exceeded for $p_w = 0.01\%$ of the time using

$$A_{0.01} = 10\log\left[\frac{Kd^{3.6}f^{0.89}(1+|\varepsilon_P|)^{-1.4}}{p_{w(0.01)}}\right] \tag{3.28}$$

For enhancement between 10 and 0 dB, we can use the following procedure:

1. Calculate the percentage of time p'_w with enhancement less than or equal to 10 dB ($E' = 10$ dB).
2. Calculate q'_e using

$$q'_e = -\frac{20}{E'}\log\left[-\ln\left(1-\frac{100-p'_w}{58.21}\right)\right] \tag{3.29}$$

3. Calculate q_s from

$$q_s = 2.05q'_e - 20.3 \tag{3.30}$$

4. Calculate q_e for the desired value of E using

$$q_e = 8+(1+0.3\times 10^{-E/20})\left[q_s+12\left(10^{-E/20}+\frac{E}{800}\right)\right]\times 10^{-0.035E} \tag{3.31}$$

5. Find the percentage of time that the enhancement E is not exceeded from

$$P_w = 58.21\left[1-\frac{1}{\exp(10^{-q_e E/20})}\right] \tag{3.32}$$

Figure 3.46 presents the distribution of the exceeded enhancement level E according to p_0.

Figure 3.47 shows, for example, the distribution of the received fields at 8 GHz in space diversity which have been recorded over a 5-day period in summer on a 28-km link located on the Persian Gulf coastline with a path inclination of 3.5 mrad. This recording shows that the distribution of the signals received simultaneously on both main and diversity antennas during a period favorable to ducting propagation is close to the model, except for the enhancement, which is more marked. For a climatic variable P_L of 90% the model reveals:

- Geoclimatic factor of 6×10^{-4}
- Multipath occurrence factor p_0 of 100

3.4.1.7 *Conversion from Average Worst Month to Average Annual Distributions* The fading and enhancement distributions for the average worst month can be converted to distributions for the average year[7] by employing the following procedure:

[7]Recommendation ITU-R P.841 also gives the conversion formulas that correspond to a generalization of relation (1.110) for various geoclimatic conditions of propagation.

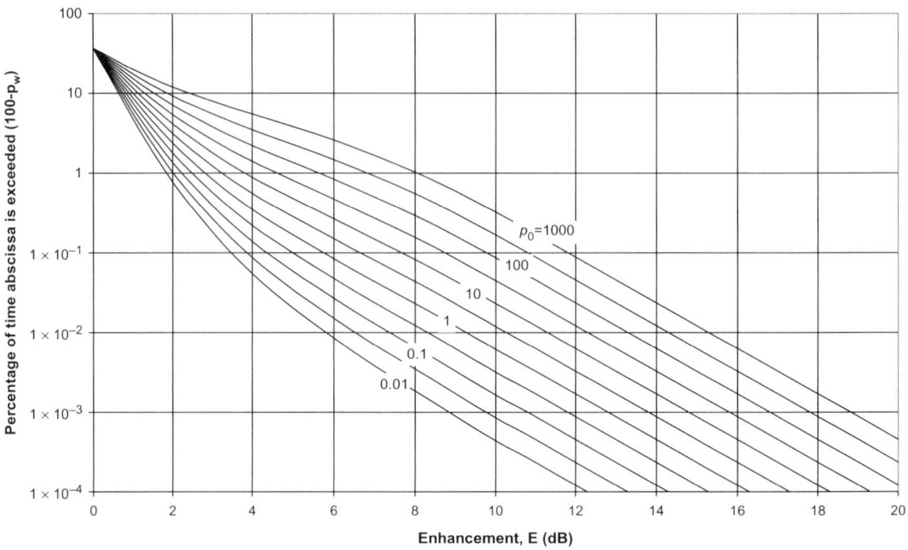

Figure 3.46 Percentage of time ($100\text{-}P_w$) that enhancement E is exceeded in an average worst month versus p_0.

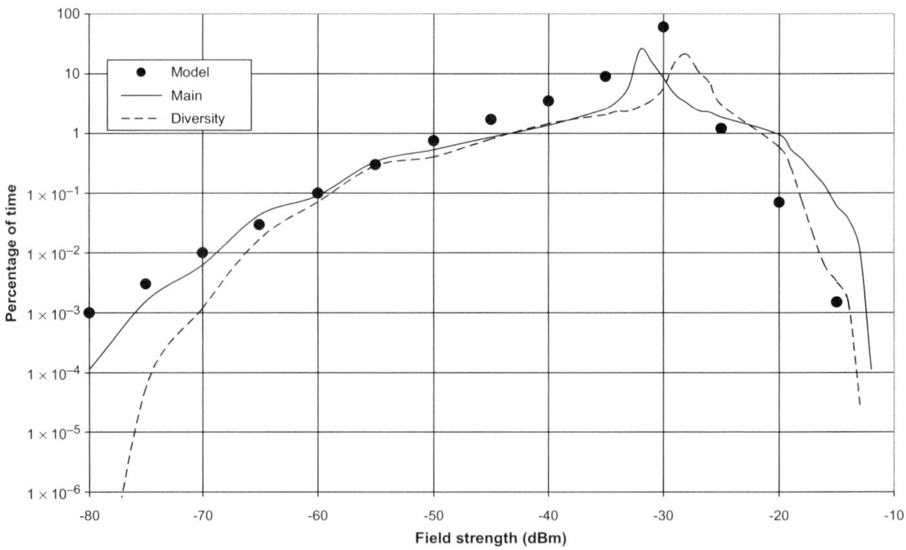

Figure 3.47 Example of field strength distribution on coastal link.

1. Calculate the percentage of time p_w that the fade depth A is exceeded in the large tail of the distribution for the average worst month from equation (3.16).
2. Calculate the geoclimatic conversion factor ΔG from

$$\Delta G = 10.5 - 5.6\log[1.1 \pm |\cos 2\xi|^{0.7}] - 2.7\log d + 1.7\log[1 + |\varepsilon_P|] \quad (3.33)$$

where $\Delta G \leq 10.8\,\text{dB}$, the $\pm$ sign is $+$ for $\xi \leq 45°$ and $-$ for $\xi > 45°$, ξ is the latitude (north or south), d is distance in kilometers, and ε_P is the inclination of the path in milliradians.

3. Calculate the percentage of time p that the fade depth is exceeded in the large fade depth tail of the distribution for the average year from

$$p = 10^{-\Delta G/10} \times p_w \quad (3.34)$$

4. Follow the method exposed in Section 3.4.1.4 replacing p_w by p in the relation (3.19) or by converting p_t to an annual value, using equation (3.34), in relation (3.21).
5. If it is required to predict the distribution of enhancement for the average year, follow the method of Section 3.4.1.6 where $A_{0.01}$ (dB) is now the fade depth exceeded for 0.01% of the time in the average year using the following procedure:
   - Calculate first p_w by inverting equation (3.34) and using $p = 0.01\%$:

$$p_{w(0.01)} = p \times 10^{\Delta G/10}$$

   - Then obtain fade depth $A_{0.01}$ exceeded for 0.01% of the time in the average year by inverting equation (3.16) and using p in place of p_w.

3.4.1.8 Prediction of Outage in Unprotected Digital Systems We saw in Section 2.3.4 that the signal can propagate between the transmitter and the receiver according to several paths that are the direct path and one or more other paths coming from a reflection on the ground or a refraction in the atmosphere and which present for that reason a certain time delay compared to the direct path. The main consequences of the superposition at the receiver of these delayed replicas are as follows:

- A severe waveform distortion and quadrature crosstalk effects
- An intersymbol or between-symbol interference, depending on whether the time delay is smaller than the symbol duration or not, which can lead to detection error

Development of digital communication systems has required an improved understanding of these effects and the means to overcome them by appropriate countermeasures such as adaptive channel equalization. The equalizer must be adaptively controlled to follow variations in transmission characteristics as propagation conditions vary, and the equalization techniques can be classified into two groups:

- Frequency domain equalization, which is designed to produce amplitude and group delay responses

- Time domain equalization, which attempts to combat intersymbol interference directly by correlating the interference that appears at the decision instant with the various adjacent symbols producing it and by adjusting tapped delay line networks to provide appropriate cancellation signals

On the other hand, the effects of relatively slow flat fading due to antenna decoupling or beam spreading, commonly referred to as defocusing, and faster frequency-selective fading due to multipath propagation must both be taken into account at the stage of link design. Number of countermeasures are available for alleviating these effects at the same time as the reduction in cross-polarization, such as a marked decrease in the effect of significant specular and diffuse surface reflections, as seen previously by the following:

- Taking advantage of obstacles along the path, such as hills, buildings, or vegetation, in order to shield the antennas from the reflection zone for all the range of effective Earth radius factor K values
- Adjusting the antenna height at one or both ends to move the reflection zone toward a poorer reflecting surface
- Reducing as much as possible the clearance in order to decrease the differential time delay and to increase the size of the interference fringes, which contribute to reducing the fading speed for the same range of variation of the effective factor K
- Optimizing one or both antenna heights so that any surface-reflected wave does not interfere destructively with the *direct* wave over the range of effective factor K
- Choosing vertical polarization, which is more advantageous than horizontal polarization on overwater paths
- Discriminating the reflected wave by a fine alignment and a positive tilt of the antennas
- Adapting the antenna size so that the variation in the launch and arrival angles of the transmitted and received waves does not induce a significant fade

Many methods of predicting the *performance of the link* are founded on the integration of various variables, such as the time delay between the *direct* ray and its replicas as well as their relative amplitude distribution, and utilize the relevant characteristics of the equipment, such as the demodulator, the equalizer, and the modules of error correction.

Signature of Receiver The method employed here is based on the signature of a receiver in which we reproduce in a laboratory, by a two-ray simulation according to the principle presented in Figure 3.48, the propagation affected

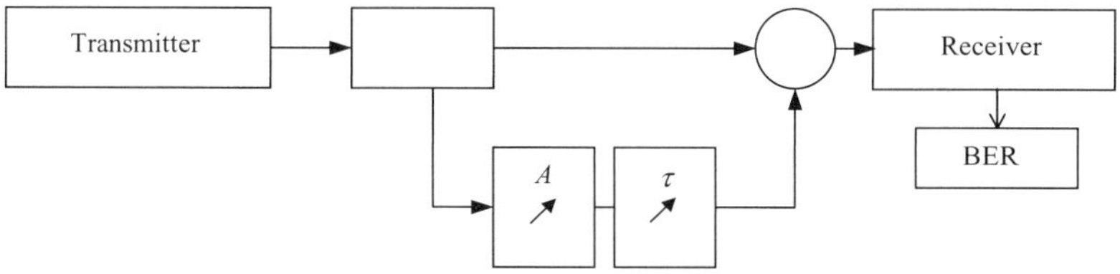

Figure 3.48 Schematic representation of propagation with echo.

by a replica of adjustable amplitude and differential time delay. The signal resulting at the receiver input has as a transfer function

$$H(\omega) = a[1 - b\exp(-j(\omega - \omega_0))\tau] \tag{3.35}$$

where a = scale factor representing amplitude of direct signal
b = amplitude of indirect signal compared to direct signal unit
τ = differential time delay between direct and indirect signals

The measurement results are presented in the shape of "M" curves while varying the level of the indirect signal compared to the direct signal by keeping constant the differential time delay τ for the same minimal value of BER (10^{-3} or 10^{-6}); the signature curves are plotted for fades with minimal phase and nonminimal phase that are in general very close. We utilize the depth of the signature B, also called notch depth, expressed in decibels:

$$B = -20\log\lambda \qquad \lambda = 1 - b \tag{3.36}$$

The standardized time delay is of 6.3 ns, but it is useful to establish a set of curves corresponding to various time delays as presented in Figure 3.49, which comes from measurements carried out on radio equipment SDH 128QAM-155 Mbits s^{-1} operating at 8 GHz for a BER of 10^{-3}. The width of the signature W represents the difference between the extreme frequencies, that is, 26 MHz.

From these curves we can deduce the maximum relative level, expressed in decibels, that the signal of echo P_R can take compared to the direct signal P_D starting from the formula

$$\frac{P_R}{P_D} = 20\log(1 - 10^{-B/20}) = 20\log b \tag{3.37}$$

Figure 3.50 presents the theoretical curve that has been obtained from the signature characteristics, illustrated by Figure 3.49, by comparison to that which corresponds to the acceptable maximum ratio between the indirect wave and the direct wave recommended by the manufacturer for a BER of 10^{-3}.

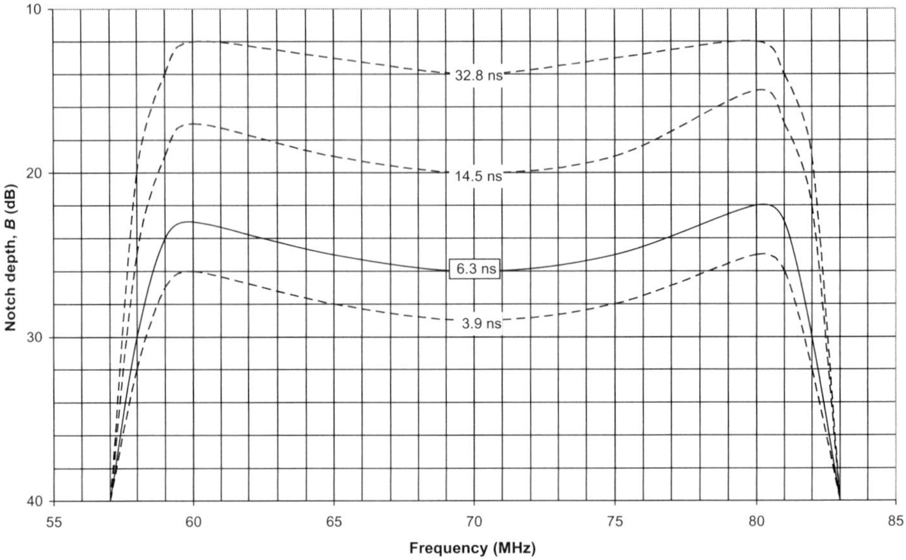

Figure 3.49 Typical signature curves for various time delays τ (BER 10^{-3}).

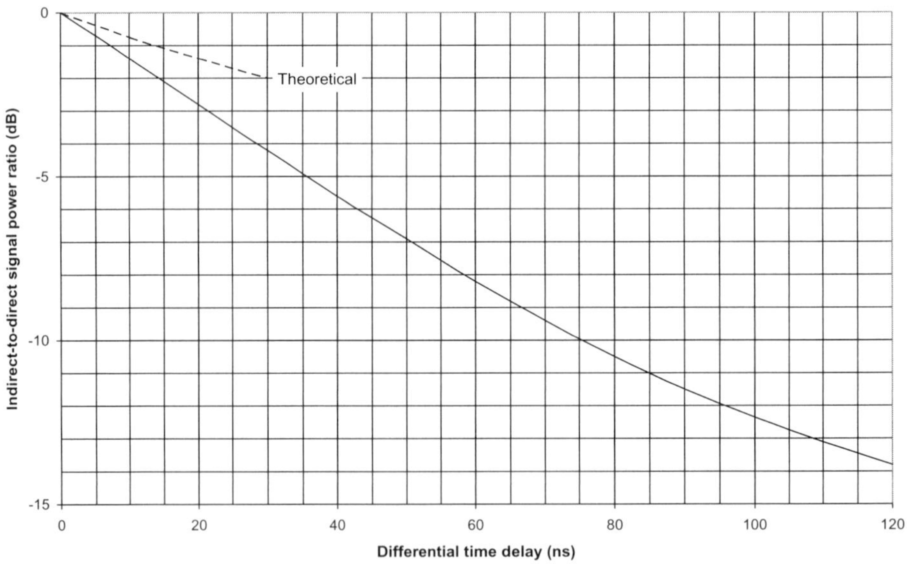

Figure 3.50 Admissible maximum indirect-to-direct signal power ratio versus differential time delay (BER 10^{-3}).

This maximum acceptable indirect-to-direct wave power ratio according to the differential time delay also makes it possible to evaluate the maximum VSWR, that is, the ratio between the reflected power and the incident power, due to mismatched feeders between radio equipment and antennas. The differential time delay due to propagation in a coaxial cable or a waveguide is on the order of 3.5–4 ns m^{-1} and must be accounted twice since the reflection takes place at the mismatched end. However, the calculation of the maximum acceptable value for the VSWR and the consecutive degradation of the total system signature of the receiver is very complex because it results from the composition of both incident and reflected waves, moving in opposite direction throughout the feeder line which contains the standing wave, under the form of a space and temporal convolution integral; Section 2.3.4.3 gives some indications about the admissible values for various modulations.

The signature curves can also be used to calculate the following:

- Standardized parameter K_n, called the surface of the signature, by the formula

$$K_n = \frac{T^2 W \lambda}{\tau} \tag{3.38}$$

where T = symbol duration (ns)
W = signature width (GHz)
λ = 1 − b
τ = reference time delay for λ (ns)

- Margin for selective fading MSF by the empirical formula

$$\mathrm{MSF} = 17.6 - 10\log\left(2W\frac{e^{-B/3.8}}{0.1584}\right) \tag{3.39}$$

where B is the notch depth in decibels.

From the signature curve for the standardized time delay of 6.3 ns, presented in Figure 3.49, we deduce the following values with B = 24 dB and W = 0.026 GHz:

$K_n \sim 0.4$	Characteristics given by manufacturer are 0.35 for BER 10^{-3} and 0.46 for BER 10^{-6}
MSF ~ 50 dB	Characteristics given by manufacturer are 47.5 dB for BER 10^{-3} and 45.5 dB for BER 10^{-6}

It is seen that the smaller the area of the signature curve, the smaller the parameter K_n and the bigger the margin for selective fading; the latter is in general higher than the thermal threshold margin of the receiver.

Probability of Outage The probability that the BER is higher than a given threshold on an unprotected channel can be obtained as follows:

- First calculate the average time delay due to the propagation, expressed in nanoseconds, starting from the formula

$$\tau_m = 0.7\left(\frac{d}{50}\right)^{1.3} \tag{3.40}$$

- Then calculate the multipath activity factor η using relation (3.26).
- Finally calculate the probability of outage due to the selective fading, P_s, as

$$P_s = 2.15\eta\left[W_M \times 10^{-B_M/20}\frac{\tau_m^2}{|\tau_{rM}|} + W_{NM} \times 10^{-B_{NM}/20}\frac{\tau_m^2}{|\tau_{rNM}|}\right] \tag{3.41}$$

where τ_m = average time delay (ns)
W_M = minimum-phase signature width (GHz)
W_{NM} = the non-minimum-phase signature width[8] (GHz)
B_M = minimum-phase notch depth (dB)
B_{NM} = non-minimum-phase notch depth (dB)
τ_{rM} = reference time delay used to obtain signature for minimal-phase fading (ns)
τ_{rNM} = reference time delay used to obtain signature for non-minimal-phase fading (ns)

3.4.1.9 Prediction of Outage in Digital Systems Using Diversity Techniques The diversity constitutes also a very efficient countermeasure to propagation distortion insofar as the signals received by different ways do not fade simultaneously when the fades are deep. The effectiveness of a fading countermeasure is usually expressed in terms of an improvement factor that is the ratio of the outage time observed for an unprotected system to that when the countermeasure is operative.

Diversity Techniques The purpose of these methods is to provide diverse signals containing the same information in which the impairments due to multipath fading are sufficiently decorrelated. The most common techniques of diversity are as follows:

- Space diversity, which is the most effective, where the transmitted signal is received on two vertically separated antennas with an appropriate spacing

[8]The signature curves of the type *M* and type *NM* are in general similar.

- Frequency diversity, where the same information is transmitted over more than one radio channel with a sufficient frequency separation
- Double-frequency and space diversity, with two transceivers, where the same information propagates over two different paths using different frequencies
- Quadruple-frequency and space diversity, which uses two transmitters and four receivers
- Angle diversity, where the diversity signal is derived from a second antenna having a different elevation angle or a second beam on the same antenna that has a different radiation pattern in the vertical dimension

Concerning space diversity, we saw in Section 3.2 that the interferometric method provides the optimized vertical separation between both receiving antennas facing both specular reflection and multipath and how a tilt upward of the latter improves the performance by discrimination of the reflected wave.

About angle diversity, we saw that a positive difference of pointing between the antennas in space diversity, by tilting upward the diversity one, for instance, makes it possible to mitigate the extreme variations of the launch and arrival angles of the direct wave due to the changes of refractivity of the atmosphere, owing to the widening effect of the combined radiation pattern, especially when the antennas are very directive, and to reduce the influence of the reflected wave.

Current systems implement diversity by using the following according to the expected severity of propagation conditions and characteristics of the transmitted signal:

- A "hitless" switch at intermediate or baseband frequencies that select the receiver which has the strongest level, the greater eye pattern opening, or the lower BER
- A combiner at radio, intermediate, or baseband frequencies that may employ equal-gain, maximum-power, or minimum-dispersion algorithms

1. *Fundamental Diversity Systems* The following figures illustrate the most used diversity systems; Figure 3.51 presents the fundamental block

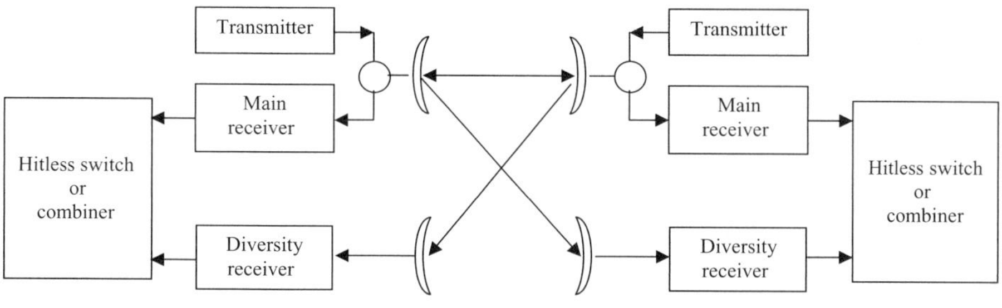

Figure 3.51 Block diagram of space diversity system.

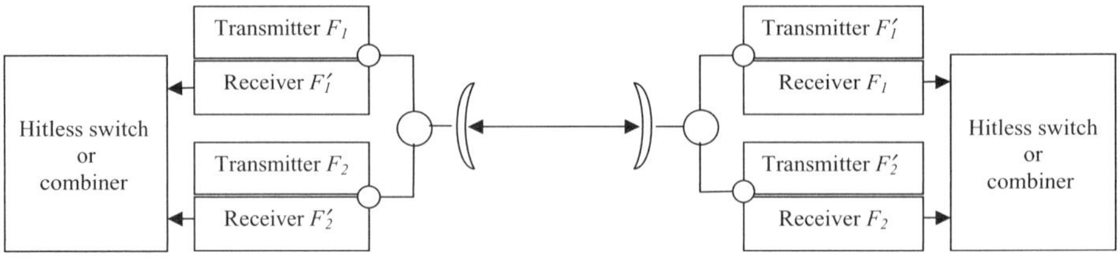

Figure 3.52 Block diagram of frequency diversity system.

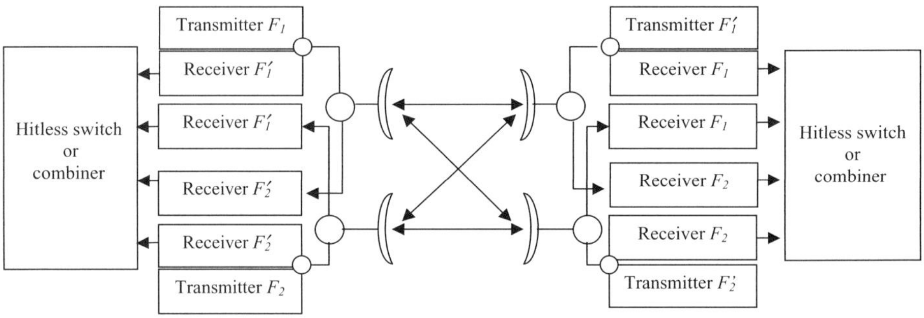

Figure 3.53 Block diagram of full-quadruple space and frequency diversity system with four receivers on two paths.

diagram of a space diversity system. Figure 3.52 presents the fundamental block diagram of a frequency diversity system where the transmitters and receivers are all connected to the same antenna at each terminal. Figure 3.53 presents the fundamental block diagram of a full quadruple space and frequency diversity system with four receivers on two paths where the transmitters are connected to both antennas.

2. *Methods of Signal Processing and Combining* We will consider that the diverse signals are Rayleigh equally distributed; such conditions of transmission suppose that transmitted power, antenna size, and path losses are the same for all signals and that they consequently have the same median field strength.

The most common techniques for signals processing in diversity include the following:

- Selection diversity, also called switching diversity, where the best signal is selected according to various criteria such as eye pattern, output level, and/or lowest BER
- Equal-gain combiner, where the signals from all the channels are multiplied by a weighting function of the same constant value and then added

- Maximal-ratio combiner, which differs from the equal-gain combiner in that the weighting functions follow a particular law based on channel measurements

The principle of combining is based on the following assumptions:

- The signal amplitudes are all in phase so that they add arithmetically while the noise amplitudes are all independent and add in an rms fashion.
- The weakest signals may contribute to the generation of the composite or combined signal, so that the resulting signal is better than the strongest of the individual signals.

Various other techniques, such as adaptive equalization in combination with space diversity, result in significant improvement compared to that obtained when adaptive equalization operates in isolation.

The properties of the random Rayleigh's variables are described in Section 1.14.3 for their individual distribution and in Section 1.11.1.3 for the combination of two variables in relation with their coefficient of correlation. The percentage of time for a given fade depth A relative to the median field strength value of a Rayleigh distributed signal is given by the relation

$$p\% = 100\left[1 - \exp\left(-\frac{10^{(A/20)^2}}{2}\right)\right] \tag{3.42}$$

and, inversely, the fade depth according to the percentage of time by the relation

$$A = 10\log\left[-2\ln\left(\frac{100 - p\%}{100}\right)\right] \tag{3.43}$$

Figure 3.54 shows the distribution of the fade depth for a single signal and the combined signal by selection of the strongest one in a double diversity according to the degree of correlation.

The diversity gain by selection of the strongest signal among a number N of channels is given by the following relation for a null correlation between the channels:

$$G = \frac{\ln(1 - P^{1/N})}{\ln(1 - P)} \tag{3.44}$$

where P is the probability that a given level is exceeded. We easily find the same improvement due to diversity using the formulas given in Section 1.11.1.3 for a null coefficient of correlation ($K = 0$) with D as the row of diversity, that is, the number of simultaneous channels; indeed, the improvement factor is worth

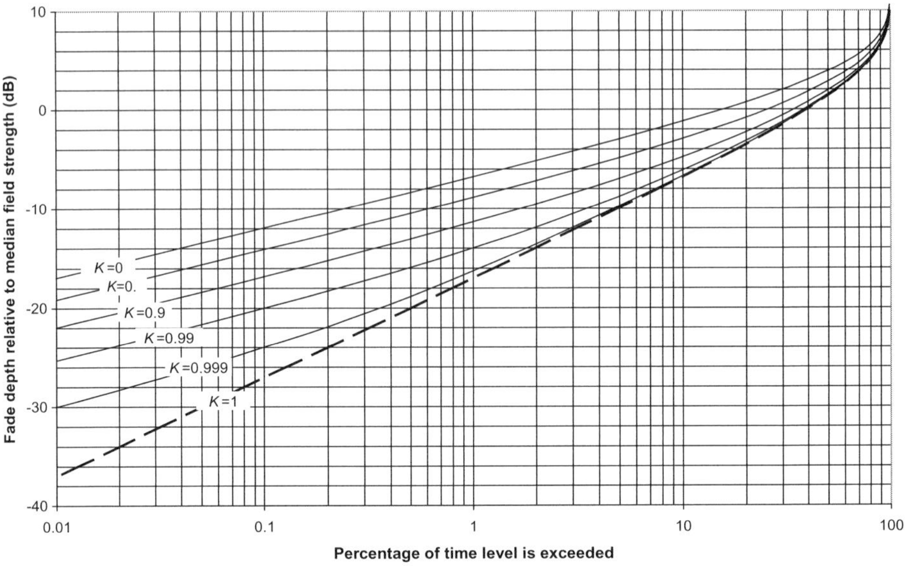

Figure 3.54 Distribution of combined signal in selection diversity system versus coefficient of correlation between two Rayleigh distributed random signals.

$$\frac{1}{P_i} \quad \text{for } D=2 \qquad \left(\frac{1}{P_i}\right)^2 \quad \text{for } D=4$$

by considering that the system $D = 4$ corresponds to the combination of two totally independent systems $D = 2$ since the signals are practically uncorrelated. Figure 3.55 shows the distribution of the fade depth of the signal combined by selection of the strongest among multiple totally independent signals according to the raw D of diversity.

The maximal-ratio combination consists of adding the various received signals S_i after attribution of a weighting coefficient α_i representative of their proper reception conditions (field strength, BER) in such a way that the combined signal S_C is given as

$$S_C = \sum_1^N \alpha_i S_i \qquad \sum_i^N \alpha_i = Cte \tag{3.45}$$

We can see that any increase of a signal level S_i results in the reduction of the other components according to their respective weight and that this type of combiner favors the strongest signals while maintaining a certain contribution from the weakest; the combined SNR is thus always higher than that of the best signal.

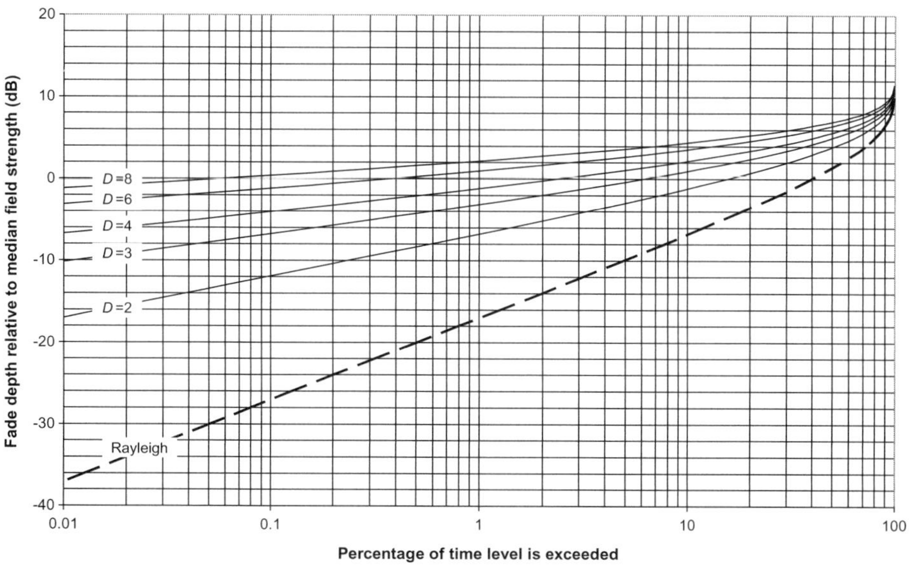

Figure 3.55 Distribution of combined signal in diversity system by selection of strongest signal versus number of totally independent Rayleigh distributed random signals.

Making use of the Schwartz inequality yields

$$\frac{S_C}{N_C} \le \sum_{i}^{N} \frac{S_i}{N_i} \tag{3.46}$$

where

$$S_i = s_i^2 \qquad N_i = \overline{n}_i^2$$

The individual Signals S_i being correlated, since they are in phase, and their proper noise N_i being uncorrelated, because of their simultaneous transmission by separate ways, it follows that

$$S_C = \left(\sum_{1}^{N} \alpha_i S_i \right)^2 \tag{3.47}$$

$$N_C = \sum_{1}^{N} \alpha_i^2 \overline{n}_i^2 \tag{3.48}$$

Figure 3.56 shows the maximal-ratio combining curves according to the percentage of time and the row of diversity compared to that of the median

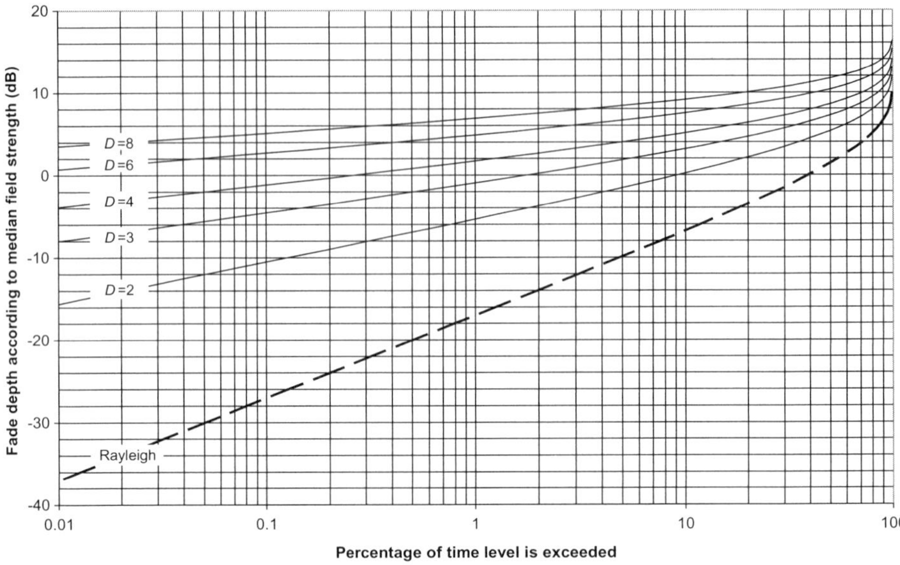

Figure 3.56 Distribution of combined signal in maximal-ratio diversity system versus number of totally independent Rayleigh distributed random signals.

individual signal. When the combiner receives identical signals in the presence of white Gaussian noise, the combined signal is composed of the sum of the amplitudes s_i of the individual signals and the combined noise by the quadratic sum of the noise powers n_i, that is,

$$\frac{S_C}{N_C} = \frac{(s_1 + \cdots + s_n)^2}{\overline{n}_i^2 + \cdots + \overline{n}_n^2} \tag{3.49}$$

In such conditions, called *static conditions*, which are generally found in the laboratory while testing the equipment, the static maximal-ratio combination gain in terms of average SNR is as follows:

Number of channels	2	3	4	6	8
Maximal-ratio static diversity gain in average SNR, dB	3	4.8	6	7.8	9

In reality, in *dynamic conditions* during propagation, when the signals are totally independent with equal mean, that is, with a null correlation and the same median value, their instantaneous levels are randomly distributed and the dynamic diversity combination gain in average SNR is given for the three considered systems by the following relations:

- Selection diversity:

$$G_{\mathrm{SNR}} = 10\log\left(\sum_{k=1}^{N}\frac{1}{k}\right) \tag{3.50}$$

- Equal-gain diversity:

$$G_{\mathrm{SNR}} = 10\log\left[1+\frac{1}{4}(N-1)\pi\right]$$

- Maximal-ratio diversity:

$$G_{\mathrm{SNR}} = 10\log N$$

Figure 3.57 illustrates, for comparison of the three systems, the diversity gain in average SNR for independently fading Rayleigh distributed signals with constant and equal rms values. We can see that the performance difference is not significant between the equal-gain and maximal-ratio diversity systems. In addition, combiners comprise coupling losses that may be higher than the gain relative to selection diversity and we have seen that the interferometric method allows us to correlate in opposition the received signals so that the combination gain would be less than expected.

Prediction of Outage in System Using Space Diversity The procedure given below applies to systems employing maximum-power combiners such as the

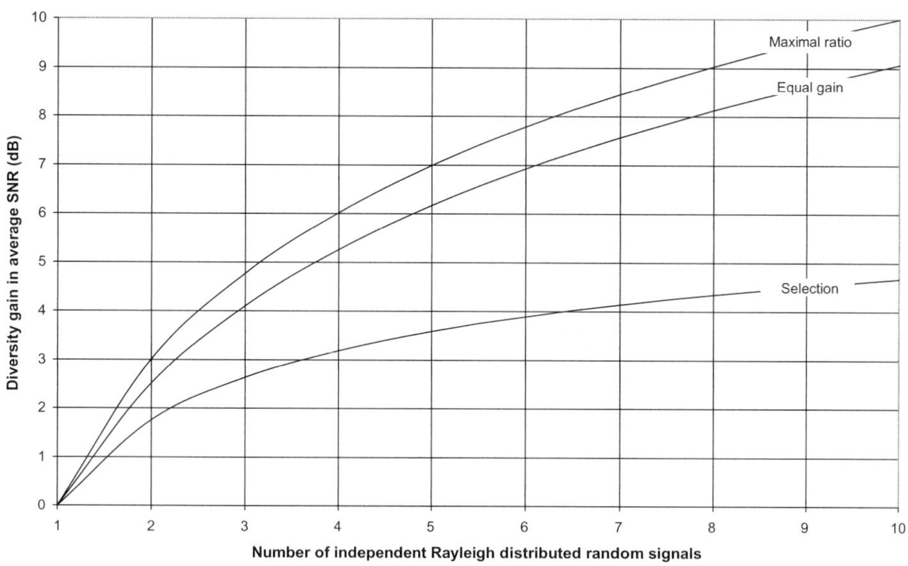

Figure 3.57 Distribution of diversity gain in average SNR versus number of totally independent Rayleigh distributed random signals.

selection of the strongest signal level; other types of combiners using both minimum distortion and maximum power dependent on radio channel evaluation may give somewhat better performance. We calculate first the multipath activity parameter η using relation (3.26) and then the square of the nonselective correlation coefficient k_{ns} by means of the formula

$$k_{ns}^2 = 1 - \frac{I_{ns} P_{ns}}{\eta} \tag{3.51}$$

where the probability of nonselective outage P_{ns} is given by relation (3.25) and the improvement factor I_{ns} can be evaluated using the following equation for narrow-band signals on an overland path by considering the flat fade margin M:

$$I_{ns} = [1 - \exp(-0.04 S^{0.87} f^{-0.12} d^{0.48} p_0^{-1.04})] 10^{(M-V)/10} \qquad V = |G_1 - G_2| \tag{3.52}$$

where p_w = percentage of time fade depth M is exceeded in average worst month, given by relation (3.19) or (3.23) as the case may be (%)
p_0 = multipath occurrence factor[9] (%)
M = flat fade margin defined by relation (3.11) (dB)
S = vertical separation center to center between receiving antennas (m)
f = frequency (GHz)
d = pathlength (km)
G_1, G_2 = gains of the receiving antennas (dB)

Relation (3.52) was based on the parameters:

Pathlength: $25 \le d \le 240\,\text{km}$
Frequency: $2 \le f \le 11\,\text{GHz}$
Antennas separation: $3 \le S \le 23\,\text{m}$

We then calculate the square of the selective correlation coefficient k_s:

$$k_s^2 = \begin{cases} 0.8238 & \text{for } r_w < 0.5 \\ 1 - 0.195(1 - r_w)^{0.109 - 0.13\log(1 - r_w)} & \text{for } 0.5 < r_w < 0.9628 \\ 1 - 0.3957(1 - r_w)^{0.5136} & \text{for } r_w > 0.9628 \end{cases} \tag{3.53}$$

[9]It is a matter of the multipath occurrence factor in percentage instead of the parameter in the recommendation.

where the correlation coefficient r_w of the relative amplitudes is given as

$$r_w = \begin{cases} 1-0.9746(1-k_{ns}^2)^{2.17} & \text{for } k_{ns}^2 \le 0.26 \\ 1-0.6921(1-k_{ns}^2)^{1.034} & \text{for } k_{ns}^2 > 0.26 \end{cases}$$

Then we calculate the nonselective outage probability with diversity P_{dns} using the formula

$$P_{dns} = \frac{P_{ns}}{I_{ns}} \tag{3.54}$$

where P_{ns} is the nonprotected outage due to nonselective fading given by relation (3.25). Lastly, we calculate the selective outage probability with diversity P_{ds} by the formula

$$P_{ds} = \frac{P_s^2}{\eta(1-k_s^2)} \tag{3.55}$$

where P_s is the nonprotected selective outage probability given by equation (3.41). The total outage probability with diversity P_d is then given by

$$P_d = (P_{ds}^{0.75} + P_{dns}^{0.75})^{1.33} \tag{3.56}$$

Prediction of Outage in System Using Frequency Diversity The method applies to a system (1 + 1), that is, with only one antenna, connected to two transmitters and two receivers at each end. We use the same procedure as for space diversity by replacing relation (3.52) for the improvement factor by the formula

$$I_{ns} = \frac{80\,\Delta f}{f^2 d} \times 10^{M/10} \tag{3.57}$$

where Δf = frequency separation (GHz) with $\Delta f / f \le 5\%$
[if $\Delta f > 0.5$ GHz, use $\Delta f = 0.5$ GHz]
f = frequency (GHz) with $2 \le f \le 11$ GHz
d = distance (km) with $30 \le d \le 70$ km
M = flat fade margin defined by relation (3.11) (dB)

Prediction of Outage in System Using Space and Frequency Diversity (Two Receivers) The nonselective fading correlation coefficient k_{ns} is given by the relation

$$k_{ns} = k_{ns,s} k_{ns,f} \tag{3.58}$$

where $k_{ns,s}$, $k_{ns,f}$ are the nonselective correlation coefficients computed for space diversity and frequency diversity using relations (3.51), (3.52) and (3.57). The following stages are the same as for the space diversity; however, it seems judicious to the author to replace formula (3.54) by the following one, not envisaged by this recommendation, in order to take account of the improvement made simultaneously by the space diversity and the frequency diversity:

$$P_{dns} = \frac{P_{ns}}{I_{ns,s} I_{ns,f}}$$

where $I_{ns,s}$, $I_{ns,f}$ are the improvement factors given by relations (3.52) and (3.57).

Prediction of Outage in System Using Space and Frequency Diversity (Four Receivers) We calculate first the multipath activity parameter η using relation (3.26) and then the diversity parameter m_{ns} by the formula

$$m_{ns} = \eta^3 (1 - k_{ns,s}^2)(1 - k_{ns,f}^2) \tag{3.59}$$

where $k_{ns,s}$, $k_{ns,f}$ are obtained in the same way as above. We next calculate the nonselective outage probability $P_{d,ns}$ with diversity from the formula

$$P_{dns} = \frac{P_{ns}^4}{m_{ns}} \tag{3.60}$$

where P_{ns} is obtained from equation (3.25). We then calculate the square of the equivalent nonselective correlation coefficient k_{ns} using the formula

$$k_{ns}^2 = 1 - \sqrt{\eta}(1 - k_{ns,s}^2)(1 - k_{ns,f}^2) \tag{3.61}$$

Finally we calculate the equivalent selective correlation coefficient k_s using the procedure for space diversity from relation (3.53). The selective outage probability with diversity P_{ds} is found from

$$P_{ds} = \left(\frac{P_s^2}{\eta(1 - k_s^2)} \right)^2 \tag{3.62}$$

where P_s is the nonprotected selective outage given by equation (3.41). Lastly, the total outage probability with diversity P_d is obtained using relation (3.56).

Prediction of Outage in System Using Angle Diversity We start by evaluating the average arrival angle μ_θ, expressed in degrees, from the formula

$$\mu_\theta = (2.89 \times 10^{-5}) G_m d \tag{3.63}$$

where G_m = average value of vertical refractivity gradient (N km^{-1})
d = pathlength (km)

When a strong ground reflection is clearly present, μ_θ can be estimated from the angle of arrival of the reflected ray in standard propagation conditions. We then calculate the nonselective reduction parameter r by means of the formula

$$r = \begin{cases} 0.113\sin\left(150\dfrac{\delta}{\Omega}+30\right)+0.963 & \text{for } q>1 \\ q & \text{for } q\le 1 \end{cases} \tag{3.64}$$
$$q = 2505\times 0.0437^{\delta/\Omega}\times 0.593^{\varepsilon/\delta}$$

where δ = angular separation between both antenna radiation patterns (deg)
ε = elevation angle of upper antenna, positive toward ground (deg)
Ω = half-power beamwidth of antenna radiation patterns (deg)

We then calculate the nonselective correlation parameter Q_0 using the formula

$$Q_0 = r\times 0.9399^{\mu_\theta}\times 10^{-24.58\mu_\theta^2}\times 2.469^{1.879(\delta/\Omega)}\times 3.615^{[(\delta/\Omega)^{1.978}(\varepsilon/\delta)]}\times 4.601^{[(\delta/\Omega)^{2.152}(\varepsilon/\delta)^2]} \tag{3.65}$$

We next calculate next the multipath activity parameter η by employing relation (3.26). The nonselective outage probability in diversity is given as

$$P_{dns} = \eta Q_0 \times 10^{M/6.6} \tag{3.66}$$

where M is the flat fade margin defined by relation (3.11) in decibels. We calculate the square of the selective correlation coefficient k_s by the formula

$$k_s^2 = 1-0.0763\times 0.694^{\mu\theta}\times 10^{23.3\mu_\theta^2}\delta(0.211-0.188\mu_\theta-0.638\mu_\theta^2)^{\Omega} \tag{3.67}$$

The selective outage probability with diversity P_{ds} is given by relation (3.55). Lastly, the total outage probability with diversity P_d is obtained by means of relation (3.56).

3.4.1.10 Prediction of Total Outage Probability The total outage probability P_{tot} corresponds to the sum of the outage probabilities due to the various uncorrelated phenomena which affect the propagation:

$$P_{\text{tot}} = \begin{cases} P_{ns}+P_s+P_{\text{rain}}+P_{\text{Xp}} & \text{for systems without diversity} \\ P_d+P_{\text{rain}}+P_{\text{Xp}} & \text{for systems with diversity} \end{cases} \tag{3.68, 3.69}$$

where P_{ns} = nonselective outage probability
P_s = selective outage probability
P_d = outage probability with diversity
P_{rain} = outage probability due to precipitation
P_{Xp} = outage probability due to XPD reduction

Prediction of Outage Probability due Clear-Air Effects This concerns the probabilities P_{ns}, P_s, and P_d, which were defined previously for systems without diversity and with diversity that are introduced into relations (3.68) and (3.69).

Prediction of Outage Probability Due to Rain Attenuation The estimated attenuation A_R due to the rain on the terrestrial paths according to latitude φ is given by relation (1.109). Reversing and solving these equations and then introducing the uniform fade margin M, we can calculate the percentage of unavailable time:

$$p\% = \begin{cases} 10^{[-0.546+\sqrt{0.298116-0.172(\log M - \log A_{0.01}+0.9208)}\,]/0.086} & \text{for } \varphi \geq 30^\circ \\ 10^{[-0.855+\sqrt{0.731025-0.556(\log M - \log A_{0.01}+1.15492)}\,]/0.278} & \text{for } \varphi < 30^\circ \end{cases} \tag{3.70}$$

The outage probability due to the rain[10] is thus equal to

$$P_{\text{rain}} = \frac{p\%}{100} \tag{3.71}$$

Prediction of Outage Probability Due to XPD Reduction The XPD can deteriorate sufficiently to cause cochannel interference and, to a lesser extent, adjacent-channel interference, especially in systems that call upon frequency reuse by employing cross-polarization techniques. The XPD reduction can occur in clear air, in conditions of multipath or reflection, and in periods of rain because of the depolarization by hydrometeors that was treated in Section 1.11.2.

1. *Reduction of XPD Due to Clear-Air Effects* The combined effect of multipath propagation and cross-polarization radiation patterns of the antennas determines the reductions in XPD for small percentages of time. To compute the effect of these reductions in link performance, we start by calculating

$$(\text{XPD})_0 = \begin{cases} (\text{XPD})_G + 5 & \text{for } (\text{XPD})_G \leq 35 \\ 40 & \text{for } (\text{XPD})_G > 35 \end{cases} \tag{3.72}$$

[10]The atmospheric attenuation A_a, included in the attenuation by rain, can be deduced from the link budget for the $r_{0.01}$ fraction of the path.

where $(\mathrm{XPD})_G$ is the manufacturer's guaranteed minimum XPD at boresight for both the transmitting and receiving antennas, that is, the minimum of the transmitting and receiving antenna boresight XPDs. We then evaluate the multipath activity parameter η by employing relation (3.26) and determine the factor

$$Q = -10\log\left(\frac{\eta k_{xp}}{P_0}\right) \tag{3.73}$$

with

$$k_{xp} = \begin{cases} 0.7 & \text{for one transmitting antenna} \\ 1 - 0.3\exp\left[-4\times 10^{-6}\left(\dfrac{S_t}{\lambda}\right)^2\right] & \text{for two transmitting antennas} \end{cases}$$

where S_t = vertical separation when two orthogonally polarized transmissions are from different antennas (m)
λ = carrier wavelength (m)

Then we can deduce the parameter C:

$$C = (\mathrm{XPD})_0 + Q \tag{3.74}$$

The probability of outage $P_{\mathrm{Xp},C}$ due to the clear-air cross-polarization reduction is given by the relation

$$P_{\mathrm{Xp},C} = P_0 \times 10^{-M_{\mathrm{XPD}}/10} \tag{3.75}$$

with the equivalent margin M_{XPD} in decibels for a given reference BER:

$$M_{\mathrm{XPD}} = \begin{cases} C - \dfrac{C_0}{I} & \text{without XPIC} \\ C - \dfrac{C_0}{I} + \mathrm{XPIF} & \text{with XPIC} \end{cases}$$

where C_0/I = carrier-to-interference ratio for reference BER
XPIC = Cross-polar interference canceler
XPIF = cross-polarization improvement factor (typically 20 dB) that gives difference in cross-polar isolation (XPI) at sufficiently large carrier-to-noise ratio (typically 35 dB) and at specific BER for systems with and without XPIC

This recommendation does not provide any indication about the improvement due to diversity concerning XPD; Section 3.6 exposes results of measurement carried out in space diversity systems which show that the improvement factor is of the same order of magnitude as for the received field fading.

2. *Prediction of XPD Outage Due to Precipitation Effects* The effects due to depolarization by hydrometeors are quantified by relation (1.130), which gives the value of $(\text{XPD})_{\text{hydro}}$, and relation (1.133) for the value of $(\text{XPD})_{\text{total}}$, which includes the share due to equipment $(\text{XPD})_E$. A rough estimate of the unconditional distribution of XPD can be obtained from a cumulative distribution of the copolar attenuation (CPA) for rain using the equiprobability relation which is valid for line-of-sight paths with small elevation angles:

$$\text{XPD} = U - V(f)\log \text{CPA} \qquad U = U_0 + 30\log f \tag{3.76}$$

where the average value of U_0 is 15 dB, f is the frequency in gigahertz, and

$$V(f) = \begin{cases} 12.8 f^{0.19} & \text{for } 8 \le f \le 20\,\text{GHz} \\ 22.6 & \text{for } 20 < f \le 35\,\text{GHz} \end{cases}$$

The outage probability results directly from the percentage of time (p%) which one calculates by means of relations (3.70) and (3.71) by replacing the uniform margin M by the attenuation A_R, which gives the value of $(\text{XPD})_{\text{rain}}$ involving the outage of the link, and proceeding as previously.

As the effects of attenuation and XPD reduction by the hydrometeors are correlated, since they are due to the same cause, we will retain the highest outage probability of both. Moreover, as the effects due to the hydrometeors are exerted simultaneously on both antennas in space diversity and/or on both frequency diversity channels, no improvement is brought by diversity.

The recommendation gives the following method:

- Calculate the attenuation $A_{0.01}$ exceeded for 0.01% of the time from relation (1.109).
- Determine the equivalent path attenuation A_{Pe} in decibels:

$$A_{Pe} = 10^{(U - C_0/\text{I} + \text{XPIF})/V} \tag{3.77}$$

  where U, V are given above and XPIF = 0 in the absence of an XPIC device.
- Determine the following parameters:

$$m = \begin{cases} 23.26\log\left(\dfrac{A_{Pe}}{0.12 A_{0.01}}\right) & \text{if } m \le 40 \\ 40 & \text{otherwise} \end{cases} \tag{3.78}$$

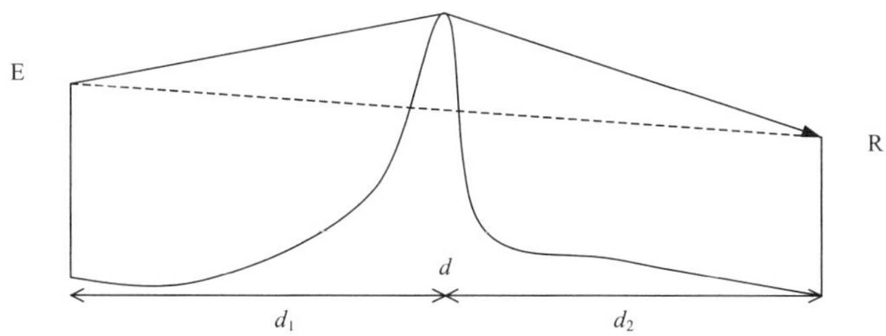

Figure 3.58 Link in diffraction on obstacle.

$$n = \frac{1}{2}(-12.7 + \sqrt{161.23 - 4m}) \quad (3.79)$$

where n must be in the range[11] –3–0.

- Evaluate the outage probability using the relation

$$P_{\mathrm{Xp},R} = 10^{n-2} \quad (3.80)$$

3.4.2 Effect Due to Obstacle

When an obstacle is on the path, the attenuation can be calculated using the relations given in Section 1.9. If this obstacle interposes permanently between the transmitter and the receiver, in conditions of superrefraction in particular, the link operates in diffraction, as shown in Figure 3.58. The median obstacle loss is then deduced from the uniform margin, but the field is in general more stable than if there were no obstacle. The reason for this stability is due to the fact that the link is shared in two sections, facing the outage due to multipath, since the top of the obstacle behaves like a source of emission with respect to the two ends.

The percentage of time for a given fade depth, given by relation (3.19) or (3.23) as the case may be, provides the following results:

For link of length d without an obstacle: p_w
For section d_1: p_{w1}
For section d_2: p_{w2}

Supposing that the effects due to multipath are completely uncorrelated on the two sections, the total percentage of time for the same fade depth becomes

[11]In some cases, especially when an XPIC device is used, values of n less than –3 may be obtained.

$$p_t = p_{w1} + p_{w2}$$

However, in relation (3.16), p_w varies according to the distance at power 3.6; then, supposing $d_1 \approx d_2$, we write

$$\frac{p_t}{p_w} \approx 2\left[\frac{1}{2}\right]^{3.6} \approx \frac{1}{6}$$

The percentage of time of fading can thus be reduced in a ratio of 6 and compensate for an obstacle loss of 8 dB deduced from the uniform margin; indeed, we saw previously that the distribution of the field is of 10 dB per decade and p_w varies as $10^{-M/10}$. It should be noted that a diffraction can also considerably improve the availability of a link while making it possible to mask a very reflective zone which would be on one of the sections; in such a case, the obstacle loss on the reflected ray is higher than that which affects the *direct* ray since it corresponds to a weaker clearance.

It is necessary, however, to take account of anomalies of propagation which can occur on diffracted links at frequencies higher than some gigahertz, such as that presented in Figure 3.59, which comprises two links in parallel, one at 2.5 GHz and the other at 7.5 GHz. The path profile presents an obstacle made up of a stripped hill surrounded by a crown of trees approximately 20 m high, and the antennas have diameters of 3 m each at station C and 2 m (2.5 GHz) and 3 m (7.5 GHz) at station D.

Figure 3.60 shows the results obtained during a recording of the fields received simultaneously at 2.5 and 7.5 GHz; we can see that the channel at

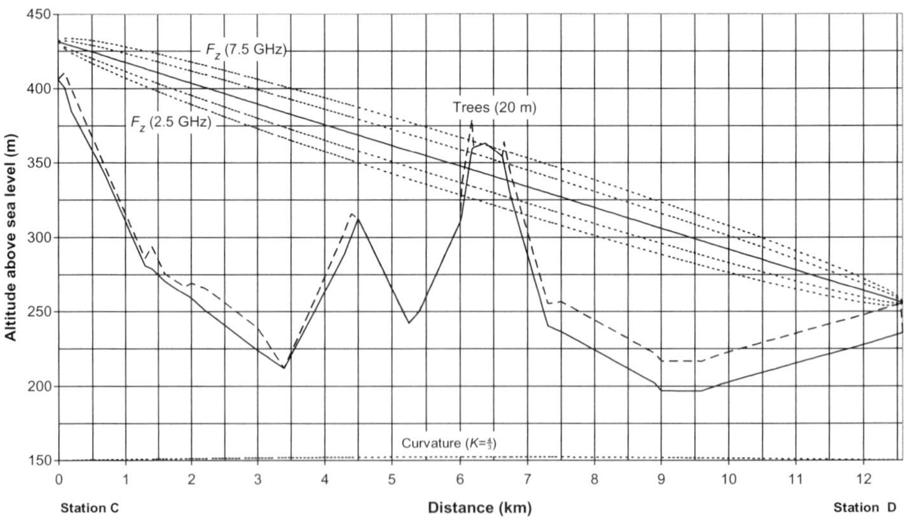

Figure 3.59 Link in diffraction at 2.5 and 7.5 GHz.

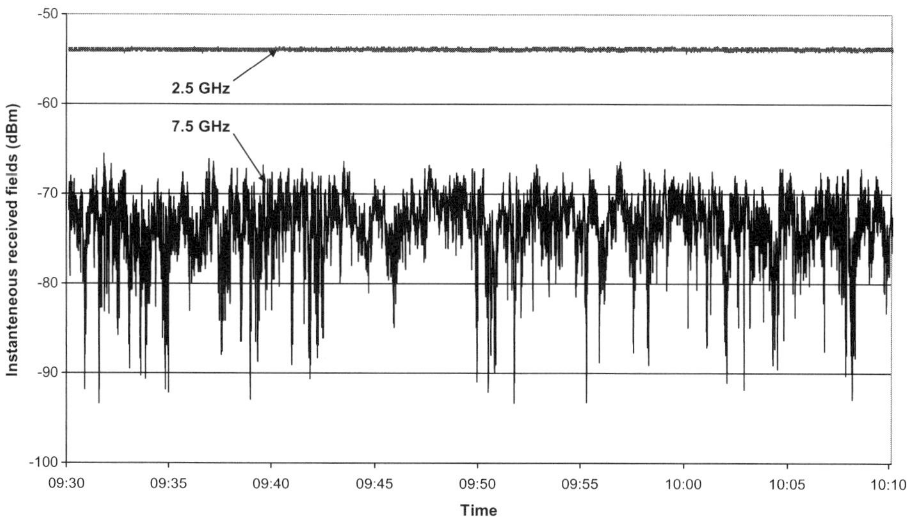

Figure 3.60 Fields received simultaneously at 2.5 and 7.5 GHz (September 8, 2004).

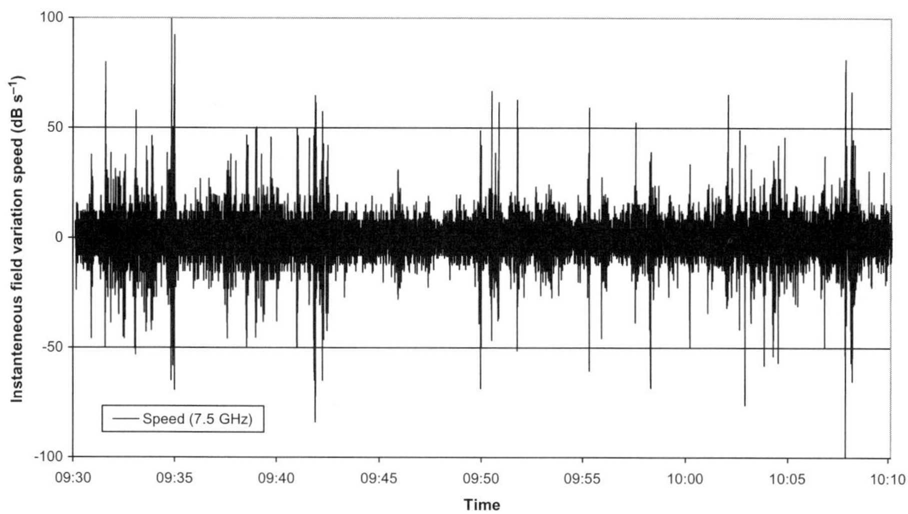

Figure 3.61 Received field variation speed at 7.5 GHz (September 8, 2004).

2.5 GHz offers a very great stability while that at 7.5 GHz presents an important lack in field strength of about 20 dB compared to the computed value as well as a very strong scintillation. Figure 3.61 shows that the variation speed of the field at 7.5 GHz can exceed 100 dB s^{-1}, whereas that at 2.5 GHz does not reach 1 dB s^{-1}.

3.4.3 Example of Link Budget

Table 3.1 summarizes the stages and results of the calculations carried out for the SDH line-of-sight microwave link of 40 km described in Section 3.2.8.

Figure 3.62 presents the distributions of the individual (P_i) and combined signals in space diversity (P_{ds}) which make it possible to visualize the corresponding improvement factors for both directions of transmission by comparison with that of totally uncorrelated signals P_d ($K = 0$).

Figure 3.63 presents the distribution of the received field enhancement that should be taken into account in calculations of radioelectric interference and to make sure that the signals remain within the limits of linearity of the receivers.

3.4.4 Evolution of Method of Prediction (ITU-R P.530-9 to P.530-12)

Further recommendations, from ITU-R P.530-9 (2001) to ITU-R P.530-12 (2007), are mainly based on the new parameters dN_1 and S_a:

dN_1 Point refractivity gradient in lowest 65 m of atmosphere not exceeded for 1% of an average year; dN_1 is provided on 1.5° grid in latitude and longitude in ITU-R P.453-9 (see Figure 3.64).

S_a Area terrain roughness[12] in meters, defined as standard deviation of terrain heights within zone of 110 km × 110 km with 30″ resolution

These data are available in tabular format from the ITU. The correct values for the latitude and longitude at the path center should be obtained by bilinear interpolation from the values for the four closest grid points.

3.4.4.1 Multipath Occurrence Factor

1. The geoclimatic factor K for the worst month becomes:
   - For a quick calculation:

$$K = 10^{-4.2-0.0029dN_1} \tag{3.81}$$

   - For a detailed link design:

$$K = 10^{-3.9-0.003dN_1} S_a^{-0.42} \tag{3.82}$$

2. The percentage of time p_w that fade depth A is exceeded in the average worst month and the multipath occurrence factor p_0 become:

[12] If $Sa < 1$ m, use the lower limit of 1 m.

TABLE 3.1 Stages and Results in SDH Line-of-Sight Microwave Links

Station A–Station B Link	ITU-R P.530-8	
Transmission station	Station A	Station B
Reception station	Station B	Station A
Frequency, GHz	8.000	8.000
Equipment type	SDH-STM1	SDH-STM1
Modulation levels	128TCM	128TCM
Bit rate, Mbits s^{-1}	155	155
Emitted power, dBm	27.0	27.0
Threshold BER 10^{-6}, dBm	–69.0	–69.0
Threshold BER 10^{-3}, dBm	–72.0	–72.0
Receiver signature width (BER 10^{-3}), MHz	26.0	26.0
Receiver signature depth (BER 10^{-3}), dB	24.0	24.0
Reference time delay, ns	6.3	6.3
Pathlength, km	40.0	40.0
Roughness, m	17.5	17.5
Free-space attenuation, dB	142.5	142.5
Mean time delay, ns	0.52	0.52
Launch/arrival angle variation, deg	+/–0.27	+/–0.27
Altitude of transmission station, m	180.0	280.0
Transmission antenna height, m	40.0	20.0
Altitude of transmission antenna, m	220.0	300.0
Altitude of reception station, m	280.0	180.0
Reception antenna height, m	20.0	40.0
Altitude of reception antenna, m	300.0	220.0
Beam slope, mrad	2.0	–2.0
Transmission specific feeder loss, dB hm^{-1}	5.8	5.8
Transmission horizontal feeder length, m	20.0	20.0
Total transmission feeder length, m	60.0	40.0
Total transmission feeder loss, dB	3.5	2.3
Reception specific feeder loss, dB	5.8	5.8
Reception horizontal feeder length, m	20.0	20.0
Total reception feeder length, m	40.0	60.0
Total reception feeder loss, dB	2.3	3.5
Transmission antenna diameter, m	2.4	2.4
Antenna efficiency, %	51.0	51.0
Half-power beamwidth, deg	±0.60	±0.60
Transmission antenna gain, dBi	43.1	43.1
Radome loss, dB	0.5	0.5
Reception antenna diameter, m	2.4	2.4
Antenna efficiency, %	51.0	51.0
Half-power beamwidth, deg	±0.60	±0.60
Reception antenna gain, dBi	43.1	43.1
Radome loss, dB	0.5	0.5
Water vapor concentration, g m^{-3}	15.0	15.0
Atmospheric attenuation, dB	0.6	0.6
Transmission branching loss, dB	1.0	1.0

TABLE 3.1 *Continued*

Station A–Station B Link	ITU-R P.530-8	
Reception Branching loss, dB	3.5	3.5
Specific transmission loss, dB	0.0	0.0
Specific reception loss, dB	0.0	0.0
Jamming threshold degradation, dB	0.0	0.0
Median obstacle loss (50%), dB	0.0	0.0
Equipment margin, dB	0.0	0.0
Median Received power, dBm	−41.2	−41.2
Flat fade margin BER 10^{-6}, dB	27.8	27.8
Flat fade margin BER 10^{-3}, dB	30.8	30.8
Topologic coefficient C_0	Hills	3.5
Latitude, deg	46.9	46.9
Fraction of path above water	0.3	0.3
Coefficient of latitude C_{lat}, dB	0.0	0.0
Coefficient of longitude C_{lon}, dB	3.0	3.0
Climatic variable P_L, %	20	20
Geoclimatic factor K	3.99×10^{-5}	3.99×10^{-5}
Multipath occurrence factor (p_0)	31.9%	31.9%
Fade depth reference (A_t), dB	27	27
Reference percentage of time (p_t)	0.0666	0.0666
Factor q'_a	2.37	2.37
Factor q_t	0.64	0.64
Factor q_a	2.30	2.30
Probability of outage BER > 10^{-3} (q_w), %	0.02657	0.02657
Nonselective outage prediction (P_{ns})	0.000266	0.000266
Multipath activity parameter (η)	0.081	0.081
Selective outage probability (P_s)	0.000025	0.000025
Worst-month outage probability for single channel, s	753.5	753.5
Space diversity, m	6.5	8.0
Multipath occurrence parameter (P_0)	0.318942	0.318942
Vertical space diversity improvement factor ($I_{ns,s}$)	30.3	36.2
Nonselective correlation coefficient ($K_{ns,s}$)	0.9492	0.9390
Correlation coefficient ($r_{w,s}$)	0.9367	0.9239
Selective correlation coefficient ($k_{s,s}$)	0.9519	0.9480
Nonselective outage probability ($P_{dns,s}$)	8.77×10^{-6}	7.34×10^{-6}
Selective outage probability ($P_{ds,s}$)	8.18×10^{-8}	7.58×10^{-8}
Total outage probability for space diversity ($P_{d,s}$)	9.39×10^{-6}	7.88×10^{-6}
Worst-month outage with space diversity, s	24.3	20.4
Frequency diversity, MHz	28	28
Frequency diversity improvement factor ($I_{ns,f}$)	1.1	1.1
Nonselective correlation coefficient ($K_{ns,f}$)	0.9983	0.9983
Correlation coefficient ($r_{w,f}$)	0.9980	0.9980
Selective correlation coefficient ($k_{s,f}$)	0.9919	0.9919
Nonselective outage probability ($P_{dns,f}$)	2.53×10^{-4}	2.53×10^{-4}
Selective outage probability ($P_{ds,f}$)	4.77×10^{-7}	4.77×10^{-7}

TABLE 3.1 *Continued*

Station A–Station B Link	ITU-R P.530-8	
Total outage probability for frequency diversity ($P_{d,f}$)	2.61×10^{-4}	2.61×10^{-4}
Worst-month outage with frequency diversity, s	677.0	677.0
Space and frequency diversity	Two receivers	Two receivers
Outage probability due to nonselective fading ($P_{dns,sf}$)	8.35×10^{-6}	6.99×10^{-6}
Correlation coefficient for nonselective fading ($K_{ns,sf}$)	0.9476	0.9374
Correlation coefficient for relative amplitudes ($r_{w,sf}$)	0.9347	0.9219
Correlation coefficient for selective fading ($k_{s,sf}$)	0.9512	0.9475
Outage probability due to selective fading ($P_{ds,sf}$)	8.07×10^{-8}	7.50×10^{-8}
Outage probability for space and frequency diversity ($P_{d,sf}$)	8.95×10^{-6}	7.52×10^{-6}
Worst-month outage for space and frequency diversity, s	23.2	19.5
ITU-T G.821 ISDN-HQ recommendation (SES / BER > 10^{-3})	0.00086%	0.00086%
ITU-T G.821 ISDN-HQ worst-month outage allocation, s	22.4	22.4
ITU-T G.821 ISDN-HQ (1/3) average-year unavailability (10*s* consecutive), s	504.9	504.9
Rain precipitation ($R_{0.001}$)	*K*	42 mm h^{-1}
Coefficient K_h, K_v	0.00412	0.00345
Coefficient A_h, A_v	1.3905	1.3797
Reduction factor of path length	0.32	0.32
Effective rain margin, dB	31.0	31.0
Atten. horiz. pol. at 0.01% average year / worst month, dB	9.5	17.9
Atten. vert. pol. at 0.01% average year / worst month, dB	7.6	14.4
Other percentage (*P* from 0.001 to 1%)	0.001%	0.001%
Atten. horiz. pol. at *P*% average year / worst month, dB	20.2	38.4
Atten. vert. pol. at *P*% average year / worst month, dB	16.3	30.9
Average-year unavailability due to rain in horiz. pol., %	<0.0002	<0.0002
Average-year unavailability due to rain in vert. pol., %	<0.0001	<0.0001
Average-year outage due to rain in horiz. pol., s	<60.0	<60.0
Average-year outage due to rain in vert. pol., s	<21.0	<21.0
Worst-month multipath outage, s	<23.2	<19.5
Worst-month availability, %	>99.99911	>99.99925

Note that frequency diversity provides poor performance in the case of this link.

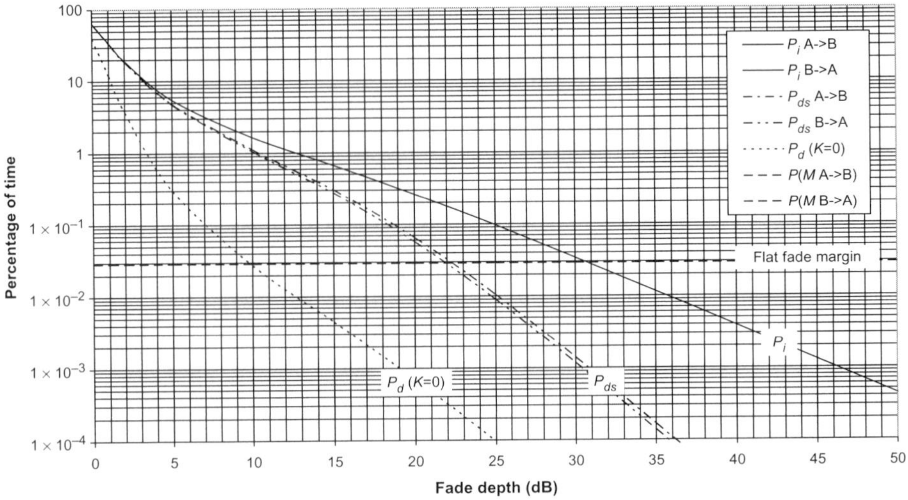

Figure 3.62 Distribution of individual and combined signals over worst month.

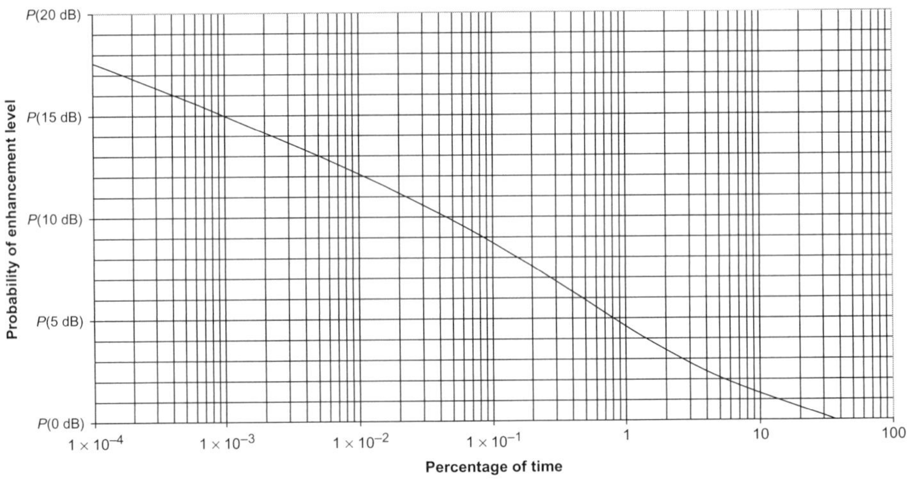

Figure 3.63 Distribution of field enhancement over worst month.

- For a quick calculation:

$$p_w = Kd^3(1+|\varepsilon_P|)^{-1.2} \times 10^{0.033f-0.001h_l-A/10} \tag{3.83}$$

$$p_0 = Kd^3(1+|\varepsilon_P|)^{-1.2} \times 10^{0.033f-0.001h_l} \tag{3.84}$$

where h_l is the altitude of the lower antenna.

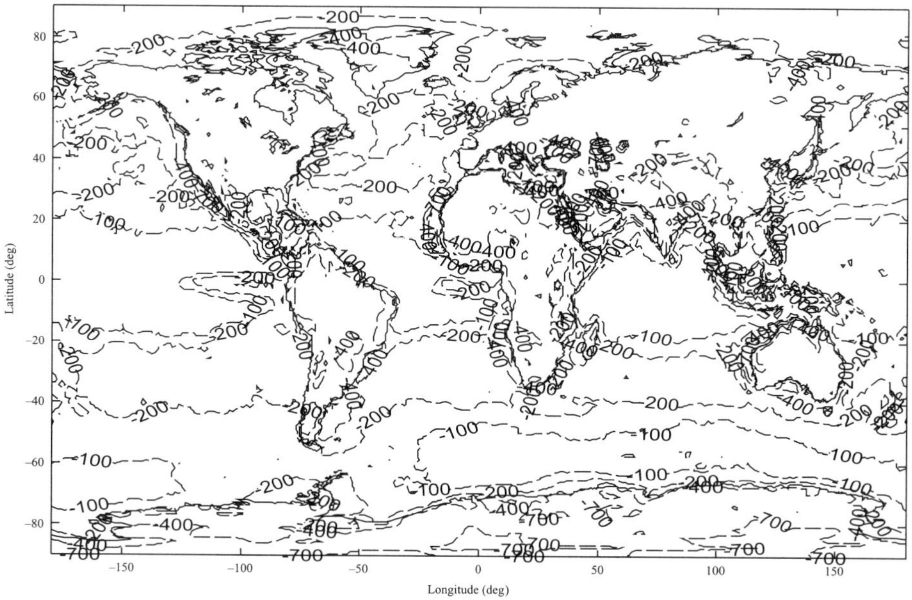

Figure 3.64 Refractivity gradient in lowest 65 m of atmosphere not exceeded for 1% of average year (ITU-R P.453-9).

- For a detailed link design:

$$p_w = Kd^{3.2}(1+|\varepsilon_P|)^{-0.97} \times 10^{0.032f-0.00085h_l-A/10} \tag{3.85}$$

$$p_0 = Kd^{3.2}(1+|\varepsilon_P|)^{-0.97} \times 10^{0.032f-0.00085h_l} \tag{3.86}$$

Equations (3.83) and (3.85) were derived from multiple regressions on fading data for 251 links in various geoclimatic regions with pathlengths in the range of 7.5–185 km, frequencies in the range of 37–450 GHz, path inclinations up to 37 mrad, altitudes from 17 to 2300 m, refractivity gradients dN_1 in the range of −860 to −150 N km^{-1}, and area terrain roughness S_a in the range of 6–850 m; the overall standard deviation of the prediction error is about 5.7 dB. These equations are also expected to be valid for frequencies to at least 45 GHz and for a lower frequency limit $f_{\min} = 15/d$ gigahertz.

Figure 3.65 represents, as an indication, the relation that has been established between the climatic variable P_L and the refractivity gradient dN_1 for a path of 40 km on smooth Earth.

Table 3.2 shows the results obtained using the formulas in the preceding example of link design where the roughness factor has been chosen as parameter S_a. The new prediction method, based on parameters dN_1 and S_a, gives about the same results than the preceding, which uses the climatic variable P_L and topologic assumptions. Note that parameters dN_1 and S_a are defined for relatively large zones and that topologic conditions may be quite different at

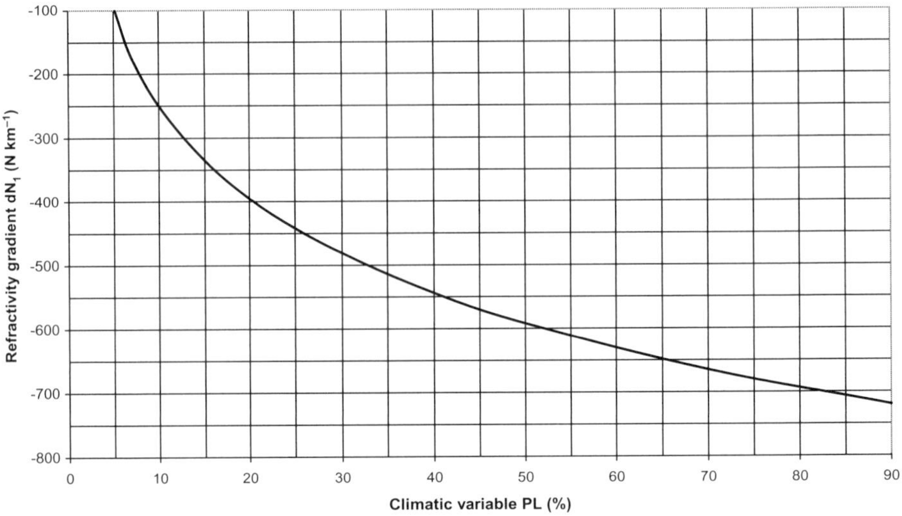

Figure 3.65 Refractivity gradient dN_1 versus climatic variable P_L.

TABLE 3.2 Results of Link Design Example

dN_1/S_a	–400	17.5
Geoclimatic factor, K		5.99E-04
Multipath occurrence factor (p_0), %	32.4	32.4
Fade depth reference (A_t), dB	27	27
Reference percentage of time (p_t), %	0.0674	0.0674
Factor q'_a	2.37	2.37
Factor q_t	0.63	0.63
Factor q_a	2.30	2.30
Probability of outage BER > 10^{-3} (p_w), %	0.02697	0.02697
Nonselective outage prediction (P_{ns})	0.000270	0.000270
Multipath activity parameter (η)	0.082	0.082
Selective outage probability (P_s)	0.000025	0.000025
Worst-month outage probability for single channel, s	764.6	764.6
Space diversity, m	6.5	8.0
Multipath occurrence parameter (P_0)	0.323745	0.323745
Vertical space diversity improvement factor ($I_{ns,s}$)	29.7	35.5
Nonselective correlation coefficient ($K_{ns,s}$)	0.9500	0.9400
Correlation coefficient ($r_{w,s}$)	0.9377	0.9251
Selective correlation coefficient ($k_{s,s}$)	0.9522	0.9483
Nonselective outage probability ($P_{dns,s}$)	9.07×10^{-6}	7.59×10^{-6}
Selective outage probability ($P_{ds,s}$)	8.32×10^{-8}	7.71×10^{-8}
Total outage probability for space diversity ($P_{d,s}$)	9.71×10^{-6}	8.15×10^{-6}
Worst-month outage with space diversity (s), s	25.2	21.1

the locus of the actual path than for the averaged whole zone; the terrain roughness given in section 3.2.3 may thus be more representative than S_a.

3.4.4.2 Conversion from Average Worst Month to Shorter Worst Periods of Time The percentage of time p_w of exceeding a deep fade A in the average worst month can be converted to a percentage of time p_{sw} of exceeding the same deep fade during a shorter worst period of time $1\,\text{h} \le T < 720\,\text{h}$ by the following relations:

- For relatively flat paths

$$p_{sw} = p_w(89.34T^{-0.854} + 0.676) \tag{3.87}$$

- For hilly paths

$$p_{sw} = p_w(119T^{-0.78} + 0.295)$$

- For hilly land paths

$$p_{sw} = p_w(199.85T^{-0.834} + 0.175)$$

3.4.4.3 Occurrence of Simultaneous Fading on Multihop Links Experimental evidence indicates that, in clear-air conditions, fading events exceeding 20 dB on adjacent hops in a multihop link are almost completely uncorrelated. For a multihop link, the total outage probability for clear-air effects can be obtained by summing the outage probabilities of the individual hops. The probability of exceeding a fade depth A can be estimated from the following formula, which has been deduced from the results of measurements on 47 pairs of adjacent line-of-sight hops operating in the 5-GHz band with pathlengths in the range of 11–97 km:

$$P_T = \sum_{i=1}^{n} P_i - \sum_{i=1}^{n-1} (P_i P_{i+1})^C \tag{3.88}$$

where P_i is the outage probability predicted for the ith hop of the total n hops and

$$C = 0.5 + 0.0052A + 0.0025(d_i + d_{i+1})$$

where $A \le 40\,\text{dB}$

$d_i + d_{i+1} \le 120\,\text{km}$ with d_i and d_{i+1} the adjacent path lengths.

$C = 1$

3.4.4.4 Optimum Choice of Antenna Heights Recommendation ITU-R P.530-11 gives a method that gives results similar to those described in Section 2.3.4 but limited to subrefraction conditions.

Pathlength Difference First we calculate the heights h_1 and h_2 above a reflection area, which may be a body of water or any consequent reflecting surface on the path of inclination angle v, as follows:

$$h_1 = h_{1G} + y_1 - y_0 + x_0 \times 10^3 \tan v$$

$$h_2 = h_{2G} + y_2 - y_0 - (d - x_0) \times 10^3 \tan v$$

where y_1, y_2 = altitudes of sites 1 and 2 above sea level (m)
h_{G1}, h_{G2} = heights of antennas at sites 1 and 2 (m)
y_0 = altitude at midpoint of reflection area above sea level (m)
x_0 = distance at midpoint of reflection area from site 1 (km)
d = pathlength (km)

For a range of effective factor K values varying from K_e (99.9%) to infinity, we calculate the distances in kilometers of the reflecting surface from sites 1 and 2 as follows:

$$d_1 = \frac{1}{2} d(1+b) \qquad d_2 = \frac{1}{2}[d(1-b)]$$

where

$$b = 2\sqrt{\frac{m+1}{3m}} \cos\left[\frac{\pi}{3} + \frac{1}{3}\arccos\left(\frac{3c}{2}\sqrt{\frac{3m}{(m+1)^3}}\right)\right]$$

$$m = \frac{d^2}{4a_e(h_1 + h_2)} \times 10^3 \qquad a_e = Ka$$

$$c = \frac{h_1 - h_2}{h_1 + h_2}$$

The pathlength difference in wavelengths between the direct and reflected rays for the same range of effective factor K values is given by the relation

$$\tau = \frac{2f}{0.3d}\left(h_1 - \frac{d_1^2}{12.74k}\right)\left(h_2 - \frac{d_2^2}{12.74K}\right) \times 10^{-3} \tag{3.89}$$

Each time the number of wavelengths τ is a positive integer as K varies, the received signal level passes through a minimum; so, if $\tau_{\max} - \tau_{\min} < 1$, the diversity protection is not necessary.

Antenna Spacing The appropriate spacing between the antennas in space diversity systems at sites 1 and 2 is given by the relations

$$\theta_1 = \frac{150d}{f(h_2 - d_2^2/(12.74K))} \qquad \theta_2 = \frac{150d}{f(h_1 - d_1^2/(12.74K))} \tag{3.90}$$

The optimum spacing of the diversity antenna at each and is then an odd multiple of half values of θ_1 and θ_2.

Frequency Separation For the same range of effective factor K values, we calculate the possible frequency separations for the main and diversity channels in frequency diversity systems from

$$\Delta f_{\min} = \frac{(7.5 \times 10^4)d}{[h_1 - d_1^2/(12.74K)][h_2 - d_2^2/(12.74K)]} \tag{3.91}$$

The possible optimum frequency separation between the main and diversity channels is then an odd multiple of $\Delta f_{\min}$.

The table below shows the difference of results between ITU-R P.530-12 and Section 2.3.4 in the case of the link design described in Section 3.2.8 for both extreme values of factor K, including ∞.

	τ			Θ_1 (m)			Θ_2 (m)			Δf_{min} (MHz)		
K	−0.7	∞	+0.7	−0.7	∞	+0.7	−0.7	∞	+0.7	−0.7	∞	+0.7
ITU-R P.530-12	23.1	11.2	3.0	3.3	5.4	14.4	9.9	12.5	17.1	17.3	357	1314
Section 2.3.4	23.1	11.2	3.1	4.2	5.4	10.1	7.7	12.5	17.4	173	357	1310

Use of Antenna Discrimination Recommendation ITU-R P.530-12 gives formulas which produce about the same effect as the method exposed in Section 3.2.10.

3.5 PROTECTION AGAINST JAMMING

The effect of a jammer on a receiver has been considered in Section 2.3.3 and the purpose here is to determine the interference levels, which are generally of two types:

- Distant interference between adjacent links and cochannels
- Local interference, which concerns coupling between equipment inside shelter and between antennas on tower

3.5.1 Calculation of Jamming Level

Figure 3.66 illustrates various jamming paths between antennas mounted on the same tower, corresponding to the local mode, and between antennas of adjacent links, corresponding to the distant mode.

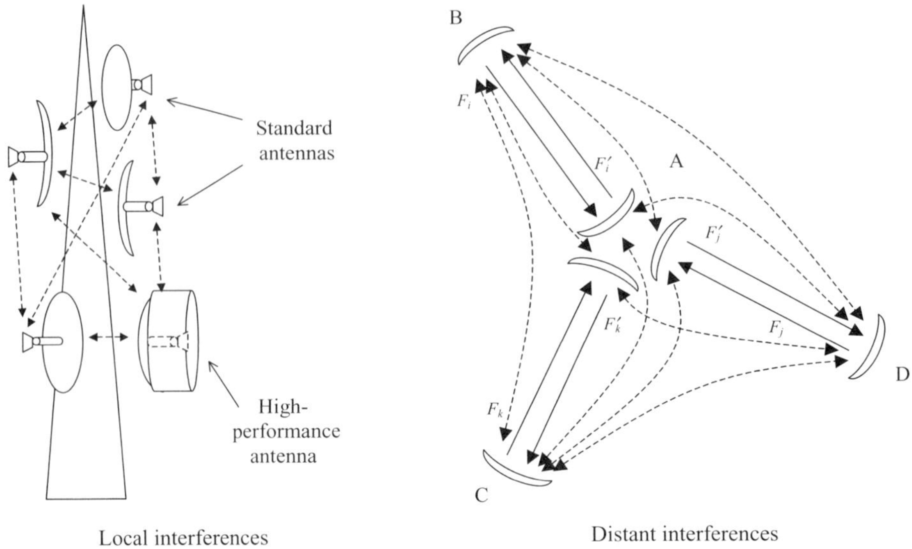

Figure 3.66 Local and distant interferences between adjacent links.

3.5.1.1 *Factor of Reduction of Interference by Antennas in Far Field* The role of the antennas is of primary importance in the study of both local and distant types of interference, in particular concerning the techniques of frequency reuse. The radiation patterns that the manufacturers provide generally are valid only for the far field, that is, beyond the Fresnel's zone. The main characteristics of the antennas which have to be taken into account in the study of jammers, as discussed in Sections 1.3 and 1.11.2, are as follows:

1. Cross-polarization isolation, which relates to the coupling on the same feed horn between two signals received in orthogonal polarization or between emission and reception
2. Cross-polarization discrimination, that is, the ratio between the component of the same signal which is received on the copolar access and that received on the cross-polar access
3. Radiation pattern, which indicates the loss of gain according to the angular separation with respect to the axis of radiation
4. Ellipticity ratio or axial ratio, which relates to the purity of polarization at emission and is in general of the same order of magnitude as the cross-polarization isolation

The high-performance (HP) antennas are distinguished from the standard (STD) antennas by a metal ring covered by an absorbing material around the reflector whose function it is to reduce the secondary lobes, particularly beyond 90°; their main lobe and their gain are practically identical but their cross-

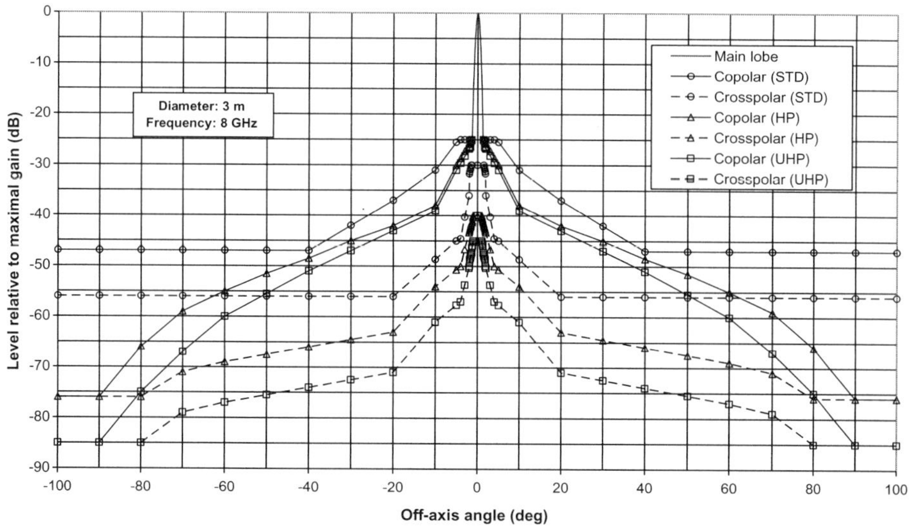

Figure 3.67 Comparative radiation patterns of STD, HP and UHP antennas.

polarization isolation and their front-to-back ratio are quite different. For example, Figure 3.67 shows the comparative radiation patterns of three types of 3-m antennas in co-polar and cross-polar configurations at 8 GHz, one of standard type and the others with high performance and ultra high performance (UHP).

3.5.1.2 Factor of Reduction of Interference by Antennas in Near Field In general, radiation pattern that is valid for the near field and allows us to determine the levels of local type interference between antennas installed on the same tower does not exist. The study here consists in calculating the free-space loss between the horns of the various antennas by regarding them as point sources and by considering the attenuation brought by the shield that the reflectors or the rings constitute as well as the possible cross-polarization discrimination. Figure 3.66 presents the configurations for antennas in the tower that one may meet for the determination of the local interferences. Generally, when the sources are visible, we employ the formula for free-space loss A_{FS} between two isotropic sources separated by a distance d by adding the cross-polarization discrimination XPD when the useful signal and the jammer are on two orthogonal polarizations:

$$A = A_{\mathrm{FS}} + \mathrm{XPD} = 20\log\frac{4\pi d}{\lambda} + \mathrm{XPD} \qquad (3.92)$$

The additional attenuation A_{Sh} brought by the shield that the reflectors or the rings of the high performance antennas constitute is difficult to evaluate, but

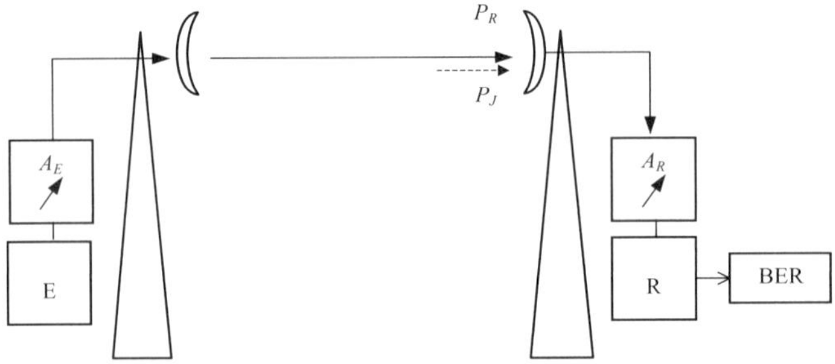

Figure 3.68 Method of two attenuators.

experiment shows that its value is generally higher than 20 dB per HP antenna. The general relation is thus obtained:

$$A = 20\log\frac{4\pi d}{\lambda} + \text{XPD} + A_{\text{Sh}} \tag{3.93}$$

Regarding indoor equipment, the jamming between channels depends on their electromagnetic compatibility and susceptibility characteristics.

3.5.2 Measurement of Jamming Level and Consecutive Degraded Threshold

The method suggested here, called the *method of two attenuators*, which measures the level of a jammer at the site affecting a receiver, has the advantages of simplicity as well as sensitivity.[13] This method is very suitable for line-of-sight links but can also be employed for other systems. It consists in indirectly measuring the degradation of the threshold of the receiver by means of the assembly presented in Figure 3.68. It is supposed that the level of the jammer P_J is largely lower than that of the useful signal, which corresponds to a necessary condition since the link would not be available if the SNR would become lower than the required C/N. The assembly consists in interposing a first variable attenuator A_E between the transmitter and the emitting antenna and a second one A_R between the reception antenna and the receiver. The test is carried out in two steps:

- The nominal value of the threshold of the receiver is measured by gradually increasing the attenuation A_R until obtaining the reference BER, attenuator A_E being minimum; in this case, the jammer is in general

[13]This indirect method has a much greater sensitivity than that of a spectrum analyzer since it is free from its noise factor and required C/N that are inherent to any receiver or system of measurement of an electromagnetic wave; the gain in terms of sensitivity can thus reach several tens of decibels.

attenuated to a sufficiently low level with respect to the reference background noise of the receiver so as not to have any more influence, and the measured value is close to the characteristic indicated by the manufacturer,

- The level of the presumably disturbed threshold is measured while placing the attenuator A_R at minimum and by increasing the attenuation at emission A_E until obtaining the reference BER; thus, the disturber does not undergo any attenuation and interferes directly with the useful signal.

The uniform margin to take into account in predicting the performance of the link must then be that which was calculated under the most favorable conditions of propagation for the disturber (field enhancement) or that which rises from the level of the degraded threshold as measured on the site.

3.6 FREQUENCY REUSE TECHNIQUES

By increasing the number of states of the carrier, the new high-level modulations make it possible to considerably improve the spectrum efficiency of the microwave links. Between two modulations employing respectively M and N states for the carrier, their respective bandwidth is in the ratio $2^{m/n}$ with $m = \log_2 M$ and $n = \log_2 N$, as shown in Figure 2.28. The various techniques of frequency reuse aim to reduce the constraints within the frequency plan while making it possible to employ again the same set of emission/reception frequencies on the collateral and adjacent connections. The spectrum efficiency of a network can thus be multiplied by several decades by using high-level modulations in association with frequency reuse techniques. The processes suggested also aim at privileging the modes of diversity that reduce the number of frequencies, such as space diversity.

3.6.1 "Back-to-Back" Process

Figure 3.69 presents two back-to-back links established with HP antennas on the central tower B. The same set of frequencies F_1–F_1' are used on the two links on both sides of the tower B; by supposing that the equipments of these links have the same features, we can easily write that the level of jamming at reception in A by the emission of B toward C is equal to

$$P_J = P_R + G_{BC} - G_{BA} - R_{FB}$$

where P_R = useful signal received at A
G_{BC} = gain of antenna of B facing C
G_{BA} = gain of antenna of B facing A
R_{FB} = front-to-back ratio of antenna of B facing C

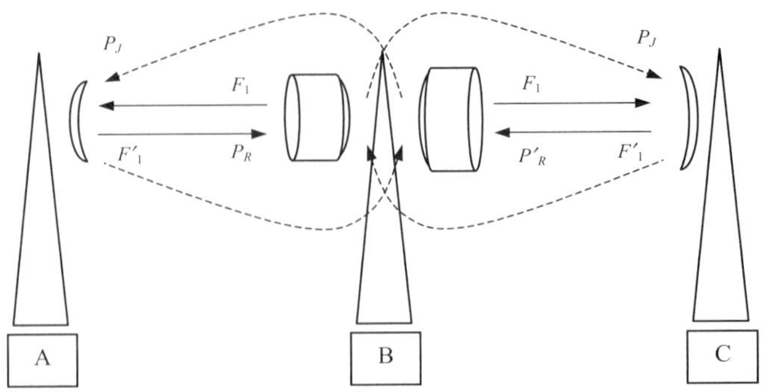

Figure 3.69 Frequency reuse with HP antennas (back to back).

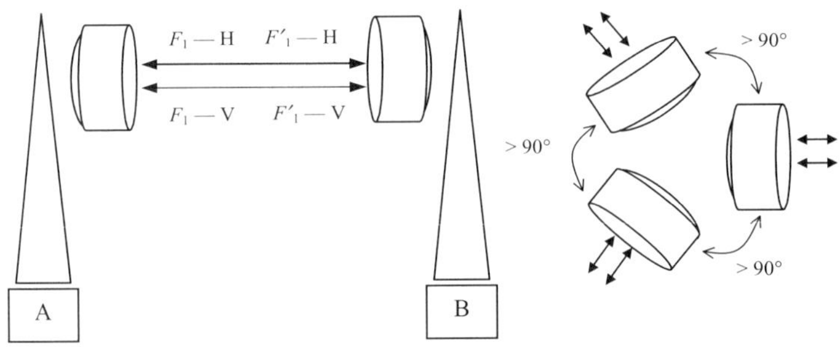

Figure 3.70 Frequency reuse with bipolar HP antennas (front to front + back to back).

From the formulas in Section 2.3.3 about interference, knowing the reference background noise N_R of the receiver, one can easily calculate the degradation of the threshold A which very strongly depends on the front-to-back ratio of the antenna toward C—thus the interest in using HP antennas. Figure 3.67 shows that the front-to-back ratio is maximum for off-axis angles higher than 90°, which makes it possible to employ again the same set of frequencies on three or even four adjacent links.

3.6.2 "Front-to-Front" Process

Figure 3.70 shows the technique that makes it possible to double the capacity of the link by using the same frequency plan for two channels on two orthogonal polarizations by means of bipolar antennas and realize up to three or four directions with the same frequencies. In order to reduce the cross-polar component, which constitutes the jammer, we generally use HP antennas with high purity of polarization and strong isolation in association with the XPIC process for the equipment. The effects of cross-polarization by the clear atmosphere and the hydrometeors can be calculated by means of the formulas in Section 3.4.1.10 for systems without diversity. The diversity of space and/or frequency does not improve the availability of the link when the depolarization is due to the hydrometeors. On the other hand, the depolarization by clear air being mainly due to multipath, the techniques of diversity provide an improvement of the same order of magnitude as that for deep fades which have the same causes as a reflection acts on the polarization of the electromagnetic waves.

3.6.3 XPD Model and Measurement

As there is no model available for diversity improvement of XPD reduction, a survey has been carried out on a 65-km SDH-XPIC link ($2 \times 155\,\text{Mbits}\,\text{s}^{-1}$) with an XPIF of 25 dB.

3.6.3.1 Test Link Characteristics The system comprises four receivers connected to both polarizations of main and diversity antennas and only one transmitter has been activated for the measurement of XPD. The table below gives the main parameters of that link. Figure 3.71 presents the path profile of that link for all possible conditions of propagation. Figure 3.72 presents an extract of the recording performed at station F of both copolar and cross-polar signals levels simultaneously with a sampling rate of 10 Hz. Figure 3.73 shows the XPD values deduced over the same period from the recording of copolar and cross-polar signals on both main and diversity bipolar antennas. The table below gives the main data about this link.

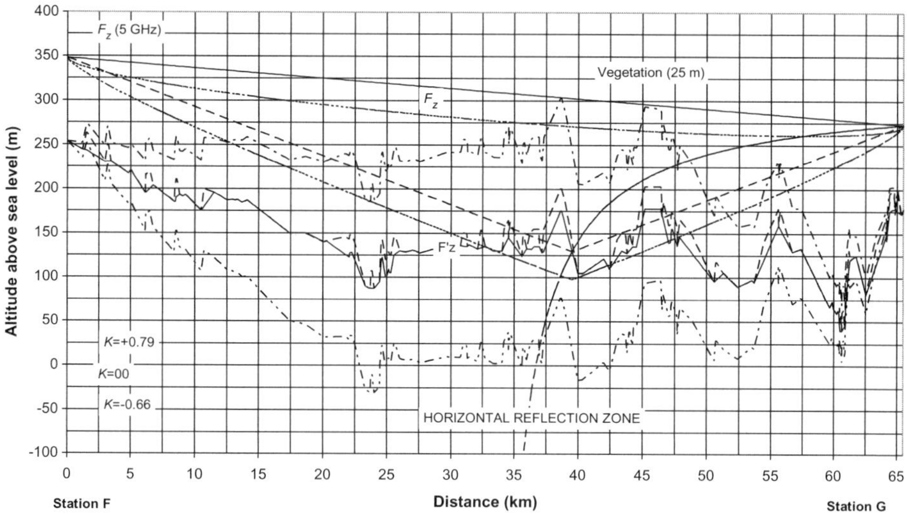

Figure 3.71 Path profile for all possible conditions of refractivity.

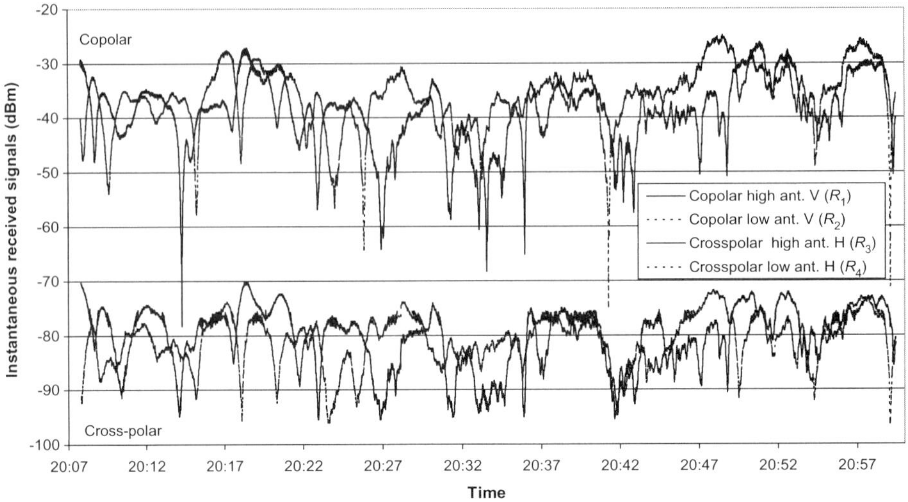

Figure 3.72 Recording of copolar and cross-polar signals (July 28, 2006).

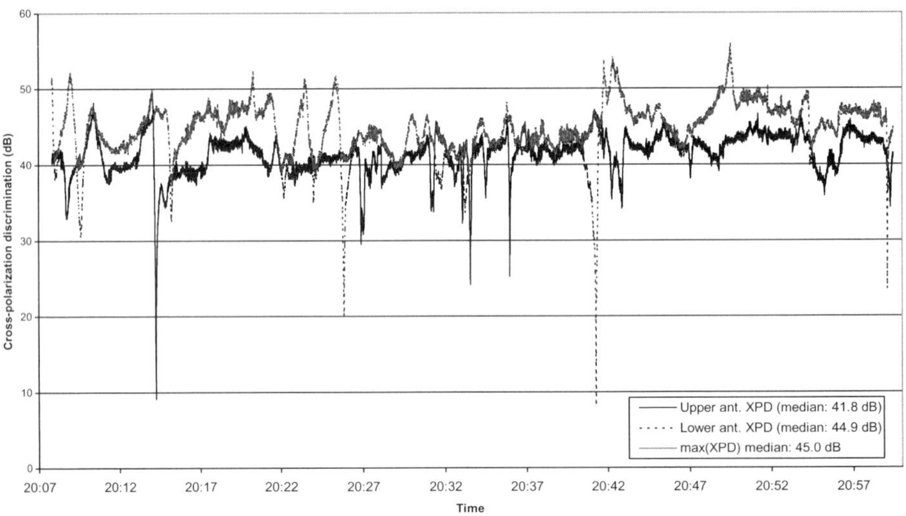

Figure 3.73 Recording of XPD on main and diversity antennas (July 28, 2006).

Frequency	5 GHz	Bit rate	2 × SDH-STM1	2 × 155 Mbits s^{-1}
Parameters	$P_L = 10\%$	$dN_1 = -250\ \mathrm{N\,km^{-1}}$	Roughness = 37.5 m	$S_a = 50$ m
Antennas (UHP)	$D = 3$ m	$G = 40.7$ dBi	$F/B = 68$ dB	XPD = 40 dB, XPI = 45 dB
Space diversity	Station F:S = 8 m	Station G:S = 14 m	Defined by the interferometric method	Defined by the interferometric method

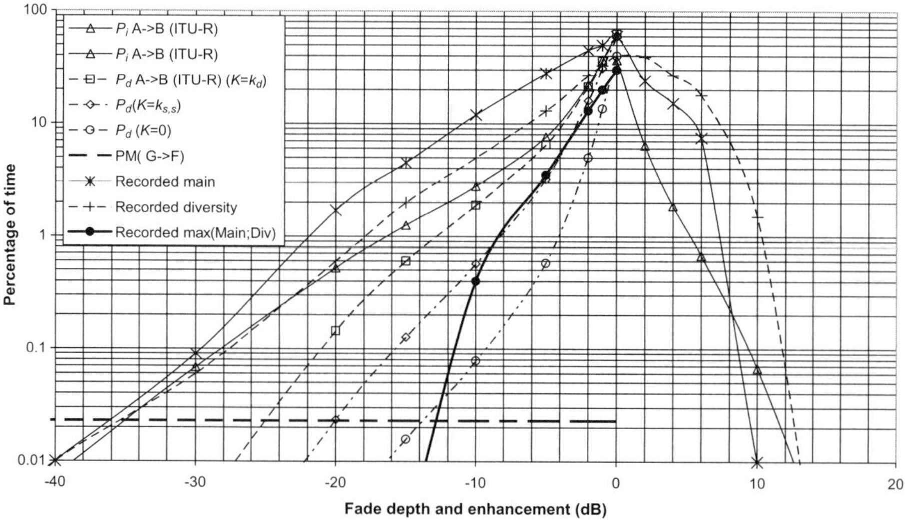

Figure 3.74 Comparative distributions of copolar signals.

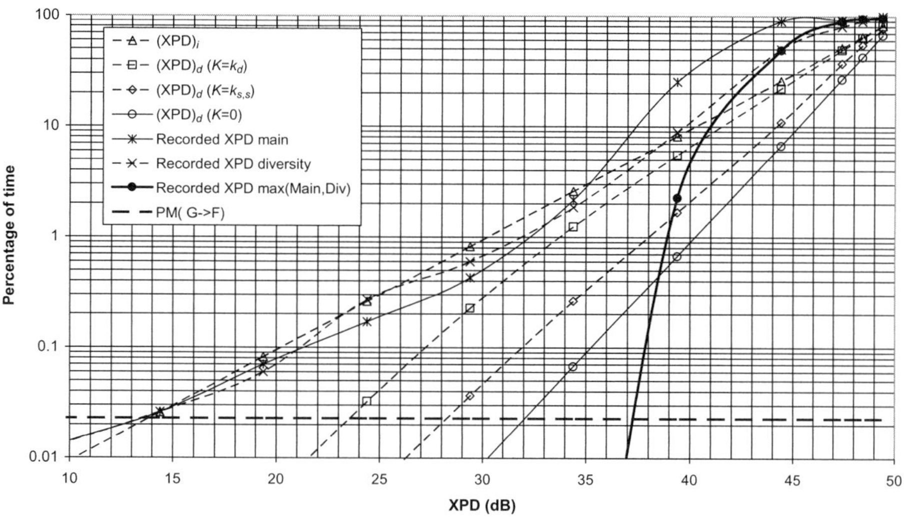

Figure 3.75 Comparative distributions of XPD.

Figure 3.74 presents the comparative distributions of both theoretical and recorded individual and combined copolar signals. Figure 3.75 presents the comparative distributions of both theoretical and recorded individual and combined XPD.

3.6.3.2 Interpretation of Recordings The distribution of the individual and combined signals as well as the improvement factor were defined according to ITU-R P.530 and compared with the new hypothesis emphasizing the following:

- Overall correlation coefficient for selective and nonselective fading ($K = k_d$) according to ITU-R P.530, which can be deduced from total outage probability without diversity ($P_i = P_{ns} + P_s$) and with diversity P_d through relation (1.114)
- Correlation coefficient for selective fading ($K = k_{s,s}$) in space diversity, which primarily refers to XPD reduction insofar as selective fading is linked with multiple reflections which also cause depolarization of electromagnetic waves
- Null coefficient of correlation ($K = 0$), which rises from correlation in opposition of signals received in diversity on antennas whose spacing is defined by interferometric method

3.6.3.3 Proposed Model When Using the Interferometric Method The recordings show that the improvement of the performance due to space diversity, when using the interferometric method, is quite better than predicted as well as for the signals than for the XPD reduction; these results suggest a new model in the case of a specularly-reflective path or in ducting conditions of propagation. This new model rests on method ITU-R P.530 for the individual fade that fits in with all the results of measurement carried out in the past and that are independent of the mode of diversity.

On the other hand, for prediction of outage in systems using diversity, the author proposes replacing the coefficients of correlation relative to the selective and the nonselective fadings by an overall null coefficient of correlation when the interferometric method is used. In this case, the percentage of time $p_d(M)$ during which a fade depth of the combined signal corresponding to the margin M is exceeded during the worst month becomes, compared to the individual percentage of time $p_w(M)$,

$$p_d(M) = 100 \times \left(\frac{p_w(M)}{100}\right)^2$$

where $p_w(M)$ is given by relation (3.19) or (3.23) as the case may be.

This new model also rests on the method in ITU-R P.530 which indicates the way of calculating the probability of interruption $P_{\text{Xp},C}$ and the equivalent margin M_{XPD} due to the XPD reduction in clear air as well as the probability of outage $P_{\text{Xp},R}$ by rain for systems without diversity. Concerning systems with diversity, unavailability due to the rain cannot be reduced because both channels are affected by the same degradation at the same time, which is not the case for outage due to multipath in clear-air conditions; the following formulas for the individual and combined outage probabilities and the equivalent combined margin seem suitable:

$$(P_{\mathrm{Xp},C})_d = (P_{\mathrm{Xp},C})_i^2 \qquad (P_{\mathrm{Xp},C})_i = P_0 \times 10^{-M_{\mathrm{XPD}}/10}$$

where P_0 is the multipath occurrence parameter and

$$(M_{\mathrm{XPD}})_d = 10\log\left(\frac{P_0}{(P_{\mathrm{Xp},C})_d}\right) = 10\log\left(\frac{P_0}{(P_{\mathrm{Xp},C})_i^2}\right)$$

3.7 COMPARISON BETWEEN VARIOUS DIVERSITY TECHNIQUES

The recordings which follow were carried out on a terrestrial link with a view to compare the techniques of diversity usually employed:

- Space diversity
- Frequency diversity
- Path and frequency diversity
- Quadruple diversity of space and frequency

We saw previously that the interferometric method allows, in the case of space diversity, to correlate in opposition the signals received on the main and diversity antennas, which confers an optimal improvement with respect to multipath and particularly in the case of reflection on water by getting an improvement factor close or even higher than the theoretical value defined by Rayleigh's distribution for a double diversity of totally independent random signals. The same result can be obtained in theory by means of the only frequency diversity, but it is practically impossible to obtain the optimal frequency separation because of the radio frequency channel arrangement in force in the frequency plan, and this process requires an additional set of frequencies, which is against the required goal for optimization of the frequency resources.

The comparative tests were carried out at the frequency of 7.5 GHz on a terrestrial link, illustrated in Figure 3.76, which is quite cleared off the main obstacle. The recording test has been performed at station T with the following data:

Path length, 102.4 km	Bit rate, 34 Mbits s^{-1}
Climatic variable P_L, 30%	Beam slope, 10.3 mrad
Roughness, 211 m	Clearance, $h/R_F(50\%) = 0.76$
Frequency separation, 77 MHz	Half-fringe size, 0.9 m
Antenna diameter, 3 m	Antenna spacing, 13.7 m

The test assembly at station T comprises two bipolar antennas in space diversity which are connected to four receivers operating two by two on both frequencies emitted by station L, one on the main antenna and the other on the diversity antenna in order to obtain a path and frequency diversity system.

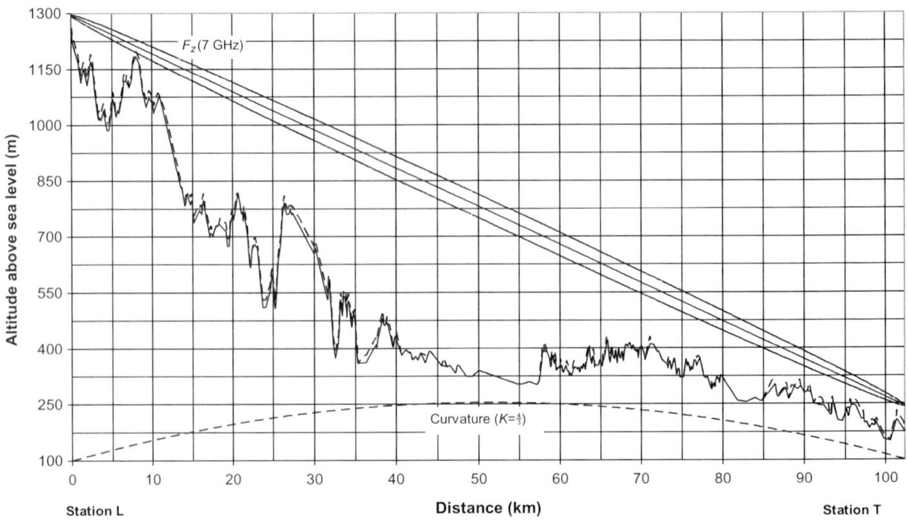

Figure 3.76 Path profile of test link (L–T).

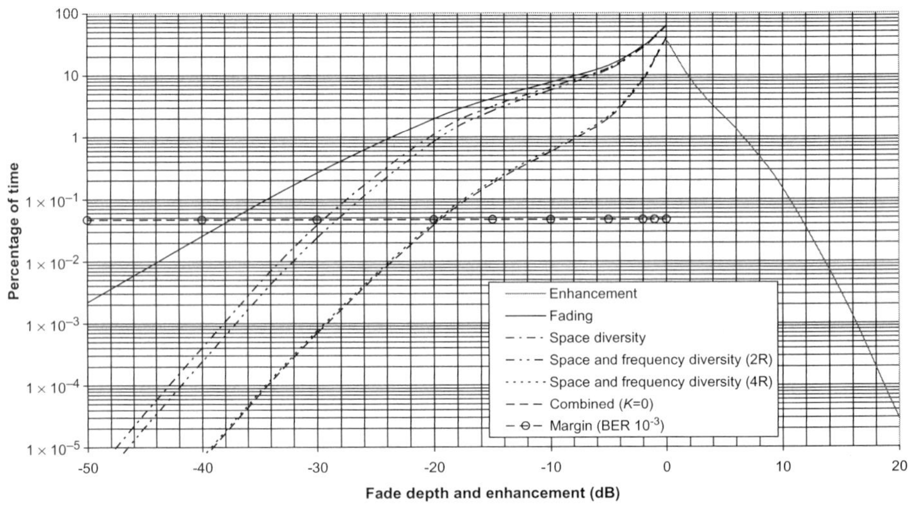

Figure 3.77 Computed distributions for individual and combined signals (worst month).

Regrouping these receivers by two or four thus makes it possible to compare the results between them and to check the predicting models for two and four receivers.

Figure 3.77 shows theoretical distributions calculated by means of the formulas in Section 3.4, such as:

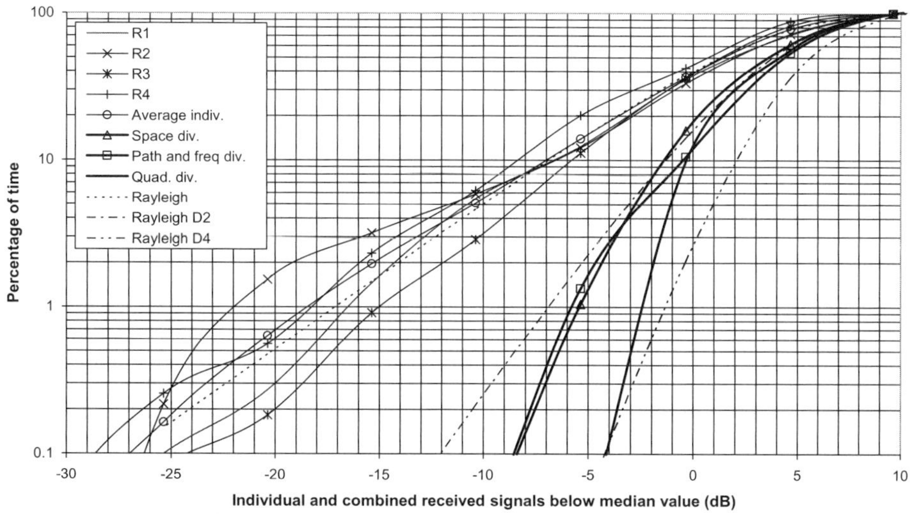

Figure 3.78 Recorded distributions of individual and combined received signals with respect to theoretical distributions (November 7, 2005–10:45 to 11:15).

- Instantaneous received signals (fading and enhancement)
- Predicted combined signal in space diversity
- Predicted combined signal in space and frequency diversity for two receivers
- Predicted combined signal in space and frequency diversity for four receivers
- Combined signal by supposing null coefficient of correlation ($K = 0$)
- Margin to threshold for BER = 10^{-3}

Figure 3.78 illustrates the recorded individual and combined received signals during a deep fading period versus the corresponding theoretical Rayleigh's distributions and shows the following:

1. Individual signals
2. Average level of individual signals
3. Individual Rayleigh's distribution
4. Combined signal in space diversity
5. Combined signal in double-path and frequency diversity (two receivers)
6. Combined Rayleigh distribution for two totally uncorrelated signals (D2)
7. Combined signal in quadruple-path and frequency diversity (four receivers)
8. Combined Rayleigh's distribution for four totally uncorrelated signals (D4)

Concerning these recordings, we can see that:

- The individual distributions follow quite perfectly the Rayleigh's distribution, which is the condition for the application of the Rayleigh's equations in the determination of the diversity gain and improvement factor exposed in Section 1.11.1.3.
- The improvement factor is close to the computed value for a null coefficient of correlation ($K = 0$) and is rather higher for the small percentages of time as well as for the double diversity, such as space diversity or path and frequency diversity with two receivers, than for a quadruple space and frequency diversity system with four receivers.
- The distribution for path and frequency diversity with two receivers is very close to that for a space diversity system only which involves totally independent signals.
- The distribution for path and frequency diversity with four receivers corresponds to that for a quadruple diversity system which involves totally independent signals.

It has to be pointed out that although the interference fringes due to horizontal reflection are integrated by the antenna, as a result of their small size at station T compared to the diameter of the antenna, the received field is subject to deep fading; multipath is therefore preponderant in the propagation.

In conclusion, although there is no reflection zone and despite significant terrain roughness, the processing of these recordings shows, by using the interferometric method, that:

1. Space diversity offers about the same improvement as a path and frequency diversity with two receivers which is close to a second-order Rayleigh's distribution for a null coefficient of correlation.
2. Path and frequency diversity with four receivers presents a higher efficiency which is close to a fourth-order Rayleigh's distribution for a null coefficient of correlation.

3.8 AVAILABILITY OF MICROWAVE LINE-OF-SIGHT SYSTEMS

Network performance in terms of quality of service, service operability, or trafficability depends on the combined aspects of availability, reliability, and maintainability of the various elements which contribute to the transmission of the information among those the microwave line-of-sight links constitute the most compound subjects. In general, the overall availability of a microwave line-of-sight link is divided into three distinct shares related to the probabilities of outage and failure:

- Hazards of propagation
- Jamming caused by interferences
- Reliability of equipment and associated materials

3.8.1 Introduction

3.8.1.1 Availability Due to Propagation Propagation performance refers to the ability of the medium to transmit the signal within intended specifications. The various causes of outage, which are mainly due to multipath and hydrometeors, thus very dependent on the climatic and topological conditions, were treated in Section 3.4. The performance objectives that are generally expressed in terms of quality of service and availability are given, for ISDNs, in Section 3.3.2.

3.8.1.2 Availability Due to Equipment There are several methods for estimating the inherent reliability of electronic circuits, equipments, and systems that are based on environmental use characteristics, piece part count, thermal and electrical stresses, subsystem repair rates, and system configuration. The ability to express reliability numerically is crucial because it enables us to concretely identify the requirements, contractual specifications, test guidelines, and performance assessment. Considering the useful lifetime of the equipment, the probability of failure remains almost constant over a given period of time; the failures can be considered as random and the equipment reliability can be predicted using analytical methods. One of the best methods is described in the handbook MIL-HDBK-217 (U.S. Department of Defense, 1992), which provides a common basis for inherent reliability predictions of a great number of electronic equipments and systems according to their component devices and conditions of use.

3.8.2 Availability of Microwave Terminal

The overall availability of a microwave line-of-sight system depends, in general, on the reliability of the following main elements:

- Equipment of transmission
- Antennas and branching devices
- Power supply and connection board, for example
- Human activities during servicing and maintenance, for instance

The inherent availability of each of these elements, that is, the ability to be in a state to perform a required function inside the system, is conditioned by their reliability, maintainability, and logistics in terms of spare parts and testing.

Evaluation of system availability lies primarily in the essential statistical parameters defined for each of its elements:

- Mean time between failures (MTBF)
- Mean time to restoration (MTTR)
- Mean repair time (MRT)
- mean downtime (MDT)

The MTBF is defined as the total functioning life of a population of an item during a specific measurement interval divided by the total number of failures within the population during that interval; it can be interpreted as the expected length of time a system will be operational between failures and used to calculate the required number of spares.

There are several methods to define the availability of an equipment according to:

- Redundancy of its main parts
- Effectiveness of mechanisms of security which come into play

The principal formulas which link together the parameters given above are, for example, as follows for a system composed of n elements:

- MTBF = $1/\Sigma_n\lambda_i$, where λ_i is the failure rate of the functional unit i among the n elements which form the system.
- Failure Rate = 1/MTBF.
- MTTR = $\Sigma_n(\lambda_i \text{ MTTR}_i)/\Sigma_n\lambda_i$, where MTTR_i is the MTTR of the functional unit i.
- Availability = MTBF/(MTBF + MTTR).
- Unavailability = 1 − Availability = MTTR/(MTBF + MTTR) ≈ MTTR/MTBF.

Figure 3.79 shows the possible configurations for the modules. The resulting availability for two serial modules, by considering for each one of them an unavailability U_i and U_j, is given by the following expression:

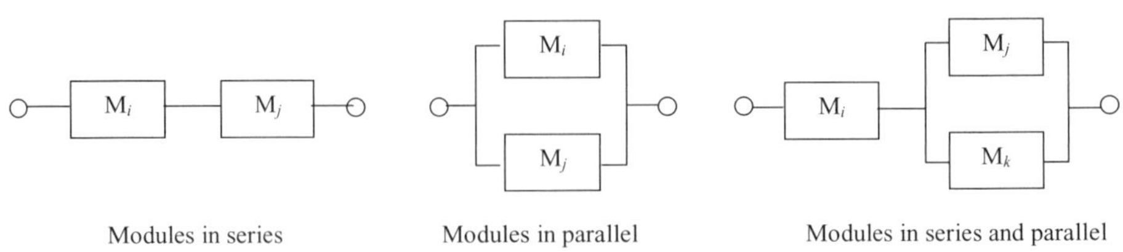

Figure 3.79 Possible module configurations.

- Availability for serial modules:

$$1 - U_i - U_j = 1 - \frac{\text{MTTR}_i}{\text{MTBF}_i} - \frac{\text{MTTR}_j}{\text{MTBF}_j}$$

- Unavailability for serial modules:

$$U_i + U_j = \frac{\text{MTTR}_i}{\text{MTBF}_i} + \frac{\text{MTTR}_j}{\text{MTBF}_j}$$

The availability for two parallel modules becomes:

- Availability for parallel modules:

$$1 - U_i U_j = 1 - \frac{\text{MTTR}_i}{\text{MTBF}_i} \frac{\text{MTTR}_j}{\text{MTBF}_j}$$

- Unavailability for parallel modules:

$$U_i U_j = \frac{\text{MTTR}_i}{\text{MTBF}_i} \frac{\text{MTTR}_j}{\text{MTBF}_j}$$

The availability for serial and parallel modules becomes, by introducing U_k for the third module:

- Availability for serial and parallel modules:

$$1 - U_i - U_j U_k = 1 - \frac{\text{MTTR}_i}{\text{MTBF}_i} - \frac{\text{MTTR}_j}{\text{MTBF}_j} \frac{\text{MTTR}_k}{\text{MTBF}_k}$$

- Unavailability for serial and parallel modules:

$$U_i + U_j U_k = \frac{\text{MTTR}_i}{\text{MTBF}_i} + \frac{\text{MTTR}_j}{\text{MTBF}_j} \frac{\text{MTTR}_k}{\text{MTBF}_k}$$

It is advisable to take into account the elements common to the chains laid out in parallel, such as the modem boards and switching circuits, which have to be regarded as elements in series in the calculation of the overall availability and must have, consequently, a very great reliability. The architecture of the whole equipment of telecommunication is thus very important in the determination of total system availability for it concerns equipment availability on the one hand and availability due to propagation on the other hand.

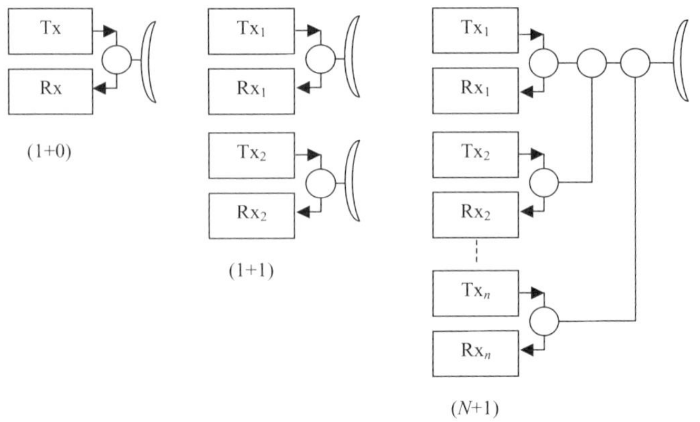

Figure 3.80 Typical radio terminal configurations.

Figure 3.80 illustrates several examples of configurations for transmitters and receivers, each pair being called transceivers, in radio relay terminals. The most current configurations for radio relay systems are as follows:

(1 + 0)	Unprotected
(1 + 1) HSB	Hot stand by
(1 + 1) HSB-S	Hot stand by with space diversity
D2-F	Frequency diversity
D2-P/F	Path and frequency diversity (two receivers)
D4-S/F	Frequency and space or path diversity (four receivers)
(N + 1)	N channels sharing a single redundant radio channel

There are other possible configurations, in particular by alternating polarizations.

3.8.3 Operational Availability of Microwave Radio Link

3.8.3.1 Operational Availability Related to Equipment The usual rules of calculation are based on the following relations where we distinguish the terminal, the units which make it up, and the link made up of two terminals:

1. Unit failure rate:

$$\lambda_i = \frac{1}{\mathrm{MTBF}_i}$$

where λ_{is} = serial modules
λ_{ip} = parallel modules

2. Terminal failure rate (global):

$$\mathrm{FR}_{TG} = \sum (\lambda_{is} + \lambda_{ip})$$

3. Terminal failure rate (degraded mode):

$$\mathrm{FR}_{TD} = \sum \lambda_{ip}$$

4. Terminal failure rate (total interruption):

$$\mathrm{FR}_{TI} = \sum \lambda_{is} + \left[\sum \lambda_{ip}\right]^2$$

5. Unavailability of channel i:

$$U_c = \mathrm{MTTR} \sum \lambda_i$$

6. Unavailability of channels i in parallel:

$$U_{cp} = \left[\mathrm{MTTR} \sum \lambda_{ip}\right]^2$$

7. Terminal unavailability:

$$U_T = \mathrm{MTTR} \sum \lambda_{is} + \left[\mathrm{MTTR} \sum \lambda_{ip}\right]^2$$

8. Terminal availability:

$$A_T = 1 - \left(\mathrm{MTTR} \sum \lambda_{is} + \left[\mathrm{MTTR} \sum \lambda_{ip}\right]^2\right)$$

9. Link availability:

$$A_L = 1 - 2\left(\mathrm{MTTR} \sum \lambda_{is} + \left[\mathrm{MTTR} \sum \lambda_{ip}\right]^2\right)$$

The elements laid out in parallel, which are redundant one of the other in a hot stand-by type configuration, can be regarded as completely independent in the field of reliability under the normal conditions of exploitation since the joint probability of failure, which is expressed by $(\lambda_i)^2$, is very weak. Their combined unavailability U_{cp} is thus the square of the product of their individual failure rate λ_i and their MTTR_i and thus we write

$$U_{cp} = [\mathrm{MTTR}_i \ \lambda_i]^2$$

Knowing the allocation for unavailability U_{cp}, we can deduce the minimum MTTR for a system made up of two elements in parallel:

TABLE 3.3 Availability Due to Microwave Radio Equipment

	MTBF (h)	Failure rate	MTBF (years)	Mode	Failures / year	MTBF / center (years)
Interface board	450,000	2.2×10^{-6}	51	Series	4	2.5
Connection board	1,200,000	8.3×10^{-7}	137	Series	2	5.0
Modem channel 1	400,000	2.5×10^{-6}	46	Parallel	5	2.0
Transceiver channel 1	200,000	5.0×10^{-6}	23	Parallel	9	1.1
Power filtering unit channel 1	1,200,000	8.3×10^{-7}	137	Parallel	2	5.0
Modem channel 2	400,000	2.5×10^{-6}	46	Parallel	5	2.0
Transceiver channel 2	200,000	5.0×10^{-6}	23	Parallel	9	1.1
Power filtering unit channel 2	1,200,000	8.3×10^{-7}	137	Parallel	2	5.0
Supervisory board	400,000	2.5×10^{-6}	46	Series	5	2.0
Service board	400,000	2.5×10^{-6}	46	Series	5	2.0
Auxiliary services board	350,000	2.9×10^{-6}	40	Series	6	1.7
Alarm display board	1,200,000	8.3×10^{-7}	137	Series	2	5.0
Branching unit	900,000	1.1×10^{-6}	103	Series	2	5.0
Power amplifier	100,000	1.0×10^{-5}	11	Series	18	0.6
Terminal (global)	25,301	4.0×10^{-5}	3			
Terminal (degraded mode)	60,000	1.7×10^{-5}	7			
Terminal (interruption of service)	43,750	2.3×10^{-5}	5			
Unavailability channel 1 or 2	2.5×10^{-5}					
Unavailability channel 1 and 2	6.3×10^{-10}					
Terminal total unavailability	9.2×10^{-6}					
Terminal total availability	99.9991%					
Link total unavailability	1.8×10^{-5}					
Link total availability	99.9982%					
Unavailability allocation ITU-R F.695	2.0×10^{-5}					
Availability allocation ITU-R F.695	99.9980%					

Note: Number of links, 100; number of terminals, 200; number of maintenance centers, 10; MTTR, 3 h.

$$\text{MTTR} \leq \frac{\sqrt{U_{cp}}}{\lambda_i}$$

Table 3.3 gives an example of the calculation of the operational availability of a microwave radio link in its various stages which makes it possible to determine, in order to comply with the total availability allocation of a network, the quantity of the necessary spare parts according to the number of terminals, centers of maintenance, and the MTTR. The network comprises 100 links and 10 maintenance centers and the computation shows that the required MTTR is 3 h in order to reach the intended availability objective. The number of necessary spare parts by center is easy to determine, knowing the estimated number of failures per year, according to the mean repair time of each kind of unit.

3.8.3.2 Operational Availability Related to Propagation We saw that the techniques of diversity, by implementing several independent channels, made it possible to considerably improve the availability of a connection with respect to the multipath but not against the hydrometeors, which affect all the channels in the same way at the same time.

Consequently, if the hydrometeors are the principal cause of the interruption or when it is about a link established in a configuration (1 + 1) HSB without diversity, the operational availability rests primarily on the equipment share and on the prediction of the outage without diversity exposed in Sections

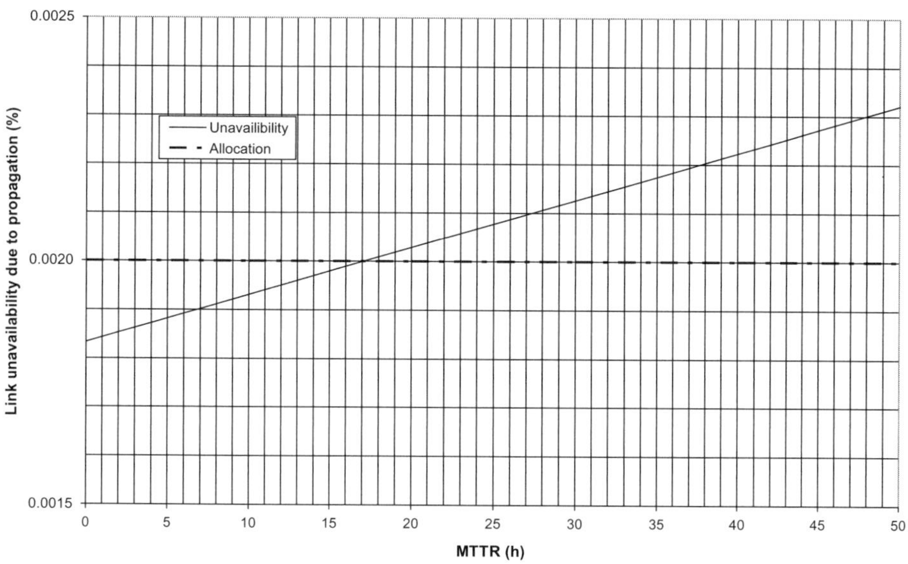

Figure 3.81 Link unavailability due to propagation during failure of redundant unit during worst month.

3.4.1.8 and 3.4.1.10. Concerning the systems using diversity, it is advisable to take into account the mechanisms which come into play when only one chain is affected by a failure or an interruption which does not relate to the others. By supposing that a channel is failing for a given period, for example, the MTTR, we can easily determine the operational availability of the link A_{Op} while utilizing the predictions of outage with diversity U_{WD} and without diversity U_{ND} by means of the expression

$$A_{Op} = 1 - \left[U_{WD} + (U_{ND} - U_{WD}) \frac{\text{MTTR}}{\text{MTBF}} \right]$$

Figure 3.81 presents the link unavailability due to propagation without diversity during failure of a redundant unit in the worst month, according to the MTTR starting from the unavailability of the equipment indicated in Table 3.3; the unavailability without diversity is supposed to be equal to the square root of the nominal availability with diversity. We can see that the required MTTR of 3h for equipment availability enables us to maintain the unavailability due to propagation below the allocated value.

CHAPTER 4

Microwave Transhorizon Systems

4.1 ENGINEERING OF TRANSHORIZON SYSTEMS

4.1.1 Introduction

In radio links established beyond the horizon, on paths generally much longer than line of sight, antennas are never visible to each other except possibly during a reduced percentage of time under extreme conditions of superrefraction. The large-scale average structure of the troposphere induces a bending of the radio waves, in relation to the general refractivity conditions, and the small-scale local structure produces the scattering that is responsible for the unexpected propagation a long way beyond the horizon. The mechanisms of propagation that then come into play at frequencies higher than 30 MHz are the diffraction on the obstacles on the surface of the Earth and the diffusion or scattering due to the presence of heterogeneities in the troposphere. The attenuation due to diffraction increases quickly with the distance, the height of the obstacles, and the frequency, as shown in the Sections 1.9 and 3.4.2, and gives way gradually to tropospheric scatter; it is thus necessary to analyze the share of each one of these mechanisms in order to predict the total propagation loss.

The propagation by tropospheric scatter, also called forward scatter or usually troposcatter, results mainly from a heterogeneous diffusion due to random irregularities or rapid variations in the refractive index of the atmosphere inside a common volume of diffusion that is delimited by the cones of both transmission and reception antenna main beams, as shown in Figure 4.1. The common volume may be located at a considerable height above the surface of Earth; however, at altitudes greater than 10 km, the troposphere is too rarefied to produce sufficient scattering and the maximum distance is thus limited to a few hundreds kilometers.

It follows that signals propagated in this manner undergo the following:

- Slow variations, ascribable to changes of general conditions of refractivity of atmosphere

Microwave Engineering: Land & Space Radiocommunications, by Gérard Barué

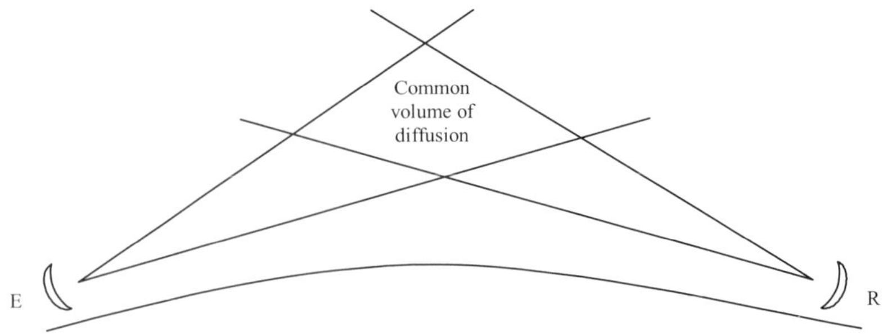

Figure 4.1 Troposcatter propagation for transhorizon transmission.

- Fast fluctuations, which are caused by displacement of zones of irregularities that are not very wide and which are due to the combination by the receiving antenna of a great number of elementary signals of same amplitude but of random phase coming from various points of common volume of diffusion

This mode of forward-scatter propagation is thus characterized by:

- Long-term median transmission loss, called hourly median value, whose evolution is daily and seasonal with a roughly Gaussian distribution (slow fading)
- Short-term instantaneous variation around this median value, which is the resultant of a number of elementary equipollent components and consequently obeys a Rayleigh distribution for periods of observation of about a minute (fast fading)

The frequency of fast fading increases in a way almost proportional to the distance and to the carrier frequency; in the decimeter bands, it can vary from a few to some hundreds of hertz.

4.1.2 Establishment of Path Profile of Troposcatter Link

The path profile of the link must be established in the plane of Earth great circle with the greatest care by using the indications given in Section 3.1; Figure 4.2 summarizes the elements to be taken into account in order to define the geometry of the path. The main geographical parameters used in the method of prediction that follows are:

a Effective Earth radius
h_{ts} Altitude above sea level of center of transmitting antenna
h_{rs} Altitude above sea level of centre of receiving antenna

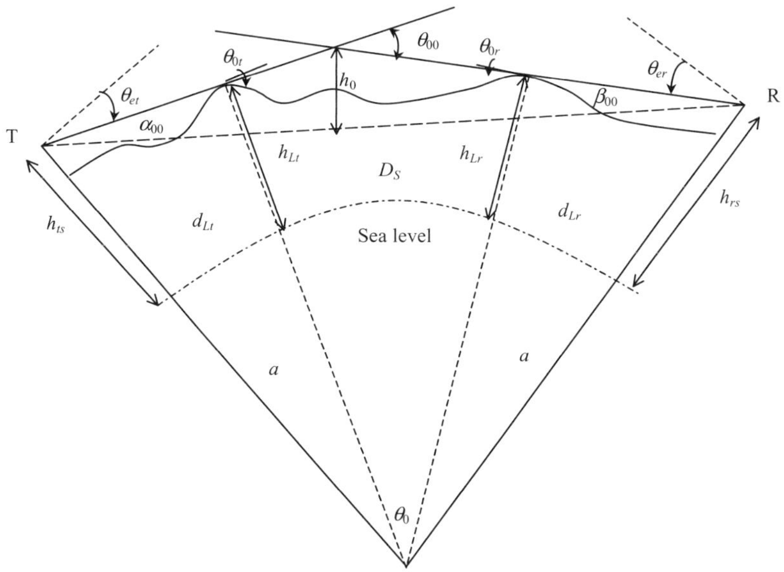

Figure 4.2 Transhorizon path geometry.

h_{Lt}	Altitude above sea level of transmitter horizon point
d_{Lt}	Distance from transmitter site to transmitter radio horizon point
h_{Lr}	Altitude above sea level of receiver radio horizon point
d_{Lr}	Distance from receiver site to receiver radio horizon point
D_S	Distance between radio horizons points
θ	Scatter angle or angular distance
θ_0	Earth center angle
θ_{et}	Takeoff horizon angle of transmitting antenna to radio horizon point
θ_{er}	Takeoff horizon angle of receiving antenna to radio horizon point
h_0	Height of intersection point of antennas boresights, also called Height of common volume
α_{00}, β_{00}	Apparent transmission and reception angles between ray to radio horizon point and chord joining both sites

4.2 METHOD OF PREDICTION

The most widespread method is extracted from technical note NBS-101 (NBS, 1966) from results obtained on more than 750 radio links operating between 40 and 10 GHz on distances of a few tens to a few hundreds of kilometers.

4.2.1 Long-Term Median Basic Transmission Loss Due to Forward Scatter

The long-term median value of the transmission loss is given, in decibels, by the formula

$$L_{bsr} = 30\log(f) - 20\log(d) + F(\theta d) + A_a - H_0 + F_0 \tag{4.1}$$

where f = frequency (MHz)
d = Earth great circle distance (km)
$F(\theta d)$ = attenuation function (dB) depending on product of scatter angle θ (rad) and pathlength d (km)
A_a = long-term median attenuation due to atmospheric absorption (dB)
H_0 = frequency gain function (dB)
F_0 = scattering efficiency correction (dB)

4.2.1.1 Effective Antenna Heights The effective heights of the antennas, h_{te} and h_{re}, are determined as follows. The average height above sea level of the surrounding terrain between each station and its radio horizon point is given by the relation

$$\bar{h}_t = \frac{1}{n-6}\sum_{i=3}^{n-3} h_{ti} \tag{4.2}$$

where h_{ti} = evenly spaced terrain elevations
n = the number of terrain data (>20)

Considering the altitudes of the station emission and reception, h_{t0} and h_{r0}, and the respective altitudes of the antennas, h_{ts} and h_{rs}, we have:

If $\bar{h}_t < h_{t0}$ (or h_{r0})

$$h_{te} = h_{ts} - \bar{h}_t \qquad h_{re} = h_{rs} - \bar{h}_t \tag{4.3a}$$

If $\bar{h}_t \geq h_{t0}$ (or h_{r0})

$$h_{te} = h_{ts} - h_{t0} \qquad h_{re} = h_{rs} - h_{r0} \tag{4.3b}$$

with h_{te}, h_{re} to be less than 1000 m.

Over a smooth spherical Earth with h_{te}, h_{re} < 1 km, the following approximate relationship exists between effective antenna heights and horizon distances:

$$h_{te} = \frac{d_{Lt}^2}{2a} \qquad h_{re} = \frac{d_{Lr}^2}{2a}$$

4.2.1.2 Effective Distance The distance where diffraction and forward-scatter transmission loss are approximately equal over a smooth Earth of effective radius 9000 km is given in kilometers by the relation

$$d_{S1} = 65\left[\frac{100}{f}\right]^{1/3} \tag{4.4}$$

The smooth Earth distance to the radio horizon points on a spherical Earth is given in kilometers by

$$d_{S0} = d - 3\sqrt{2h_{te}} - 3\sqrt{2h_{re}} \tag{4.5}$$

where h_{re} = transmitting effective antenna height (m)
h_{re} = receiving effective antenna height (m)
f = frequency (MHz)

It has been observed that the long-term variability of hourly medians is greatest on average for values of d only slightly greater than the sum of d_{s1} and d_{s0}. The equivalent distance d_e is arbitrarily calculated using the following formulas with the lengths expressed in kilometers:

$$d_e = \begin{cases} \dfrac{130}{1 + (d_{S1} - d_{S0})/d} & \text{for } d_{S0} \le d_{S1} \\ 130 + d_{S0} - d_{S1} & \text{for } d_{S0} > d_{S1} \end{cases} \tag{4.6}$$

4.2.1.3 Angular Distance and Elevation Angles to Radio Horizon Points In determining the angular distance, we first calculate the elevation angles θ_{et} and θ_{er} of the antenna boresight to the radio horizon points relative to the horizontal line at each site. The horizon point is the one for which the radio horizon elevation angle is a maximum; when the trial values are negative, the maximum is the value nearest zero. The radio horizon elevation angles may be measured with surveying instruments in the field or determined directly on the path profile; the antenna axis must be slightly above. The uncorrected angular distance θ_{00}, which is the minimum diffraction angle or scattering angle expressed in radians between the axis to the radio horizon points in the Earth great circle plane passing by both stations, and the transmission and reception elevation angles to the horizon points are given by the relations

$$\theta_{00} = \theta_0 + \theta_{et} + \theta_{er} = \frac{d}{a} + \theta_{et} + \theta_{er} = \frac{D_S}{a} + \theta_{0t} + \theta_{0r} = \alpha_{00} + \beta_{00} \tag{4.7}$$

where[1]

$$\theta_0 = \frac{d}{a} \tag{4.8}$$

$$\theta_{et} = \frac{h_{Lt} - h_{ts}}{d_{Lt}} - \frac{d_{Lt}}{2a} \qquad \theta_{er} = \frac{h_{Lr} - h_{rs}}{d_{Lr}} - \frac{d_{Lr}}{2a} \tag{4.9}$$

$$\theta_{0t} = \theta_{et} + \frac{d_{Lt}}{a} \qquad \theta_{0r} = \theta_{er} + \frac{d_{Lr}}{a} \tag{4.10}$$

$$D_S = d - d_{Lt} - d_{Lr} \tag{4.11}$$

with the altitudes expressed in meters and the distances in kilometers and where a is the effective Earth radius in kilometers defined by relation (1.83) or illustrated in Figure 1.46. For a smooth Earth,

$$\theta_{et} = -\sqrt{2\frac{h_{te}}{a}} \qquad \theta_{er} = -\sqrt{2\frac{h_{tr}}{a}}$$

The apparent transmission and reception angles are defined by the formulas

$$\alpha_{00} = \frac{d}{2a} + \theta_{et} + \frac{h_{ts} - h_{rs}}{d} \tag{4.12}$$

$$\beta_{00} = \frac{d}{2a} + \theta_{er} + \frac{h_{rs} - h_{ts}}{d} \tag{4.13}$$

These angles are positive for beyond-horizon paths.

To allow for the effects of the nonlinear refractivity gradient, α_{00} and β_{00} are modified by corrections $\Delta\alpha_0$ and $\Delta\beta_0$ to give the angles α_0 and β_0, whose sum is the angular distance θ and whose ratio defines the path asymmetry factor S. Here, $\Delta\alpha_0$ and $\Delta\beta_0$ are functions of the angles θ_{0t} and θ_{0r} and of the distances d_{st} and d_{sr} from each horizon point to the crossover of the radio rays:

$$\alpha_0 = \alpha_{00} + \Delta\alpha_0 \qquad \beta_0 = \beta_{00} + \Delta\beta_0 \tag{4.14}$$

The distances d_{st} and d_{sr} are approximated as

$$d_{st} = \frac{d\beta_{00}}{\theta_{00}} - d_{Lt} \qquad d_{sr} = \frac{d\alpha_{00}}{\theta_{00}} - d_{Lr} \tag{4.15}$$

[1]The terms $d_{Lt}/(2a)$ and $d_{Lr}/(2a)$ represent the downward slope of the horizon at the emission and the reception according to the distance, which was defined in Section 1.8.5

Over a smooth Earth $d_{st} = d_{sr} = D_s/2$. If θ_{0t} or θ_{0r} is negative, we compute

$$d'_{st} = d_{st} - |a\theta_{0t}| \qquad d'_{sz} = d_{sr} - |a\theta_{0r}|$$

Then we substitute d'_{st} for d_{st} and d'_{sr} for d_{sr} and read $\Delta\alpha_0$ and $\Delta\beta_0$ on Figure 4.3 using θ_{0t} or $\theta_{0r} = 0$. Figure 4.3 shows $\Delta\alpha_0$ or $\Delta\beta_0$ as a function of θ_{0t} or θ_{0r} and d_{st} or d_{sr}; the curves have been drawn for the reference surface refractivity $N_S = 301$ and for θ_{0t}, $\theta_{0r} < 50$ mrad and d_{st}, $d_{sr} < 400$ km. For values of N_S other than 301, these values must be multiplied by $C(N_S)$:

$$\begin{aligned}\Delta\alpha_0(N_S) &= C(N_S)\Delta\alpha_0(301)\\ \Delta\beta_0(N_S) &= C(N_S)\Delta\beta_0(301)\end{aligned} \tag{4.16}$$

where $C(N_S) = (1.3N_S^2 - 60N_S)\times 10^{-5}$. Figure 4.4 illustrates the function $C(N_S)$ according to the surface refractivity N_S as defined by relations (1.76) and (1.77). Then, the angular distance or scatter angle is given by the relation

$$\theta = \alpha_0 + \beta_0 \tag{4.17}$$

The path asymmetry factor S is calculated as

$$S = \frac{\alpha_0}{\beta_0} \tag{4.18}$$

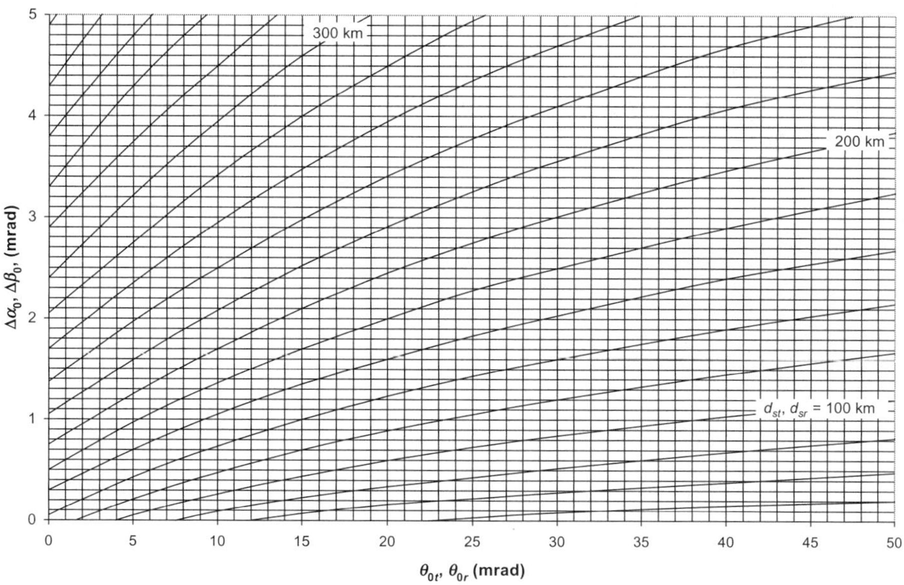

Figure 4.3 $\Delta\alpha_0$ or $\Delta\beta_0$ as function of θ_{0t} or θ_{0r} and d_{st} or d_{sr}.

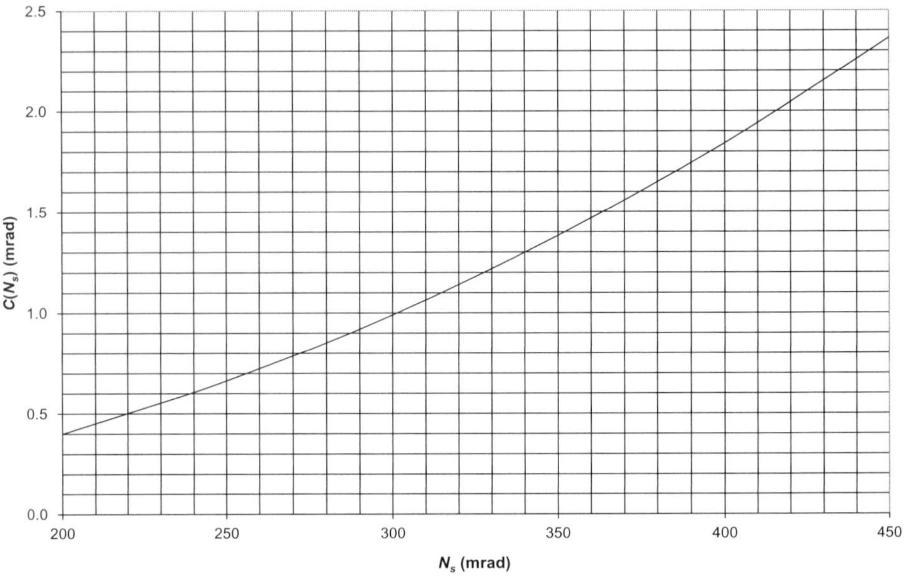

Figure 4.4 Correction factor $C(N_S)$ versus N_S.

For small θ_{0t} and θ_{0r}, no correction is required for values d_{st} and d_{sr} less than 100 km; when both $\Delta\alpha_0$ and $\Delta\beta_0$ are negligible, we write

$$\theta \approx \theta_{00} = \alpha_{00} + \beta_{00} \qquad S \approx \frac{\alpha_{00}}{\beta_{00}}$$

4.2.1.4 Attenuation Function The attenuation function $F(\theta d)$ for the reference surface refractivity $N_S = 301$ and for a path asymmetry factor S ranging between 0.75 and 1 is given by the following relations in decibels:

$$F(\theta d, 301) = \begin{cases} 135.82 + 0.33\,\theta d + 30\log(\theta d) & \text{for } 0.01 \le \theta d \le 10 \\ 129.5 + 0.212\,\theta d + 37.5\log(\theta d) & \text{for } 0.01 \le \theta d \le 70 \\ 119.2 + 0.157\,\theta d + 45\log(\theta d) & \text{for } \theta d > 70 \end{cases} \quad (4.19)$$

For any other value of N_S and replacing S by $1/S$ when $S > 1$, we have

$$F(\theta d, N_S) = F(\theta d, 301) - 0.1(N_S - 301)\exp\left(-\frac{\theta d}{40}\right) \quad (4.20a)$$

When $\theta\, d > 10$ and $S \neq 1$, we use the corrected formula

$$F(\theta d, N_S) = F(\theta d, 301) - 0.1(N_S - 301)\exp\left(-\frac{\theta d}{40}\right) - 0.37(\theta d - 10)^{0.83} \times 10^{-2S} \quad (4.20b)$$

Figure 4.5 presents the transmission loss function $F(\theta d)$ where θ is expressed in radians and d in kilometers for an asymmetry factor equal to 1 and various values of N_S. Figure 4.6 shows the variation of $F(\theta d)$ according to the asymmetry factor S for $N_S = 301$.

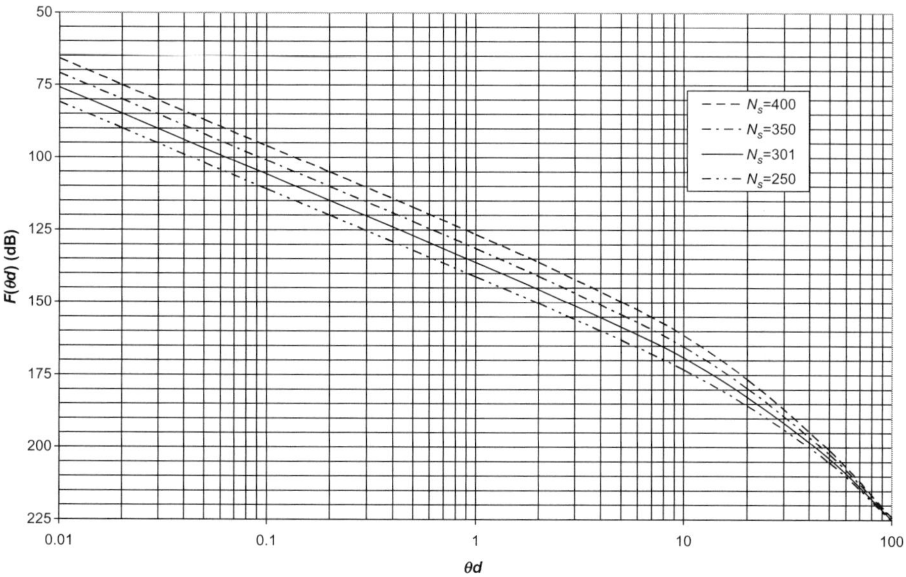

Figure 4.5 Attenuation function $F(\theta d)$ for $S = 1$ and $N_S = 250$–400.

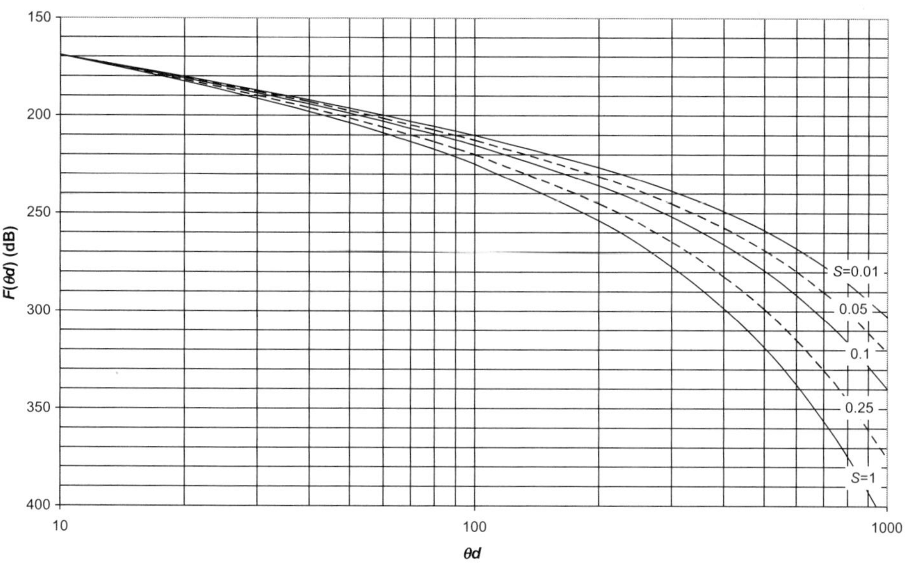

Figure 4.6 Attenuation function $F(\theta d)$ for $S \neq 1$ and $N_S = 301$.

4.2.1.5 Long-Term Median Attenuation Due to Atmospheric Absorption

As shown in previous chapters, the principal effects of the atmosphere are as follows:

- Median absorption by vapor and oxygen according to frequency and distance, presented in Figure 1.73, for propagation above a smooth Earth between 10-m antenna heights
- Attenuation by clouds, negligible below 6 GHz, which depends on frequency, water content (in grams per cubic meter), and temperature
- Attenuation and depolarization by rain, treated in Section 1.11
- Sky noise reradiation by oxygen, water vapor, and rain, which causes an increase in noise temperature of the system of reception and was treated in Section 1.13.

The median absorption by water vapor and oxygen according to frequency and path length can be calculated using the relations given in Section 1.10 and in particular by relation (1.105). The total absorption attenuation within a cloud can be written, in decibels per kilometer, as

$$A_c = KM$$

where K = attenuation coefficient depending on frequency and temperature
M = liquid water content ($\text{g}\,\text{m}^{-3}$)

The attenuation by ice clouds is much weaker than by water clouds.

Figure 4.7 presents the total absorption attenuation caused by water clouds at various temperatures and expressed in decibels per kilometer per grams per cubic meter of liquid water content.

4.2.1.6 Frequency Gain Function

The frequency gain function H_0, which expresses the reduction of the effectiveness of the low part of the common volume, where scattering efficiency is great, when the frequency decreases because of destructive interference of the direct wave and the wave reflected by the ground in front of the antennas, is negligible for

$$\frac{h_{te}}{\lambda} > 4\frac{a}{d} \qquad \text{and} \qquad \frac{h_{re}}{\lambda} > 4\frac{a}{d}$$

and does not exceed 25 dB except when the antennas are very low. The function H_0 depends on effective antenna heights in terms of wavelength, path asymmetry, and function η_S given by the relation

$$\eta_S = 0.5696 h_0 [1 + (0.031 - 2.32 N_S \times 10^{-3} + 5.67 N_S^2 \times 10^{-6}) \exp(-3.8 h_0^6 \times 10^{-6})] \tag{4.21}$$

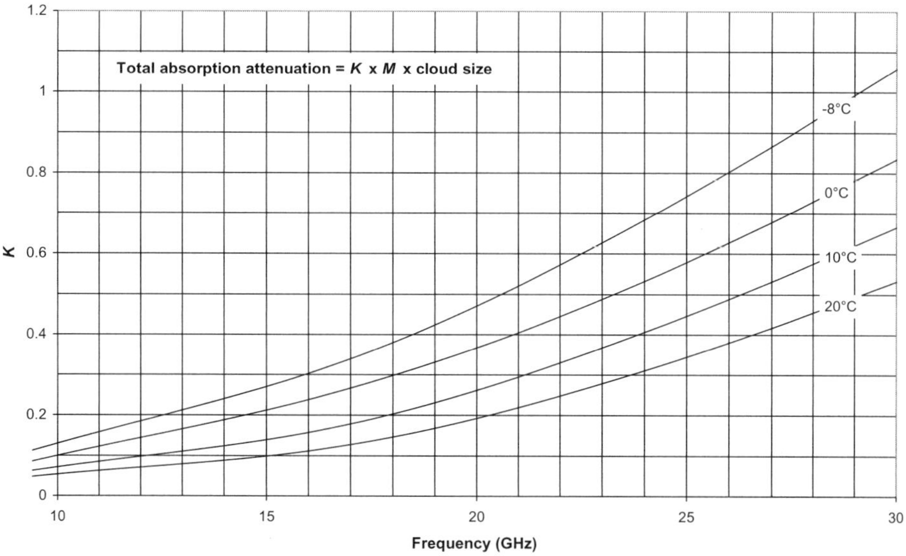

Figure 4.7 Absorption by water clouds.

$$\text{for } h_0 > 10\,\text{km}: \quad \eta_S = 0.5696\,h_0$$

where h_0, the height of the intersection point of the rays to the horizon points in kilometers, is given as

$$h_0 = \frac{S\theta d}{(1+S)^2} \tag{4.22}$$

The function η_S is presented in Figure 4.8. If $\eta_S \geq 1$, H_0 is given by

$$H_0 = \frac{1}{2}[H_0(r_1) + H_0(r_2)] + \Delta H_0 \tag{4.23}$$

where the function $H_0(r)$ can be read in Figure 4.9 or obtained by the relations

$$H_0(r) = \begin{cases} \dfrac{19.5}{r^{0.48}} - 0.7 + \dfrac{14\eta_S}{5} & \text{for } r \leq 0.5 \\[2ex] \dfrac{19.2}{r^{0.77}} - 2 + \left[\dfrac{14}{r^{0.7}} - \dfrac{2}{r^3} - \dfrac{r^{1.5}}{100}\right]\dfrac{\eta_S}{5} & \text{for } r > 0.5 \end{cases}$$

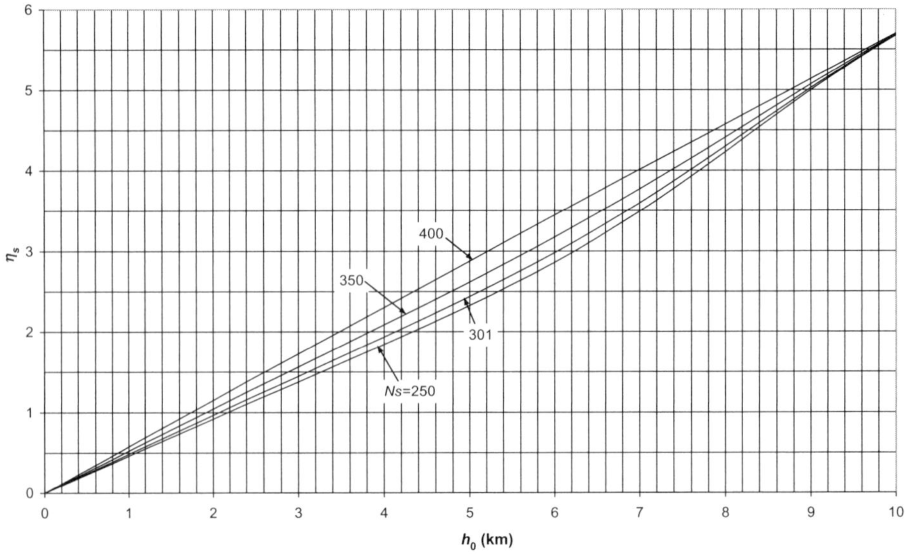

Figure 4.8 Function $\eta_S(h_0)$.

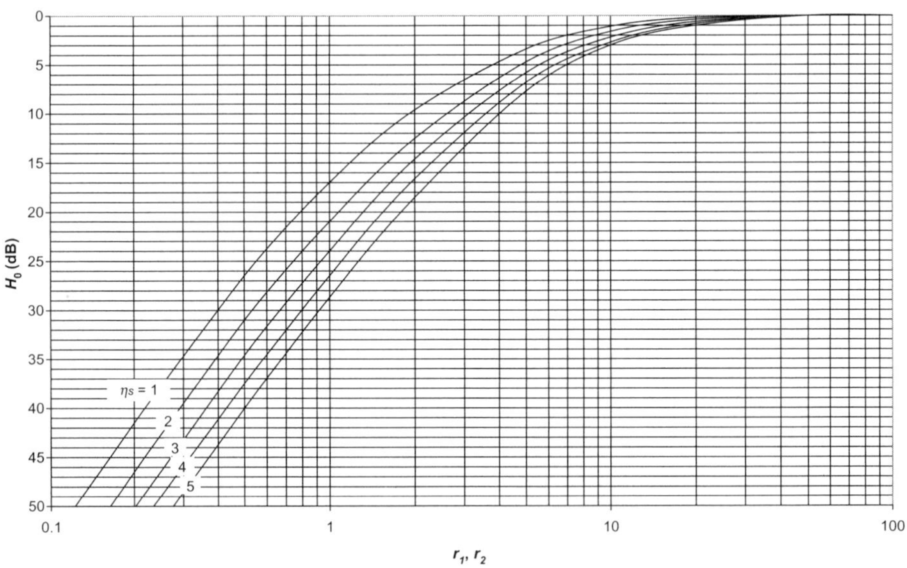

Figure 4.9 Frequency gain function $H_0(r)$.

where

$$r_1 = \frac{4\pi\theta h_{te}}{\lambda} \qquad r_2 = \frac{4\pi\theta h_{re}}{\lambda}$$

and

$$\Delta H_0 = 6(0.6 - \log \eta_S)\log S \log q \qquad q = \frac{r_2}{S r_1} \tag{4.24}$$

The correction term ΔH_0 is null for $\eta_S = 4$, $S = 1$, or $q = 1$ and reaches the maximum value of 3.6 dB for highly asymmetrical paths when $\eta_S = 1$. The following limits should be applied in determining ΔH_0:

If $\eta_S > 5$, H_0 keeps the computed value for $\eta_S = 5$.
If $S \geq 10$ or $q \geq 10$, $S = 10$ or $q = 10$.
if $S \leq 0.1$ or $q \leq 0.1$, $S = 0.1$ or $q = 0.1$.
If $\Delta H_0 \geq \frac{1}{2}[H_0(r_1) + H_0(r_2)]$, $H_0 = H_0(r_1) + H_0(r_2)$.
If ΔH_0 makes H_0 negative, $H_0 = 0$.

4.2.1.7 Scattering Efficiency Correction The correction term F_0, which allows for the reduction of scattering efficiency at great heights in the atmosphere, is given by the relation

$$F_0 = 1.086 \frac{\eta_S}{h_0}(h_0 - h_1 - h_{Lt} - h_{Lr}) \tag{4.25}$$

where

$$h_1 = \frac{S D_S \theta}{(1+S)^2} \qquad D_S = d - d_{Lt} - d_{Lr}$$

with the heights and distances expressed in kilometers and the angular distance in radians. The scattering efficiency correction F_0 exceeds 2 dB only when the distances and the antenna heights are such that h_0 is higher than h_1 by more than 3 km.

4.2.2 Median Long-Term Transmission Loss Variation

The variability of troposcatter radio transmission depends upon the changes in the atmosphere and the complex relationships between various propagation mechanisms. Long-term power fading, identified with the variability of the hourly median values of the transmission loss, is usually due to slow changes in average atmospheric refractivity, degree of atmospheric stratification, or intensity of refractive index turbulence. Just beyond line of sight, fading rate and fading range depend in a very complex manner on the relative importance of various propagation mechanisms and we can say that:

- An increase of the atmospheric refractive index increases long-distance diffraction or forward-scatter range but may lead to multipath fading problems over short paths.
- Increased turbulence of the atmosphere generates a structure composed of vertices or blobs that may result in either an increase or a decrease of the transmission loss depending on the geometry of the path.
- Increased stratification generates a structure composed of many small reflecting layers, which favors propagation by reflection and sometimes the guiding of energy by ducts or superrefractive layers.

Ordinarily, a diffraction signal fades slowly and the fades are of relatively short duration and very deep while a tropospheric forward-scatter signal shows a tendency to go into relatively deep fades, with durations from less than a minute to hours, and exhibits the rapid and severe fading characteristic of the Rayleigh distribution around the hourly median field strength.

4.2.2.1 Median Long-Term Transmission Loss The predicted median long-term transmission loss, exceeded by half of all hourly medians in a given climatic region, is given by the expression

$$L(50\%) = L_{bsr} - V(50\%, d_e) \tag{4.26}$$

where the climatic function $V(50\%,d_e)$ is given by the following relation according to the effective distance d_e:

$$V(50\%, d_e) = [c_1 d_e^{n1} - f_2(d_e)]\exp[-c_3 d_e^{n3}] + f_2(d_e) \tag{4.27}$$

where

$$f_2(d_e) = f_\infty + [f_m - f_\infty]\exp[-c_2 d_e^{n2}]$$

The effective distance d_e was introduced to take into account a certain number of factors, such as the effective height of the antennas and the radio frequency used; it is always higher than the geographical distance d and we employ the method described hereafter to determine it.

We distinguish eight standard climates with their respective coefficients to calculate relation (4.27) (Table 4.1):

- Climate 1: *Continental Temperate* For the regions localized between 30° and 60° of the northern or southern latitude and characterized by an

TABLE 4.1 Coefficients of Standard Climatic Zones for $V(50\%,d_e)$

Climate		c_1	c_2	c_3	n_1	n_2	n_3	f_m	f_∞
1	Continental temperate	1.59×10^{-5}	1.56×10^{-11}	2.77×10^{-8}	2.32	4.08	3.25	3.9	0
2	Maritime temperate overland	1.12×10^{-4}	1.26×10^{-20}	1.17×10^{-11}	1.68	7.30	4.41	1.7	0
3	Maritime temperate oversea	**4×10^{-5}**	**2.9×10^{-13}**	**8×10^{-9}**	**2.1**	**4.61**	**3.65**	7	3.2
4	Maritime subtropical oversea	**6×10^{-5}**	5.89×10^{-18}	**2.1×10^{-7}**	**2**	**6.75**	**3**	**6.15**	**1.7**
5	Maritime subtropical overland[a]								
6	Desert, Sahara	**1.5×10^{-5}**	2.76×10^{-14}	**3×10^{-13}**	**2.25**	4.82	**5.1**	**8.3**	**8.4**
7	Equatorial	3.45×10^{-7}	3.74×10^{-12}	6.97×10^{-8}	2.97	4.43	3.14	1.2	−8.4
8	Continental subtropical	1.59×10^{-5}	1.56×10^{-11}	2.77×10^{-8}	2.32	4.08	3.25	3.9	0
—	Polar[b]								

Note: The coefficients that appear in bold type have been modified compared to the values of origin in order to improve the coincidence with the standard curves.

[a]Climate 5 is removed; for the hot and wet climates we will use the values of climate 4 and for the coastal zones those of climate 3.

[b]A ninth climatic region, polar, is yet to be defined.

annual mean N_s of 320 N-units with an annual range of monthly variation from 20 to 40 N-units; pronounced diurnal and seasonal changes in propagation are expected to occur.

- Climate 2: *Maritime Temperate Over Land* For the regions localized between 20° and 50° of the northern or southern latitude near the sea, where prevailing winds unobstructed by mountains carry moist maritime air inland, and characterized by a rather small annual mean N_s of 320 N-units with an annual range of monthly variation from 20 to only 30 N-units.
- Climate 3: *Maritime Temperate Over Sea* For the coastal and over-sea areas with the same general characteristics as those for climate 2; the distinction made is that a radio path with both horizons on the sea is considered to be an over-sea path.
- Climate 4: *Maritime Subtropical Over Land* For the regions localized between 10° and 30° of the northern or southern latitude, usually on low lands near the sea with definite rainy and dry seasons, and characterized by an annual mean N_s of 370 N-units with an annual range of monthly variation from 30 to 60 N-units.
- Climate 5: *Maritime Subtropical Over Sea* For the regions corresponding to the conditions observed in coastal areas with the same range of latitude as climate 4; the curves for this climate were based on an inadequate amount of data and have been deleted and it is suggested that the curves for climates 3 or 4 be used, selecting whichever seems more applicable to each specific case.
- Climate 6: *Desert, Sahara* For the regions characterized by an annual mean N_s of 280 N-units with year-round semiarid conditions and an annual range of monthly variation from 20 to 80 N-units.
- Climate 7: *Equatorial* Can be observed in areas ranging between 20° of the northern and southern latitudes and is characterized by an annual mean N_s of 360 N-units and an annual range of monthly variation from 0 to 30 N-units; such areas undergo monotonous heavy rains and high average summer temperatures.
- Climate 8: *Continental Subtropical* Refers to the areas ranging between 20° and 40° of the northern latitute with an annual mean N_s of about 320 N-units and an annual range of monthly variation from 60 to 100 N-units; it is a hot climate, typified by the Sudan and monsoons, with seasonal extremes of winter drought and summer rainfall.

A *continental polar* climate may also be defined; temperatures are low to moderate all year-round and the annual mean N_s is about 310 N-units with an annual range of monthly variation from 10 to 40 N-units.

Figure 4.10 presents the climatic function $V(50\%,d_e)$ established for the various standard climates by applying the coefficients in Table 4.2 in relation 4.27.

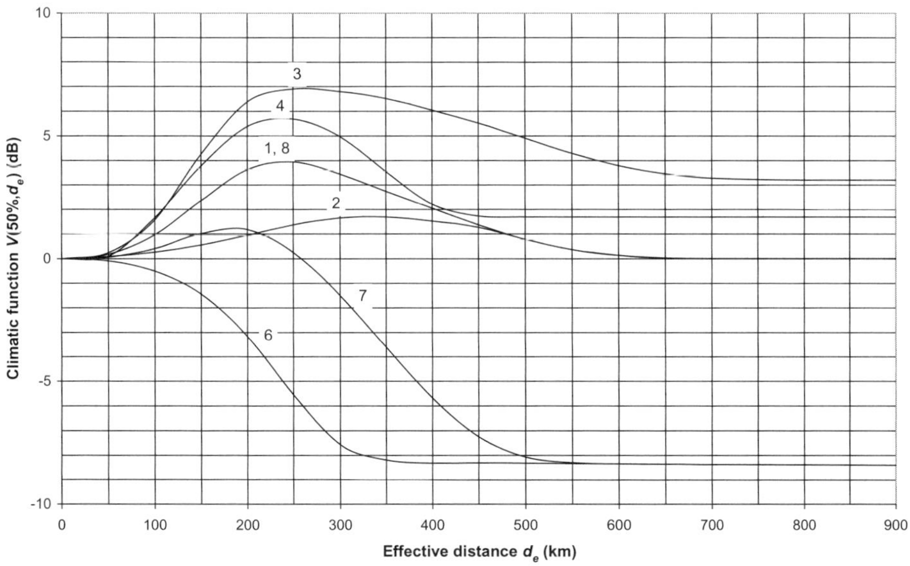

Figure 4.10 Climatic function $V(50\%,d_e)$

4.2.2.2 Transmission Loss Not Exceeded for p% of Time

The estimated value of the transmission loss not exceeded for a given percentage $p\%$ of the hourly medians is defined by the relation

$$L(p\%) = L(50\%) - Y(p\%, d_e) \tag{4.28}$$

where $Y(p\%,d_e)$ is the variability of $L(p\%)$ relative to its long-term median value $L(50\%)$ obtained by relation (4.26) for a specified climatic region and a given effective distance. The variations according to various percentages of the time thus enable us to predict the important fades that should be taken into account in the estimation of long-term availability as well as the significant enhancement of the field which intervenes in the study of reciprocal jamming between adjacent connections.

The amplitude of variation of the function $Y(p\%,d_e)$ depends on the frequency and is given by the relation

$$Y(p\%, d_e) = Y(p\%, d_e, f_0)\, g(p\%, f_{\mathrm{MHz}}) \tag{4.29}$$

where $g(p\%,f_{\mathrm{MHz}})$ = frequency factor defined for each climate
f_0 = frequency of reference

Figure 4.11 presents the variation of the frequency factor $g(10\%,f_{\mathrm{MHz}})$ in a continental temperate climate during 10% of time with respect to $f_0 = 100\,\mathrm{MHz}$. Figure 4.12 presents the variation of the frequency factor $g(90\%,f_{\mathrm{MHz}})$ in a

TABLE 4.2 Coefficients of Standard Climatic Zones for $Y(p\%,d_e)$

Climate	$Y(p\%,f_{0MHz})$	c_1	c_2	c_3	n_1	n_2	n_3	f_m	f_∞
1	$Y(10\%)$ 100 MHz	3.56×10^{-2}	9.85×10^{-8}	1.5×10^{-11}	1.13	2.8	4.85	10.5	5.4
	$Y(10\%)$ >1000 MHz[a]	**3.4×10^{-2}**	**7.3×10^{-8}**	**4.9×10^{-11}**	**1.16**	**2.83**	**4.6**	**11**	**6**
	$Y(90\%)$ 100 MHz	9.48×10^{-3}	5.7×10^{-11}	5.56×10^{-6}	1.33	3.96	2.44	8.2	3
	$Y(90\%)$ >1000 MHz	**4.5×10^{-3}**	**5×10^{-11}**	**5.56×10^{-6}**	**1.53**	**3.96**	**2.4**	**8.2**	**3.5**
2	$Y(10\%)$, bands I–II	**6.93×10^{-3}**	1.57×10^{-7}	**1.3×10^{-11}**	1.52	**2.8**	5.04	**12.5**	11
	$Y(10\%)$, band III	3.6×10^{-2}	3.19×10^{-8}	6.91×10^{-18}	1.11	2.96	7.14	12.5	11
	$Y(10\%)$, bands IV–V	6.28×10^{-4}	3.19×10^{-8}	6.06×10^{-12}	1.92	2.96	5.05	13	12.5
	$Y(90\%)$, bands I–II	**1×10^{-3}**	1.68×10^{-12}	8.07×10^{-6}	1.7	**4.55**	2.36	**9.5**	3.5
	$Y(90\%)$, band III	**9.7×10^{-4}**	2.68×10^{-14}	**1.2×10^{-16}**	1.74	5.29	6.82	**10.3**	3.5
	$Y(90\%)$, bands IV–V	**7×10^{-5}**	1.93×10^{-15}	2.81×10^{-4}	2	**5.85**	**1.6**	**12**	4.5
3	$Y(10\%)$, bands I–II	**1×10^{-2}**	1.04×10^{-11}	1.42×10^{-5}	1.38	4.42	2.27	16	13
	$Y(10\%)$, band III	**1.3×10^{-3}**	**5×10^{-1}**	**8.5×10^{-8}**	1.94	**2.5**	**3.2**	**18**	**14.5**
	$Y(10\%)$, bands IV–V	**1.6×10^{-2}**	2.4	6.92×10^{-15}	1.31	0	5.78	**19.5**	**13**
	$Y(90\%)$, bands I–II	**1.7×10^{-3}**	**5.25×10^{-9}**	**1.6×10^{-8}**	1.75	**3.28**	**3.67**	**14**	3.5
	$Y(90\%)$, band III	**1.1×10^{-3}**	**6.5×10^{-9}**	**1.15×10^{-8}**	1.9	**3.28**	3.67	**13**	**4.1**
	$Y(90\%)$, bands IV–V	1.25×10^{-3}	6.57×10^{-16}	1.49×10^{-9}	1.72	5.96	3.84	**10.5**	4
4	$Y(10\%)$	4.33×10^{-2}	7.13×10^{-11}	1.19×10^{-12}	1.09	3.89	4.93	17.5	13.6
	$Y(90\%)$	7.24×10^{-3}	4.26×10^{-15}	1.12×10^{-6}	1.35	5.41	2.56	12.7	8.4
6	$Y(10\%)$	6.09×10^{-2}	1.36×10^{-5}	3.18×10^{-11}	1.08	1.84	4.6	15.1	6
	$Y(90\%)$	3.19×10^{-2}	5.66×10^{-8}	7.39×10^{-11}	1.14	2.76	4.4	11.4	3.3
7	$Y(10\%)$	**5.7×10^{-3}**	**1.1×10^{-4}**	5.22×10^{-17}	**1.38**	**1.38**	**6.9**	**8.4**	**1.6**
	$Y(90\%)$	**8.2×10^{-3}**	**8.9×10^{-5}**	**6×10^{-17}**	**1.29**	**1.55**	**7.1**	**10.4**	**2.8**
8	$Y(10\%)$	1.01×10^{-2}	2.2×10^{-7}	3.9×10^{-9}	1.46	2.67	3.78	**15**	**8.7**
	$Y(90\%)$	**3.9×10^{-3}**	1.08×10^{-9}	**1.5×10^{-10}**	**1.5**	3.49	4.48	**10.8**	**3.9**

Note: The coefficients that appear in bold type have been modified compared to the values of origin in order to improve the coincidence with the standard curves.

[a] $Y(10\%)$ and $Y(90\%)$ for >1000 MHz can be obtained by multiplying the values for 100 MHz in decibels by 1.1.

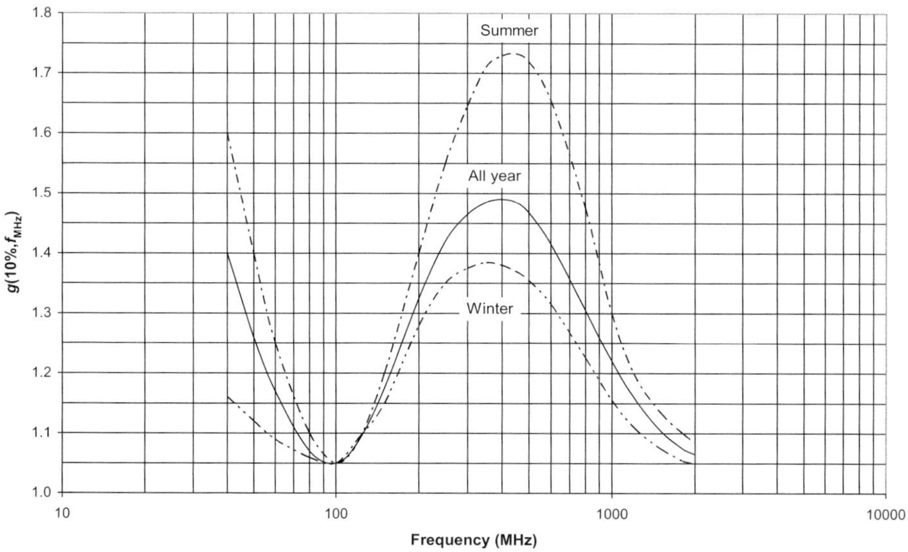

Figure 4.11 Frequency factor $g(10\%,f_{\mathrm{MHz}})$ for continental temperate climate (f_0 = 100 MHz).

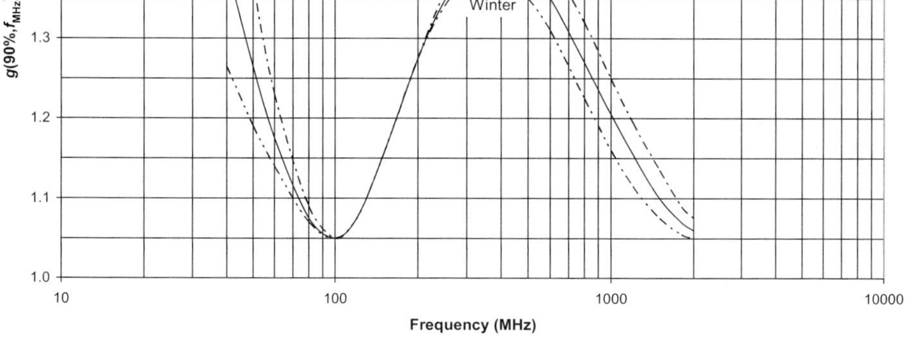

Figure 4.12 Frequency factor $g(90\%,f_{\mathrm{MHz}})$ for continental temperate climate (f_0 = 100 MHz).

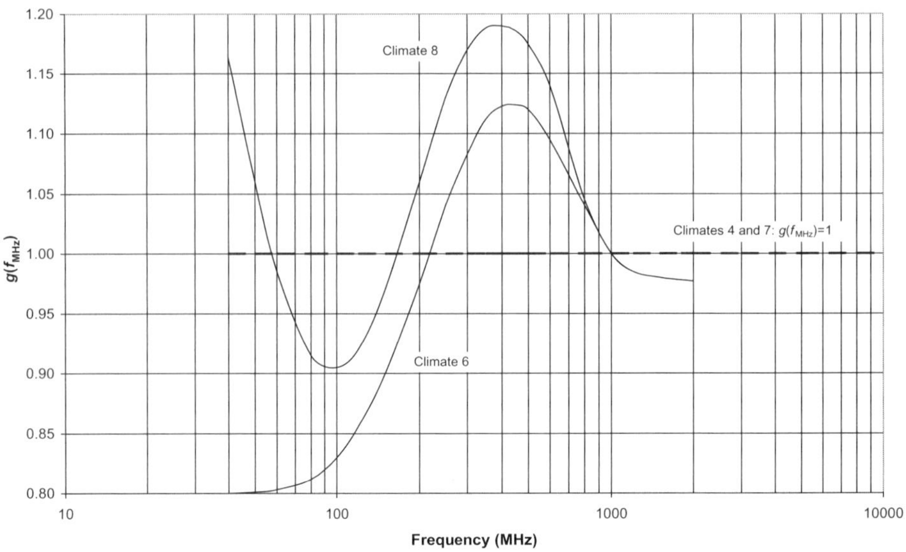

Figure 4.13 Frequency factor $g(f_{MHz})$ for climates 4, 6, 7, and 8 ($f_0 = 1000\,MHz$).

continental temperate climate during 90% of the time compared to the reference frequency $f_0 = 100\,MHz$. Figure 4.13 presents the frequency factor $g(f_{MHz})$ that can be used for the climates maritime subtropical over land, desert, equatorial, and continental subtropical with respect to reference frequency $f_0 = 1000\,MHz$.

In the continental temperate climate, the transmission loss is often nearly lognormally distributed; the standard deviation may be as much as 20 dB for short transhorizon paths where the mechanisms of diffraction and forward scatter are about equally important. When a propagation path in a maritime temperate climate is over water, a lognormal distribution may be expected from $L(50\%)$ to $L(99.9\%)$, but considerably higher fields are expected for small fractions of time when pronounced superrefraction and ducting conditions are present.

Data from the United States, Germany, and France provide the basis for predicting long-term power fading in a continental temperate climate; more than half a million hourly median values of basic transmission loss recorded over some 200 paths were used in developing these estimates.

The empirical curves $g(p\%,f_{MHz})$ should not be regarded as an estimate of the dependence of long-term variability on frequency but represent only an average of many effects, some of which being frequency sensitive.

The apparent frequency dependence is a function of the relative dominance of various propagation mechanisms and this in turn depends on the climate, the hour of day, the season, and the particular types of terrain profiles for which data are available; for example, a heavily forested low-altitude path will usually show greater variability than that observed over a treeless high-altitude prairie and this effect is frequency sensitive.

An allowance for the year-to-year variability is also included. The variability $Y(p\%,d_e)$ of the transmission loss not exceeded during 10 and 90% of the time compared to the long-term median $L(50\%)$ is given for the various climates considered by using relation (4.27) with the coefficients given in Table 4.2. The passage to $p\%$ of the time is obtained by multiplying the values for $Y(10\%)$ and $Y(90\%)$ by the coefficients which appear in the tables below and which are Gaussian most of the time:

$Y(10\%) \rightarrow +1.28\sigma$	$Y(0.01\%,d_e)$	=	$+3.719\ /\ 1.28\ Y(10\%,d_e)$	=	$+2.91\ Y(10\%,d_e)$
	$Y(0.1\%,d_e)$	=	$+3.09\ /\ 1.28\ Y(10\%,d_e)$	=	$+2.41\ Y(10\%,d_e)$
	$Y(1\%\ d_e)$	=	$+2.327\ /\ 1.28\ Y(10\%,d_e)$	=	$+1.82\ Y(10\%,d_e)$
	$Y(99\%,d_e)$	=	$-2.327\ /\ 1.28\ Y(90\%,d_e)$	=	$-1.82\ Y(90\%,d_e)$
$Y(90\%) \rightarrow -1.28\sigma$	$Y(99.9\%,d_e)$	=	$-3.09\ /\ 1.28\ Y(90\%,d_e)$	=	$-2.41\ Y(90\%,d_e)$
	$Y(99.99\%,d_e)$	=	$-3.719\ /\ 1.28\ Y(90\%,d_e)$	=	$-2.91\ Y(90\%,d_e)$

Climate	0.01%	0.1%	1%	99%	99.9%	99.99%
1	**3.33**	**2.73**	**2**	1.82	2.41	2.91
2 (BI–BII)	2.91	2.41	1.82	1.82	2.41	2.91
2 (BIII)	—	—	—	1.82	2.41	2.91
2 (BIV–BV)	—	—	—	1.82	2.41	2.91
3 (BI–BII)	—	—	—	1.82	2.41	2.91
3 (BIII)	—	—	—	1.82	2.41	2.91
3 (BVI–BV)	—	—	—	1.82	2.41	2.91
4	—	—	—	1.82	2.41	2.91
6	2.91	2.41	1.82	1.82	2.41	2.91
7	2.91	2.41	1.82	1.82	2.41	2.91
8	2.91	2.41	1.82	1.82	2.41	2.91

Figures 4.14–4.25 present $Y(p\%,\ d_e)$ for the various standard climates considered.

4.2.2.3 *Estimation of Prediction Error* The hourly median values of the received field strength at terminals during a fraction $p\%$ of all hours over a large number of randomly different propagation paths can be considered as very nearly normally distributed with a median and a standard deviation denoted by $\sigma_c(p\%)$. This path-to-path variability may assume a random distribution of all parameters which are not taken into account in the prediction method for a hypothetical path. For a specific propagation path, it is calculated in accordance with the method of prediction using a given set of parameters $(d, f, \theta, \ldots)$. In such a case, $\sigma_c(p\%)$ represents the standard error of prediction

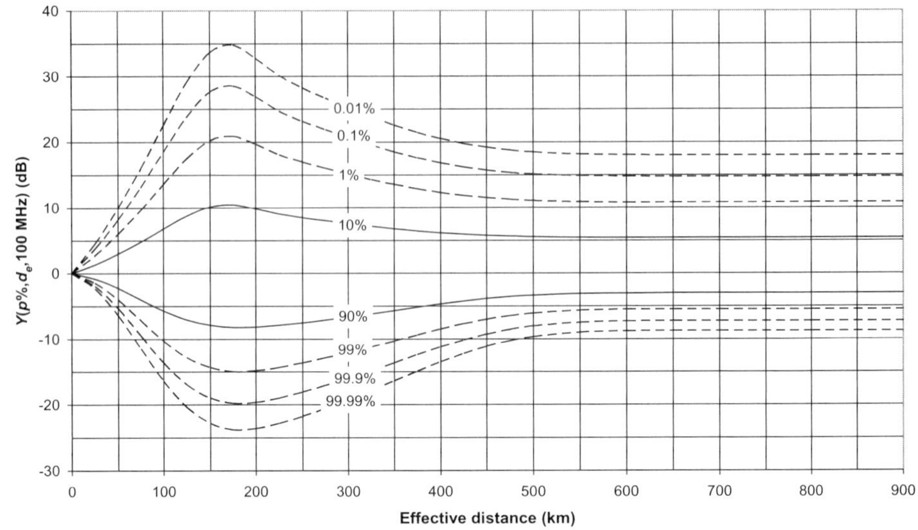

Figure 4.14 Climate 1: continental temperate ($f_0 = 100$ MHz).

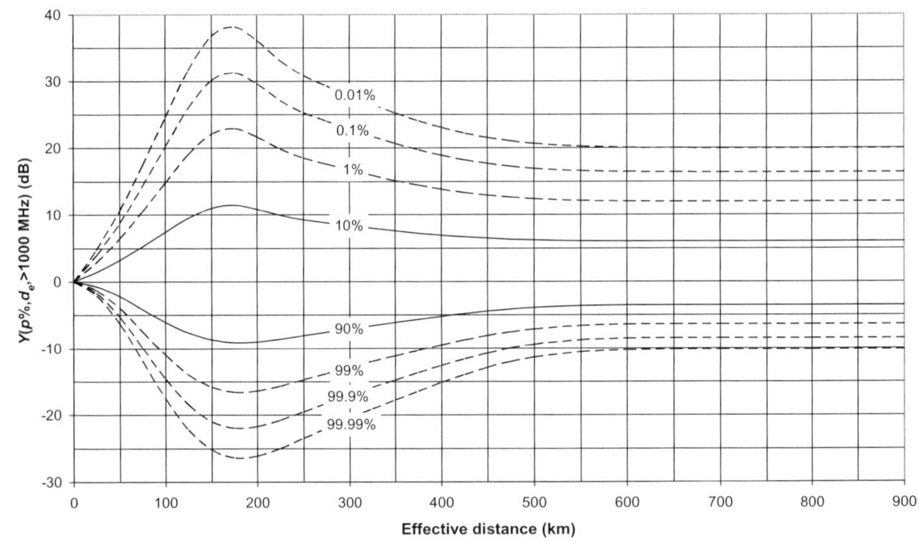

Figure 4.15 Climate 1: continental temperate (>1000 MHz).

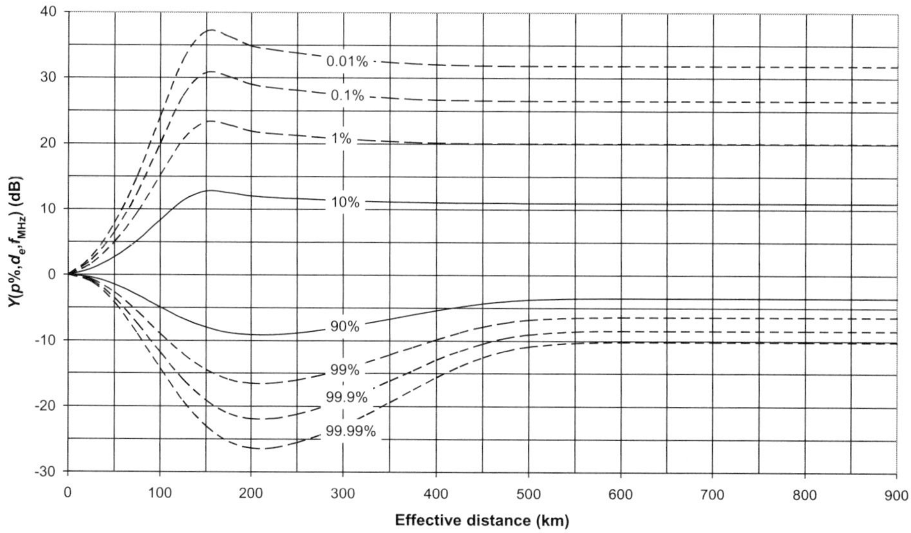

Figure 4.16 Climate 2: maritime temperate over land (40–100 MHz).

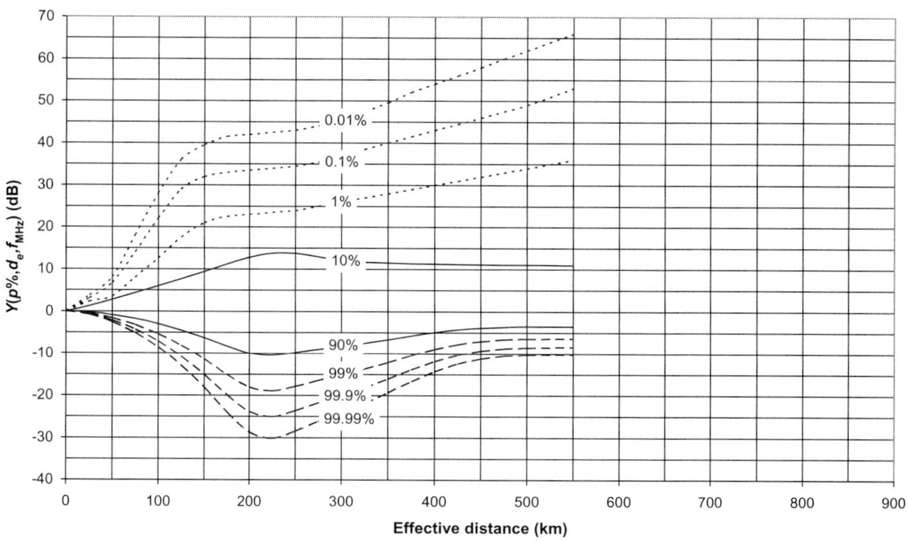

Figure 4.17 Climate 2: maritime temperate over land (150–250 MHz).

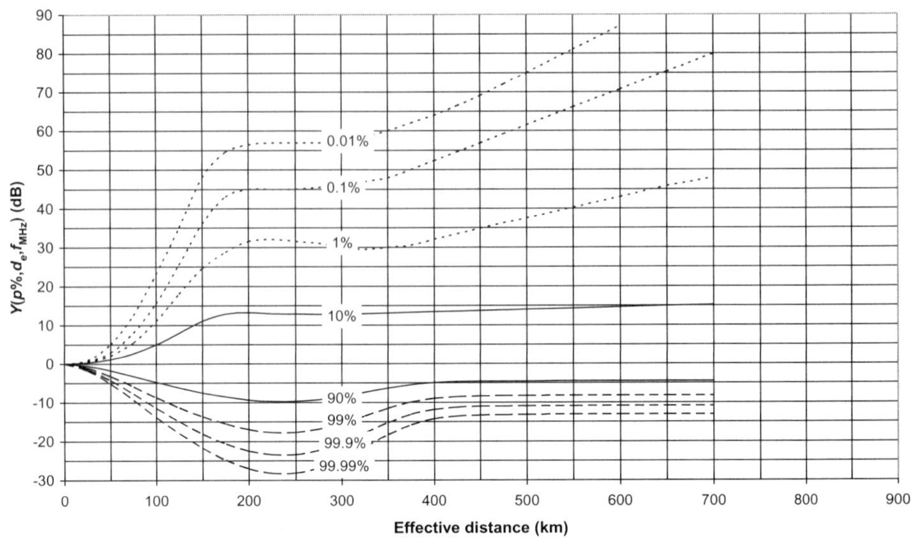

Figure 4.18 Climate 2: maritime temperate over land (450–1000 MHz).

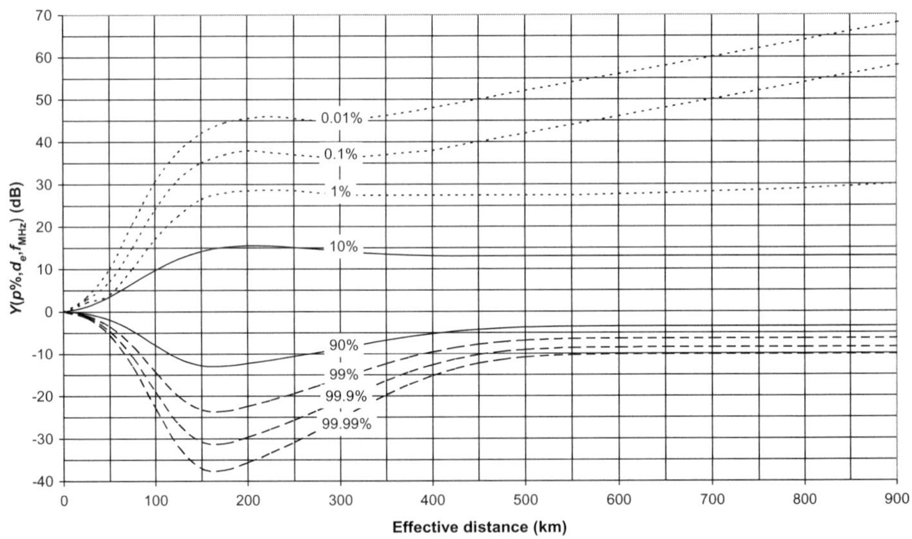

Figure 4.19 Climate 3: maritime temperate over sea (40–100 MHz).

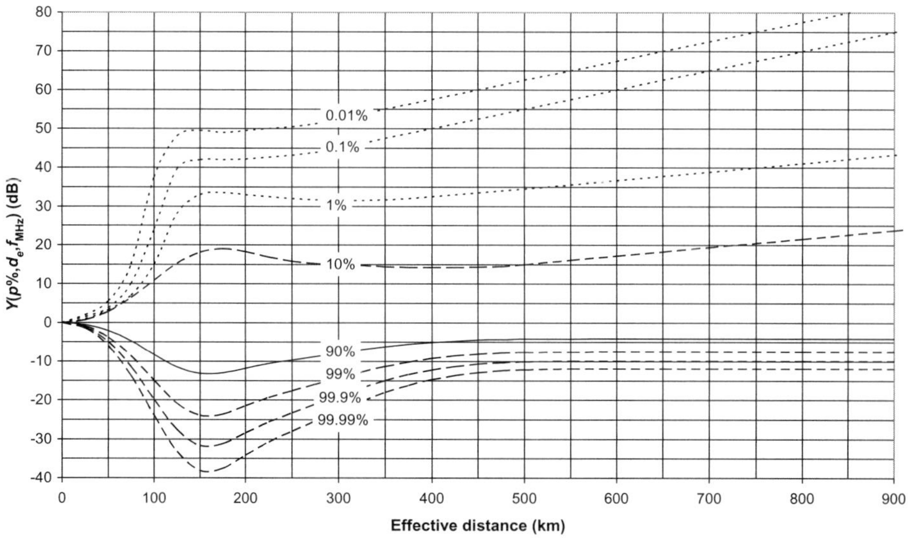

Figure 4.20 Climate 3: maritime temperate over sea (150–250 MHz).

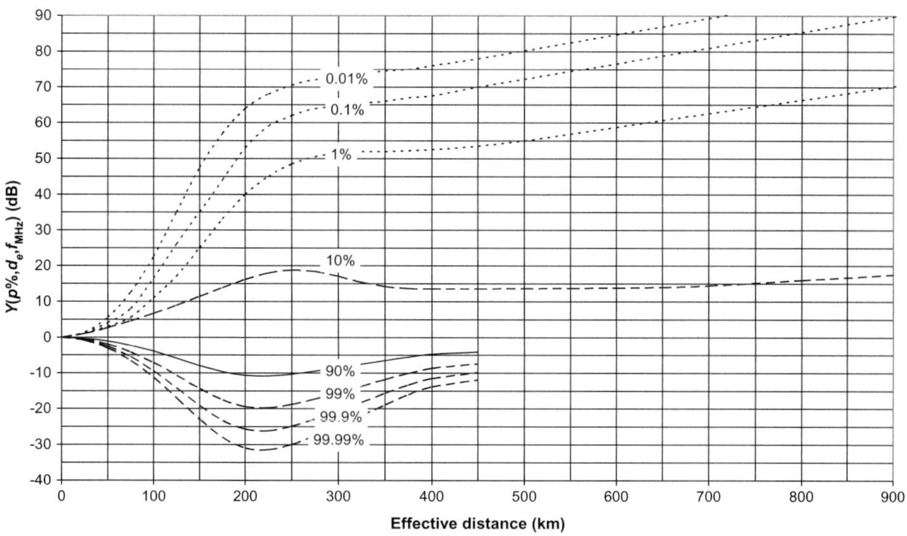

Figure 4.21 Climate 3: maritime temperate over sea (450–1000 MHz).

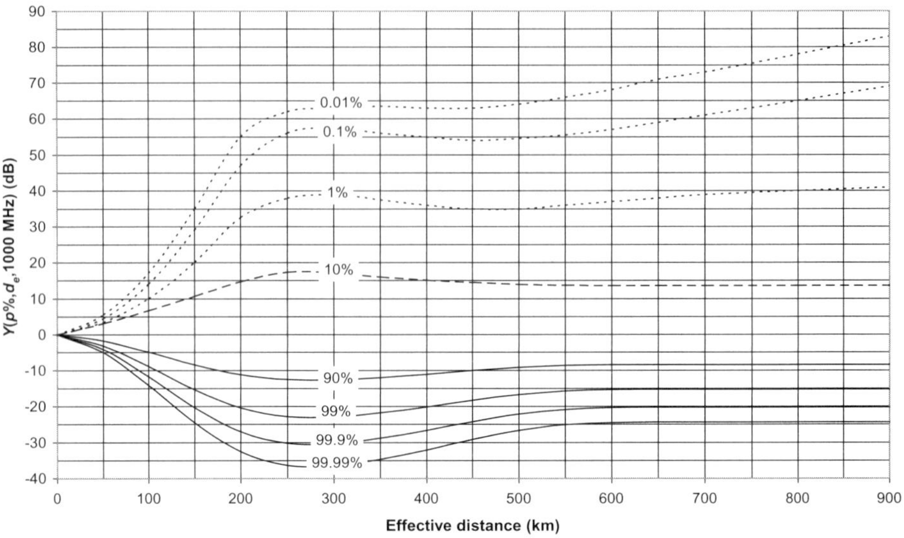

Figure 4.22 Climate 4: maritime subtropical over land (f_0 = 1000 MHz).

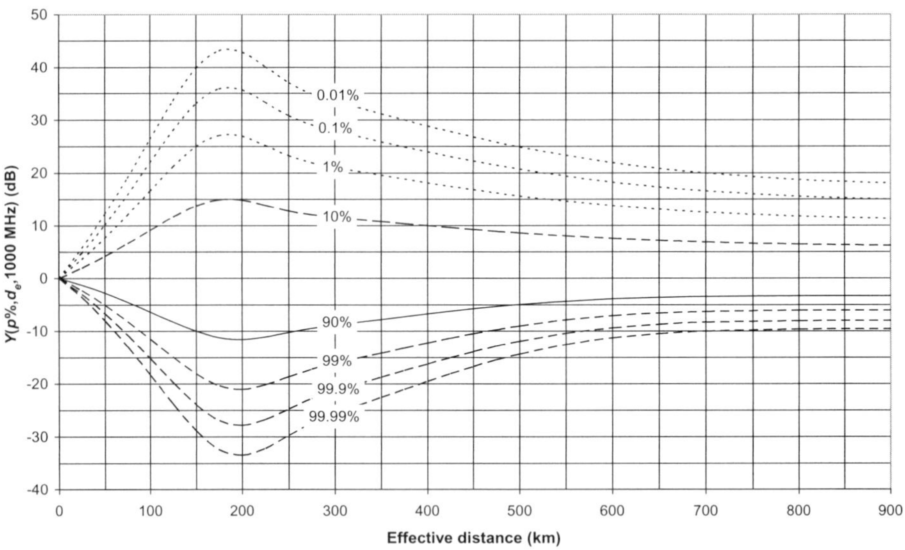

Figure 4.23 Climate 6, desert (f_0 = 1000 MHz).

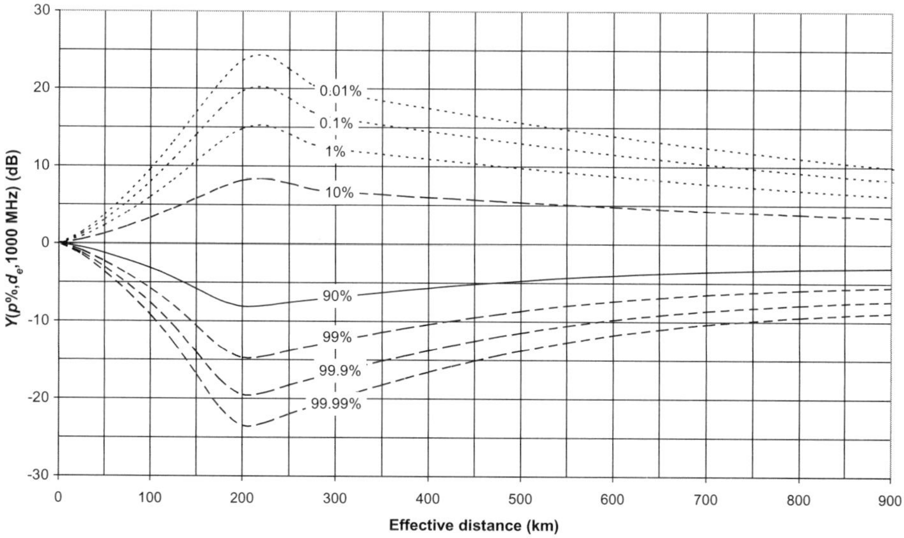

Figure 4.24 Climate 7, equatorial (f_0 = 1000 MHz).

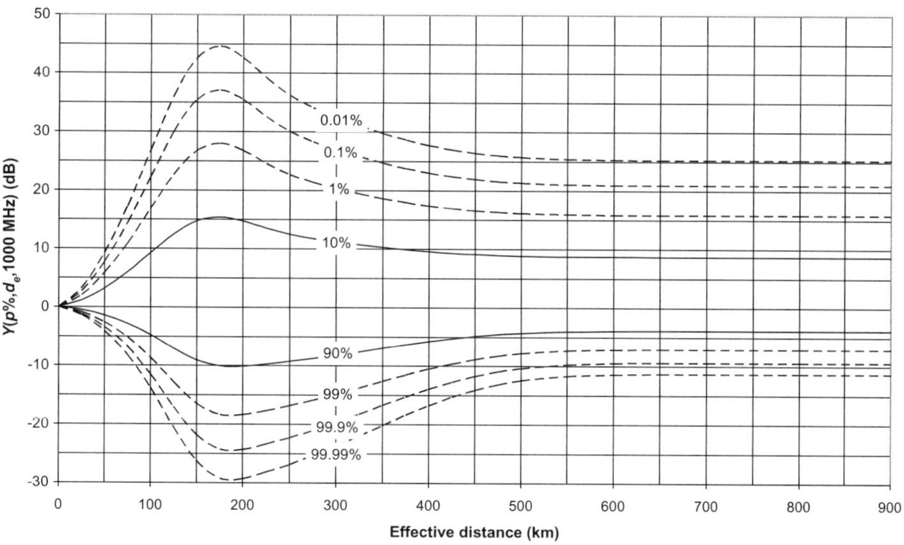

Figure 4.25 Climate 8: continental subtropical (f_0 = 1000 MHz).

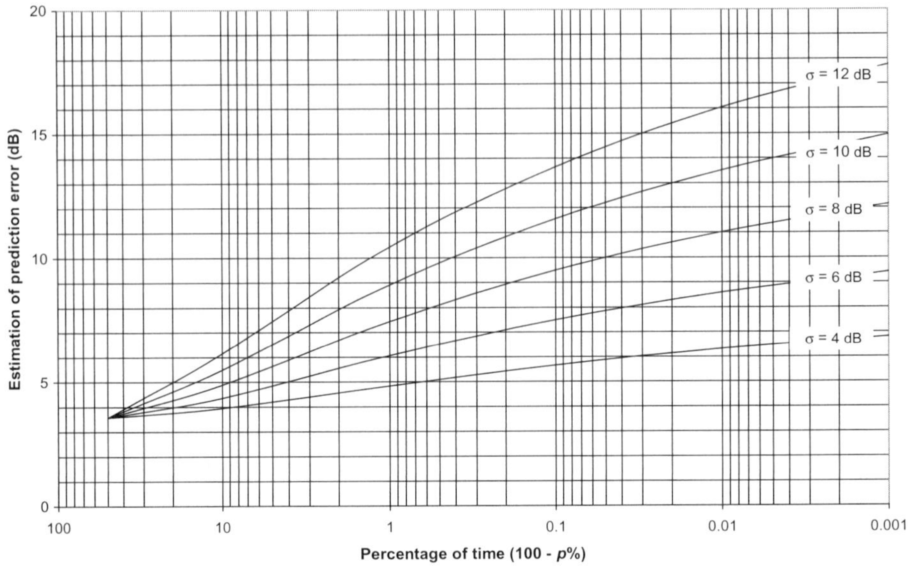

Figure 4.26 Estimated prediction error.

and can be estimated from presently available transhorizon transmission loss data as

$$\sigma_c(p\%) = \sqrt{12.73 + 0.12Y^2(p\%)} \tag{4.30}$$

where $Y(p\%)$ is the variability of the long-term Gaussian distribution which can be given by the following relation valid for $50\% < p\% < 99.999\%$:

$$Y(p\%) \approx -5.31\left[1.167\left(\frac{\ln 100 - \ln p(\%)}{2}\right)^{0.14} - 1\right]\sigma$$

where σ is the standard deviation $[Y(84\%)$ or $Y(16\%)]$. It has to be noticed that $Y(50\%) = 0 \rightarrow \sigma_c(50\%) = 3.6\,\text{dB}$. Figure 4.26 illustrates the prediction error according to the percentage of time[2] $(100 - p\%)$ and the standard deviation of the Gaussian distribution of the transmission loss.

4.2.2.4 Service Probability For noise-limited service of at least grade g and time availability p, the service probability is the probability that, due to prediction error, the hourly median transmission loss is higher than the maximum allowable value, that is,

$$L(p\%) - L_M(g) > 0$$

[2]For graphical convenience.

Assuming that the estimation error lies mainly on the path variance $\sigma_c^2(p)$, or mean-square prediction error, it is convenient to represent the service probability Q as a function of the standard normal deviation which has a mean of zero and a variance of unity:

$$z_M(p\%) = \frac{L_M - L(p\%)}{\sigma_c(p\%)}$$

where L_M is the transmission exceeded a fraction $(100 - p\%)$ of the time with a probability Q. The probability of service can then be expressed in terms of the error function or the normal distribution:

$$Q[z_M(p\%)] = \frac{\exp(-z_M^2(p\%)/2)}{\sqrt{2\pi} \times 0.661 z_M(p\%) + 0.339\sqrt{z_M^2(p\%) + 5.51}} \tag{4.31}$$

with

$$z_M(p\%) = \frac{Y_M - Y(p\%)}{\sigma_e(p\%)}$$

where Y_M is the maximum allowable value. Figure 4.27 shows the service probability $(1 - Q)$ versus percentage of time $(100 - p\%)$ for various values of Y_M considering a standard deviation of 10dB for the distribution. It is then

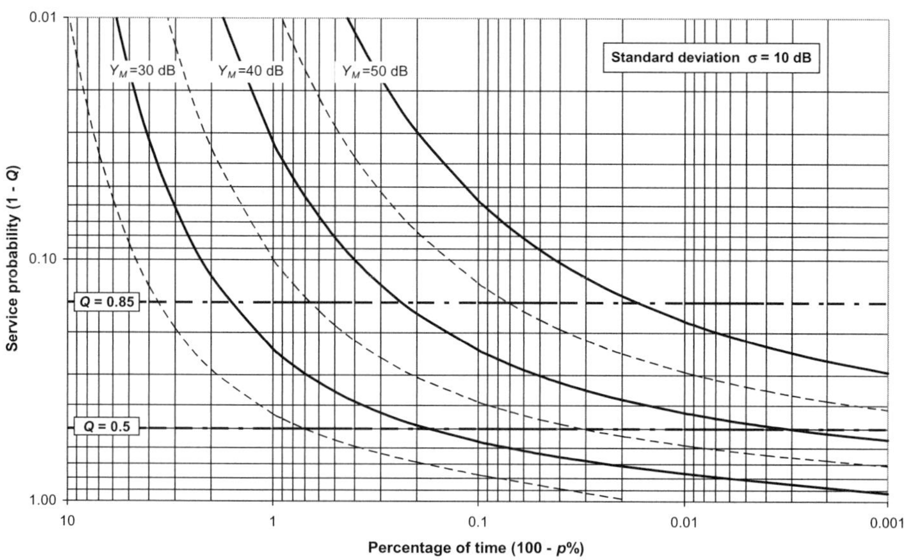

Figure 4.27 Service probability versus percentage of time.

possible to express the availability, in terms of percentage of time, according to the service probability by using the following steps:

1. Calculate the standard normal deviation:

$$z_M(p\%) = -5.31\left[1.167\left(-\frac{1}{2}\ln Q\right)^{0.14} - 1\right] \tag{4.32}$$

2. Calculate the variability:

$$Y(p\%) = \frac{Y_M - \sqrt{Y_M^2 - [1 - 0.12Z_M^2(p\%)][Y_M^2 - 12.73Z_M^2(p\%)]}}{1 - 0.12Z_M^2(p\%)} \tag{4.33}$$

3. Calculate the corresponding availability:

$$p\% = 100 \times \left(1 - \frac{\exp\{-(1/2)[Y(p\%)/\sigma]^2\}}{\sqrt{2\pi}\,0.661|Y(p\%)/\sigma| + 0.339\sqrt{[Y(p\%)/\sigma]^2 + 5.51}}\right) \tag{4.34}$$

Figure 4.28 shows the percentage of time of unavailability ($100 - p\%$) according to the service probability for various values of Y_M considering a standard deviation of 10 dB for the distribution. In relation with the link budget, the maximum allowable value Y_M corresponds to the minimum received field strength for a given BER below the median long-term transmission loss given by relation (4.26).

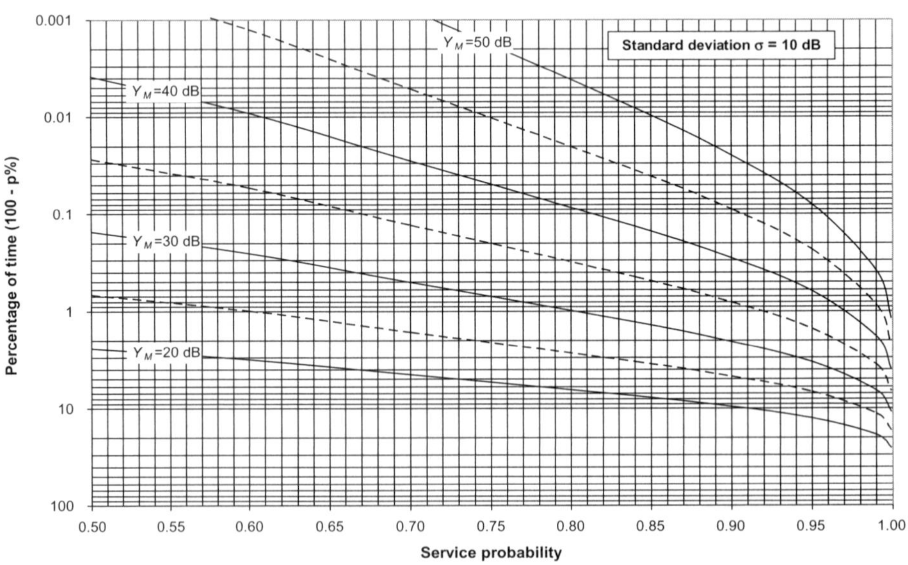

Figure 4.28 Percentage of time versus Service probability.

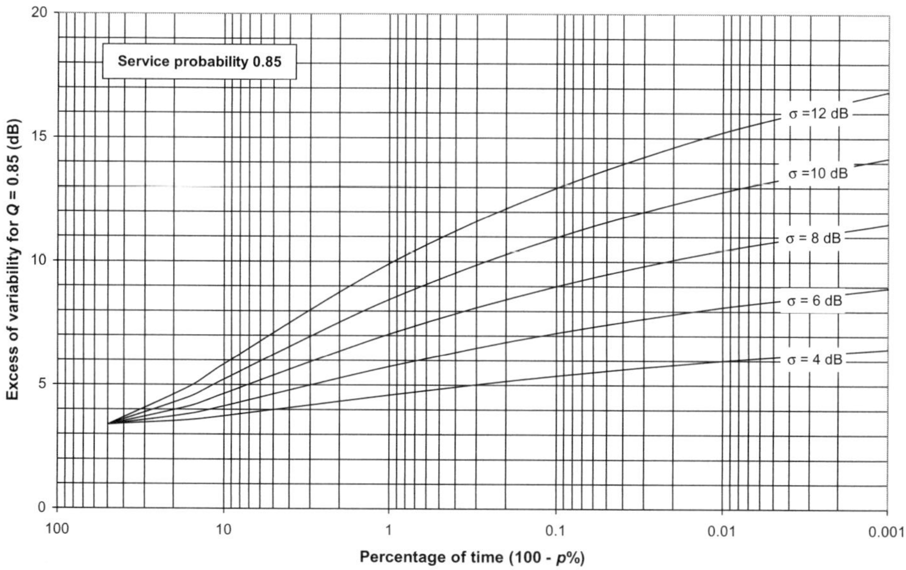

Figure 4.29 Excess of variability for $Q = 0.85$.

It is also possible to compute the variability of the signal for a given service probability in relation with the variability for $Q = 0.5$ by introducing the relation

$$L_Q(p\%) = L_{Q=0.5}(p\%) + z_M \sigma_c \tag{4.35}$$

Figures 4.29 and Figure 4.30 show respectively as an example the excess of the variability of the attenuation of propagation and the unavailability for $Q = 0.85$ relative to $Q = 0.5$; such value of 0.85 means that among 100 paths using the same parameters 85 would be equal or exceed specifications.

4.3 LINK BUDGET

The hourly median power received according to the percentage of time over the average year is expressed by the following relation in decibels:

$$P_R = P_E + G_E + G_R - L_F - L_{gp} - L(p\%) \tag{4.36}$$

where P_R = hourly median received power (dBW)
P_E = emitted power (dBW)
G_E = transmitting antenna gain (dBi)
G_R = receiving antenna gain (dBi)
L_F = total branching loss (dB)
L_{gp} = multipath coupling loss (dB)

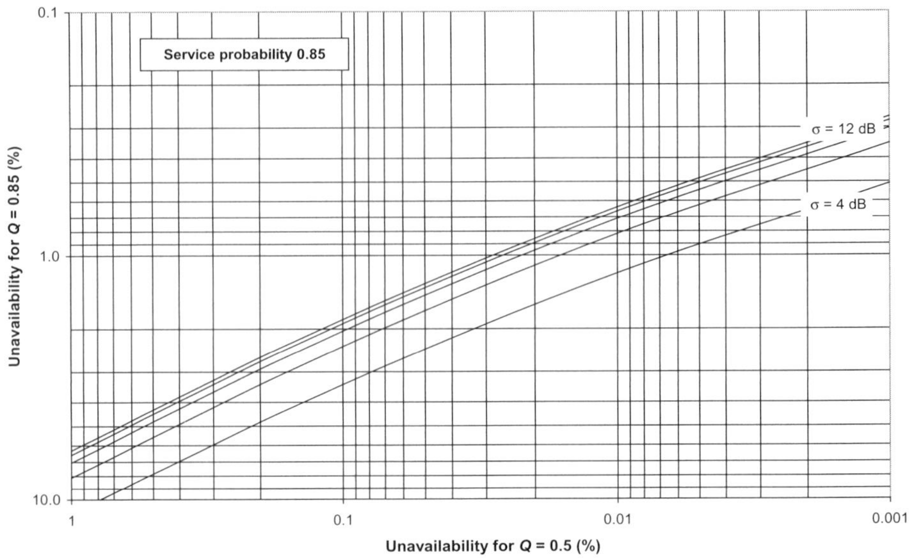

Figure 4.30 Excess of variability for $Q = 0.85$ for various standard deviation values.

$L(p\%)$ = estimated value of transmission loss not exceeded for given percentage of time $p\%$ defined by relation (4.28)

4.3.1 Multipath Coupling Loss

The multipath coupling loss L_{gp} between the antennas and the medium of propagation, which is also called antenna-to-medium coupling loss or loss-in-path antenna gain, results in an apparent reduction of the gain of both reception and emission antennas. The transmission of the wave energy being done by reradiation of all the heterogeneities illuminated by the transmitting antenna, the common volume of diffusion behaves like a large-sized incoherent source instead of a point source. In consequence, the received wave is not a plane wave and the apparent gain of the receiving antenna is lower than its gain in free space; according to the reciprocity theorem, the transmitting antenna undergoes the same loss in gain. In addition, with high-gain antennas, the density of scatterers is uniform, and when the gain of the antennas increases, the illumination increases proportionally to the inverse of the square of the beamwidth, but, because of the reduction in size of the common volume, the number of illuminated scatterers decreases as the inverse of the cube of the beamwidth and so there is an overall gain loss proportional to the inverse of the beamwidth. The loss in path antenna gain can be calculated using the relation

$$L_{gp}=\begin{cases}\dfrac{1+0.72\,[\log(S\mu)]^{1.22}}{1+1.1\,[\log(S\mu)]^{1.22}}\times 8.55\,[0.7+\log(\hat{n}+0.07)]^2 & \text{for } 0.1<\hat{n}\le 10\\[2ex] \dfrac{1+0.72\,[\log(S\mu)]^{1.22}}{1+1.1\,[\log(S\mu)]^{1.22}}[29\log(\hat{n}+0.07)-4.3] & \text{for } \hat{n}>10\end{cases} \tag{4.37}$$

where the asymmetry factor S is given by relation (4.18) and the factor μ by

$$\mu=\frac{\alpha_E}{\alpha_R} \tag{4.38}$$

where α_E = half beamwidth at half power of transmitting antenna
α_R = half beamwidth at half power of receiving antenna

considering the following:

- If $S\mu < 1$, take $1 / S\mu$.
- The factor $\hat{n}$ is given by the relation

$$\hat{n}=\frac{n+0.03v}{f(v)} \tag{4.39}$$

where

$$n=\begin{cases}\dfrac{\alpha_{00}}{\alpha_E} & \text{if } S\mu\ge 1\\[2ex] \dfrac{\beta_{00}}{\alpha_R} & \text{if } S\mu<1\end{cases} \tag{4.40}$$

$$v=\frac{\eta_S}{2}\quad [\eta_S \text{ given by relation (4.21)}] \tag{4.41}$$

$$f(v)=\frac{1.36+0.116v}{1+0.36e^{-0.56v}} \tag{4.42}$$

Figure 4.31 presents the function $f(v)$. Figure 4.32 presents the function L_{gp} according to factor $\hat{n}$ and $S\mu$ that is obtained using relation (4.37).

4.3.2 Influence of Transverse Directivity

When the antennas are pointed using the traditional radioelectric process, which consists in seeking the maximum received field in turn at each end of the link, it appears that under certain conditions of propagation and for

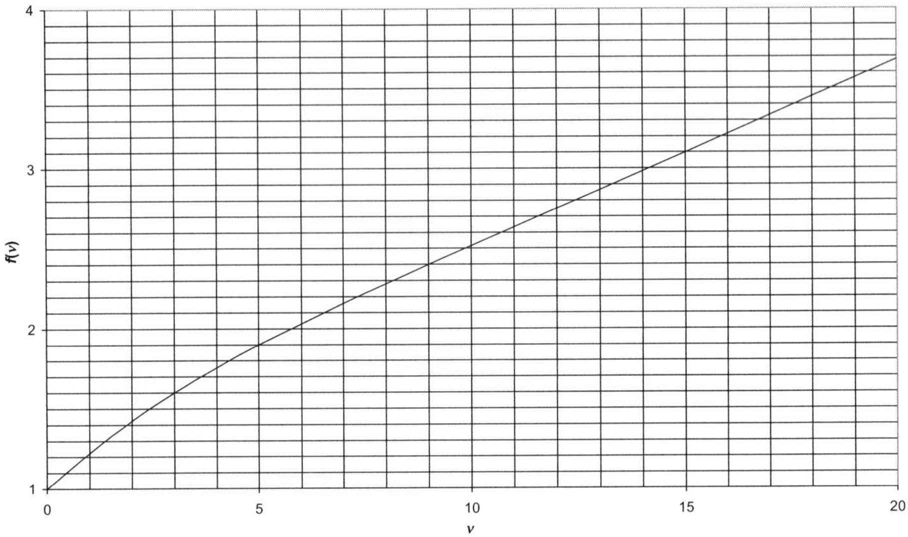

Figure 4.31 Function $f(v)$.

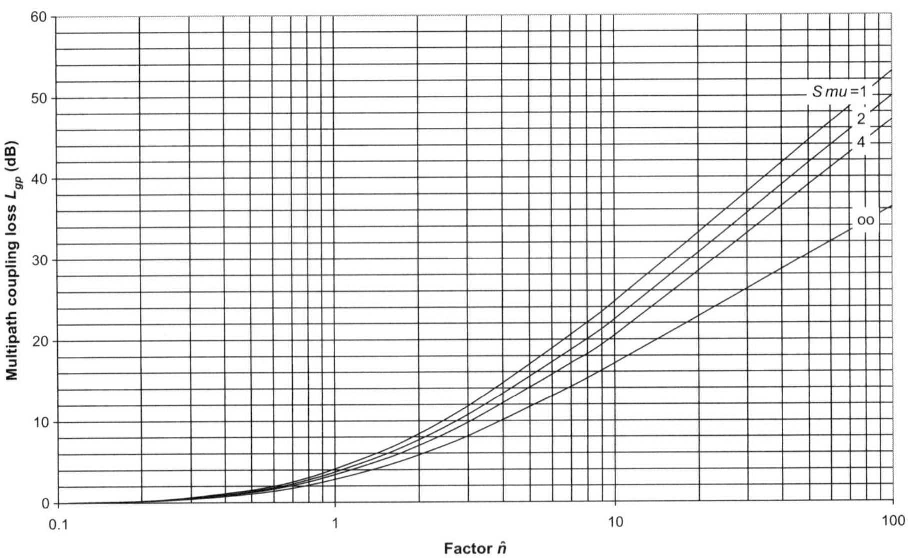

Figure 4.32 Multipath coupling loss L_{gp} versus factor.

antenna gain higher than a certain value there is no convergence toward the nominal azimuth; consequently, the link is established on a shifted common volume of diffusion with a significant degradation of the performance.

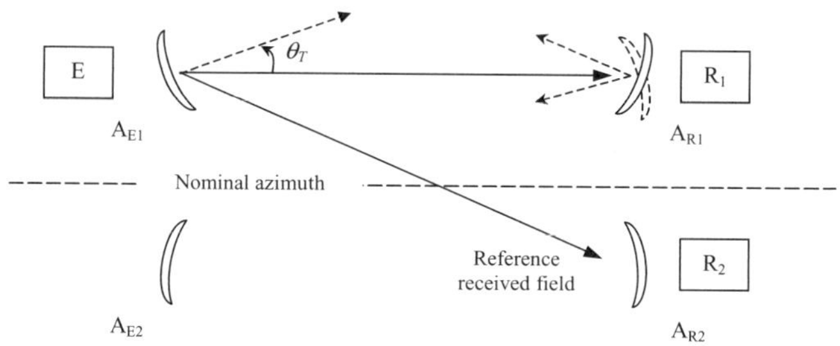

Figure 4.33 Classical radioelectric pointing process.

The radioelectric process of pointing generally employed is illustrated in Figure 4.33 which presents a traditional configuration of double diversity of space or of quadruple diversity of space and frequency. In this process, the wave emitted by transmitter E is received simultaneously by receivers R_1 and R_2, which are assigned respectively to antenna A_{R1}, which is in the progress of pointing, and antenna A_{R2}, which remains fixed; the field received by receiver R_2 thus constitutes a reference with which one permanently compares that which is received by receiver R_1 in order to free from the fluctuations of the long-term propagation attenuation while the fast fluctuations in the short-term are integrated by a time-constant on the order of 1 min.

Let us suppose now that the transmitting antenna A_{E1} is pointed so that the field at the station of reception is sufficient to carry out the operations of pointing but is shifted an angle θ_T compared to the nominal azimuth which is in the Earth great circle plane passing by both stations, which is generally the case when the link was established following a prepointing carried out in an imprecise way, for instance by means of a compass, as is usual.

Figure 4.34 shows the various radiation patterns obtained at the reception for different off-axis incremental positions θ_T of the transmitting antenna, corresponding to an angular separation of 1°, on a 82-km troposcatter link operating at the frequency of 5 GHz on a smooth Earth in a semidesert climate and using antennas of 3 m.

We see in this example that a misalignment of the transmitting antenna by 1° toward the right (transmission 1° Right) causes a displacement of the common volume of diffusion about 0.5° compared to the nominal azimuth and the path loss is then penalized by 5–6 dB; it is thus necessary to carry out several successive radioelectric pointings by alternating the emission and the reception to converge toward the nominal azimuth since the shift of the common volume decreases by half during each operation.

In the same way, Figure 4.35 shows the various radiation patterns obtained, with an angular separation of 0.5°, on a 167-km troposcatter link operating on a smooth Earth in the same conditions as above and using antennas of 8 m as described in Section 4.4.4. We see in Figure 4.35, where the values of the reference field level for each curve are indicated as conditions of propagation at

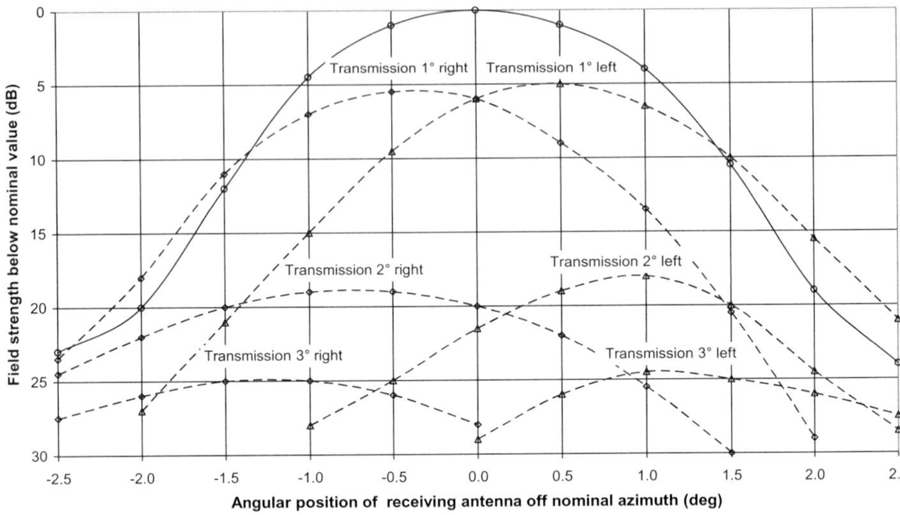

Figure 4.34 Transverse directivity effects on 3-m antennas. Radiation patterns of receiving antenna for various incremental positions of transmitting antenna (5 GHz) on 82-km troposcatter link.

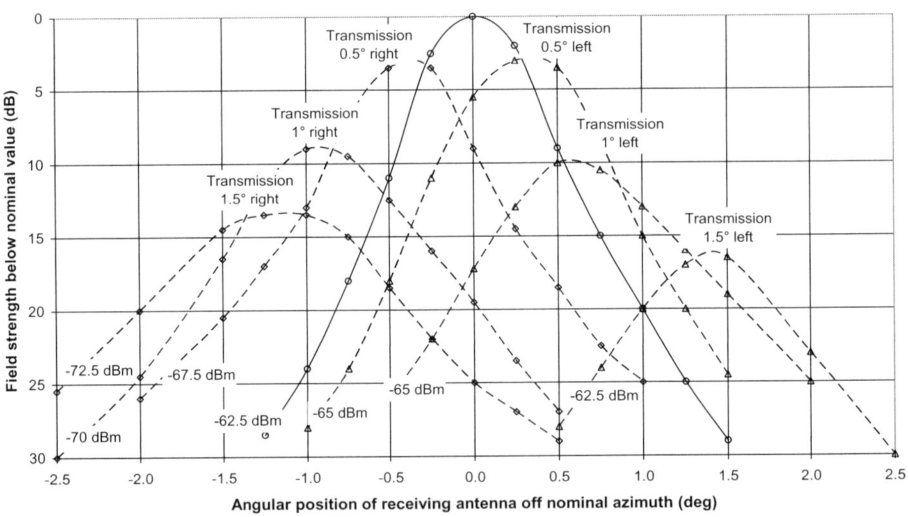

Figure 4.35 Transverse directivity effects on 8-m antennas. Radiation patterns of receiving antenna for various incremental positions of transmitting antenna (5 GHz) on 167-km troposcatter link.

the time, that the shift of the common volume is of the same order of magnitude as the shift angle of the transmitting antenna; convergence toward nominal azimuth cannot thus be obtained any more by successive approximations as previously but requires that the operation of pointing be repeated for various incremental positions of the transmitting antenna in order to locate the maximum maximorum by using an iterative process.

This method of radioelectric pointing by iteration, which consists of aligning the transmitting antenna instead of the receiving antenna, remains valid in all the cases; it indeed allows also to locate possible sources of transverse diffraction in the vicinity of the path in zones where the relief is accentuated or which comprise disseminated obstacles.

In order to explain this phenomenon of propagation due to the diffusion by the medium in all directions horizontally as well as vertically, called *transverse directivity* by the author in 1977, and to measure its effects, the following tests were carried out and are illustrated in Figures 4.36–4.38, representing a top view of the link and the recordings for various conditions of propagation.

Figure 4.36a shows the case where both antenna emission and reception are pointed on a shifted common volume owing to a shift angle θ_T of the transmitting antenna relative to the nominal azimuth to which corresponds

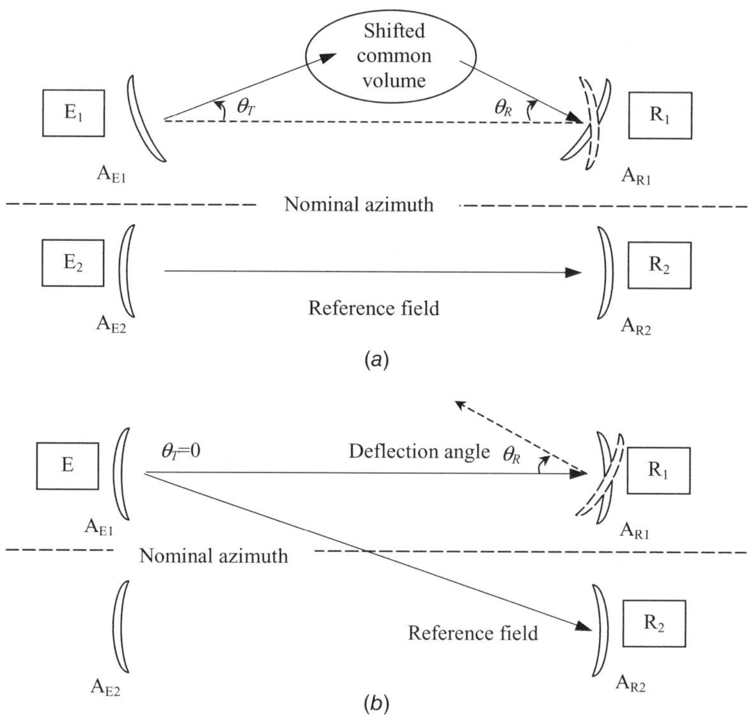

Figure 4.36 (*a*) Measurement of transverse directivity loss. (*b*) Measurement of off-axis deflection loss.

the shift angle θ_R of the receiving antenna that results from the research of the maximum of power; the difference between the field that is received in this configuration and the reference field corresponds to the transverse directivity loss.

Figure 4.36b shows the manner usually employed to measure in the field the radiation pattern of the receiving antenna while varying the shift angle θ_R whereas the transmitting antenna is aligned on the nominal azimuth ($\theta_T = 0$); thus the difference between the received field and the reference field represents the off-axis deflection loss.

Figure 4.37 shows the respective values of the losses due to the transverse directivity and off-axis deflection, according to the angular variation, that were measured using the 8-m antennas on the 167-km link for various conditions of propagation represented by the level of the reference field in a range from –45 to –80 dBm. It has to be noticed that:

- The transverse directivity loss is always lower than the off-axis deflection loss in the case of high-gain antennas, whatever the conditions of propagation are, which explains why there cannot be convergence toward the nominal azimuth and the link can be established only on a shifted common volume; the process of pointing per iteration thus proves essential to optimize link performance.
- In the case of a very high field strength, such as that of –45 dBm, which corresponds to conditions of propagation of the specular type, the deflection loss is close to the theoretical diagram of the antenna in free space.

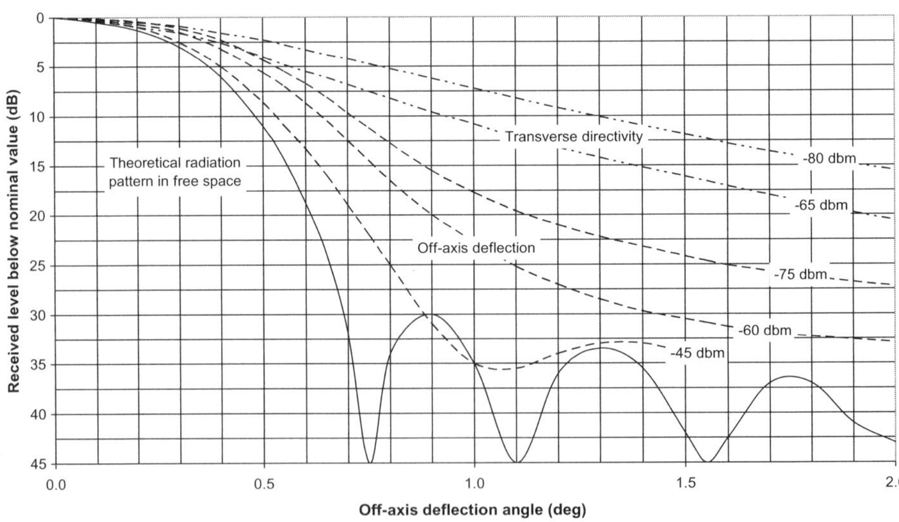

Figure 4.37 Transverse directivity and off-axis deflection losses: antennas 8 m, frequency 5 GHz, pathlength 167 km on smooth Earth.

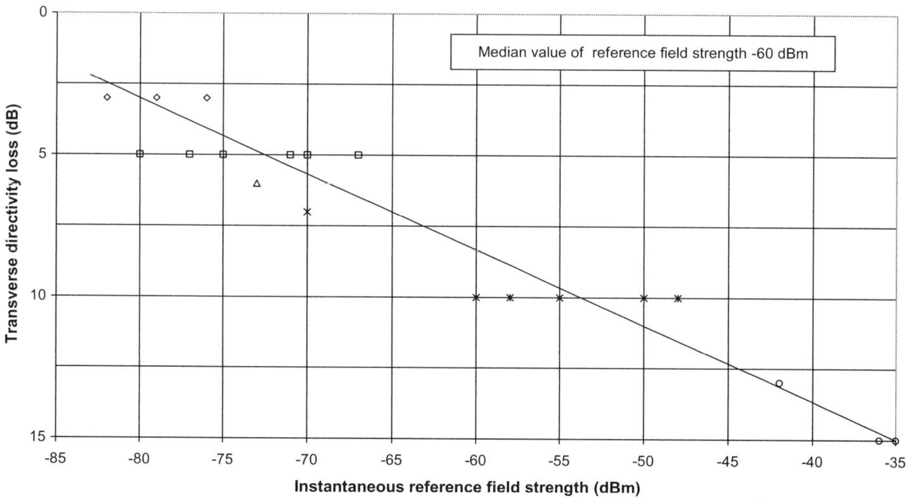

Figure 4.38 Distribution of transverse directivity loss versus conditions of propagation: antennas 8 m, frequency 5 GHz, pathlength 167 km on smooth Earth; shift angle 0.75°.

On the other hand, similar measurements carried out with 3-m antennas on the 82-km link under median conditions of propagation showed that the losses due to transverse directivity and off-axis deflection were of the same order of magnitude and relatively close to the theoretical diagram of radiation, which explains why convergence on the nominal azimuth can be obtained by successive approximations when the antennas have a low gain.

Figure 4.38 presents the distribution of the transverse directivity loss according to the instantaneous reference field strength that was obtained for an azimuth shift angle of 0.75° for both 8-m transmitting and receiving antennas on the 167-km link. We observe a roughly linear distribution which manifests itself by a permanent additional loss as well as by greater variability of the received field but also by a reduced standard deviation; the transverse directivity loss decreases with the reference field level in the proportion $\frac{1}{2}$; this property is made profitable when transverse angle diversity is used.

4.3.2.1 Manual Process for Pointing Antennas The process for manual pointing of the antennas suggested here makes it possible to be free, whatever the size of the antennas and the conditions of propagation may be, from the effects of transverse directivity and to align them on an axis very close to the nominal azimuth; the elevation angle must have been, beforehand, fixed just above the radio horizon by means of an inclinometer.

- First, insofar as the received field level obtained after the phase of prepointing is sufficient, we need to locate the first maximum (1), which corresponds to the initial position of the transmitting antenna, by employing the method described above and illustrated in Figure 4.33.

- Then we shift the transmitting antenna by an incremental value $\Delta\theta_T$, in relation with the radiation pattern of the antenna and the required degree of accuracy, and seek a new maximum (2) which corresponds to this new position and results in the variation $\Delta\theta_R$ at reception side.
- The operation is started again, putting on a graph the measurements obtained successively until one exceeds the maximorum, which corresponds to the nominal azimuth.
- The final pointing is then carried out on the transmitting antenna settled on the optimum value, which can be easily determined graphically as shown in Figure 4.39.
- The two other antennas, which are used for space diversity, can finally be pointed in only one operation on those that were thus optimized at the two ends of the link.

Figure 4.39 illustrates the iterative process that makes it possible to easily locate the nominal azimuth for the transmitting antenna by tracing on the graph the locus of the successive maxima 1, 2, 3, 4, 5, ... which have been obtained at the reception side. It has to be pointed out that it is not necessary to carry out the same operation by reversing the direction of transmission since the process makes it possible to point at the same time the emission and reception antennas.

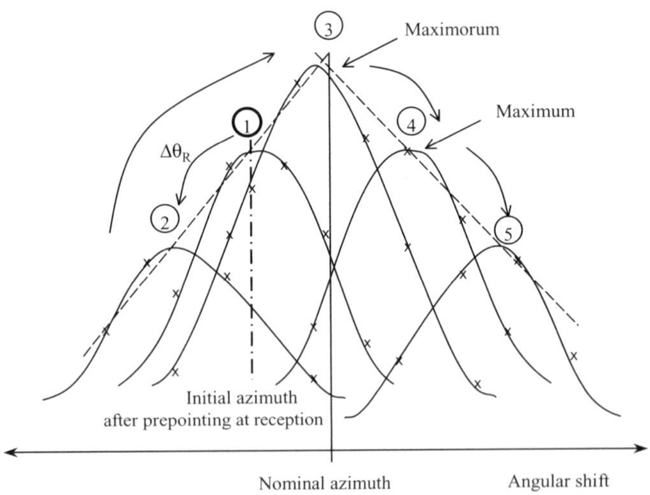

Figure 4.39 Manual process for pointing antennas by using a graph.

4.3.2.2 Automatic Process for Pointing Antennas With tactical links, where an implementation as fast as possible is sought after, antennas are generally actuated by jacks in order to facilitate the pointing. Antennas at each end can be prepointed with a slow sweep on one side and a fast sweep on the other so that the antenna moving fast has the time to go outward and return by the time the antenna moving slowly passes in the vicinity of the nominal azimuth; the sweeping of both antennas is then stopped as soon as the signal is captured. The sweep angle of the antennas can be several tens of degrees, which is what enables to quickly find the opposite station, of which one knows the position only in a given sector, and to avoid having to employ preliminary telecommunications as well as precise processes of localization of the stations and materialization of the nominal azimuth; the relationship between the scanning rates depends on the beamwidth delimited by the antenna pattern.

The automatic process suggested here is thus particularly adapted to the single-antenna configuration where space diversity would be replaced, for example, by angle diversity. In this case, the instantaneous received field that is measured on the antenna in the course of pointing cannot be compared any more with the reference field that was previously received on the space diversity antenna that remained fixed in order to compensate for the long-term fluctuations. Figure 4.40 illustrates the process of automatic research for the localization of the opposite station and the capture of the signal in the case of a single antenna configuration at each end. In order to be as free as possible from the long-term slow variations of the received signal and to preserve good accuracy, it is necessary to carry out the pointing operation in a minimum of time by calling upon a double algorithm:

- A phase of research by successive approximations of each maximum corresponding to the various incremental positions of the transmitting antenna
- A phase of determination of the nominal azimuth by the iterative process as described above

In small-size antennas, the automatic process can be limited to the first algorithm of successive approximations since, as we have seen above, the shift angle for each maximum at the reception is less than the shift angle of the transmitting antenna.

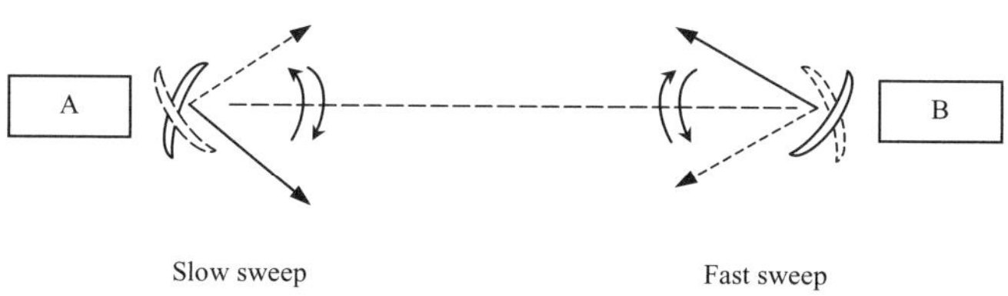

Figure 4.40 Process of automatic signal research and capture.

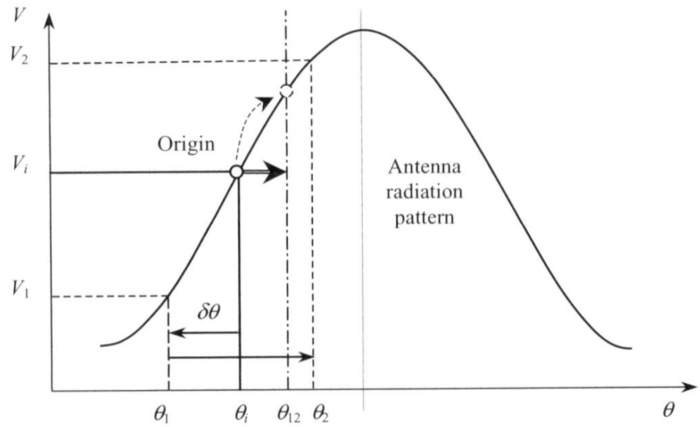

Figure 4.41 Sequence of readings of a set of received signal levels.

Process by Successive Approximations The process by successive approximations comprises a succession of sequences of readings of the level of the received signal for various angular positions of the receiving antenna by alternating the direction of transmission between each sequence. The process consists in creating sets of values of the received signal level for incremental positions $\pm\delta\theta$ of the receiving antenna with respect to a position of origin.

In Figure 4.41, which presents the reception antenna radiation pattern, we see that starting from an initial given angular position θ_i, corresponding to the position of capture at the issue of the process of signal research, we obtain for an incremental angular displacement $-\delta\theta$ the level[3] V_1 which corresponds to the position θ_1 and then for a displacement $+\delta\theta$ the level V_2 which corresponds to the position θ_2. Using a calculator, we then determine by interpolation the median position θ_{12} of the first set using the formula

$$\theta_{12} = \frac{\theta_1 + \theta_2}{2} + \delta\theta \frac{\|V_2 - V_1\|}{V_M} \tag{4.43}$$

where V_M is a characteristic which represents the maximum amplitude of the radiation pattern and $\|V_2 - V_1\|$, the absolute value of $V_2 - V_1$ with the sign of $(V_2 - V_1)/(\theta_2 - \theta_1)$. The same operation is then carried out at the station at the opposite end of the link by reversing the direction of transmission, that is, the antenna that was previously pointed emits in its turn on the position θ_{12}.

We see in Figure 4.42 that the first set of values gives the angular position θ_{A12} for station A, with station B transmitting on the initial position θ_{Bi}, then that station A, while emitting in its turn on the angular position θ_{A12}, makes it possible to obtain the position θ_{B12} at station B, which, while emitting again on

[3]The value V, which represents the field strength, is supposed to be linear.

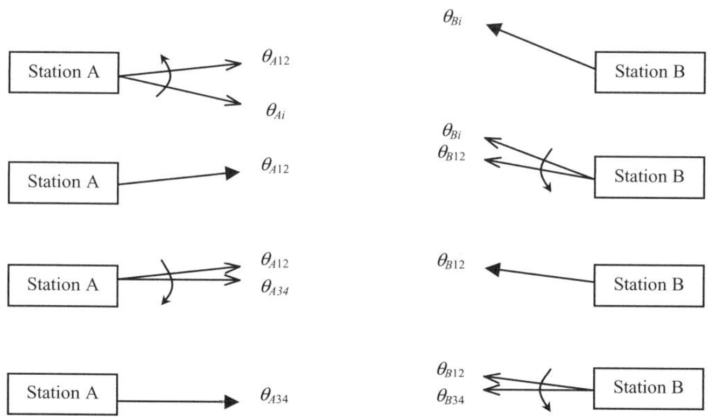

Figure 4.42 Sequences of pointing by successive approximations.

the position θ_{B12}, allows us to determine θ_{A34} at station A, and so on. It is necessary to take care to reverse the direction of displacement $\delta\theta$ in order to compensate for a possible variation of the path attenuation between the readings of two sets of measurements.

A few sets of measurements are generally enough to determine an accurate position for both antennas and the process of pointing can stop with this phase when the antennas are of small dimensions or in the case of links using the diffraction mode.

Process by Iteration We have seen that when antennas are large sized, convergence of the receiving antenna is not carried up on the nominal azimuth but onto a shifted common volume of diffusion which is defined by the off-axis angular position of the transmitting antenna; thus the preceding phase made it possible to obtain only a maximum without converging toward the maximorum.

Research of the maximorum, which corresponds to the alignment of the antennas on the nominal azimuth according to the Earth great circle, then needs a second algorithm to seek for several successive maxima corresponding to incremental positions $\delta\theta$ of the transmitting antenna, according to the process illustrated in Figure 4.39, this time keeping the same direction of transmission since it is a question of optimizing the pointing of the transmitting antenna. Then the calculator determines the direction of the nominal azimuth starting from the values of the received field that have been obtained respectively on each maximum and the angular position of the transmitting antenna which corresponds to them.

If one affects, for example, analogical values V_{Mi} to the levels of the fields that are received on the different maxima, which were obtained on the corresponding angular positions θ_{Mi} of the transmitting antenna, the optimal azimuth of the latter can be given by the relation

$$\theta_{\text{optimal}} = \frac{\sum_{i=1}^{n} \theta_{Mi} V_{Mi}}{\sum_{i=1}^{n} V_{Mi}} \tag{4.44}$$

The experiment showed that the complete process of pointing did not exceed half an hour with the following durations for each phase:

- Phase of research and capture <5 mn
- Determination of each maximum <3 mn

The final pointing accuracy which was obtained compared to the nominal azimuth was about a tenth of the half-power beamwidth of the antennas. The same process can be applied to optimize the elevation angle in tactical applications.[4]

4.3.3 Fast Fading Speed

We saw previously that the received field level underwent simultaneously slow variations of Gaussian distribution, corresponding to the hourly median attenuation related to the radioclimatic conditions at the moment, and fast fluctuations around this hourly median value that obey a Rayleigh distribution during intervals of time of a few minutes. The fading speed is an important factor from the point of view of the autocorrelation function in the time which determines the transfer function of the transmission channel.

Figure 4.43 shows an example of distributions over a 1-min period with time constants of 5, 0.3, and 0.015 s compared to the theoretical distribution. We can see that with a time constant of 0.015 s the distribution is very close to the theoretical Rayleygh law, with a variation of 10 dB per decade for the tail of the distribution, and that a time constant much higher than 5 s is required to integrate the instantaneous fluctuations in order to measure the median value of the field; the typical value employed for the time constant in this case is of about 1 min.

4.3.4 Reception in Diversity

Owing to the heterogeneity of the atmosphere, transmitted signals are not correlated in the short term outside a spatial cell that presents a limited time-variable width, height, and depth as well as a limited correlation frequency bandwidth. Therefore, if the signals are separated in paths or in frequency by more than their relevant correlation distance, they are decorrelated at the reception. In addition, to account for the importance of the instantaneous fluctuations of the received signal, it is necessary to use diversity techniques

[4]For each 0.1° of antenna misalignment, there is an additional loss of ~1 dB in azimuth and ~3 dB in elevation.

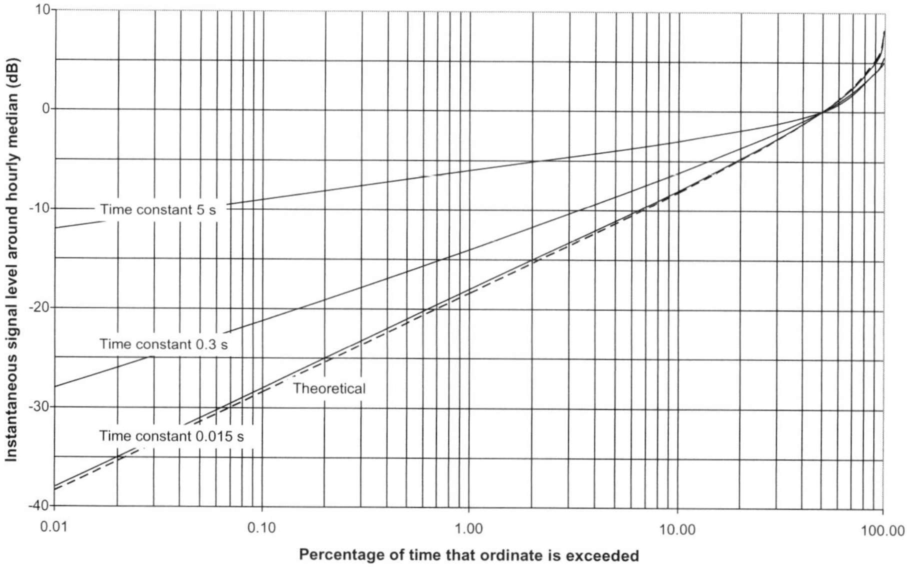

Figure 4.43 One-minute distribution of fast fading.

that consist in multiplying the number of channels by seeking the weakest possible correlation between them; the same information is thus transmitted on several paths and frequencies whose statistical fluctuations can be regarded as independent.

Section 3.4.1.9 set out the available diversity techniques in propagation conditions following a Rayleigh's distribution as well as their comparative performances. It is appropriate, now, to analyze the behavior of a receiver when it is subjected to such conditions of propagation; we will consider, for example, the case of a traditional modulation of 2DPSK (Differential phase shift keying) type that is generally used for links of this kind. The characteristic of the BER according to the carrier-to-noise ratio *C/N* of the receiver, if only the white Gaussian noise intervenes, is given by the formula

$$\mathrm{BER} = \frac{0.5}{\exp(10^{0.1(C/N)})} \tag{4.45}$$

In the case of a propagation affected by Rayleigh fading, the value of BER according to the *C/N* of a diversity system of reception operating in maximal-ratio combining depends on the row *D* of diversity employed and is given by the relation

$$\mathrm{BER} = \frac{0.5}{(1 + 10^{0.1(C/N)})^{D}} \tag{4.46}$$

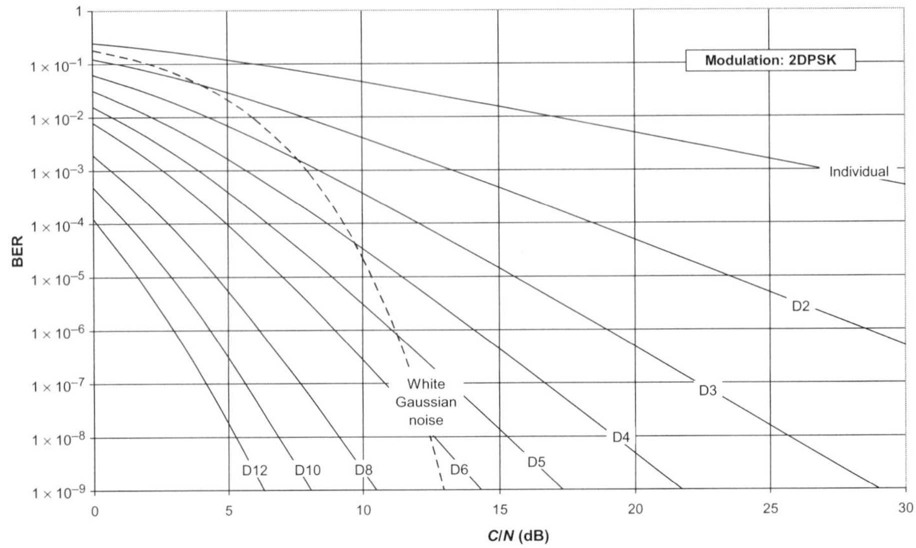

Figure 4.44 Characteristic *C/N*(BER) according to row of diversity in maximal-ratio combining system.

Figure 4.44 shows the characteristic of the BER according to the *C/N* of a single receiver affected by the only AWGN and those of systems of reception operating in maximal-ratio combining according to the row of diversity employed up to 12. The characteristics of the BER in the presence of Rayleigh fluctuations are much more spread out than the traditional characteristics obtained in an AWGN only line-of-sight radio link. The reception on a single receiver is practically not exploitable and it is only when starting from the diversity row of order 4 that one approaches the characteristic in Gaussian mode; however, it is seen that, for a BER ranging between 10^{-3} and 10^{-6}, the variation of the *C/N* in D4 is about 8 dB for only 3 dB in the Gaussian mode, which explains why troposcatter links generate a BBER higher than that of line-of-sight radio links. In contrast, increasing the diversity row beyond D4 provides high BER performances.

4.3.4.1 Space Diversity The suitable spacing between both reception antennas for frequencies higher than 1000 MHz is given by the following relations from ITU-R Rep.238 (where D is the antenna diameter in meters):

- Horizontal space diversity:

$$\Delta_H = 0.36\sqrt{D^2 + 4l_H^2} \qquad l_H = 20\,\text{m} \tag{4.47}$$

- Vertical space diversity[5]:

$$\Delta_H = 0.36\sqrt{D^2 + l_Z^2} \qquad l_Z = 15\,\text{m} \tag{4.48}$$

4.3.4.2 Frequency Diversity The suitable value for the frequency spacing, expressed in megahertz, is given by the following relation from ITU-R Rep.238:

$$\Delta f = \frac{1.44 f}{\theta d}\sqrt{D^2 + l_Z^2} \tag{4.49}$$

where f = frequency (MHz)
θ = angular distance (mrad)
d = distance (km)
D = antenna diameter (m)
l_Z = 15 m

4.3.4.3 Quadruple Space and Frequency Diversity Figure 4.45 shows the distribution of the instantaneous individual and combined received fields by selection diversity which was obtained on a 235-km troposcatter link at 5 GHz, established in a continental temperate climate as described in section 4.4.5, in a quadruple space and frequency diversity system using 7-m antennas horizontally spaced by 50 m, a frequency spacing of 112 MHz, and an emitted power of 1 kW; these results have to be related to Figure 3.55.

The period of observation has been chosen during real troposcatter conditions of propagation which result in Rayleigh distributed instantaneous individual received fields, as shown by the curves in Figure 4.45 related to the theoretical distributions. Figure 4.46 shows the evolution of the coefficients of correlation that have been observed respectively in space diversity and frequency diversity over the considered 7-min period.

4.3.4.4 Angle Diversity This mode of diversity consists in using antennas with separated beams horizontally and/or vertically so as to create several common volumes side by side and/or superposed; the angular separation (squint angle) is optimum when about the beamwidth. The process of vertical angle diversity should be applied only at the reception, by preserving the emission on the lower beam, in order to limit the transmission loss on the higher volume, as shown in Figure 4.47.

[5]With regard to vertical space diversity, it is necessary to consider the risk of gain loss of the higher antenna due to reflections on the ground, such as that evoked in Section 2.3.4.1, as well as the reduction of frequency gain H_0, given in Section 4.2.1.6, which becomes negligible beyond a certain effective height of the antennas; Section 1.7.4 also provides indications on the limitation of the zone of reflection which favors the antennas close to the ground.

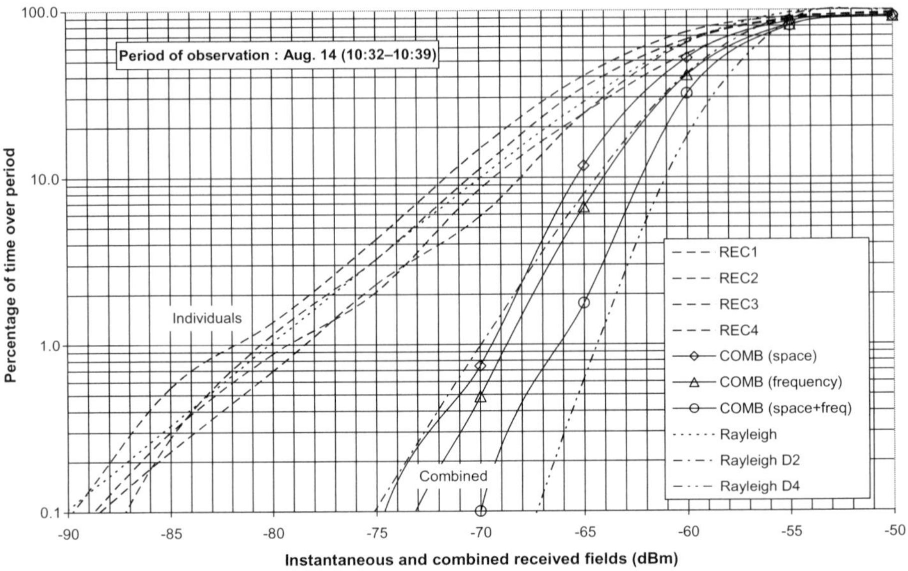

Figure 4.45 Instantaneous individual and combined received fields by selection of strongest.

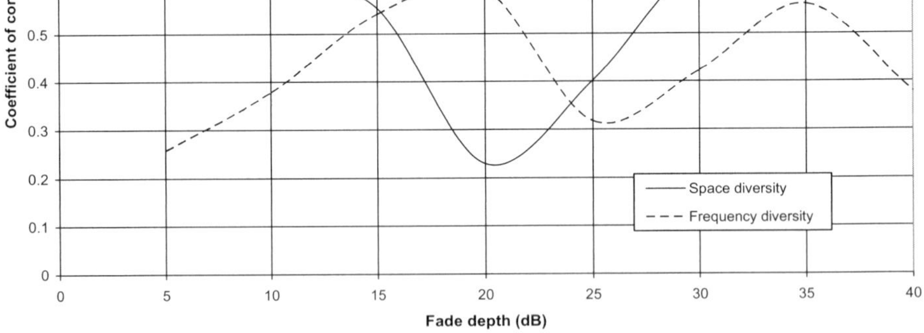

Figure 4.46 Coefficients of correlation in space and frequency diversity.

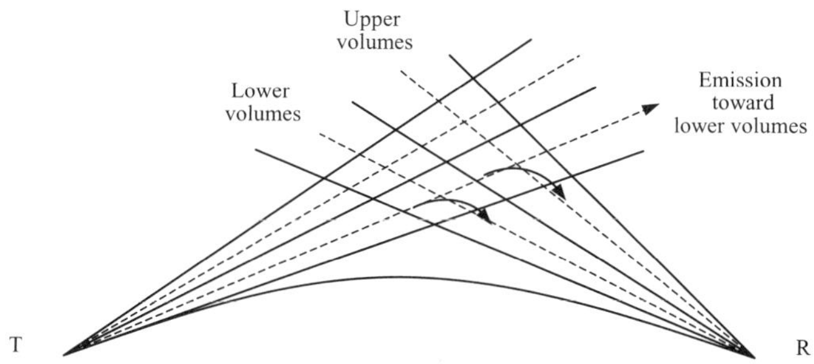

Figure 4.47 Reception in vertical angle diversity.

Measurements showed that a separation higher than approximately 6 mrad between the beams gives a coefficient of correlation that can be lower than 0.4 with results comparable with those of a diversity of frequency (ITU-R Rep.238). Moreover, it has been observed on the 167-km link established in a semidesert climate with 8-m antennas that, as we will see, in section 4.4.4, the higher path presented:

- A lower additional loss than that which is predicted by the attenuation function when introducing a more important angular distance
- A weaker standard deviation of the long-term hourly median attenuation of propagation than that of the lower path

These results can be explained by the fact that the asymmetry factor is more important on the higher path than on the lower one on this type of profile. At last, a description of a very efficient system using angle and frequency diversity of row 6, by means of a single three-beam antenna at each end, is presented in section 4.4.6 as well as the results that have been experienced on an 180-km link in a continental temperate climate.

4.3.4.5 Polarization Diversity According to ITU-R Rep.238, polarization is well preserved in the transmission and the fading of the signal does not depend on the polarization used; thus there is no diversity effect if the radio path is the same for two different polarizations.

4.3.5 Link Availability

The margin for a given BER and a given row of diversity D is equal to

$$M(\text{BER}, D) = P_R - N_R - \frac{C}{N}(\text{BER}, D) \qquad (4.50)$$

where P_R is the hourly median received field strength given by relation (4.36), N_R is the channel reference background noise given by relation (2.50), and

$$\frac{C}{N}(\mathrm{BER}, D) = 10\log\left[\sqrt[D]{\frac{0.5}{\mathrm{BER}}} - 1\right] \tag{4.51}$$

It is thus a question of comparing this hourly median margin $M(\mathrm{BER},D)$, expressed in decibels, with the estimated variation of the transmission loss $Y(p\%,d_e)$ not exceeded during $p\%$ of the time, which is given by relation (4.29). As $Y(p\%,d_e)$ is Gaussian, its distribution function is given by relation (1.144):

$$F(x) \approx \frac{\exp(-x^2/2)}{\sqrt{2\pi}\,(0.661|x| + 0.339\sqrt{x^2 + 5.51})} \qquad x = \frac{M(\mathrm{BER}, D)}{\sigma}$$

where σ is the standard deviation of the distribution, expressed in decibels, that we deduce from the tables and formulas in Section 4.2.2 or from the curves of the variability according to the effective distance for each standard climate; for example, we can record the value of $Y(99.9\%,d_e)$ and divide it by the Gaussian coefficient 3.09 to obtain it.

4.3.6 Coherence Bandwidth of Troposcatter Channel

We saw previously that the received signal results from the composition of a great number of elementary components that are transmitted by diffusion within the common volume to the transmitting and receiving antennas and whose phase is random; the resulting signal is for that reason modulated in amplitude and phase. The coherence bandwidth B_C, which characterizes the selectivity of the tropospheric channel or the bandwidth of the transmissible spectrum, is related to the differential time delay of propagation τ of the rays in the volume of diffusion that is delimited by the radiation patterns of the antennas by the relation

$$\frac{1}{B_C} = \tau \tag{4.52}$$

Figure 4.48 presents the geometric elements which come into play for determination of the differential time delay, where θ is the angular distance; β_E, β_R are the elevation angles at transmission and reception sites, and α_T is the half-power total beamwidth of the antennas. When the frequency bandwidth B_S occupied by the signal approaches this limit value B_C, strong distortions of amplitude and phase occur, called "multipath time dispersion", that generate an important degradation of the quality which comes from, in particular, significant interference between the successive digital pulses or intersymbol interference. The in-phase response is studied starting from the response in group delay, which is defined by the relation

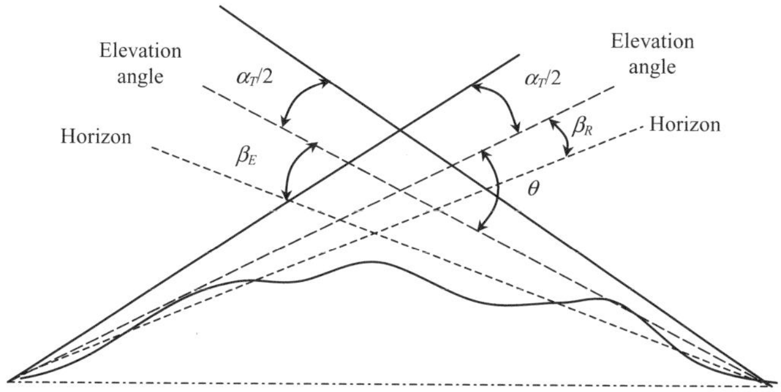

Figure 4.48 Geometric elements of troposcatter channel.

$$T_G = \frac{d\varphi}{d\omega} \qquad \omega = 2\pi f \tag{4.53}$$

Generally, the receiver is equipped with an autoadaptive equalizer which makes it possible to correct these effects and to admit a reduction of the transmission bandwidth corresponding to approximately half of the Nyquist bandwidth.

4.3.6.1 Model of Sunde The maximum differential time delay, expressed in seconds, between the higher path and the lower path is given by the relation

$$\Delta = 2\tau = \frac{d\alpha_T}{4c}(\theta + \alpha_T) \qquad \theta = \beta_E + \beta_R + \frac{d}{a} \tag{4.54}$$

where d = distance (km)
c = speed of light (~$3 \times 10^5\,\text{km s}^{-1}$)
α_T = half-power beamwidth of antennas (rad)
β_E, β_R = elevation angles above horizon of antennas (rad)

By supposing that the path differential time delay τ compared to the average path time delay is equally distributed, the coherence bandwidth B_C, expressed in hertz, becomes

$$B_C = \frac{c}{d\alpha_T(\theta + \alpha_T)} \tag{4.55}$$

We thus obtain an average coherence bandwidth, based on geometric considerations, which does not provide any indication about its evolution in the course of time.

4.3.6.2 Model of Rice The model of Rice supposes that the contribution of heterogeneities decreases starting from the center of the common volume according to a Gaussian law in three dimensions; the coherence bandwidth resulting from this model has as a value

$$B_C = \frac{2c\sqrt{3}}{\pi d \alpha_T \theta} \tag{4.56}$$

4.3.6.3 Model of Collin This method aims at defining the frequency bandwidth inside which the variations of the response in amplitude occur, on average, in the same direction according to the time. It calls upon an autocorrelation function which expresses the degree of correlation existing between the amplitudes of the fields that are received at two frequencies of difference Δf; the coherence bandwidth is, by definition, the value of the variation δf for which the coefficient of autocorrelation is 0.4.

As the conditions of propagation evolve over time, we consider the coherence bandwidth as an average value over a period of 1 min that fluctuates during the hours of the year; it thus presents a distribution function with a median value (50% of the time) and values exceeded for various percentages $p\%$ of the hours of the year.

The relations that have been established for the determination of the median value and the standard deviation of the coherence bandwidth, proceeding from measurements carried out on an experimental link and statistical results published for about 15 links established in temperate climate, are as follows:

$$B_{C(50\%)} = \frac{9.87 L_{gp}}{f\theta^2 d^{1.65}} \quad \text{(expressed in MHz)} \tag{4.57}$$

$$\frac{B_{C(50\%)}}{B_{C(84\%)}} = \frac{1.883}{(\theta d)^{0.17}} \tag{4.58}$$

where L_{gp} = multipath coupling loss (dB) defined by relation (4.37)
f = frequency (GHz)
θ = angular distance (rad)
d = pathlength (km)

and the conditions of validity are as follows:

Climate	Continental temperate ($N_S \approx 320$)
Terminal altitude	<200 m
Pathlength	$135 < d < 270$ km
Frequency	$0.9 < f < 7.6$ GHz
Angular distance	$1.7 < \theta\, d < 11$
Multipath coupling loss	$L_{gp} \geq 0.4\, f$

The standard deviation, expressed in decibels, is given by

$$\sigma = 20\log\left(\frac{B_{C(50\%)}}{B_{C(84\%)}}\right) \tag{4.59}$$

We can thus deduce from it the value not exceeded of the coherence bandwidth for $p\%$:

$$B_{C(p\%)} = B_{C(50\%)} \times 10^{-(\sigma C_G/20)} \tag{4.60}$$

where C_G is the Gaussian coefficient given in Section 4.2.2.2. For a given minimal bandwidth B_S, acceptable for the signal, the availability of the channel in terms of coherence bandwidth is obtained using relation (1.144) with

$$x = \frac{20\log(B_{C(50\%)}/B_S)}{\sigma} \tag{4.61}$$

The model can be connected to the climatic characteristic $V(50\%,d_e)$, in order to extend it to the other standard climates, by the relation

$$B_{C(50\%)} = \frac{6.3L_{gp}}{f\theta^2 d^{1.65}} \times 10^{[V(50\%,de)/20]} \tag{4.62}$$

where $V(50\%,d_e)$ is given by relation (4.27) or Figure 4.10.

4.3.7 Combination of Diffraction and Troposcatter Transmission Losses

On relatively short links, the two modes of propagation can coexist; the median value of the received field then becomes

$$P_C = P_1 + 10\log[1 + 10^{(P_2 - P_1)/10}] \tag{4.63}$$

where P_1 and P_2 represent the diffracted and scattered fields. Figure 4.49 shows the result of the combination of the hourly median received fields power according to the relationship between their respective power.

Concerning the instantaneous fields, mechanisms of propagation differ for each of the two modes, as shown in Section 3.4.2 for diffraction and Section 4.3.3 for scattering. The instantaneous resulting signal can thus be affected by important fluctuations according to the difference between the respective magnitude and path time delay of both components in relation to the wavelength. Under such conditions, frequency diversity or vertical space diversity may prove more effective than space horizontal diversity, the latter not introducing any difference in length between the paths of propagation.

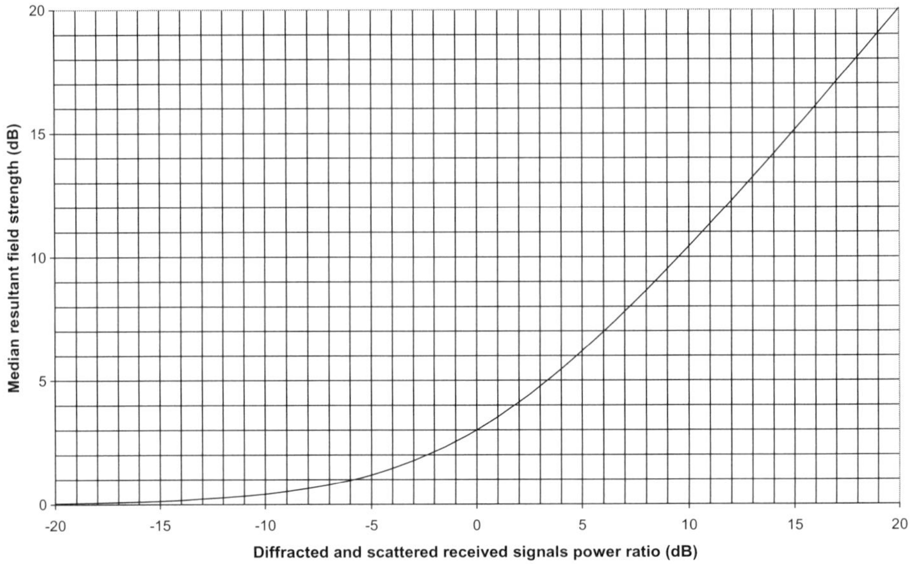

Figure 4.49 Combination of diffracted and scattered received signals.

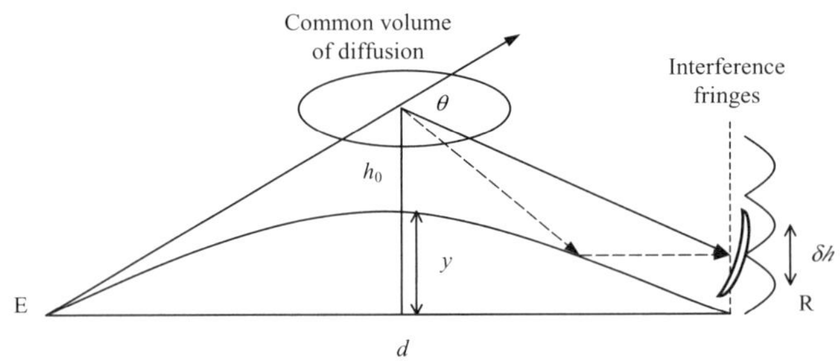

Figure 4.50 Geometry of forward-scatter link affected by reflection on ground at reception side.

4.3.8 Influence of Reflection on Ground

As we saw in Section 2.3.4, a reflection on the ground or in the low layers of the atmosphere produces interference fringes which are the result of the combination of direct and reflected waves; under these conditions, the resulting signal cannot be regarded as an equal-phase plane wave.

In the case of a forward-scatter link, the reflection can occur both at the emission side and the reception side since the antennas have in common the volume of diffusion, which explains why the frequency gain H_0 depends on the effective height of the emitting–receiving antennas. Figure 4.50 shows the geometry of a link whose antenna is affected by a reflection on the ground at the reception side. For example, for a link established on a smooth Earth, we

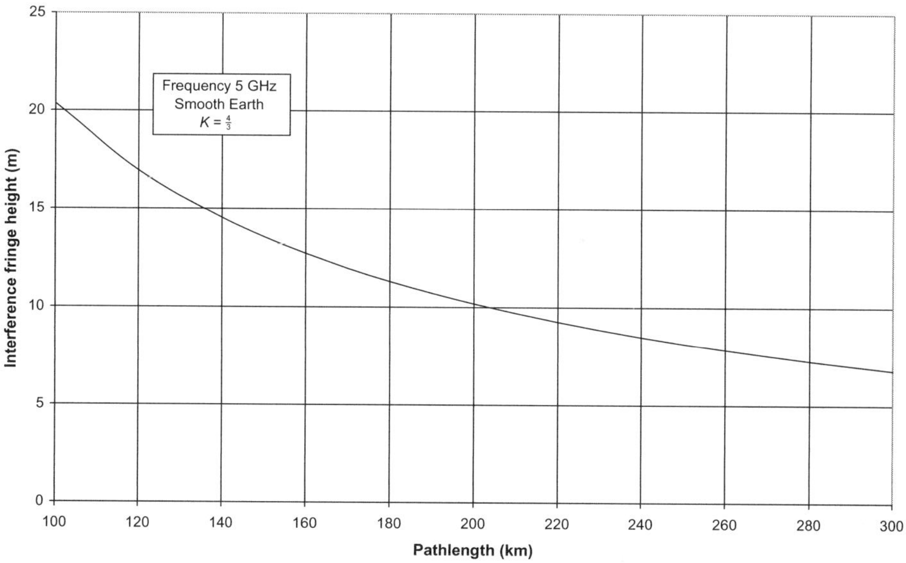

Figure 4.51 Interference fringe height versus pathlength over smooth Earth.

can calculate the height of the interference fringe according to the average height h_0 of the volume of diffusion from relation (2.67):

$$\delta h = \frac{\lambda d}{2(h_0 - y)} \qquad h_0 = \frac{S\theta d}{(1+S)^2}$$

where

$$S = 1 \quad \text{since link is symmetrical}$$

$$\theta \approx \theta_0 = \frac{d}{a}$$

$$y = \frac{d^2}{8a}$$

We then write

$$\delta h \approx 4\frac{\lambda a}{d} \tag{4.64}$$

Figure 4.51 presents the variation of the height of the interference fringes according to the pathlength for a frequency of 5 GHz and $K = \frac{4}{3}$. We can see that the height of the interference fringes can be of the same order of magnitude or lower than the usual size of the antennas, which means that the phase of the received signal may be reversed completely between the lower edge and the higher edge of the antenna, just as the level of field strength may vary between a maximum and a minimum; under such conditions, the received signal cannot be regarded as an equal-phase plane wave.

Calculation of the antenna gain loss due to reflection on the ground is very complex because it results from the integration by the antenna of the elementary components resulting from all the points of the common volume of diffusion that take part in the propagation. It is consequently desirable to mask as much as possible the antennas with respect to the reflection on the ground, as well as in the low layers of the atmosphere, by placing them close to the ground or by using antennas of sufficient diameter so that discrimination of the direct wave from the reflected one is effective.

4.3.9 Conversion of Annual Statistics to Worst-Month Statistics

Recommendation ITU-R P.841 gives further detailed information for the conversion of annual statistics to worst-month statistics according to the geoclimatic conditions of the zone; relations (1.110) can be used in general.

4.4 EXAMPLES OF TRANSHORIZON LINKS

4.4.1 Knife-Edge Diffraction Path

Figure 4.52 presents an 104-km link established in a semidesert climate that comprises a sharp stripped hill as the main obstacle close to one end as well as a rounded hill also deprived of vegetation as the secondary obstacle near the other end; the method of Deygout thus appears most suitable to apply. The radioelectric axis of the 3-m antennas was placed 5 m above the ground at both ends, the frequency employed was 4.7 GHz, and the emitted power was 1 kW.

We will consider an annual average value for the refractive index N_0 of 290, to which corresponds, according to relations (1.76) and (1.82), a surface refractivity N_S and a median value of the coefficient K:

$$N_S = 282.7 \qquad K(50\%) = 1.29$$

- *Free-Space Loss* The free-space loss is given by relation (1.49):

$$A = 20\log\left(\frac{4\pi \times 104{,}000\,\text{m} \times 4.7\,\text{GHz}}{0.3}\right) = 146.2\,\text{dB}$$

The atmospheric attenuation, given by relations (1.104) or (1.105), is $A_a \approx 0.8\,\text{dB}$.

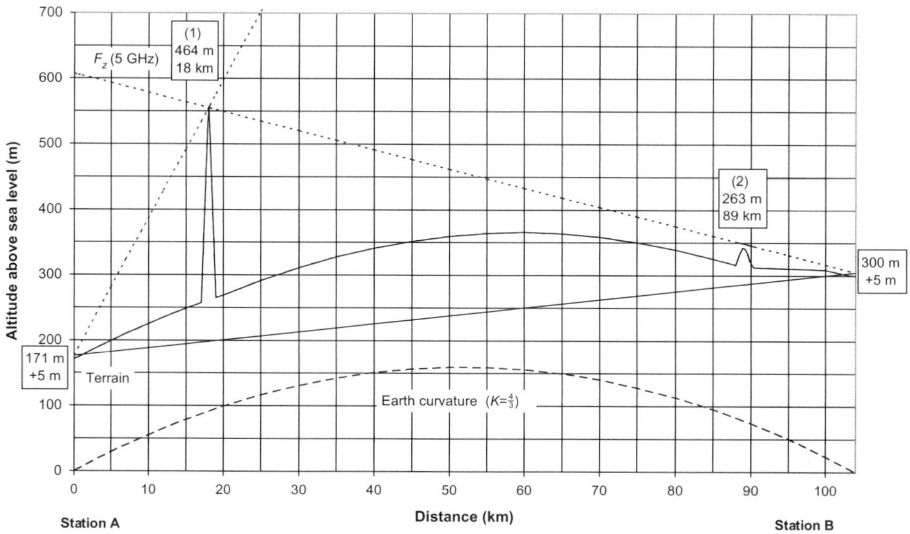

Figure 4.52 Path profile of knife edge diffraction link.

- *Main Obstacle Loss (1)* The diffraction loss due to a rounded obstacle is given by relation (1.97). We start by determining the height of the obstacle in the following way:

  (a) Altitude of obstacle 1 above chord joining both sites:

$$H_{C1} = 176\,\text{m} + (305\,\text{m} - 176\,\text{m})\frac{18\,\text{km}}{104\,\text{km}} = 198\,\text{m}$$

  (b) Earth curvature for $K(50\%) = 1.29$ using relation (1.84):

$$y_1 = \frac{18\,\text{km} \times 86\,\text{km}}{2 \times 1.29 \times 6378\,\text{km}} = 0.094\,\text{km} = 94\,\text{m}$$

  from which the height of the main obstacle is $h_1 = 464\,\text{m} - 198\,\text{m} + 94\,\text{m} = 360\,\text{m}$.

  (c) Determination of angle of diffraction by neglecting occultation distance D_S for sharp edge:

$$\theta_1 \approx \arcsin\left(\frac{360\,\text{m}}{18{,}000\,\text{m}}\right) + \arcsin\left(\frac{360\,\text{m}}{86{,}000\,\text{m}}\right) = 0.0242\,\text{rad} \Rightarrow 24.2\,\text{mrad}$$

The equivalent radius of curvature R of the top of the obstacle was estimated at 500 m approximately; the attenuation calculated using relation (1.97) has as a value

$$A_{OB1}(50\%) = 45.7\,\text{dB}$$

- *Secondary Obstacle Loss (2)* This obstacle consists of a rounded hill that presents a variation of height of about 30 m over a distance of 1500 m and whose maximum altitude was evaluated at 263 m. The purpose is now to determine the height of this obstacle compared to the line which connects the top of the main obstacle (1) to station B as previously.

  (a) Altitude of obstacle 2 above chord joining both sites:

$$H_{C2} = 305\,\text{m} + (464\,\text{m} - 305\,\text{m})\frac{15\,\text{km}}{86\,\text{km}} = 332.7\,\text{m}$$

  (b) Earth curvature for $K(50\%) = 1.29$ using relation (1.84):

$$y_2 = \frac{15\,\text{km} \times 71\,\text{km}}{2 \times 1.29 \times 6378\,\text{km}} = 0.0647\,\text{km} = 64.7\,\text{m}$$

  from which the height of the secondary obstacle is determined:

$$h_2 = 263\,\text{m} - 332.7\,\text{m} + 64.7\,\text{m} = -5\,\text{m}$$

  We can estimate the equivalent radius of curvature of the secondary obstacle starting from the relation of the circle in the following way:

$$R = \frac{(750\,\text{m})^2}{2 \times 30\,\text{m}} \approx 10{,}000\,\text{m}$$

  The obstacle attenuation due to the secondary obstacle can be roughly estimated, for a clearance $h/R_{F1} \sim -0.1$ in Figure 1.70, at an intermediate value between a knife edge and an irregular terrain, that is,

$$A_{OB2}(50\%) = 8\,\text{dB}$$

  The total median loss between stations A and B for the diffraction mode is thus established as

$$A(50\%)_{\text{diffraction}} = 146.2\,\text{dB} + 0.8\,\text{dB} + 45.7\,\text{dB} + 8\,\text{dB} = 200.7\,\text{dB}$$

  The budget link then becomes:

Emitted power	+60 dBm
Cumulated antenna gain	+81.6 dBi
Total branching loss	–5 dB
Attenuation of propagation	–200.7 dB
Median received field strength $P_R(50\%)$	–64.1 dBm

It is interesting to determine the maximum value of the diffraction loss for the extreme conditions of subrefraction; Figure 1.54 indicates a value of 0.92 for K_{min}, from which the new values for the obstacle heights to be taken into account are:

(a) Main obstacle (1):

$$y_1 = \frac{18\,\text{km} \times 86\,\text{km}}{2 \times 0.92 \times 6378\,\text{km}} = 0.132\,\text{km} = 132\,\text{m}$$

from which the maximum height of the main obstacle is

$$h_1 = 464\,\text{m} - 198\,\text{m} + 132\,\text{m} = 398\,\text{m}$$

and the maximum diffraction loss which corresponds to it is

$$A_{OB1}(99.9\%) = 47.5\,\text{dB}$$

(b) Secondary obstacle (2):

$$y_2 = \frac{15\,\text{km} \times 71\,\text{km}}{2 \times 0.92 \times 6378\,\text{km}} = 0.0907\,\text{km} = 90.7\,\text{m}$$

from which the maximum height of the secondary obstacle is

$$h_2 = 263\,\text{m} - 332.7\,\text{m} + 90.7\,\text{m} = 21\,\text{m}$$

The attenuation calculated using relation (1.97) has as a value

$$A_{OB2}(99.9\%) = 16\,\text{dB}$$

that is very close to that in Figure 1.70, while regarding as previously an obstacle halfway between knife edge and irregular terrain, for a clearance $h/R_{F1} \sim 0.4$.

The maximum diffraction loss[6] then becomes

$$A(99.9\%)_{\text{diffraction}} = 146.2\,\text{dB} + 0.8\,\text{dB} + 47.5\,\text{dB} + 16\,\text{dB} = 210.5\,\text{dB}$$

and the minimum field strength is

$$P_R(99.9\%) = -75.9\,\text{dBm}$$

[6] Obstacle 2 has a great influence on the diffraction loss according to the variation of the vertical gradient of the refractive index.

- *Troposcatter Loss* The elevation angle at stations A and B can be determined using relation (4.9):

$$\theta_{et} = \frac{0.464\,\text{km} - 0.176\,\text{km}}{18\,\text{km}} - \frac{18\,\text{km}}{2 \times 1.29 \times 6378\,\text{km}} = 0.0149\,\text{rad} \Rightarrow 14.9\,\text{mrad}$$

$$\theta_{er} = \frac{0.464\,\text{km} - 0.305\,\text{km}}{86\,\text{km}} - \frac{86\,\text{km}}{2 \times 1.29 \times 6378\,\text{km}} = -0.0034\,\text{rad} \Rightarrow -3.4\,\text{mrad}$$

The angular distance[7] is given by relation (4.7):

$$\theta = \frac{104\,\text{km}}{1.29 \times 6378\,\text{km}} + 0.0149\,\text{rad} - 0.0034\,\text{rad} = +0.0242\,\text{rad} \Rightarrow +24.2\,\text{mrad}$$

The calculation for a probability of service of 0.5 gives the following values:

$$L(50\%) = 224.1\,\text{dB}$$

$$\sigma = \begin{cases} 13.2\,\text{dB} & \text{for field enhancement} \\ 9\,\text{dB} & \text{for fading} \end{cases}$$

$$P_R(50\%) = -93.6\,\text{dBm} \quad \text{for median value of scattered field}$$

Figure 4.53 shows the results of a 24-h recording which was carried out with a time constant of 1 min; if a value close to the diffracted field for the diurnal period is found, when the sun causes a homogenization of the atmosphere by thermal convection, the night period is strongly disturbed by the field enhancement due to specular reflections in elevated ducts that are frequent in the desert zone and by multipath on the longest part of the weak slope of the link, which can be predicted using the method described in Section 3.4.

4.4.2 Rounded-Edge Diffraction Path

Figure 4.54 presents the path profile of a 60-km link established in frequency diversity in a continental temperate climate with a rounded obstacle of 2 km radius; the size of the antennas is 4.5 m and the emitted power is 1 kW. The predicted median received field strength is –45.6 dBm, with an obstacle loss of 43.4 dB; Figure 4.55 shows a one-week recording with a time constant of 1 min of the received signals presenting a ducting period during which a deep fading of about 20 dB occurred.

[7]The diffusion angle is equal to the diffraction angle previously calculated.

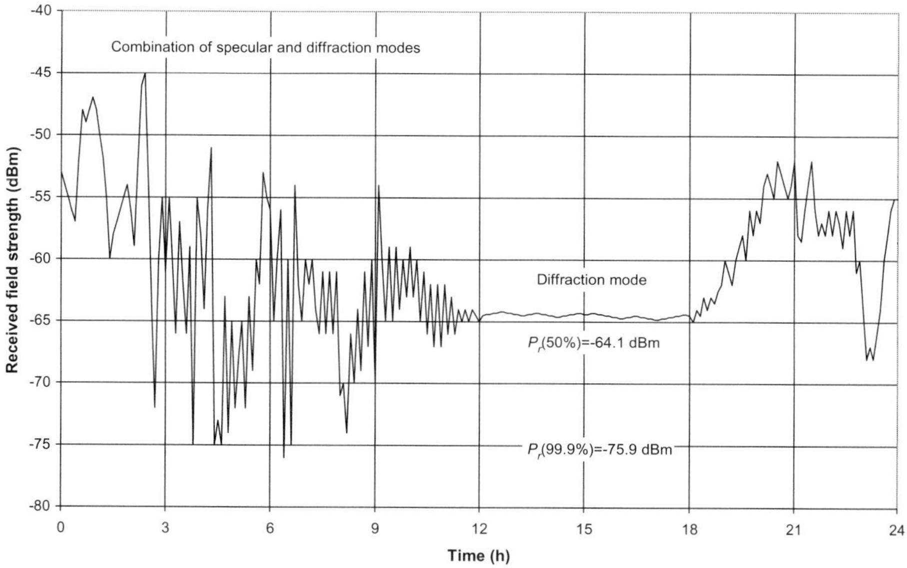

Figure 4.53 Received field strength recording in knife edge diffraction.

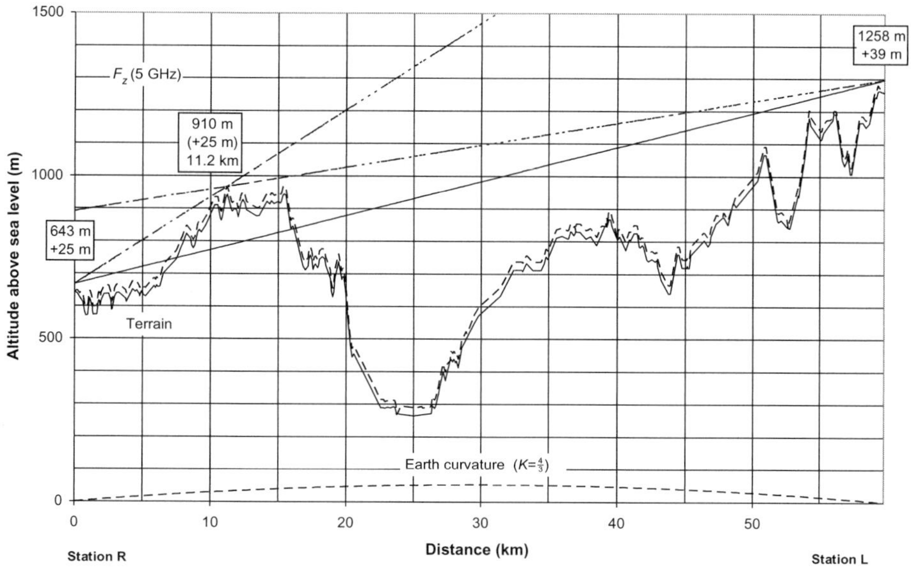

Figure 4.54 Path profile of link with a rounded-edge obstacle.

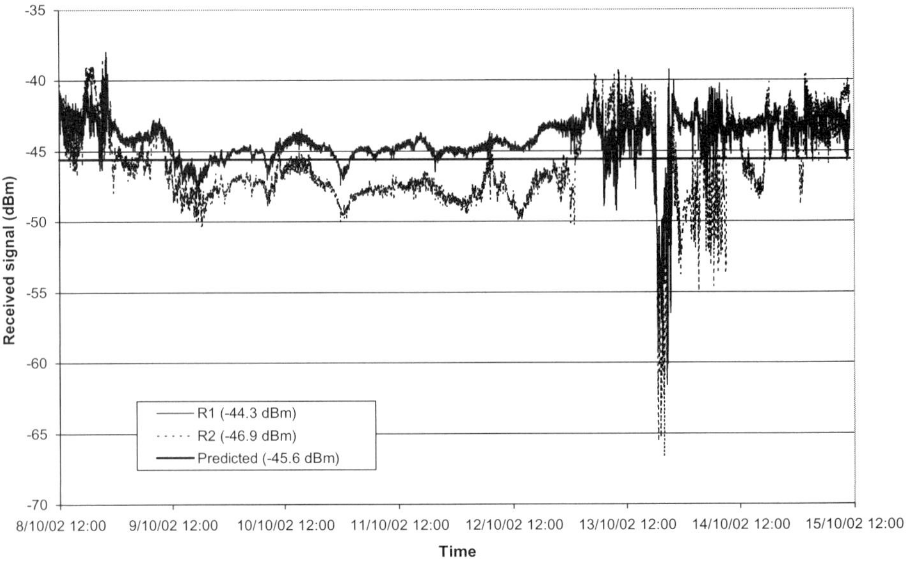

Figure 4.55 Received fields recorded with time constant of 1 min.

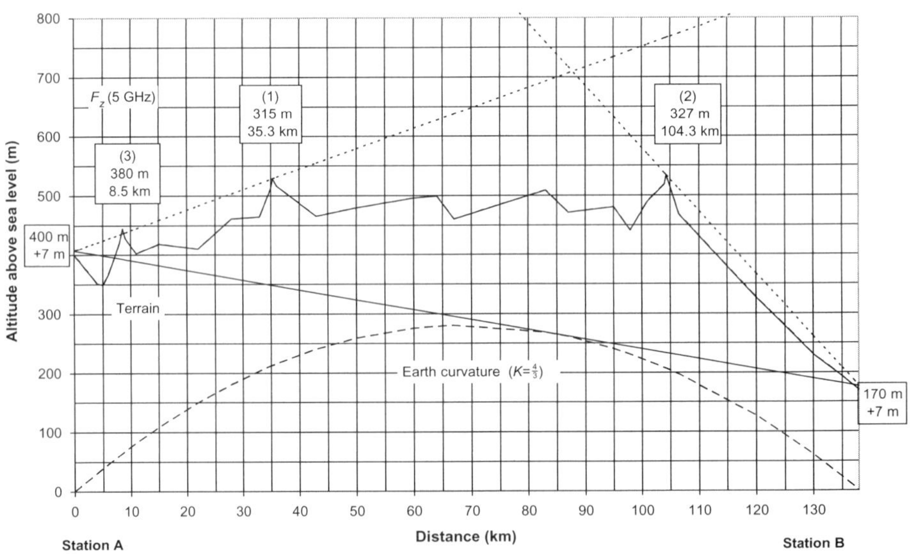

Figure 4.56 Path profile of double-diffraction link.

4.4.3 Double-Diffraction Path

Figure 4.56 presents the path profile of an 137.9-km link established in a semi-desert climate comprising two main obstacles 1 and 2 which consist of transverse edges whose altitudes are similar and a secondary obstacle 3; these three obstacles, like the whole profile, are deprived of vegetation. Thus the method of Epstein and Peterson will be employed here and the budget link is established in the manner which follows. The axis of the 8-m antennas was placed 7 m above the ground at both ends, the frequency employed was 4.7 GHz, and the emitted power was 1 kW.

- *Free-Space Loss* Attenuation of propagation: $A = 20\log[0.3/(4\pi \times 137{,}900\,\text{m} \times 4.7\,\text{GHz})] = 148.7\,\text{dB}$

  Atmospheric attenuation: $A_a = 1.1\,\text{dB}$
- *First Main Obstacle Loss (1)* Obstacle 2 is the source of emission, and we can determine the following:

  (a) Altitude of obstacle 1 above chord joining both sites:

$$H_{C1} = 407\,\text{m} + (327\,\text{m} - 407\,\text{m})\frac{35.3\,\text{km}}{104.3\,\text{km}} = 379.9\,\text{m}$$

  (b) Earth curvature for $K(50\%) = 1.29$ using relation (1.84):

$$y_1 = \frac{35.3\,\text{km} \times 69\,\text{km}}{2 \times 1.29 \times 6378\,\text{km}} = 0.148\,\text{km} = 148\,\text{m}$$

  from which the height of the first main obstacle is

$$h_1 = 315\,\text{m} - 379.9\,\text{m} + 148\,\text{m} = 83.1\,\text{m}$$

  The equivalent radius of curvature R of the top of the obstacle was estimated at 300 m; the attenuation calculated using relation (1.97) has as a value

$$A_{OB1}(50\%) = 23.6\,\text{dB}$$

- *Second Main Obstacle Loss (2)* Obstacle 1 is the source of emission, and we can determine, by considering an apparent radius of the top of the obstacle of about 100 m,

$$H_{C2} = 315\,\text{m} + (177\,\text{m} - 315\,\text{m})\frac{69\,\text{km}}{102.6\,\text{km}} = 222.2\,\text{m}$$

$$y_2 = \frac{69\,\text{km} \times 33.6\,\text{km}}{2 \times 1.29 \times 6378\,\text{km}} = 0.1408\,\text{km} = 140.8\,\text{m}$$

$$h_2 = 327\,\text{m} - 222.2\,\text{m} + 140.8\,\text{m} = 245.6\,\text{m}$$

  The attenuation brought by obstacle 2 has as a value

$$A_{OB2}(50\%) = 34.3\,\text{dB}$$

The correction term is

$$A_C = 10\log\left(\frac{(35.3\,\text{km} + 69\,\text{km})(69\,\text{km} + 33.6\,\text{km})}{69\,\text{km} \times 137.9\,\text{km}}\right) = 0.5\,\text{dB}$$

- *Secondary Obstacle Loss (3)* Obstacle 1 is the source of the emission, and we can determine, by considering an apparent radius of the top of the obstacle of about 100 m,

$$H_{C3} = 407\,\text{m} + (315\,\text{m} - 407\,\text{m})\frac{8.5\,\text{km}}{35.3\,\text{km}} = 384.8\,\text{m}$$

$$y_3 = \frac{8.5\,\text{km} \times 26.8\,\text{km}}{2 \times 1.29 \times 6378\,\text{km}} = 0.0138\,\text{km} = 13.8\,\text{m}$$

$$h_3 = 380\,\text{m} - 384.8\,\text{m} + 13.8\,\text{m} = 9\,\text{m}$$

The attenuation brought by obstacle 3 has as a value

$$A_{OB3}(50\%) = 11.6\,\text{dB}$$

The total median loss for the diffraction mode is thus established with the following value:

$$\begin{aligned} A(50\%)_{\text{diffraction}} &= 148.7\,\text{dB} + 1.1\,\text{dB} + 23.6\,\text{dB} + 34.3\,\text{dB} + 11.6\,\text{dB} + 0.5\,\text{dB} \\ &= 219.8\,\text{dB} \end{aligned}$$

The link budget of transmission then becomes

Emitted power	+60 dBm
Cumulated antenna gain	+98.6 dBi
Total branching loss	–5 dB
Attenuation of propagation	–219.8 dB
Diffraction median received field strength	$P_R(50\%) = -66.2$ dBm

The diffracted field calculated for $K_{\min}$ is

$$P_R(99.9\%) = -76.7\,\text{dBm}$$

- *Troposcatter Loss* The elevation angle at stations A and B can be determined using relation (4.9):

$$\theta_{et} = \frac{0.380\,\text{km} - 0.407\,\text{km}}{8.5\,\text{km}} - \frac{8.5\,\text{km}}{2 \times 1.29 \times 6378\,\text{km}} = -0.0037\,\text{rad} \Rightarrow -3.7\,\text{mrad}$$

$$\theta_{er} = \frac{0.327\,\text{km} - 0.177\,\text{km}}{33.6\,\text{km}} - \frac{33.6\,\text{km}}{2 \times 1.29 \times 6378\,\text{km}} = +0.0024\,\text{rad} \Rightarrow +2.4\,\text{mrad}$$

The calculation gives the following results for a probability of service of 0.5:

$$L(50\%) = 220.5\,\text{dB}$$

$$\sigma = \begin{cases} 12.9\,\text{dB} & \text{for field enhancement} \\ 9\,\text{dB} & \text{for fading} \end{cases}$$

$$P_R(50\%) = -75.3\,\text{dBm} \quad \text{for median value of scattered field}$$

The scattered field is weaker by about 10 dB than the diffracted field.

Figure 4.57 presents a recording of the received field strength over one period of 24 h on this link where we observe the effects ascribable to the various modes of propagation, in particular the specular mode during the night, and to their combination. It will be noticed that the measured value of the diffracted field is relatively close to the one calculated above.

4.4.4 Troposcatter Link on Smooth Spherical Earth

Figure 4.58 presents the path profile of an 167-km link established in a semi-desert climate at 4.7 GHz on a smooth spherical Earth using 8-m antennas

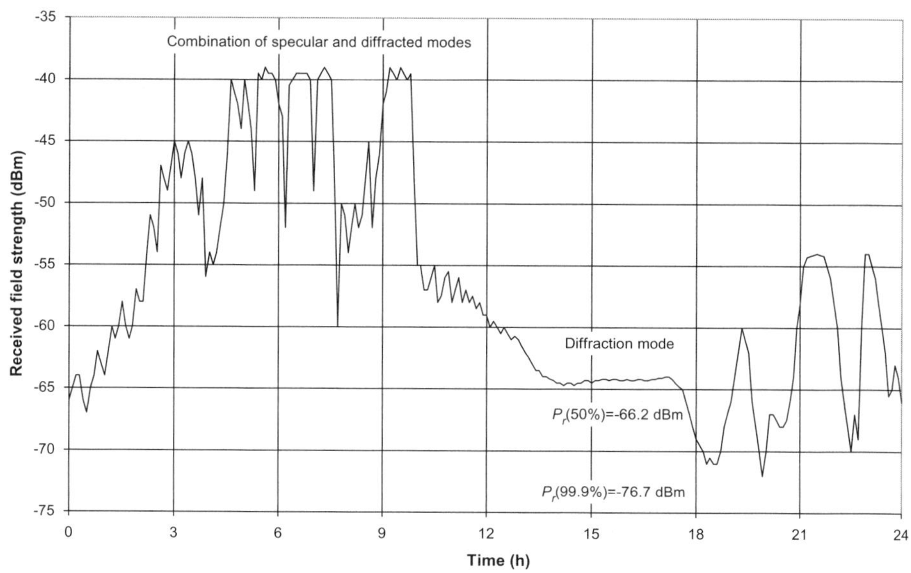

Figure 4.57 Received field recording with time constant of 1 min.

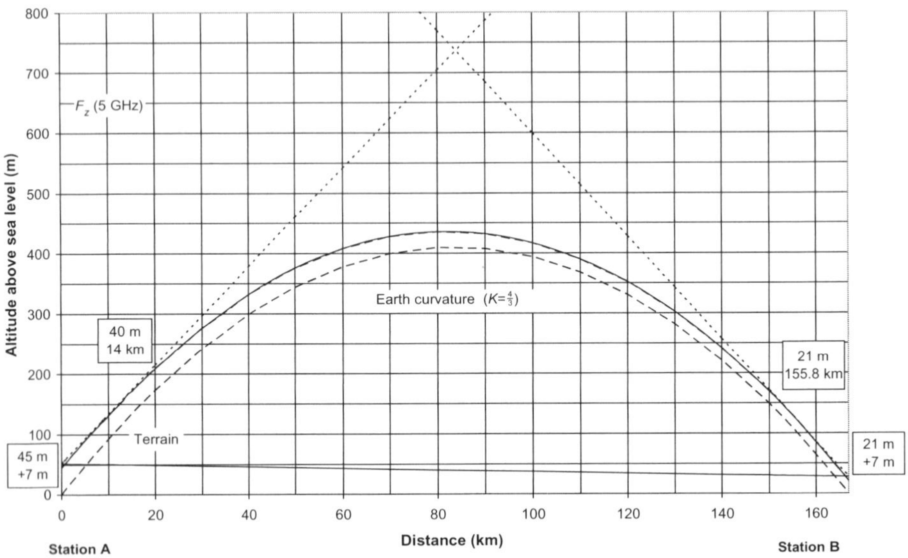

Figure 4.58 Path profile of troposcatter link on smooth Earth.

placed at a height of 7 m, with an emitted power of 1 kW to which corresponds the test related in Section 4.3.2. Using the relations given in Section 4.2, for a desert climate, we obtain the following results:

$N_{S0} = 290$	$N_S = 288.8$	$a = 8320.9\,\text{km}$	$K = 1.30$
$\theta_0 = 0.020\,\text{rad}$	$\theta_{et} = -1.7\,\text{mrad}$	$\theta_{er} = -1.3\,\text{mrad}$	B_c (Sunde) = 6.9 MHz
$\theta = 0.0171\,\text{rad}$	$\theta d = 2.84\,\text{rad km}$	$f(\theta\, d) = 151.5\,\text{dB}$	B_c (Rice) = 11.8 MHz
$A_a = 1.5\,\text{dB}$	$H_0 = 0\,\text{dB}$	$F_0 = 0\,\text{dB}$	$L_{bsr} = 218.7\,\text{dB}$
$\bar{h}_t = 40.8\,\text{m}$	$\bar{h}_r = 21\,\text{m}$	$h_{te} = 11.3\,\text{m}$	$h_{re} = 7\,\text{m}$
$d_{S0} = 141.3\,\text{km}$	$d_{S1} = 18.0\,\text{km}$	$d_e = 253.3\,\text{km}$	$V(50\%, d_e) = -5.7\,\text{dB}$
$L(50\%) = 224.4\,\text{dB}$	$\sigma(16\%) = 11.2\,\text{dB}$	$\sigma(84\%) = 7.9\,\text{dB}$	
$\alpha_{00} = 8.5\,\text{mrad}$	$\beta_{00} = 8.6\,\text{mrad}$	$S = 0.99$	$h_0 = 711\,\text{m}$
$\eta_S = 0.34$	$\nu = 0.17$	$f(\nu) = 1.04$	$\mu = 1$
$S_\mu = 1.01$	$n = 1.76$	$\hat{n} = 1.7$	$L_{gp} = 7.3\,\text{dB}$
$B_c(50\%) = 11.4\,\text{MHz}$	$\sigma = 4\,\text{dB}$	$B_c(99.9\%) = 2.8\,\text{MHz}$	

These data make it possible to calculate the characteristic values of the received field with a service probability of 0.5, 0.67, and 0.85 corresponding respectively to $\frac{1}{2}$, $\frac{2}{3}$, and $\frac{5}{6}$:

Q	0.5	0.67	0.85
$P_R(50\%)$	−78 dBm	−79.4 dBm	−81.5 dBm
$P_R(99.9\%)$	−102.4 dBm	−106.8 dBm	−112.3 dBm
$P_R(0.1\%)$	−43.4 dBm	−48.6 dBm	−56.1 dBm

Figure 4.59 shows the distribution of the hourly median of the received field that has been recorded over a two-year period with a time constant of 1 min. The recorded distribution for each year is compared with those calculated for the desert and for the continental temperate climates. Comparison with the distribution defined by the method of prediction for the desert climate shows that:

- The hourly median of the received field strength and that which corresponds to the small percentages of time are higher.
- The standard deviation is greater.
- The value measured for 99.9% of the time, which is the most significant for the dimensioning of the means to implement in order to obtain the required availability, is very close to that which was calculated.

This variation from the prediction is ascribable to the enhancement of the received field caused by the specular mode that occurs frequently at night in the desert zones.

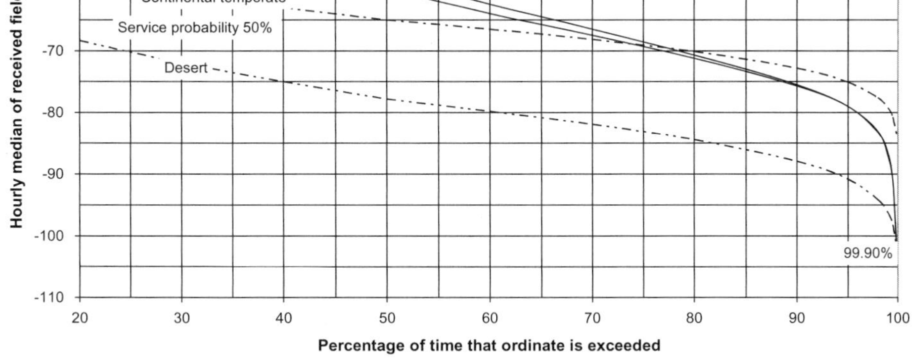

Figure 4.59 Distribution of hourly median over period of two years.

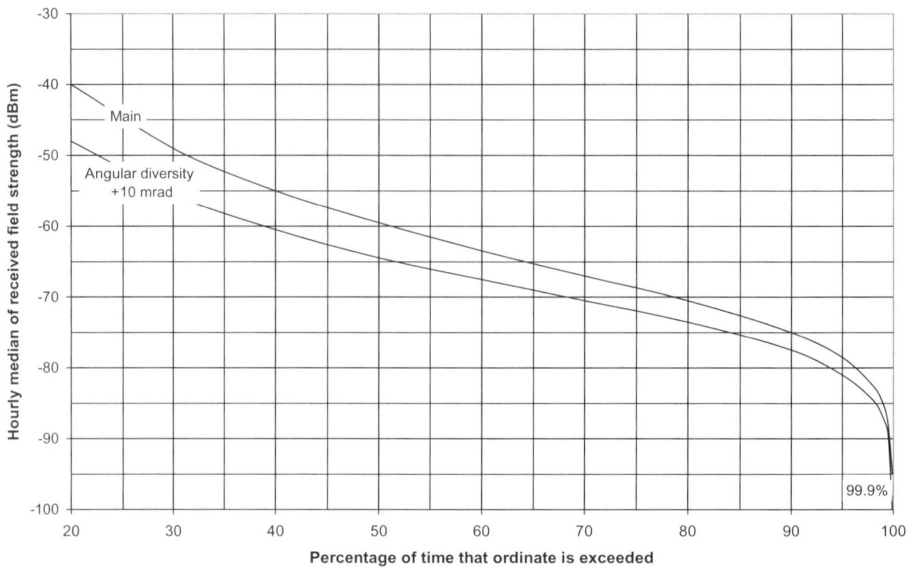

Figure 4.60 Reception in vertical angle diversity over one-year period.

We can also observe that:

- The annual variability is small over these two years of recording, with a median value of about 2 dB, especially for the high percentages of time.
- The seasonal variation between spring, summer, autumn, and winter did not exceed ±2.5 dB compared to the annual distribution.

Figure 4.60 presents the results of measurement that have been carried out over a one-year period while tilting upward, on an experimental basis, the elevation angle of one of the two antennas in horizontal space diversity at the reception side by an angle of 10 mrad. This test showed the viability of the vertical angle diversity provided it is carried out at the reception side only and the angular separation does not exceed about 10 mrad with respect to the nominal elevation angle.

Figure 4.61 shows the variation measured in winter time over three consecutive periods of one week each for a tilt upward in elevation of respectively 5, 10, and 15 mrad.

4.4.5 Combined Troposcatter and Diffraction Link

Figure 4.62 shows the path profile of the 235-km link discussed in Section 4.3.4.3 that presents both scattering and diffraction modes of propagation. Detailed computation of the budget link is given in Table 4.3 for a continental

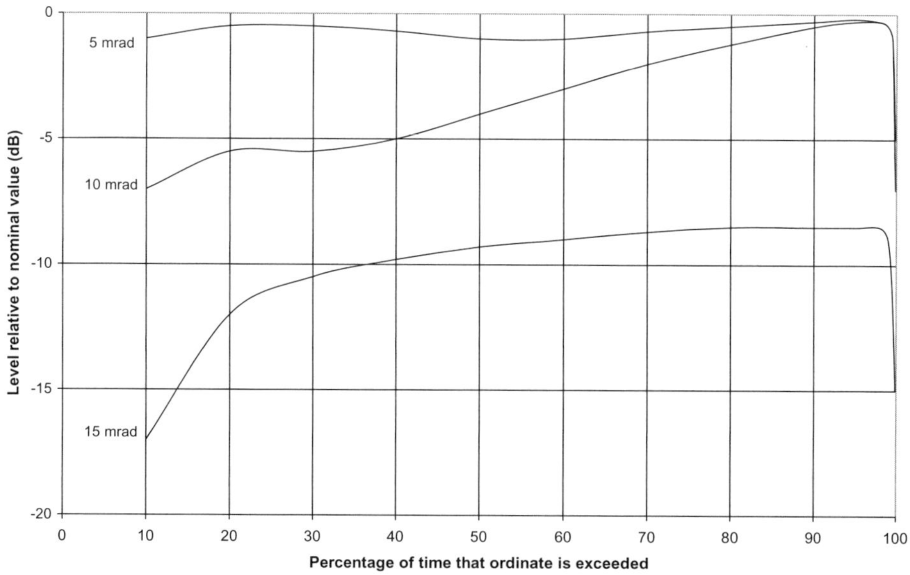

Figure 4.61 Field strength relative to nominal value according to tilt angle.

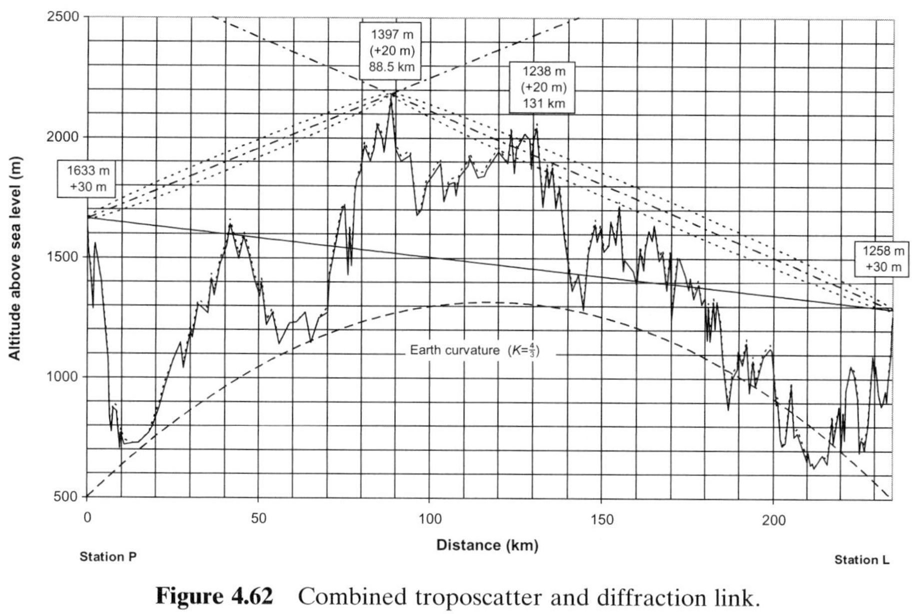

Figure 4.62 Combined troposcatter and diffraction link.

TABLE 4.3 Computation of Budget Link for Continental Temperate Climate

Station P—Station L	Technical Note 101 (NBS, 1966)	
Transmission station	Station P	Station L
Reception station	Station L	Station P
Frequency, GH$_z$	4.650	4.650
Equipement	TROPO D4	TROPO D4
Modulation	2CPFSK	2CPFSK
Bit rate, kbits s^{-1}	2048.0	2048.0
Emitted power, dBm	60.0	60.0
Threshold BER 10^{-5}, dBm	−91.0	−91.0
Threshold BER 10^{-3}, dBm	−93.0	−93.0
Noise figure, dB	5.0	5.0
Frequency bandwidth, MHz	3.25	3.25
Diversity row	4	4
C/N (BER 1×10^{-5}), dB	11.4	11.4
C/N (BER 1×10^{-3}), dB	5.7	5.7
Antenna low-noise preamplifier, dB	25.0	25.0
Noise figure, dB	2.0	2.0
Pathlength, km	235.3	235.3
Altitude of transmission station (h_{t0}), m	1633.0	1258.0
Transmission antenna height (h_{te}), m	30.0	30.0
Altitude of transmission antenna (h_{ts}), m	1663.0	1288.0
Altitude of reception station (h_{ro}), m	1258.0	1633.0
Reception antenna height (h_{re}), m	30.0	30.0
Altitude of reception antenna height (h_{rs}), m	1288.0	1663.0
Transmission specific feeder loss, dB hm^{-1}	2.8	2.8
Transmission feeder length, m	30.0	105.0
Total transmission feeder length, m	60.0	135.0
Total transmission feeder loss, dB	1.7	3.8
Reception specific feeder loss, dB hm^{-1}	2.8	2.8
Reception feeder length, m	105.0	30.0
Total reception feeder length, m	135.0	60.0
Total reception feeder loss, dB	3.8	1.7
Transmission antenna diameter, m	7.0	7.0
Antenna efficiency, %	55.0	55.0
Half-power beamwidth, deg	±0.32	±0.32
Transmission antenna gain, dBi	48.1	48.1
Radome loss, dB	0.5	0.5
Reception antenna diameter, m	7.0	7.0
Antenna efficiency, %	55.0	55.0
Half-power beamwidth, deg	±0.32	±0.32
Reception antenna gain, dBi	48.1	48.1
Radome losses, dB	0.5	0.5

TABLE 4.3 *Continued*

Station P—Station L	Technical Note 101 (NBS, 1966)	
Total E/R branching loss, dB	1.0	1.0
Specific loss, dB	0.0	0.0
Jamming threshold degradation, dB	0.0	0.0
Other losses, dB	0.0	0.0
Equipment margin, dB	0.0	0.0
Means refractivity at see level (N_0)	310	310
Surface refractivity (N_s)	265.2	265.2
Effective earth radius (a)/factor K, km	$8020.3/K = 1.26$	
Earth center angle (θ_0), mrad	29.3	29.3
Takeoff horizon elevation angle of transmitting antenna (θ_{et}), mrad	–8.3	–6.8
Takeoff horizon elevation angle of receiving antenna (θ_{er}), mrad	–6.8	–8.3
Angular distance (θ_{00})	14.3	14.3
(θ_d)/f(θ_d), rad km/dB	3.35 /	156.0
Atmospheric absorption, dB	1.8	1.8
Frequency gain (H_0), dB	0.0	0.0
Scattering efficiency correction (F_0), dB	–1.0	–1.0
Long-term median transmission loss (L_{bsr}), dB	219.4	219.4
Average altitude to radio horizon emission/reception (h_{tm}, h_{rm}), m	938.9	718.4
Effective antenna height emission/reception (h_{te}, h_{re}), m	724.1	569.6
Smooth earth distance to radio horizons (d_{s0}), km	19.9	19.9
Smooth earth diffraction distance (d_{s1}), km	18.1	18.1
Effective distance (d_e), km	131.8	131.8
Climate type / climate function V (50%, d_e), dB	1	1.8
Median long-term transmission loss L (50%), dB	217.6	217.6
Frequency factor g (0.1%, f_{MHz})/g(0.9%,f_{MHz})	1.05	1.05
Long-term median variability at 0.1% Y(0.1)/90% Y(0.9), dB	9.6	7.7
Long-term transmission loss variation at 0.001% Y(0.001) / 99.9% Y(99.9), dB	26.3	18.5
Standard deviation Y(0.16) / Y(0.84), dB	8.5 /	6.0
α_{00}, mrad	8.0	6.3
β_{00}, mrad	6.3	8.0
Asymmetry factor (S)	1.27	0.79
Intersection height of rays to radio horizons (h_0), m	826.7	826.7
$\eta_s(h_0)$	0.38	0.38
v	0.19	0.19
$f(v)$	1.04	1.04
μ	1.00	1.00

TABLE 4.3 *Continued*

Station P—Station L	Technical Note 101 (NBS, 1966)	
$S\mu$ (or $1/S\mu$)	1.27	1.27
n	1.42	1.42
$\hat{n}$	1.4	1.4
L_{gp}, dB	5.7	5.7
Service probability/Gaussian coefficient	0.5	0.0
Antenna long-term troposcatter median reception power $P_r(50\%)$, dBm	–70.8	–72.9
Radio equipment lont-term reception power, dBm	–49.6	–49.6
Reception system global noise figure, dB	2.1	2.0
Flat Fade margin (BER 10^{-5}), dB	23.1	21.0
Flat Fade margin (BER 10^{-3}), dB	25.1	23.0
Reception system global reference noise, dBm	–106.8	–106.8
C/N, dB	36.0	33.9
Diversity margin (BER 10^{-5}), dB	24.6	22.5
Diversity margin (BER 10^{-3}), dB	30.3	28.2
Propagation time delay, ns	91	91
Coherence bandwidth (Sunde), MHz	4.4	4.4
Median coherence bandwidth (Collin), MHz	7.2	7.2
Coherence bandwidth standard deviation (Collin), MHz	3.7	3.7
Coherence bandwidth at 99.9% (Collin), MHz	1.9	1.9
Coding factor	1.00	1.00
Filtering coefficient/Nyquist bandwidth	0.50	0.50
Radio spectrum, MHz	4.096	4.096
Effective bandwidth, MHz	1.024	1.024
Availability due to coherence bandwidth, %	99.99977	99.99977
Availaibility for BER 10^{-5}/year, %	99.99796	99.99135
Availaibility for BER 10^{-3}/year, %	99.99998	99.99988
Cumulated unavailable time due to troposcatter/year, s	6.6	38.5
ITU-R no-quality allowance of ISDN (DM/BER > 10^{-6}), %	0.03765	0.03765
ITU-R no-quality allowance of ISDN (SES/BER > 10^{-3}), %	0.00508	0.00508
ITU-R allowance cumulated no-quality time for ISDN, s	131.7	131.7
ITU-R unavailable time for ISDN (10-s consecutives), s	742.6	742.6
P_L/K	30% /	1.31×10^{-5}
Free-space attenuation, dB	153.1	153.1
Clearance main obstacle transmission loss, dB	–11.01	36.8
Main obstacle radius /rounded-edge additional loss	0.00 km /	0.0 dB

TABLE 4.3 *Continued*

Station P—Station L	Technical Note 101 (NBS, 1966)	
Second obstacle transmission loss, dB	0.0	26.0
Diffraction total transmission loss, dB	216.0	216.0
Antenna diffraction long-term median reception power $P_r(50\%)$, dBm	−63.5	−65.6
Flat fade margin (BER 10^{-3}), dB	32.4	30.3
Worst-month cumulated outage time, s	<12355.2	<25668.9
Worst-month availability for BER 10^{-3}, %	>99.52333	>99.00969
Combined troposcatter and diffraction received field strength at antenna port, dBm	−62.8	−64.9
Rain rate ($R_{0.01}$), mm h^{-1}	46	46
Coefficients K_h, K_v	0.00110	0.00100
Coefficients A_h, A_v	1.21500	1.17500
Pathlength reduction factor	0.07	0.07
Attenuation due to rain for H/V polarization at 0.01% of average year, dB	1.9	1.5
Attenuation due to rain for H/V polarization at 0.01% of worst month, dB	3.6	2.8
Total availability over average year (BER 10^{-5}), %	>99.99796	>99.99135
Total availability over average year (BER 10^{-3}), %	>99.99997	>99.99987
Total availability over worst month (BER 10^{-5}), %	>99.98702	>99.95426
Total availability over worst month (BER 10^{-3}), %	>99.99966	>99.99880

temperate climate and for a positive minimal value k_{min} of 1 not exceeded for 99.9% of the year and resumed by the following results:

	Troposcater	Diffraction		Combination
$K(50\%) = 1.33$	—	$h_1 = 660.8$ m	$h/R_{F1} = 11.01$	
$A_{OB1}(K = 1.33)$	—	36.8 dB		
	—	$h_2 = 139.6$ m	$h/R_{F2} = 3.15$	
$A_{OB2}(K = 1.33)$	—	26 dB		
$A_{OB,total}(K = 1.33)$	—	62.8 dB		
$K(99.9\%) = 1$	—	$h_1 = 913.5$ m	$h/R_{F1} = 15.22$	
$A_{OB1}(K = 1)$	—	39.7dB		
	—	$h_2 = 225.8$ m	$h/R_{F2} = 5.09$	
$A_{OB2}(K = 1.33)$	—	30.1 dB		
$A_{OB,total}(K = 1.33)$	—	69.8 dB		
$P_R(50\%)$	−70.8 dBm	−63.5 dBm	—	−62.8 dBm
$Y(84\%)$	6 dB	2.2 dB		
$P_R(99.9\%)$	−88.5.4 dBm	−70.4 dBm	—	−70.4 dBm
$Y(16\%)$	8.5 dB			
$P_R(0.1\%)$	−43.7 dBm			

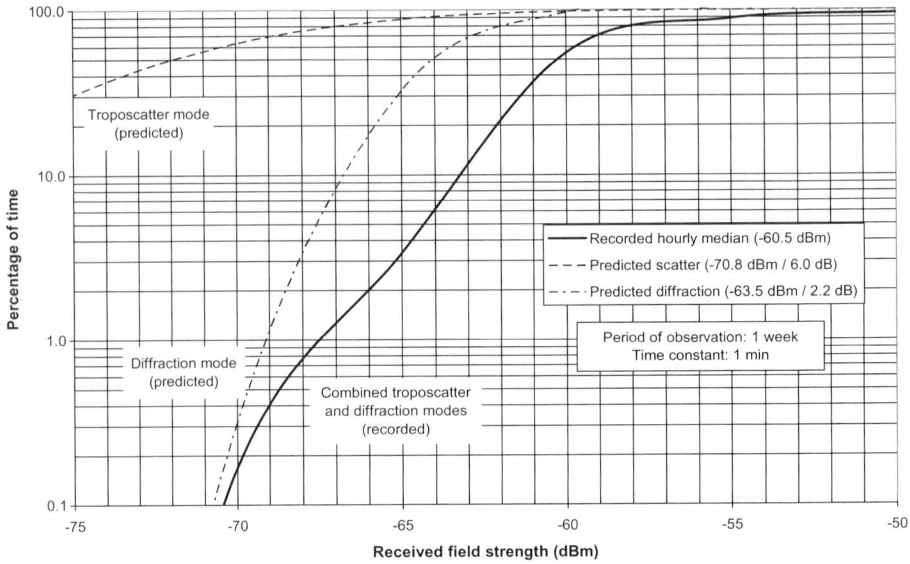

Figure 4.63 Long-term distribution of received field over period of one week.

Figure 4.63 illustrates the long-term distribution of the field strength recorded with a 1-min time constant over a one-week period and Figure 4.45 the short-term distribution of the instantaneous and combined fields; as the diffraction mode is predominant, the standard deviation of the long-term distribution is less than that predicted for the troposcatter mode.

Figure 4.64 shows the distribution of the average field strength over the one-week period using a time constant of 1 min.

Figure 4.65 shows a 1 min recording of the instantaneous individual and maximal-ratio combined received fields using a frequency sampling rate of 10 Hz.

4.4.6 Angle and Frequency Diversity Troposcatter Link

Figure 4.66 presents the path profile of an 180-km link that was carried out in a continental temperate climate by means of a single three-beam 5.6-m antenna at each end in order to obtain the same performance as a classical quadruple space and frequency diversity.

The goals of the tests of propagation, carried out at 5 GHz, were to optimize the coupling in both transverse and vertical angle diversity between the antennas and the propagation medium, by limiting the radiation losses at the emission and at the reception, and to seek the lowest possible correlation between the various channels.

Figure 4.67 presents the schematic one-way configuration of both terminal equipment which involves two transmitters at frequencies f_1 and f_2 and six receivers connected two by two to the three beams.

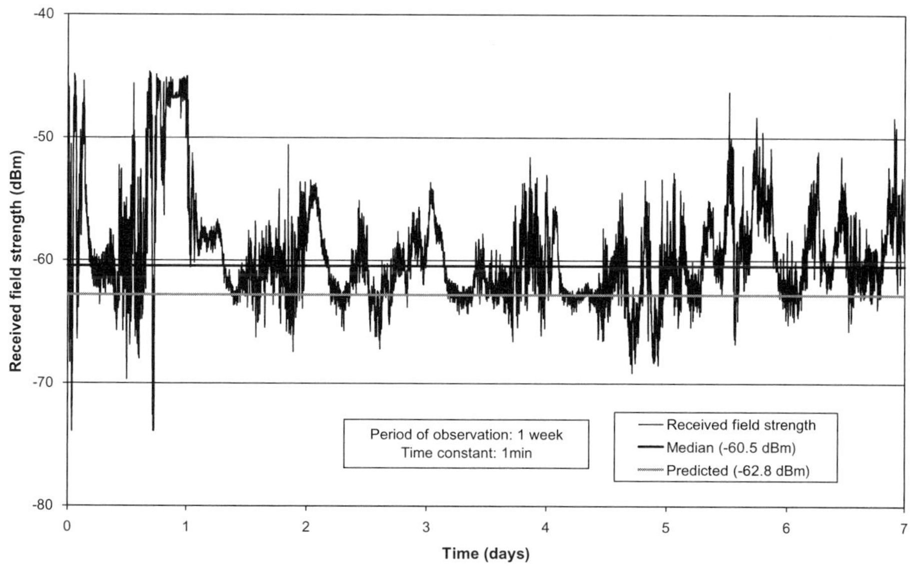

Figure 4.64 Long-term average received field strength (August 14 to 21, 2002).

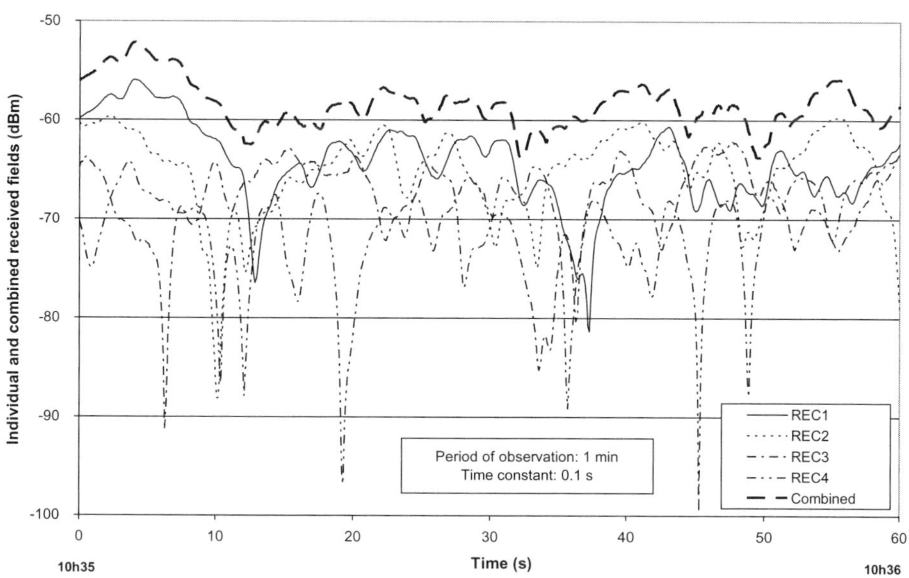

Figure 4.65 Short-term instantaneous individual and combined signals (August 14, 2002–10:35 to 10:36).

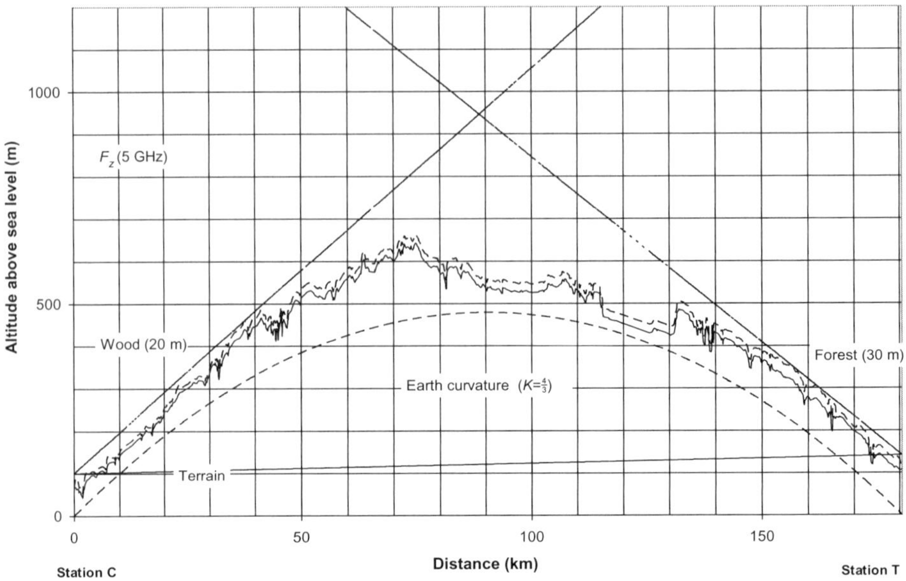

Figure 4.66 Path profile of angle and frequency diversity link.

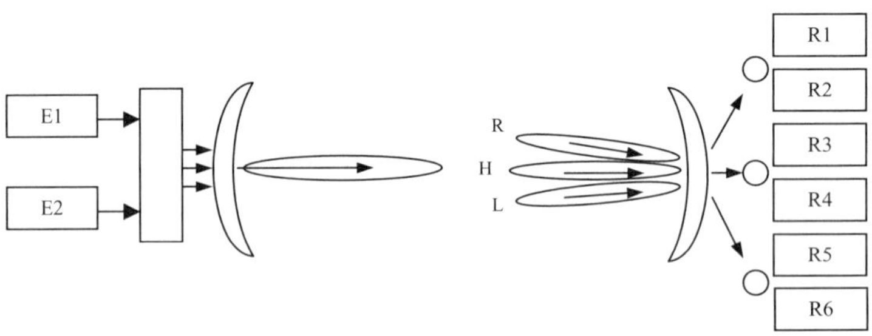

Figure 4.67 Schematic configuration of angle and frequency diversity.

Figure 4.68 shows the distribution of the individual instantaneous signals received on the left lobe (L), the right lobe (R), and the high lobe (H) and those which correspond to their combination in angle diversity (LRH) for each frequency and for both frequencies; these distributions are compared to the theoretical ones of Rayleigh for totally independent random variables.

Figure 4.69 shows the variation relative to the hourly median of individual and combined signals in angle diversity and angle/frequency diversity representing the diversity gain.

Figure 4.70 shows the values that were experienced for the diversity improvement factor by combining the signals in various manners in order to reveal the effects of angle diversity, frequency diversity, and full sixfold angle and frequency diversity.

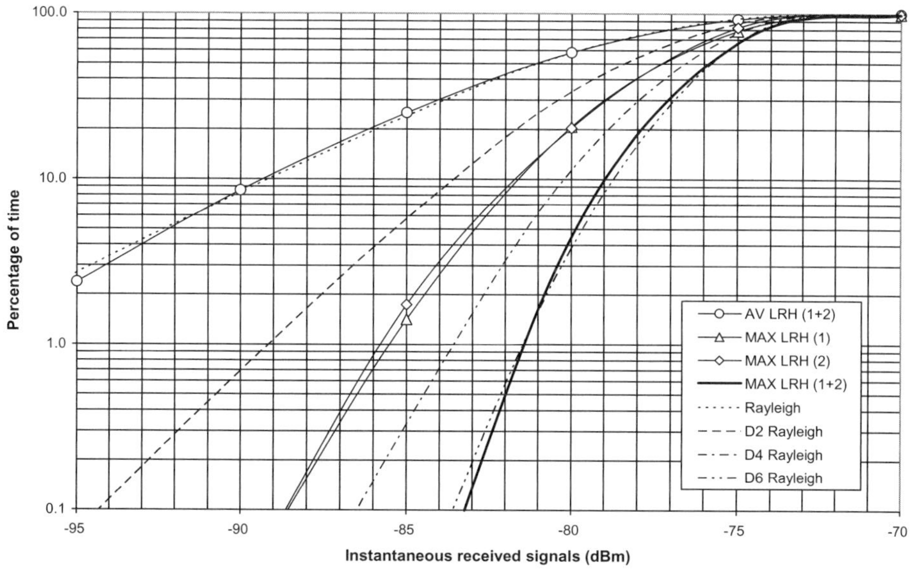

Figure 4.68 Comparative distribution between recorded and theoretical instantaneous individual and combined received signals (November 5, 2005–12:00 to 13:00).

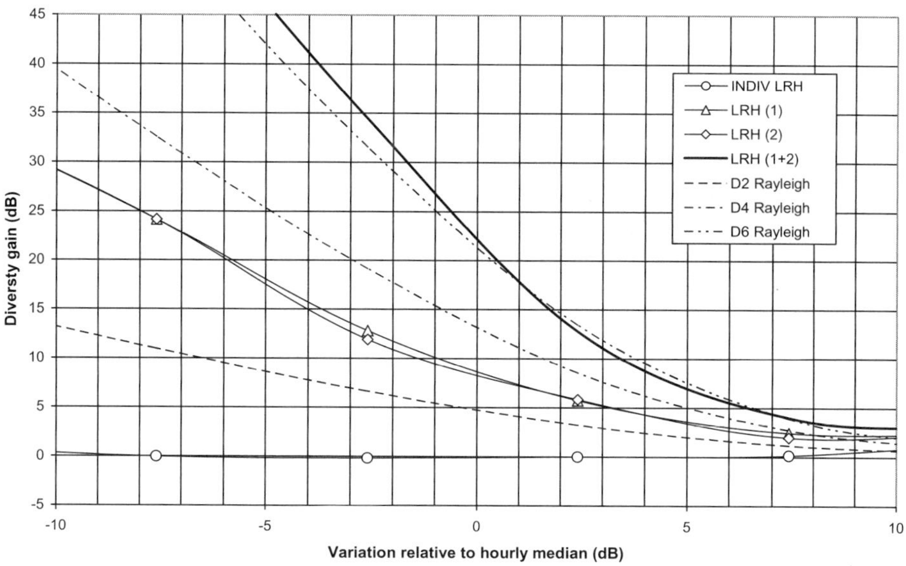

Figure 4.69 Diversity gain (November 5, 2005–12:00 to 13:00).

Figure 4.71 shows a recording of the six received signals over a period of five days with a time constant of 1 min.

Figure 4.72 shows the distribution of the sum of the received signal power combined in angle diversity at each frequency and in full sixfold angle and frequency diversity compared to the average of the individual fields.

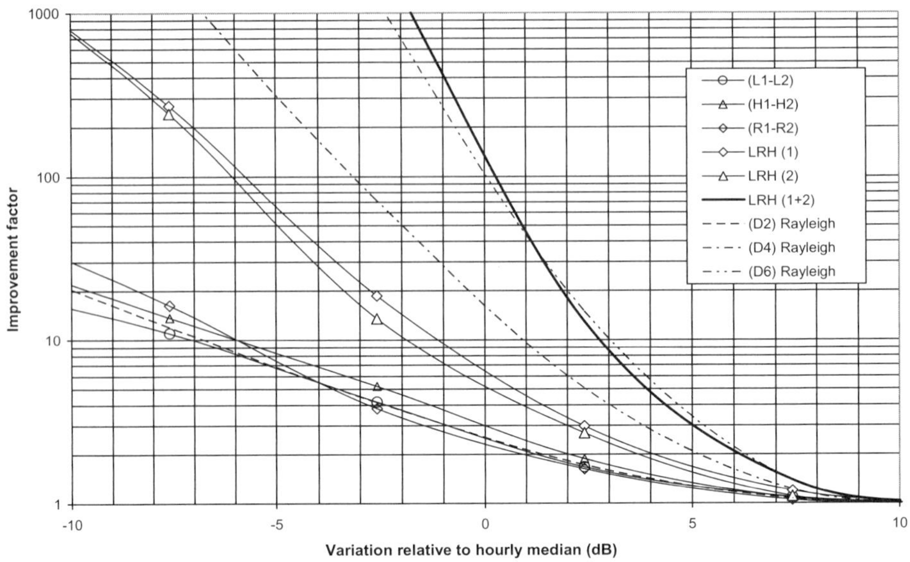

Figure 4.70 Improvement factor (November 5, 2005–12:00 to 13:00).

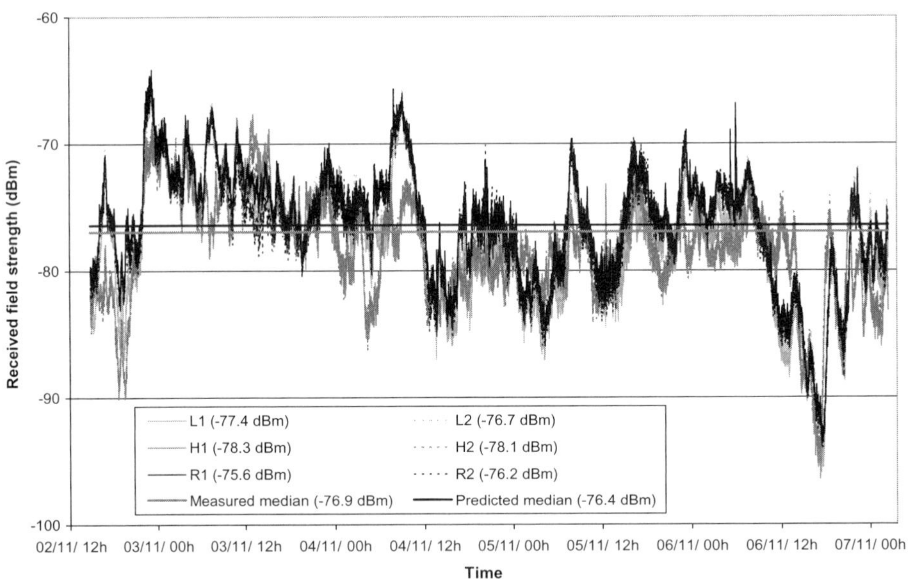

Figure 4.71 Recording of individual received signals (November 2 to 7, 2005).

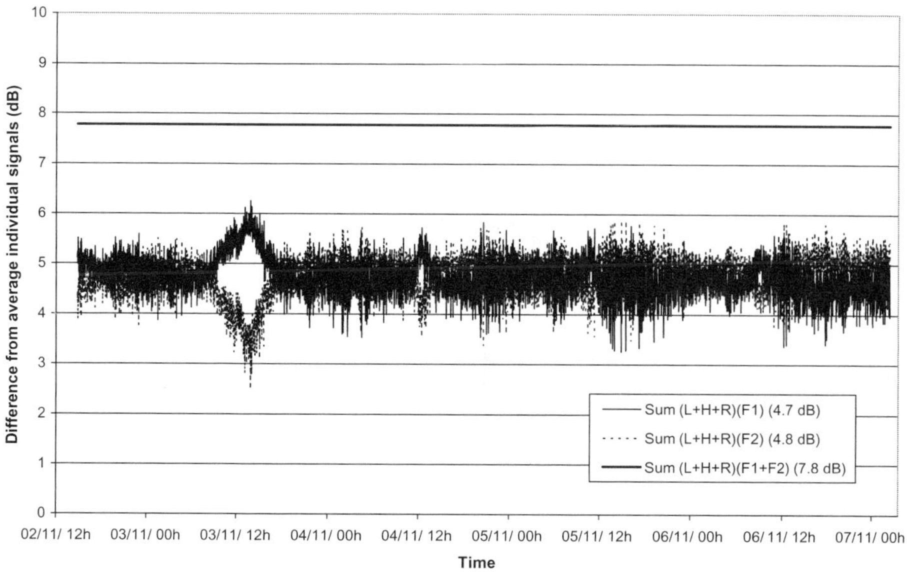

Figure 4.72 Recording of sums of received signals (November 2 to 7, 2005).

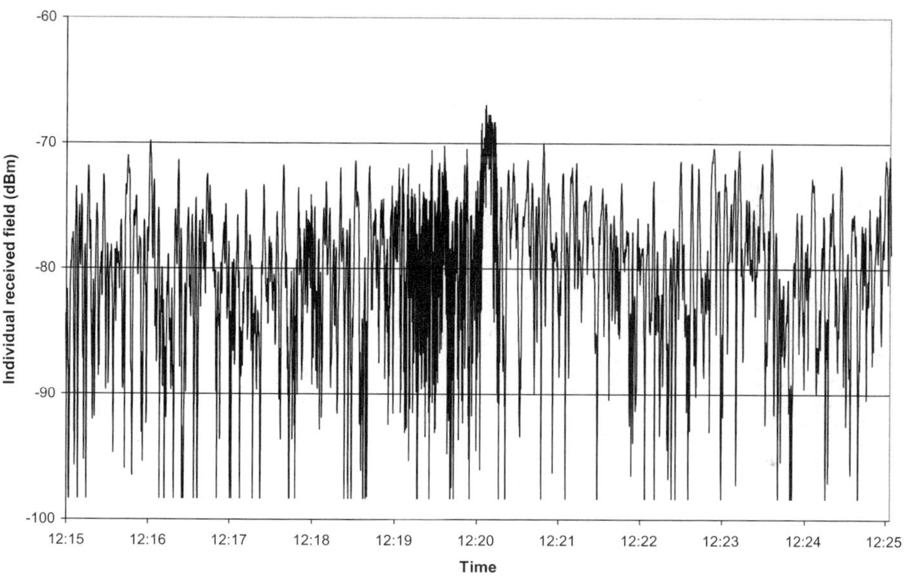

Figure 4.73 Recording of single instantaneous received field (November 5, 2005).

Figure 4.73 shows the amplitude of variation over a period of 10 min of one of the received fields.

Figure 4.74 shows the speed of fluctuation of that instantaneous field over the same period that can reach 250 dB s^{-1}.

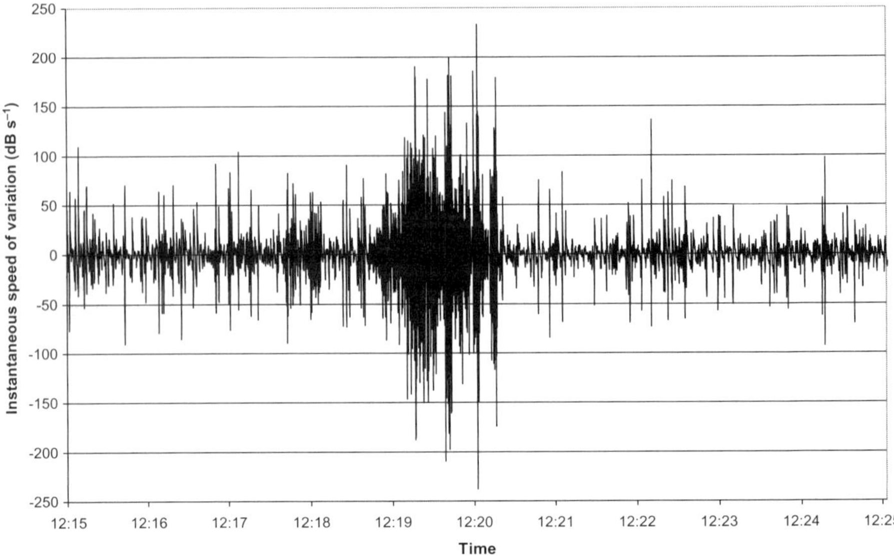

Figure 4.74 Speed of fluctuation of instantaneous received field (November 5, 2005).

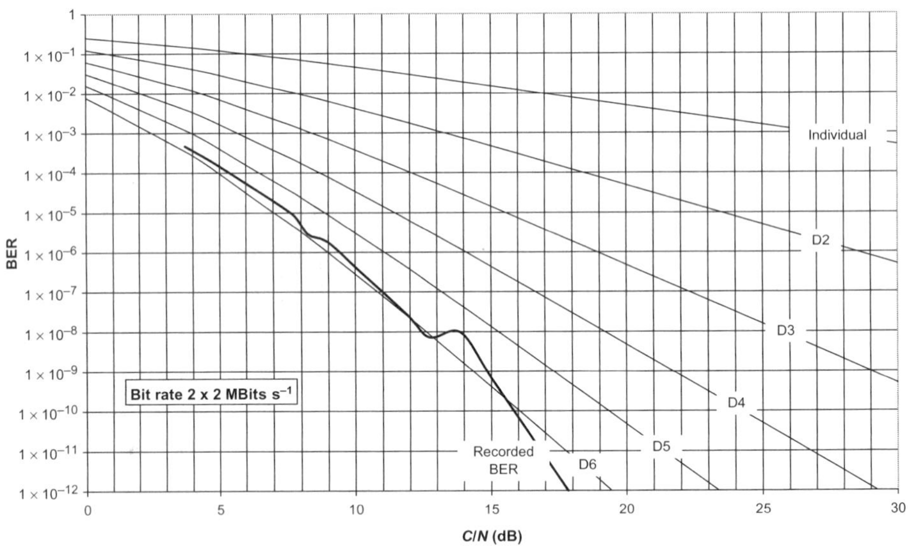

Figure 4.75 BER recorded distribution (November 5, 2005).

Figure 4.75 shows the distribution of the BER in full sixth-order diversity compared with the theoretical distributions for a null coefficient of correlation.

Following is an interpretation of the measurement results:

1. Figure 4.68 shows that the individual received fields obey Rayleigh's law, that is, the distribution of the signals combined in angle diversity on each of the two frequencies is quite close to the third-order Rayleigh's distribution for totally independent random variables and the distribution of the signals combined in full sixfold angle and frequency diversity is also very close to theoretical sixth order.
2. Figure 4.69 presents the diversity gains that correspond to the combination of the received fields in three-beam angle diversity at each frequency and in full sixfold angle and frequency diversity relative to the individual Rayleigh's distribution.
3. Figure 4.70 shows that the corresponding improvement factors are in concordance with the theoretical values of Rayleigh.
4. Figure 4.71 shows that all of the received field levels are equipollent and that their median is very close to the predicted value.
5. Figure 4.72 shows that, compared to the individual average level, the median power of the sum of the fields combined in three-beam angle diversity on each frequency as well as the sum of the whole of the signals present values close to the theoretical ones of respectively 4.8 and 7.8 dB.
6. Figure 4.75 shows that the distribution of the measured BER corresponds to the theoretical sixth-order diversity and the recovering loss of the three main lobes of the receiving antenna is practically compensated by comparison with the theoretical fourth-order distribution in space and frequency diversity that is considered as the reference.

The process of three-beam angle diversity thus presents performance that is comparable with space diversity; it makes it possible to reduce the importance of the infrastructure at the cost of a weak increase of the terminals which include two more receivers only. Moreover, by offering characteristics equivalent to theoretical sixth-order diversity, we see that the availability, dependent on the high values of the BER, is slightly reduced, but the quality of the link is clearly improved for the low values of the BER.

4.4.7 Digital Adaptive Modem

In general, microwave links are planned in order to obtain a certain availability at a given fixed bit rate; consequently, except for a small percentage of time, the terminals are designed in order to achieve performances that correspond to the most unfavorable conditions of propagation, that is, from 0.1 to 0.001% for troposcatter and line of sight. It is obvious that the bit rate could be much higher during the favorable conditions of propagation that occur most of the time provided that the modem is adaptive. Figure 4.76 shows the distribution of the instantaneous bit rate according to propagation data from the recording of Figure 4.71 by using an adaptive tropo modem 2–16 Mbits s^{-1};

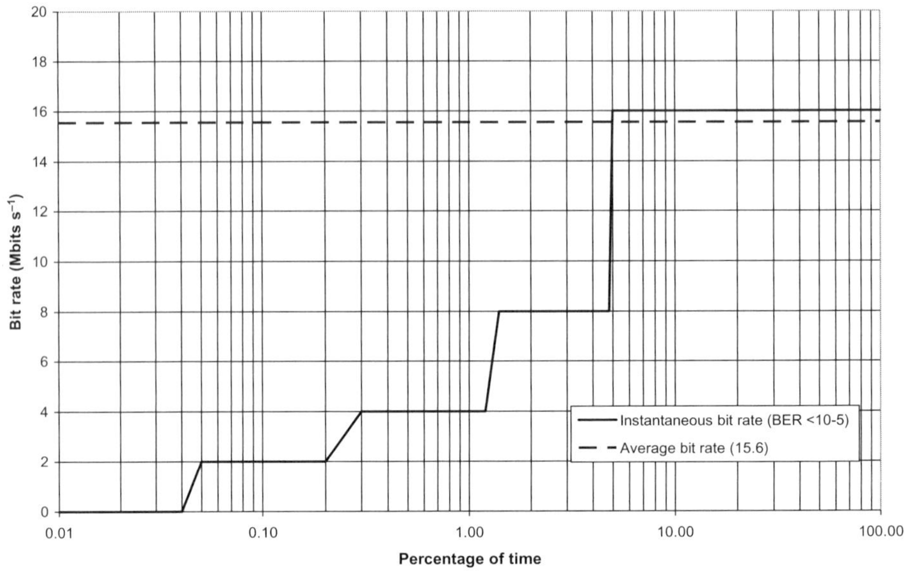

Figure 4.76 Bit rate distribution using adaptive tropo modem.

we see that the possible average bit rate is of the order of 15.6 Mbits s^{-1} whereas the terminals have been initially designed for a BER < 10^{-5} at 2 Mbits s^{-1} for 99.95% of the average year. Adaptation of the bit rate to the conditions of propagation is facilitated by the fact that they correspond practically to the slow Gaussian fading which is the same for both directions.

As impairments occur frequently in troposcatter propagation, generating severe disturbance of the transmitted signal when bulk encryption is used, a good solution consists in employing ATM packet switching with integrated ciphering; in fact, the encrypted cells are much less vulnerable and the recovery time of the complete transmission chain would thus be considerably reduced. To keep the same channeling and take account of the coherance bandwidth, which implies that the signal occupies the same frequency bandwidth whatever the bit rate may be, the modem can use various modulations levels from BPSK to 16QAM for instance. In ATM, it is also possible to introduce the time diversity using an appropriate time delay in frequency diversity so as to reduce still the overall correlation coefficient between signals.

4.5 OTHER MODELS OF PREDICTION

In general, the following methods give more pessimistic results than the method in technical note 101 (NBS, 1966) and should consequently be taken into account on a purely conservatory basis.

4.5.1 Method I of ITU-R

This method, simpler to use, is given by ITU-R Rep.238 and ITU-R Rec.617; it is founded on observations that were made at frequencies ranging between 200 and 4 GHz. The hourly median transmission loss exceeded for 100 − p% of the time over the year, expressed in decibels, can be calculated using the relation

$$L(p) = M + 30\log(f) + 10\log(d) + 30\log(\theta) + N(H,h) + L_C - G_E - G_R - C(p)Y(90) \quad (4.65)$$

where f = frequency (MHz)
d = pathlength (km)
θ = angle that makes radio horizon rays in Earth great circle plane for median atmospheric conditions for particular climate (mrad)

and $N(H,h)$ is a term accounting for the height at the base of the common volume according to the formula

$$N(H, h) = 20\log(5 + \gamma H) + 4.343\gamma h$$

where

$$H = \frac{\theta d}{4} 10^{-3} \quad \text{(km)}$$

$$h = \frac{\theta^2 a}{8} 10^{-6} \quad \text{(km)}$$

and a is the effective Earth radius in kilometers. The meteorological and atmospheric structure parameters M, γ, respectively, with different values in different climates are as follows:

Climate	M (dB)	γ (km^{-1})
Continental temperate	29.73	0.27
Maritime temperate, over land	33.2	0.27
Maritime temperate, over sea	26	0.27
Mediterranean		
Continental subtropical	29.73	0.27
Maritime subtropical	19.3	0.32
Desert	38.5	0.27
Equatorial	39.6	0.33
Polar		

The aperture-to-medium coupling loss L_C in decibels is given by the relation

$$L_C = 0.07\exp[0.055\,(G_E + G_R)] \tag{4.66}$$

and is presented in Figure 4.77.

The coefficient depending on q, $C(p)$, is given below:

p	50%	90%	99%	99.9%	99.99%
$C(p)$	0	1	1.82	2.41	2.9

The transmission loss difference relative to the median for 90% of times, $Y(90)$, is given as

$$Y(90) = L(50) - L(90)$$

and is calculated as follows:

- For climates continental subtropical, continental temperate, and maritime temperate, over land,

$$Y(90) = -2.2 - (8.1 - 2.3\times10^{-4} f)\exp(-0.137h)$$

- For climate maritime temperate, over sea,

$$Y(90) = -9.5 - 3\exp(-0.137h)$$

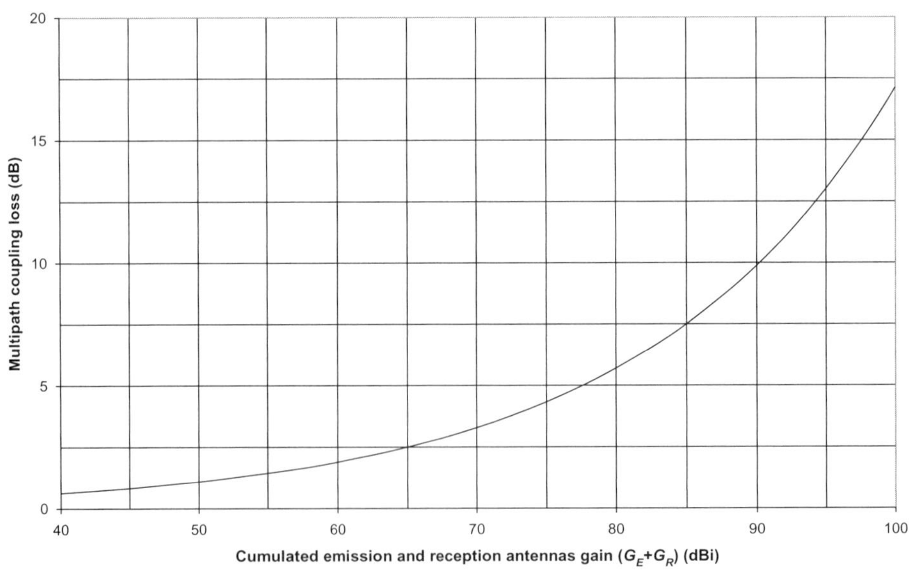

Figure 4.77 Aperture-to-medium coupling loss (ITU-R Rep.238).

- For climates equatorial, maritime subtropical, and desert, $Y(90)$ can be predicted from the graph of Figure 4.78 according to distance d_S, which is the difference between the pathlength and the sum of the distances from the antennas to the radio horizon points given approximately by

$$d_S = \frac{\theta a}{1000}$$

Figures 4.79–4.82 present curves that give the difference between the worst-month basic transmission loss and the annual basic transmission loss.

Figure 4.83 presents the basic isotropic propagation attenuation over the average year $L(50\%)$ at 1 GHz for the various climates according to relation (4.65) considering a smooth Earth and null elevation angles.

4.5.2 Method II of ITU-R

Recommendation ITU-R Rep.238 presents along with method II a simplified former version of technical note 101 (NBS, 1996); the median value in the long term of the transmission loss is given roughly by the following formula with the same units as in expression (4.1):

$$L(50\%) = 30\log(f) - 20\log(d) + F(\theta d) - G_T - G_R + L_C - V(50\%, de) \quad (4.67)$$

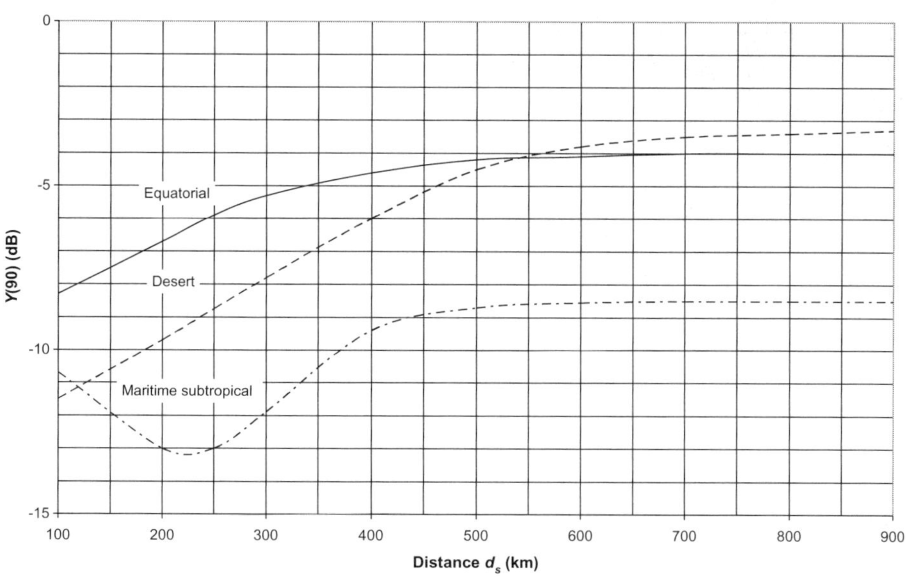

Figure 4.78 $Y(90)$ for equatorial, desert, and maritime subtropical climates.

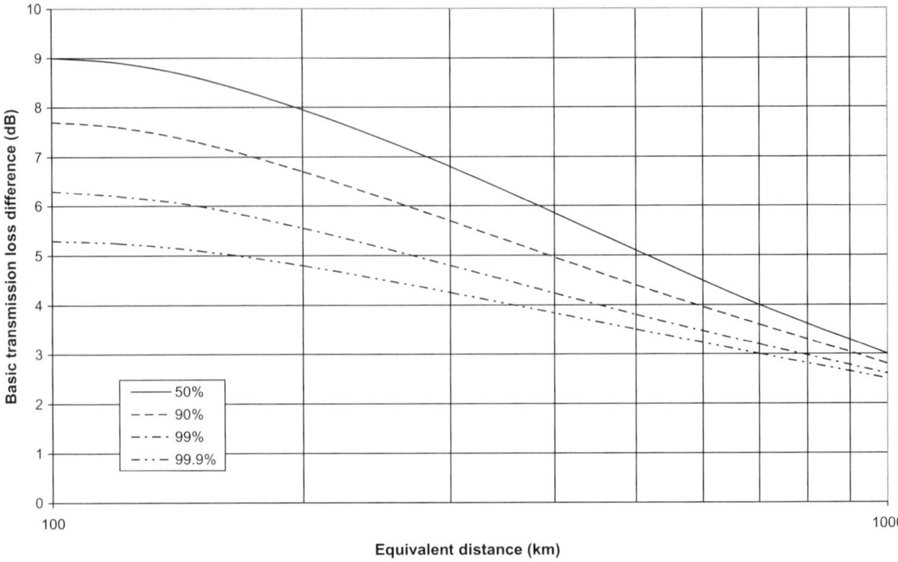

Figure 4.79 Difference between worst-month basic and annual basic transmission loss at 1 GHz, temperate climate (ITU-R Rep.238, method I).

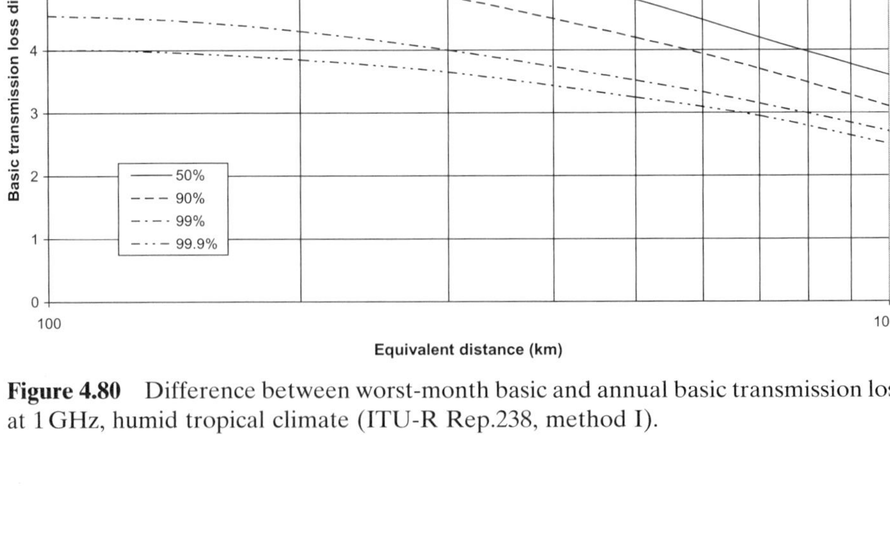

Figure 4.80 Difference between worst-month basic and annual basic transmission loss at 1 GHz, humid tropical climate (ITU-R Rep.238, method I).

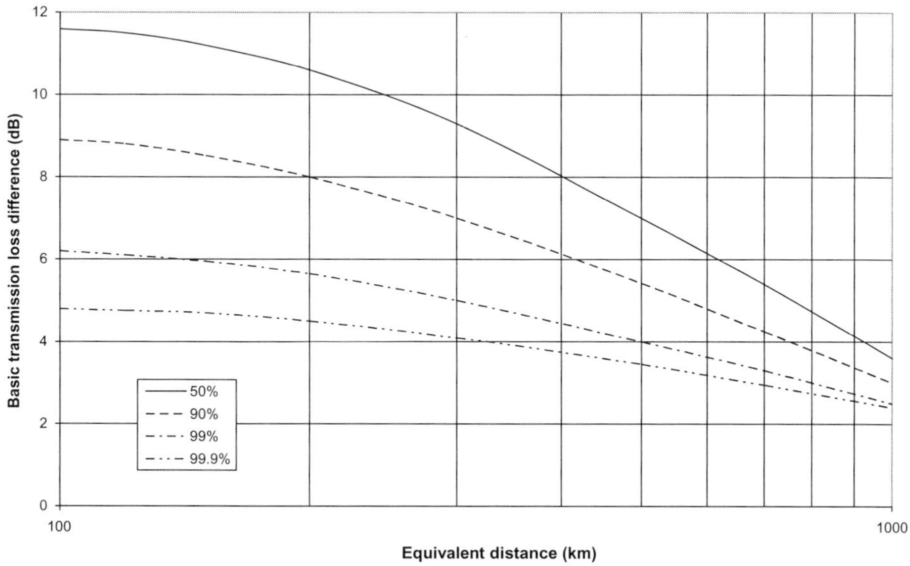

Figure 4.81 Difference between worst-month basic and annual basic transmission loss at 1 GHz, desert climate (ITU-R Rep.238, method I).

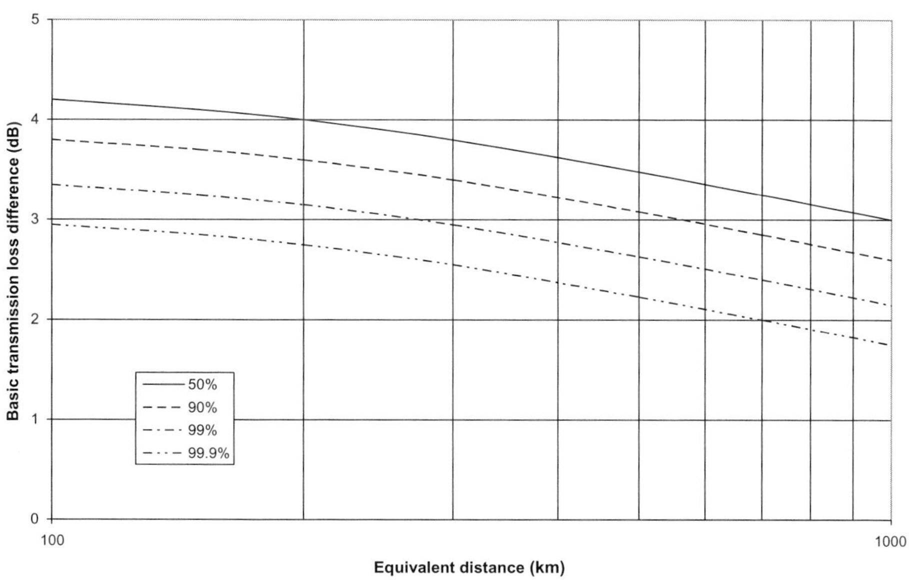

Figure 4.82 Difference between worst-month basic and annual basic transmission loss at 1 GHz, equatorial climate (ITU-R Rep.238, method I).

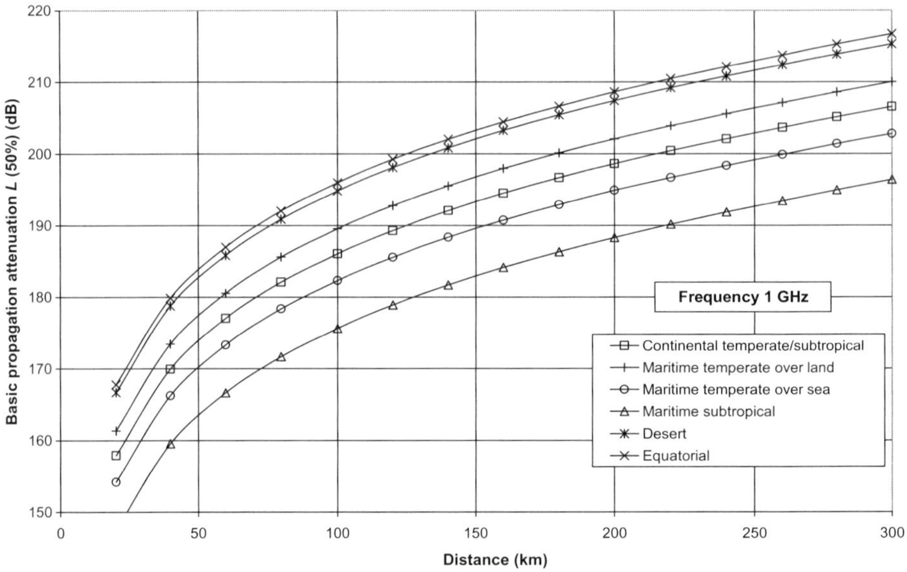

Figure 4.83 Basic propagation attenuation $L(50\%)$ at 1 GHz.

where G_E = emission antenna gain (<50 dBi)
G_R = reception antenna gain (<50 dBi)
L_C = multipath coupling loss (dB) given by relation (4.66) and presented in Figure 4.77

The characteristic $V(50\%,d_e)$ is obtained using relation (4.27) or from Figure 4.10 through the following formulas to calculate the equivalent distance:

$$d_L = 4.1\left(\sqrt{h_{te}} + \sqrt{h_{re}}\right) \qquad d_{se} = 302 f^{-1/3}$$

$$d_e = \begin{cases} \dfrac{130\,d}{d_L + d_{se}} & \text{for } d \le d_L + d_{se} \\ 130 + d - (d_L + d_{se}) & \text{for } d > d_L + d_{se} \end{cases}$$

The transmission loss thus calculated is more important because of the higher value for the multipath coupling loss of the antennas by comparison with the formulas given in Section 4.3.1 concerning the method in technical note 101 (NBS, 1966).

4.5.3 Method III of ITU-R

This method (by Boithias and Battesti), described in ITU-R Rep.238, rests on families of curves established in Europe and Africa according to the equivalent distance and represented by:

- Figures 4.84–4.87 for the various standard climates, which give statistics for the worst month of the transmission loss between isotropic antennas at 1 GHz for certain percentages of time and for null angles of elevation
- Figures 4.88–4.91 for the statistics for the annual basic transmission loss

For frequencies other than 1 GHz, the correction factor is given by

$$f_C = 30\log\left(\frac{f}{1000}\right)$$

When the elevation angles are not null, the equivalent distance is given by the formula

$$d_e = d + 8.5\,(\theta_1 + \theta_2)$$

where d = pathlength (km)
θ_1 = transmission elevation angle (mrad)
θ_2 = the reception elevation angle (mrad)

The following term must be added to the calculated transmission loss:

$$20\log\left(\frac{d}{d_e}\right)$$

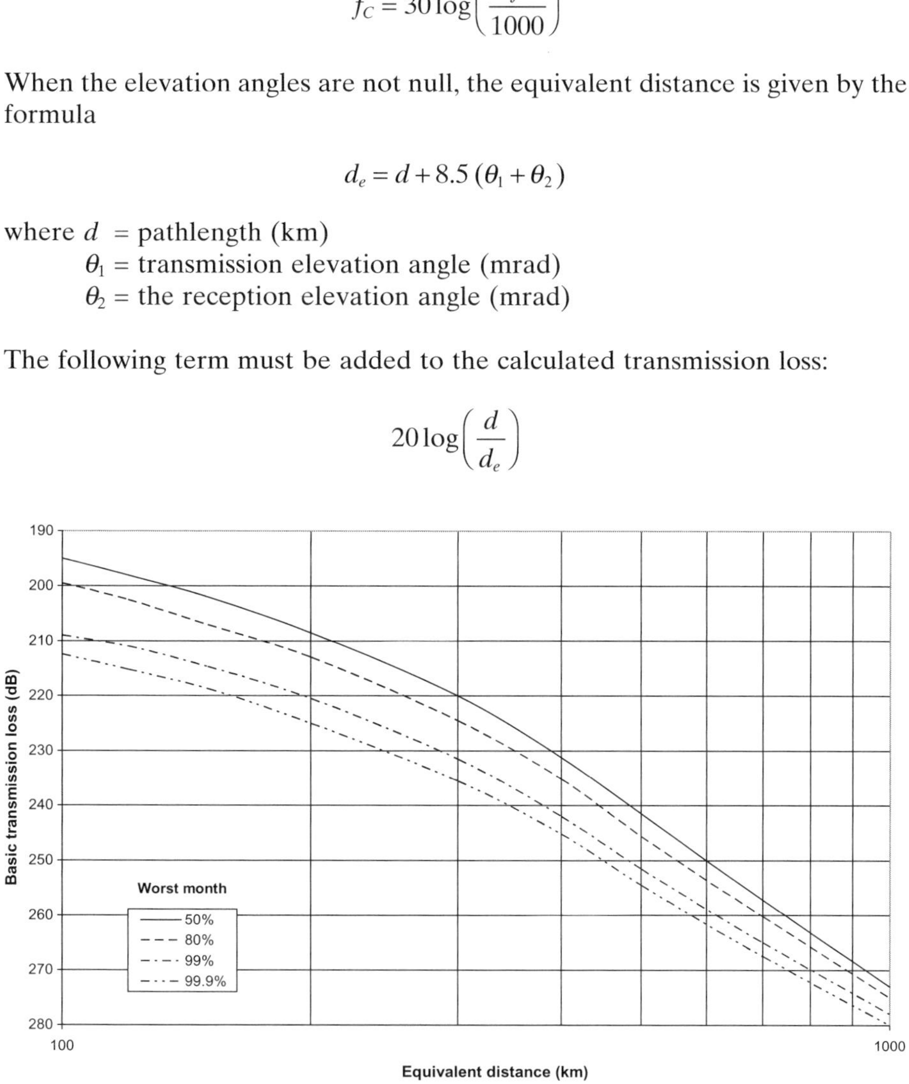

Figure 4.84 Worst-month basic transmission loss for tropospheric scatter at 1 GHz, continental temperate and maritime temperate, over-land climate (ITU-R Rep.238, method III).

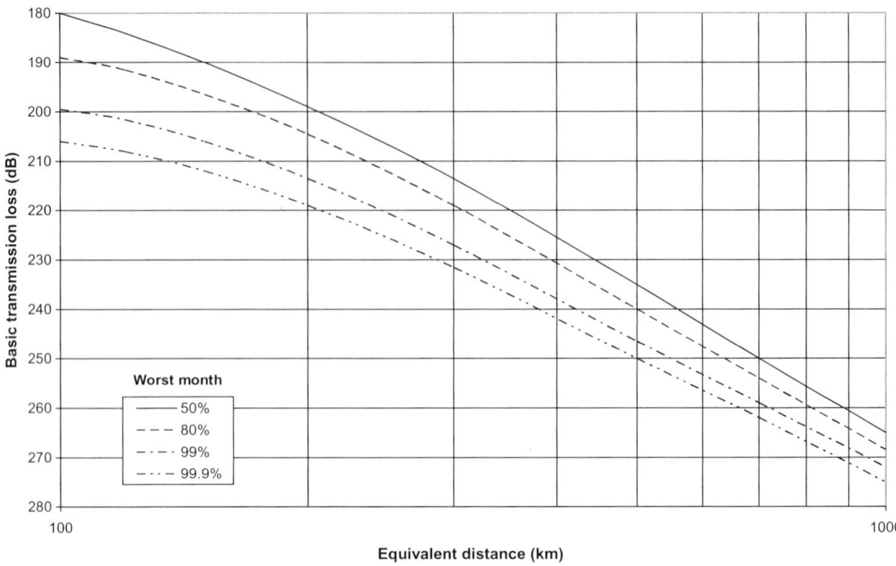

Figure 4.85 Worst-month basic transmission loss for tropospheric scatter at 1 GHz, maritime subtropical climate (ITU-R Rep.238, method III).

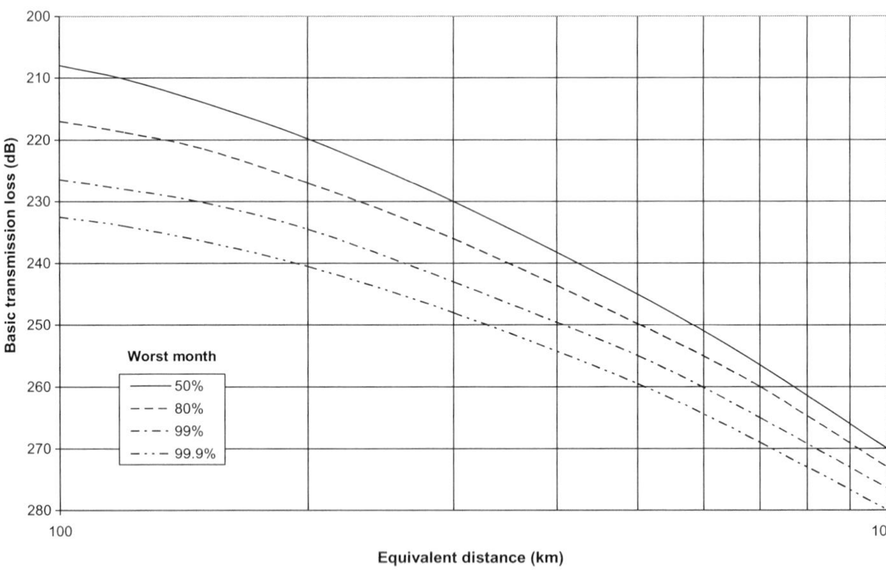

Figure 4.86 Worst-month basic transmission loss for tropospheric scatter at 1 GHz, desert climate (ITU-R Rep.238, method III).

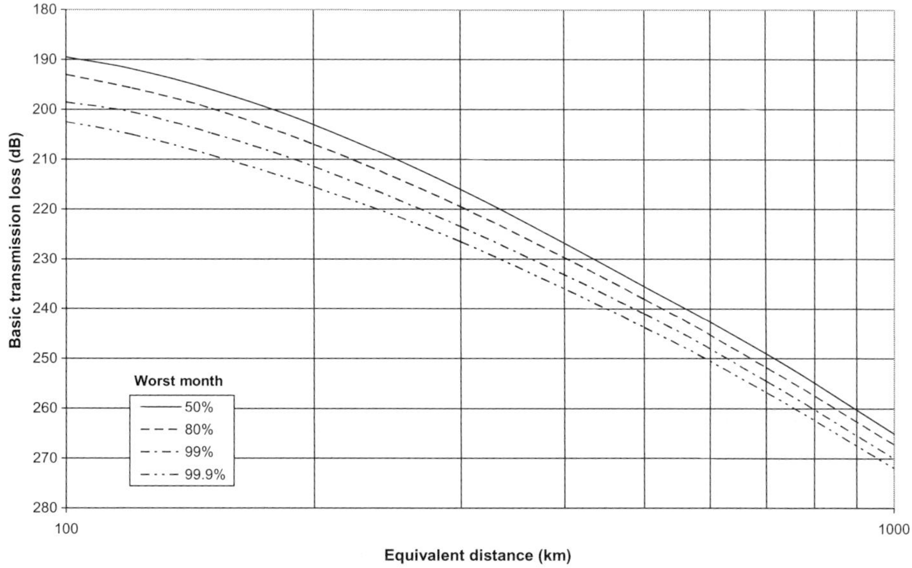

Figure 4.87 Worst-month basic transmission loss for tropospheric scatter at 1 GHz, equatorial climate (ITU-R Rep.238, method III).

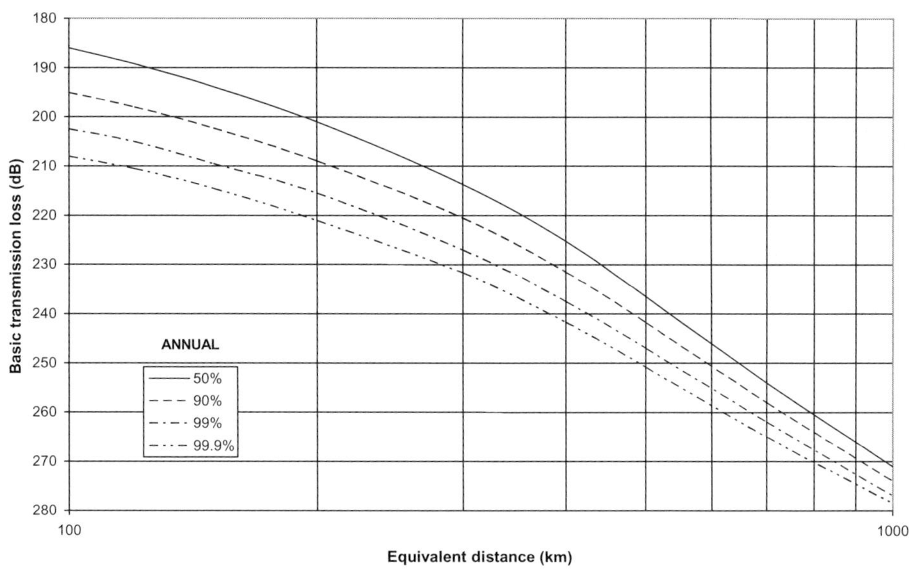

Figure 4.88 Annual basic transmission loss for tropospheric scatter at 1 GHz, temperate climate (ITU-R Rep.238, method III).

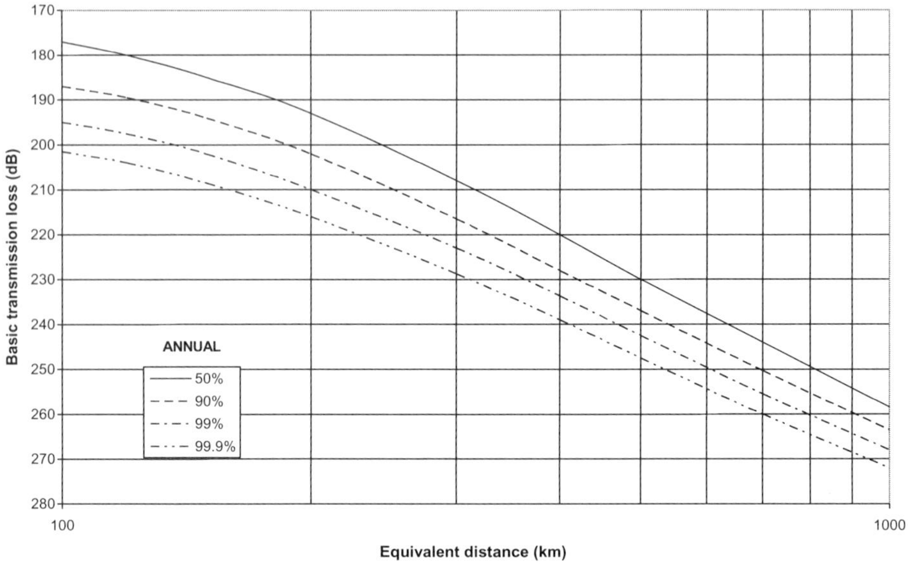

Figure 4.89 Annual basic transmission loss for tropospheric scatter at 1 GHz, humid tropical climate (ITU-R Rep.238, method III).

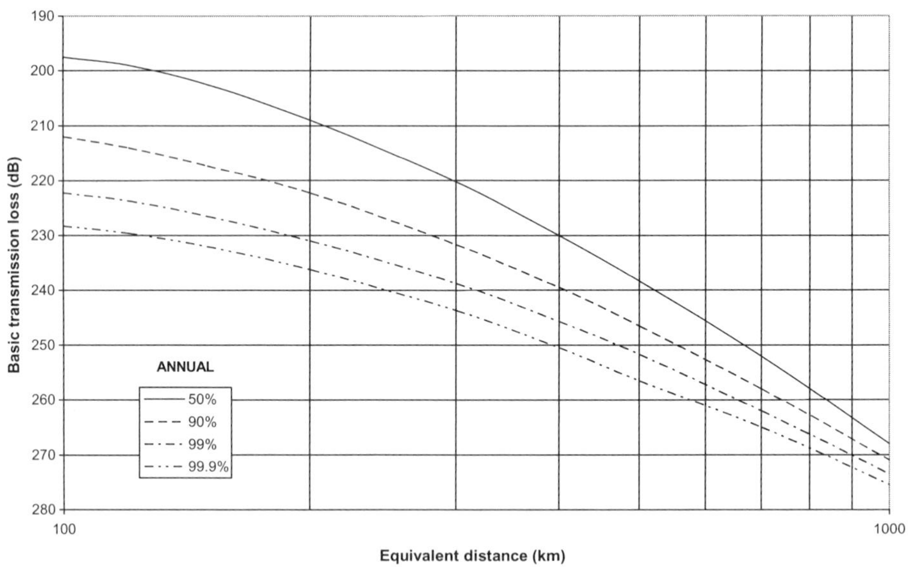

Figure 4.90 Annual basic transmission loss for tropospheric scatter at 1 GHz, desert climate (ITU-R Rep.238, method III).

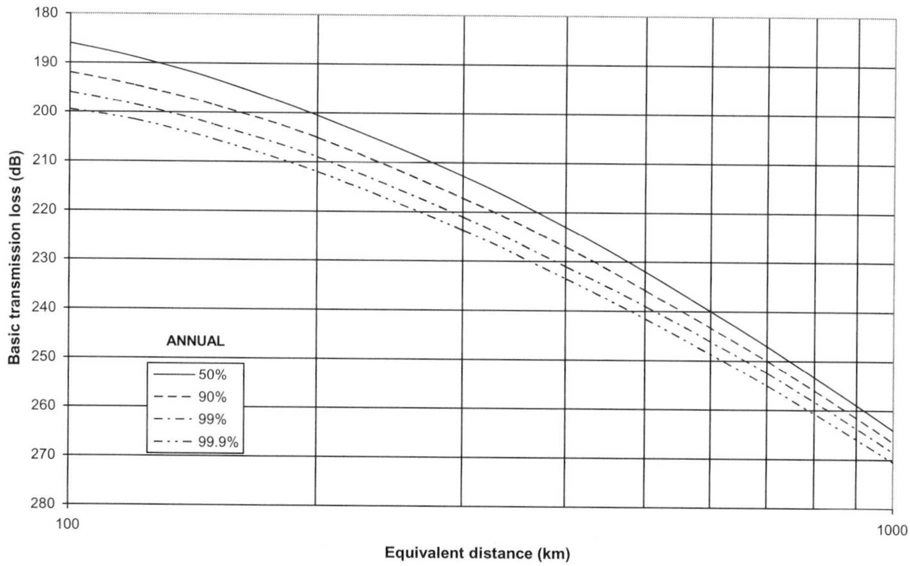

Figure 4.91 Annual basic transmission loss for tropospheric scatter at 1 GHz, equatorial climate (ITU-R Rep.238, method III).

Finally, for calculation of the actual transmission loss, we must use the total effective gain of the antenna emission and reception, expressed in decibels, which is given by the following formula or is represented as in Figure 4.92:

$$G_E + G_R - L_C = (G_E + G_R)\exp\left\{-\frac{[(G_E + G_R)/148]^4}{1+[(G_E + G_R)/148]^4}\right\} \tag{4.68}$$

4.5.4 Seasonal and Diurnal Variations in Transmission Loss

In temperate climates, the transmission loss varies annually and diurnally; the monthly median losses tend to be higher in winter than in summer and the diurnal variations are most pronounced in summer. In dry, hot desert climates, the attenuation reaches a maximum in summer; the annual variations of the monthly medians for median-distance paths exceed 20 dB while the diurnal variations are very large. In equatorial climates, the annual and diurnal variations are generally small. In monsoon climates, the maximum values of N_S occur during the wet season but the minimum attenuation is between the wet and the dry seasons.

4.6 TOTAL AVAILABILITY OF TROPOSCATTER LINKS

Section 3.8 treats the aspects of the operational availability of microwave links, particularly line-of-sight links where only Rayleigh fading occurs. Concerning

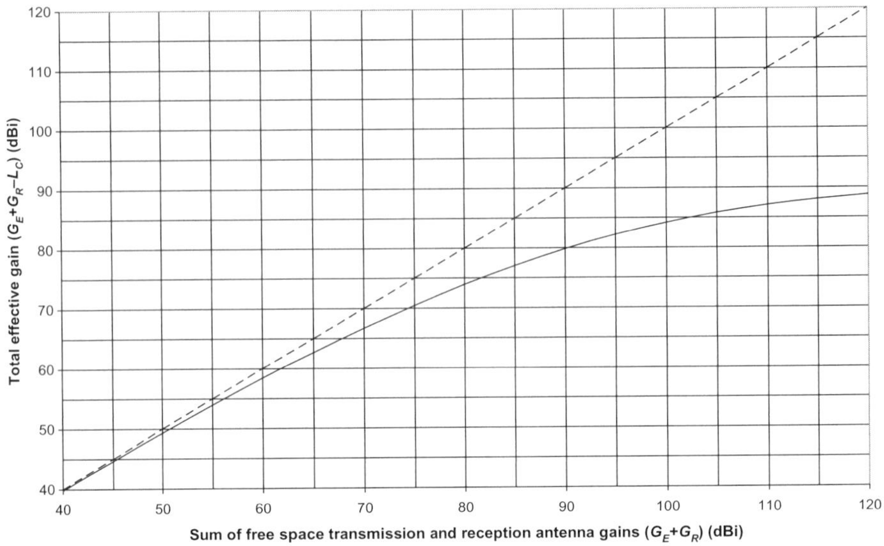

Figure 4.92 Total effective gain (Boithias and Battesti).

troposcatter links, it is necessary to consider the properties of propagation, which is more complex, as well as the mechanisms specific to systems of diversity.

4.6.1 Physical Principles of Propagation in Troposcatter Mode

We saw that the propagation in troposcatter mode obeys simultaneously a Gaussian distribution in the long term, characterized by a hourly median value and a standard deviation, and a Rayleigh's distribution in the short term which results in deep fading of the instantaneous fields around the median with a slope of about 10 dB per decade. The maximal-ratio combiner has as a function to create, from the independent individual signals collected simultaneously by the various receivers operating in diversity, a combined signal whose characteristic *C/N* according to the BER depends on the row of diversity on the assumption of a low correlation between the received signals. The degradation of the performance of the link in terms of operational availability does not obey a linear law because the phenomena of propagation that come into play as well as the mechanisms of combination in diversity are not linear.

Consequently, the calculation of the real unavailability over the average year, which results from a temporary degradation *X* in decibels of the *C/N* of one among the multiple channels, must necessarily rest on a statistical model utilizing simultaneously:

1. The characteristic *C/N*(BER) for a maximal-ratio combination (Figure 4.44)

2. A parameter which expresses the influence of the breakdown according to the role of the failing element in the transmission chain
3. The unavailability over the average year according to the standard deviation of the Gaussian distribution in the long term
4. The MTBF and MTTR of the equipment in question

Because of the extreme variability of the propagation loss, calculation of the impact of temporary degradation is related to the average year, which refers to a period much longer; a breakdown of any equipment during a favorable period would have, of course, a negligible consequence on the overall performance of the link in the long term, contrary to the same event occurring during an unfavorable period.

4.6.2 Model of Calculation of Operational Availability

The model rests on the combination of four laws:

- The first law makes it possible to determine $\Delta[C/N(\mathrm{BER})]$ from the row of degraded diversity D_d compared to the nominal row of diversity D_{nom} and acceptable BER:

$$\Delta[C/N(\mathrm{BER})] = 10\log[10^{\log(\mathrm{BER}/2)/D_d} - 1] - 10\log[10^{\log(\mathrm{BER}/2)/D_{\mathrm{nom}}} - 1]$$

- The second law allows us to determine $\Delta[C/N]$ starting from the degradation X [0; ∞] of the C/N due to the breakdown of the considered equipment:

$$\Delta[C/N] = [1 - 10^{-(X/10)}]\Delta[C/N(\mathrm{BER})]$$

- The third law allows us to calculate the unavailability over the average year $U_Y(\%)$ as if it were about a permanent degradation by considering the nominal availability $P_{\mathrm{nom}}(\%)$ [50% ; 99.999%] and the standard deviation σ in decibels of the distribution:

$$U_Y(\%) \approx 100 \times \frac{\exp(-A^2/2)}{\sqrt{2\pi}(0.661|x| + 0.339\sqrt{x^2 + 5.51})}$$

where

$$A = 5.31\left[1.167\left(\frac{\ln 100 - \ln P_{\mathrm{nom}}(\%)}{2}\right)^{0.14} - 1\right] + \frac{\Delta[C/N]}{\sigma}$$

- The fourth law serves to calculate the real unavailability over the average year U_R:

$$U_R = I_{\text{nom}} + [U_Y - U_{\text{nom}}]\frac{\text{MTTR}}{\text{MTBF}}$$

Table 4.4 shows, for example, the influence of the same breakdown of a transmitter or a low-noise amplifier (LNA) on the typical configurations D4 and D6 illustrated in Figure 4.93.

Figure 4.94 shows how to determine graphically the degradation $\Delta[C/N(\text{BER})]$ of C/N in the case of a breakdown which reduces by two the row of diversity such as the failure of one transmitter over two in D4 and D6 systems for BER 10^{-5}.

Figure 4.95 presents the effective degradation $\Delta[C/N]$ according to a degradation X of C/N for various values of $\Delta[C/N(\text{BER})]$ defined previously.

TABLE 4.4 Operational Unavailability (BER > 10^{-5}) for MTTR/MTBF = 0.01

	Configuration D4		Configuration D6	
	%	Hours	%	Hours
Nominal performances	0.05	4.4	0.05	4.4
Transmitter breakdown	0.1	8.8	0.1	8.8
LNA breakdown	0.1	8.8	0.062	5.4

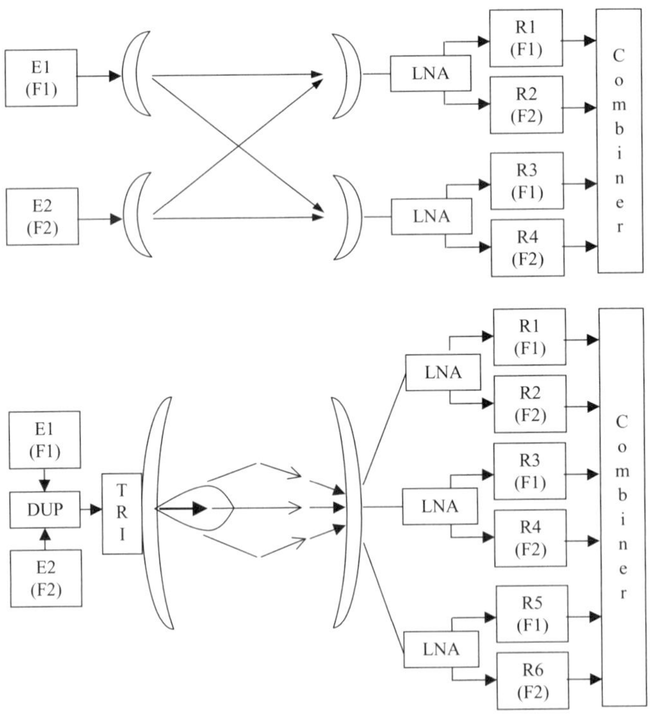

Figure 4.93 Configurations D4 and D6.

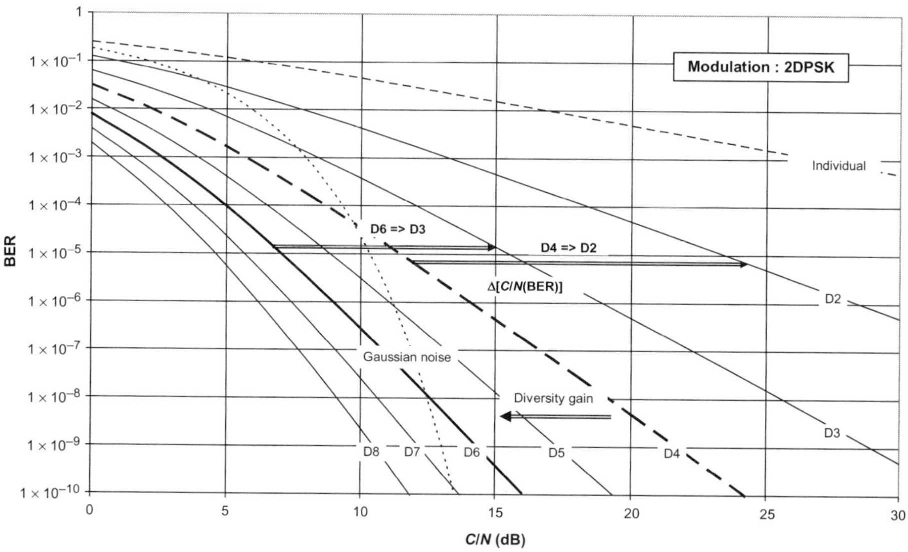

Figure 4.94 Determination of $\Delta[C/N(BER)]$ for a given BER in a maximal-ratio combining system

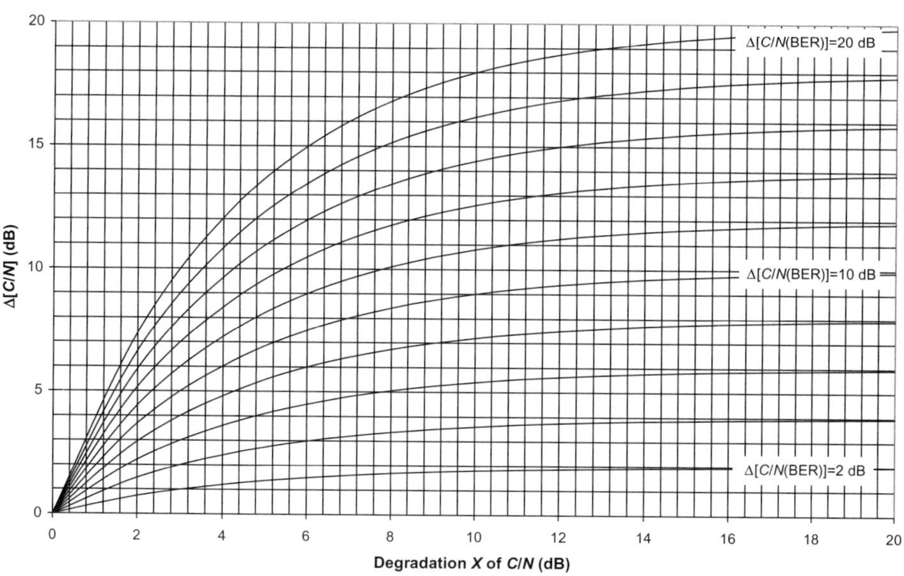

Figure 4.95 Determination of $\Delta[C/N]$ versus degradation X of C/N and $\Delta[C/N$ (BER)].

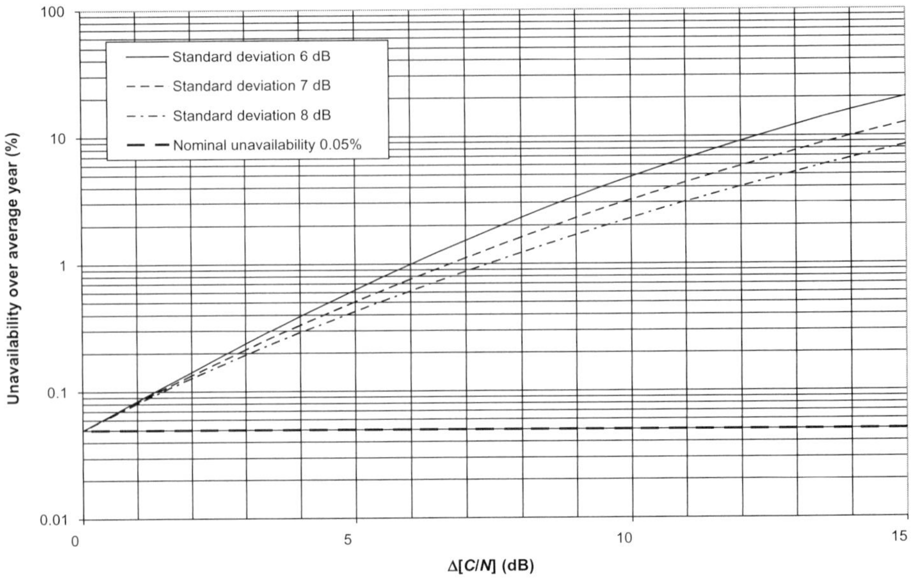

Figure 4.96 Unavailability over average year for permanent degradation $\Delta[C/N]$.

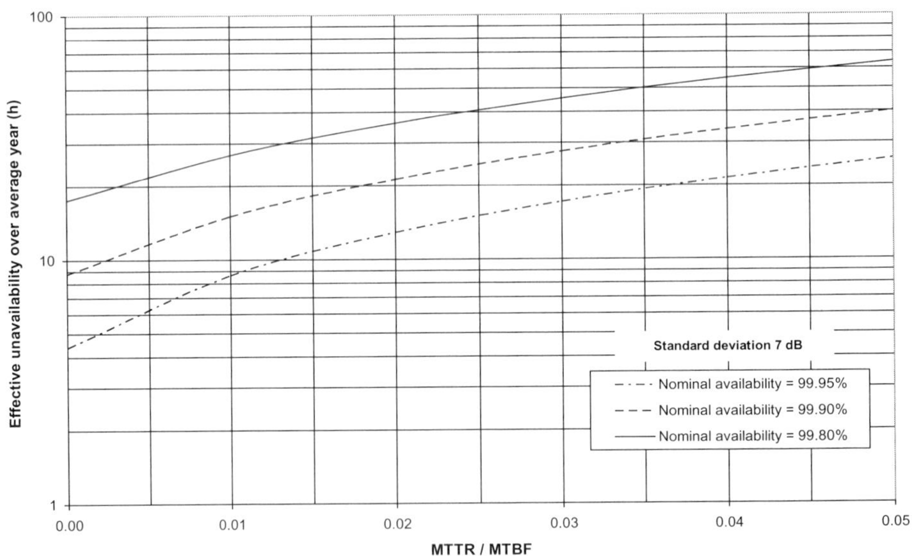

Figure 4.97 Effective unavailability according to $MTTR/MTBF$.

Figure 4.96 presents, for instance, determination of the unavailability over the average year $U_Y(\%)$ corresponding to a permanent degradation $\Delta[C/N]$ for various standard deviations of the Gaussian distribution and considering a nominal unavailability of 0.05%.

Figure 4.97 presents the effective unavailability over the year in relation with the ratio MTTR/BTBF of the component in failure in the case of one transmitter among two.

CHAPTER 5

Satellite Communications

5.1 SPACE GEOMETRY OF SATELLITE SYSTEM

5.1.1 Introduction

This chapter provides useful indications for determination of satellite orbits, such as the period of revolution, the distance with respect to the center of Earth or to an Earth station, and the travel speed of the satellite on its orbit. It is devoted, essentially, to the systems of telecommunication using a geostationary satellite, but the principles remain valid for the systems comprising a moving satellite.

5.1.2 General Characteristics of Orbits

Calculation of a satellite orbit rests, on the one hand, on Kepler's laws and, on the other hand, on universal gravitation, Newton's law. Figure 5.1 presents the relative positions of Earth and the satellite in its orbit with the following notations:

r	Distance between centers of gravity of Earth and satellite
a	Half large axis of ellipse
c	Distance between center of Earth and that of ellipse
M	Mass of Earth
m	Mass of satellite
α	Angular distance of satellite relative to perigee

5.1.2.1 Kepler's Laws

1. The trajectory of a planet in space, called its orbit, is an ellipse described in a plane which contains the center of the sun placed at one of the focus (1602).
2. The vector which connects the sun to the planet sweeps equal surfaces in equal times (1605).

Microwave Engineering: Land & Space Radiocommunications, by Gérard Barué

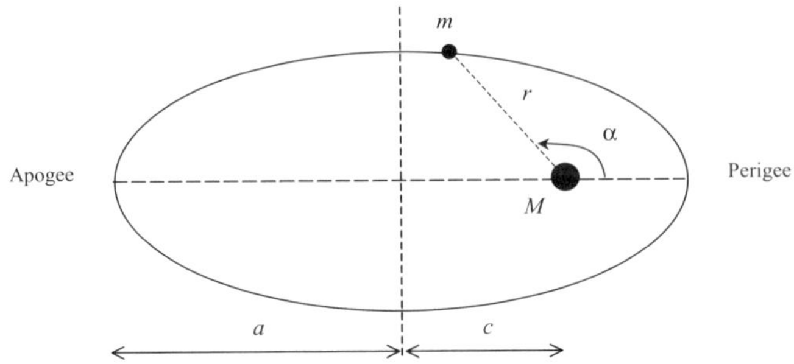

Figure 5.1 Representation of satellite orbit.

3. The ratio of the square of the period of revolution around the sun to the cube of half of the large axis of the ellipse is identical for all planets (1618).

5.1.2.2 *Newton's Law* Two bodies of mass M and m attract each other with a force F that is proportional to their mass and inversely proportional to the square of the distance r that separates them:

$$F = G\frac{Mm}{r^2} \tag{5.1}$$

where G $= 6.672 \times 10^{-11}$ ($m^3 kg^{-1} s^{-2}$) is the gravitational constant
$M = 5.974 \times 10^{24}$ (kg) is the mass of Earth

5.1.2.3 *Determination of Orbits* We will consider that:

1. The mass m of the satellite is negligible with respect to the mass M of Earth.
2. The movements take place in a vacuum.
3. There is no other interaction than the attraction between Earth and the satellite.
4. Earth is spherical and homogeneous.

The shape of the ellipse is characterized by its eccentricity e, which is given by the relation

$$e = \frac{c}{a}$$

The following formula connects the distance r between the center of Earth and the satellite, the half large axis a, the eccentricity e, and the angular distance α between the direction of the perigee and that of the satellite:

$$r = a\frac{1-e^2}{1+e\cos\alpha} \tag{5.2}$$

The period of revolution T is given by the relation

$$T = 2\pi\sqrt{\frac{a^3}{GM}} \tag{5.3}$$

The speed of the satellite on its orbit is given by the relation

$$V = \sqrt{GM\left(\frac{2}{r}-\frac{1}{a}\right)} \tag{5.4}$$

The table below provides the characteristics of various circular low, medium, and geostationary orbits:

Altitude (km)	Radius (km)	Period (s)	Speed ($m\,s^{-1}$)
250	6,628	5,370	7,755
1,000	7,378	6,307	7,350
2,500	8,878	8,325	6,700
10,000	16,378	20,860	4,933
35,786	42,164	86,164	3,075

5.1.2.4 Origins of Disturbances of Orbits The disturbances of the orbits have the following principal origins:

- Earth is not a point mass and presents a certain asymmetry, called oblateness, as well as a certain heterogeneity
- Attraction by the sun and moon
- Pressure of radiation of the sun, or solar wind, which produces forces on surfaces of the satellite that are exposed to the sun

Consequently these multiple influences cause, among others, a drift in longitude and modification of the slope of the orbit or its eccentricity, which oblige operators to periodically reposition the satellite on its nominal orbit.

5.1.3 Types of Satellite Systems

There are several types of satellite systems according to the orbit used:

- Elliptical orbits, inclined at an angle of 64° with respect to the equatorial plane, which are employed for communication with mobiles

- Circular low Earth orbit (LEO), with near 90° inclination, which are utilized for observation or for communication in constellations of several tens of satellites
- Circular medium Earth orbit (MEO), at an altitude of about 10,000 km and an inclination of about 50°, which requires a constellation comprising less satellites
- Circular orbits with zero inclination (equatorial orbit) and geostationary

5.1.3.1 Geometry of Low and Medium Circular Orbits The circular orbits are characterized by a null eccentricity $e = 0$, which yields $c = 0$ in the relations above; in such systems, the satellite moves faster in its trajectory as the distance from Earth decreases and Earth stations have to be equipped with tracking devices for the antenna and to face high Doppler effects on the transmitted signal.

Consider, for example, the equatorial circular orbit in Figure 5.2 with a satellite S and an Earth station T by introducing the following parameters:

$\mathbf{w}$	Angular velocity of satellite ($2\pi/T$ rad s^{-1})
$\mathbf{v}$	Velocity of satellite on its orbit
$\mathbf{v_R}$	Radial velocity of satellite toward Earth station
$\mathbf{W}$	Angular velocity of rotation of Earth ($2\pi/86{,}400$ s rad s^{-1})
H_S	Altitude of satellite
θ	Elevation angle at Earth station

We write

$$\begin{aligned} \gamma &= \arcsin\left(\frac{R}{R+H_s}\cos\theta\right)+\theta-\frac{\pi}{2} \\ v_R &= \sqrt{R^2+(R+H_s)^2-2(R+H_s)R\cos\gamma}- \\ &\quad \sqrt{R^2+(R+H_s)^2-2(R+H_s)R\cos(\gamma-w\pm W)} \end{aligned} \tag{5.5}$$

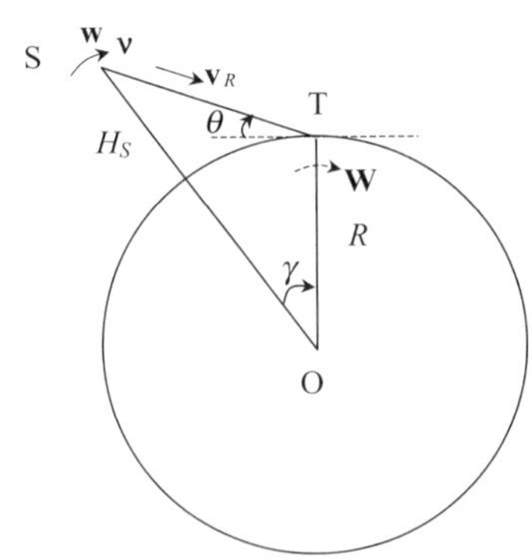

Figure 5.2 Equatorial circular orbit.

Figure 5.3 presents the period of revolution of the satellite on a circular orbit according to its altitude. Figure 5.4 shows the radial speed of a satellite on an equatorial circular orbit toward an Earth station located in the same plane according to its altitude.

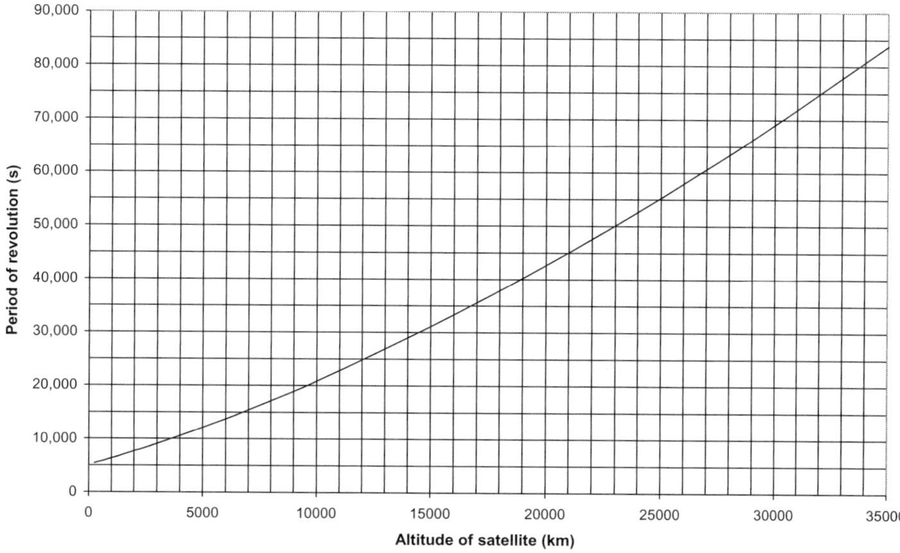

Figure 5.3 Period of revolution of a satellite on a circular orbit.

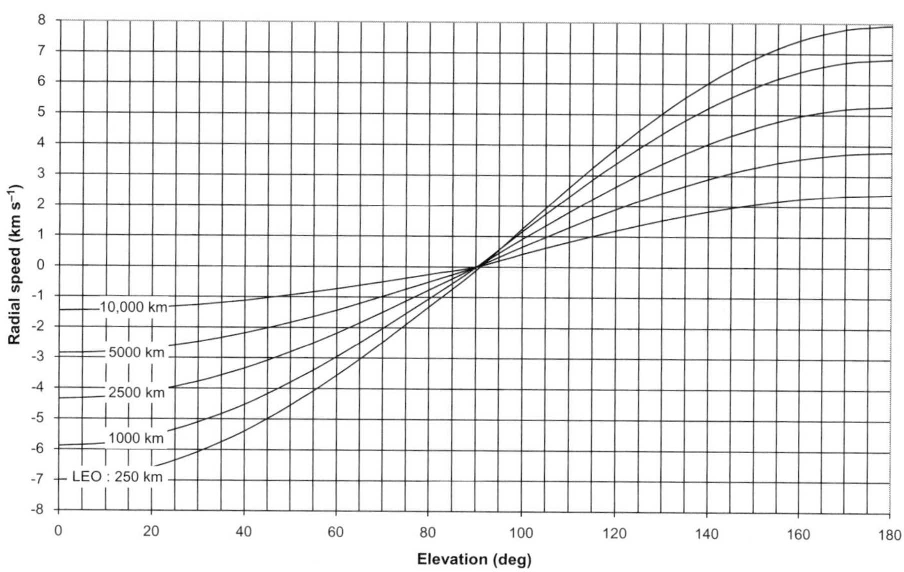

Figure 5.4 Radial speed for satellite on equatorial circular orbit.

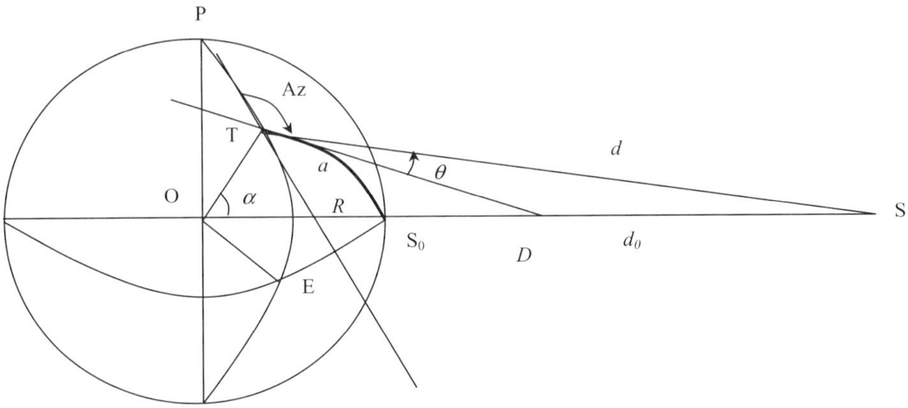

Figure 5.5 Geometric elements of geostationary satellite.

5.1.3.2 Geometry of Connection Using Geostationary Satellite Taking into account the very long distance from the geostationary satellite to Earth and for reasons of convenience, the azimuth and the elevation of the antenna of the Earth station are defined by reference to the geosphere. Figure 5.5 presents the geometry of the connection in three dimensions, where the following appear:

Satellite S
Earth station T
Geographical North Pole P
Subsatellite point S_0
Distance from satellite to Earth, d_0
Meridian arc PTE passing by Earth station
Arc of large circle a passing by T and S_0 and corresponding geocentric angle α
Azimuth Az
Elevation angle θ
Distance d between satellite and Earth station
Distance D between satellite and center of Earth

Geosynchronism Distance for Satellite The radius of the circular orbit, or geosynchronism distance, is calculated by writing the equality between the force of gravitation and the centrifugal force:

$$G\frac{Mm}{D^2} = m\omega^2 D \tag{5.6}$$

where m = mass of satellite, which is negligible compared to M (kg)
D = geosynchronism distance with respect to center of Earth (m)
ω = angular velocity of Earth (rad s^{-1}), $=2\pi/T$

and the duration of the sidereal day is $T = 86{,}164$ s, which yields

$$D = \sqrt[3]{\frac{GM}{\omega^2}} = 42{,}164.2\ \text{km} \tag{5.7}$$

The terrestrial equatorial radius having as a value

$$R = 6378.1\ \text{km}$$

The distance from the satellite to the surface is given as:

$$d_0 = 35{,}786.1\ \text{km}$$

The geocentric angle is given by

$$\cos\alpha = \cos(L_T - L_s)\cos(l_T - l_S)$$

Azimuth from Earth Station Let L_T and l_T be the longitude and latitude of the Earth station and L_S and l_S the longitude and latitude of the satellite (all in degrees). We can then write the fundamental relation of the spherical triangle applied to TPS_0:

$$\cos\alpha = \cos\text{TS}_0 = \cos\text{TP}\cos\text{PS}_0 + \sin\text{TP}\sin\text{PS}_0\cos\text{T}\hat{\text{P}}\text{S}_0$$

from which, replacing the complementary arcs,

$$\alpha = \arccos[\sin l_T \sin l_S + \cos l_T \cos l_S \cos(L_T - L_S)] \tag{5.8}$$

we get

$$\cos\text{PS}_0 = \cos\text{PT}\cos\text{TS}_0 + \sin\text{PT}\sin\text{TS}_0\cos\text{P}\hat{\text{T}}\text{S}_0$$

That is,

$$\sin l_S = \sin l_T \cos\alpha + \cos l_T \sin\alpha \cos A_Z$$

which yields

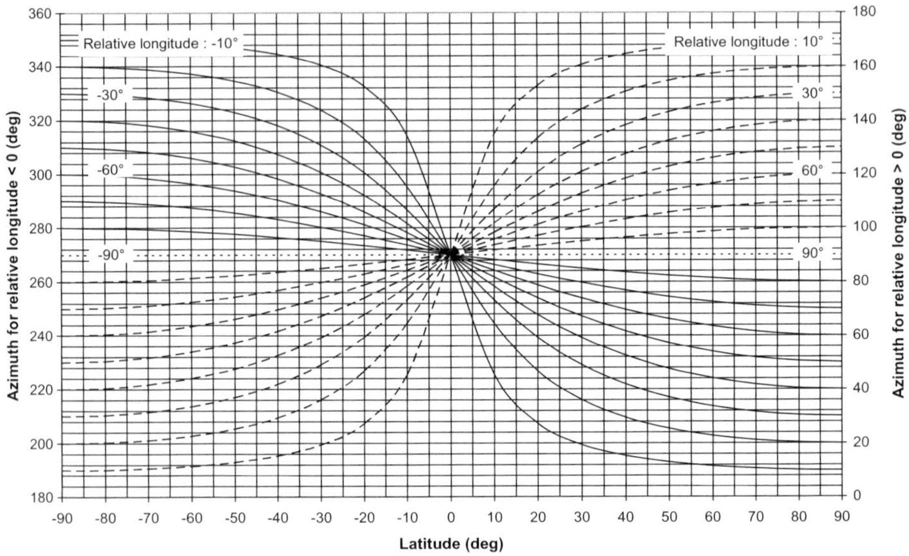

Figure 5.6 Azimuth of satellite from Earth station.

$$A_Z = \begin{cases} \arccos\left[\dfrac{\sin l_S}{\cos l_T \sin\alpha} - \dfrac{\tan l_T}{\tan\alpha}\right] & \text{for } L_T - L_S < 0 \\ 360° - \arccos\left[\dfrac{\sin l_S}{\cos l_T \sin\alpha} - \dfrac{\tan l_T}{\tan\alpha}\right] & \text{for } L_T - L_S > 0 \end{cases} \tag{5.9}$$

The satellite being geostationary, $l_S = 0$, from which we get the simplified expressions

$$\alpha = \arccos[\cos l_T \cos(L_T - L_S)] \tag{5.10}$$

$$A_Z = \begin{cases} \arccos\left[-\dfrac{\tan l_T}{\tan\alpha}\right] & \text{for } L_T - L_S < 0 \\ 360° - \arccos\left[-\dfrac{\tan l_T}{\tan\alpha}\right] & \text{for } L_T - L_S > 0 \end{cases} \tag{5.11}$$

Figure 5.6 presents the azimuth according to latitude and relative longitude.

Distance between Satellite and Earth Station Figure 5.7 presents the geometry of the connection in the plane containing the satellite, the Earth station, and the center of Earth. In triangle OTS, we can write the fundamental relation

$$d^2 = R^2 + D^2 - 2RD\cos\alpha$$

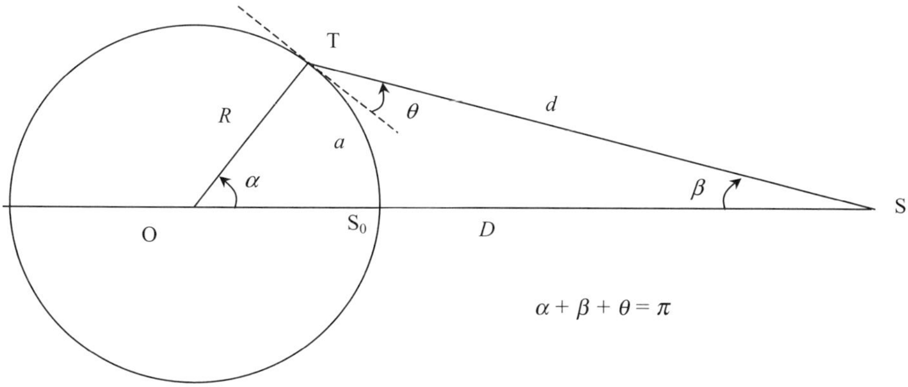

Figure 5.7 Representation in plane containing arc of Earth great circle.

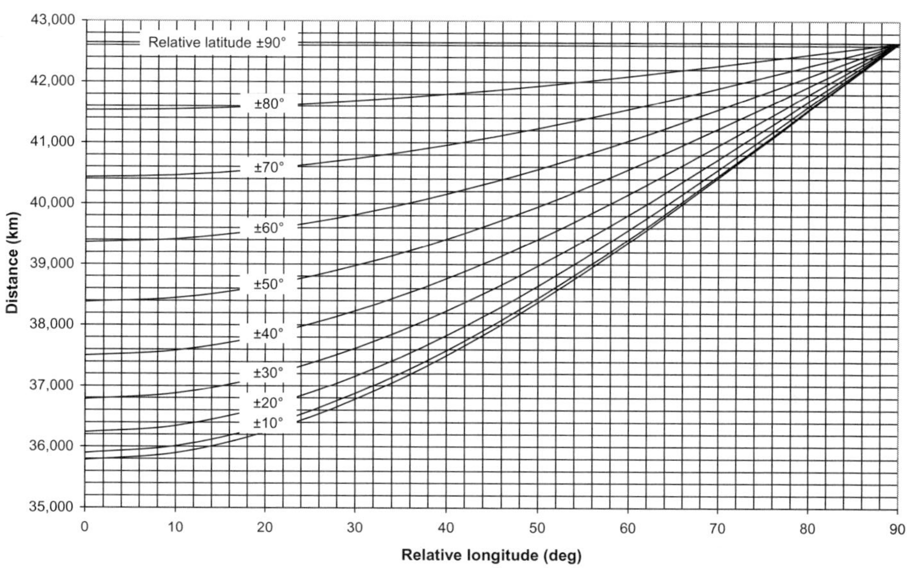

Figure 5.8 Distance between satellite and Earth station.

which yields

$$d = \sqrt{R^2 + D^2 - 2RD\cos\alpha} = 35{,}786.1\sqrt{1 + 0.42[1 - \cos(L_T - L_S)\cos(l_T - l_S)]} \tag{5.12}$$

Figure 5.8 presents the distance between the satellite and the Earth station according to the longitudc and the relative latitude.

Elevation Angle from Earth Station In the same way, we write

$$\frac{d}{\sin\alpha}=\frac{R}{\sin\beta}=\frac{D}{\sin(\pi/2+\theta)}=\frac{D}{\cos\theta}$$

which yields

$$\theta=\arccos\left[\frac{D}{d}\sin\alpha\right]=\arccos\left[\frac{D}{R}\sin\beta\right] \tag{5.13}$$

Figure 5.9 presents the elevation angle from the Earth station according to the relative longitude and latitude.

Nadir Angle from Satellite

$$\beta=\arcsin\left[\frac{R}{d}\sin\alpha\right] \tag{5.14}$$

Figure 5.10 shows the nadir angle, which corresponds to the angle at the level of the satellite between the vertical and the Earth station according to the relative longitude and latitude.

Optical Limit of Visibility The optical limit of visibility corresponds to the null angle of elevation:

$$\theta = 0^\circ \qquad \alpha = 81.3^\circ \qquad \beta = 8.7^\circ \qquad \mathrm{d} = 41{,}679\,\mathrm{km}$$

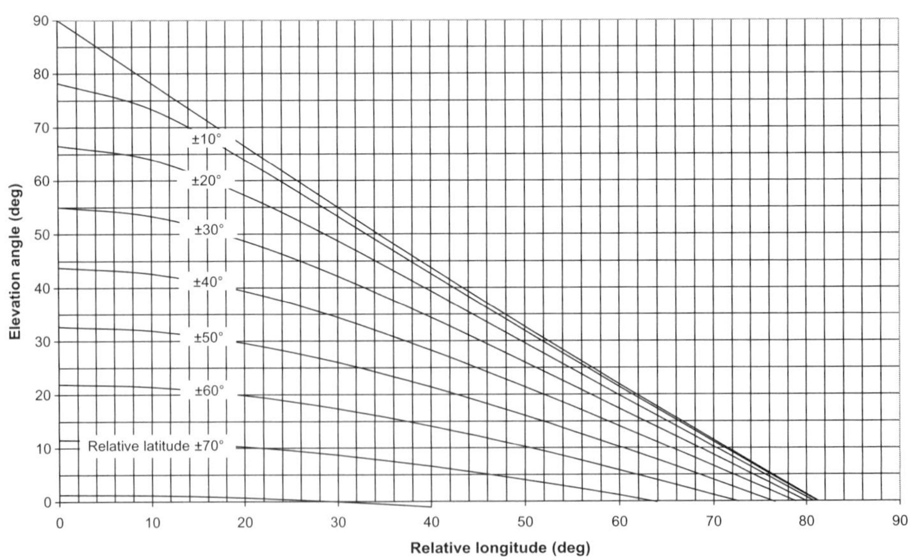

Figure 5.9 Elevation angle of satellite from Earth station.

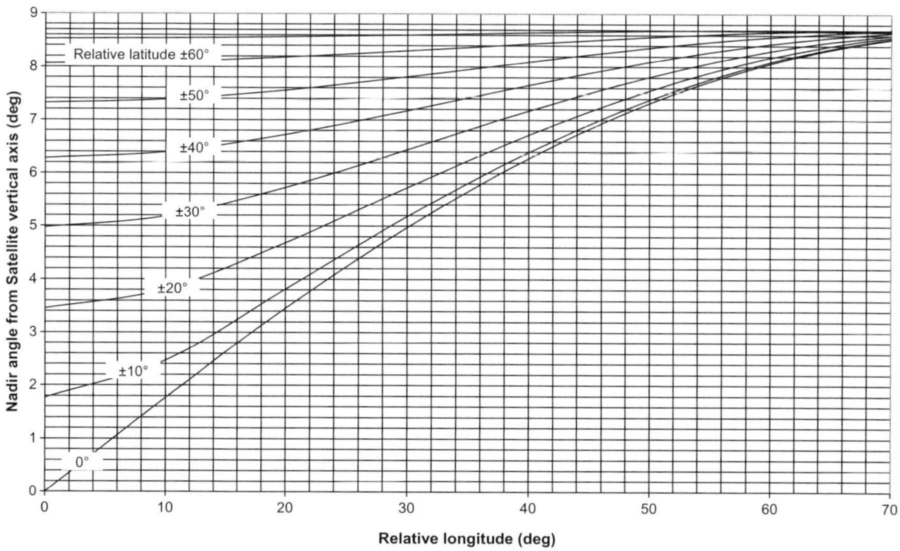

Figure 5.10 Satellite nadir angle relative to vertical axis.

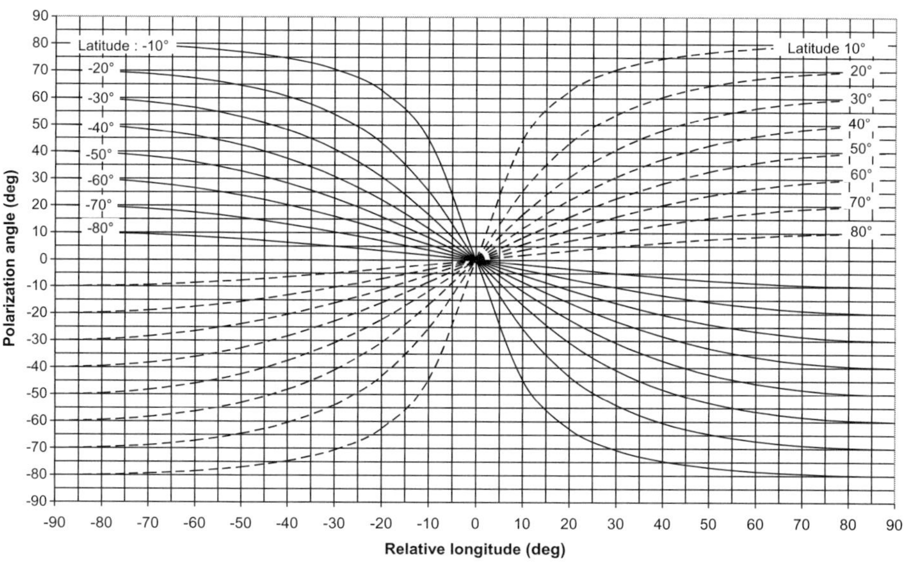

Figure 5.11 Polarization angle from meridian plane.

Angle of Polarization By supposing that the plane of polarization at the satellite is confused with the meridian plane, the angle of polarization at the Earth station is given by Figure 5.11 and the relation

$$\psi = \arctan\left[\frac{\sin(L_T - L_S)}{\tan(l_T - l_S)}\right] \tag{5.15}$$

5.2 CONFIGURATION OF SATELLITE COMMUNICATION SYSTEM

A satellite communications system comprises three main segments:

- Space segment, which involves one or several active and spare satellites operating separately or organized in a constellation
- Ground segment, which consists of all the traffic Earth stations of different size depending on the type of service considered
- Control system, which concerns all the ground facilities for the control and monitoring of the satellites and for the management of traffic and associated resources onboard

Figure 5.12 illustrates a connection by satellite between two Earth stations A and B for only one direction of propagation, which involves:

- Repeater onboard satellite
- Uplink from Earth station A to satellite
- Downlink from satellite to Earth station B

We generally use a higher frequency band for the uplink path than for the downlink path; it is indeed easier to compensate for the attenuation by the atmosphere and the hydrometeors at the level of the Earth stations by using larger antennas or amplifiers of higher power than at the level of the repeater.

5.2.1 General Characteristics of Repeater

The repeater constitutes to some extent the keystone of the system; it can be either of the following:

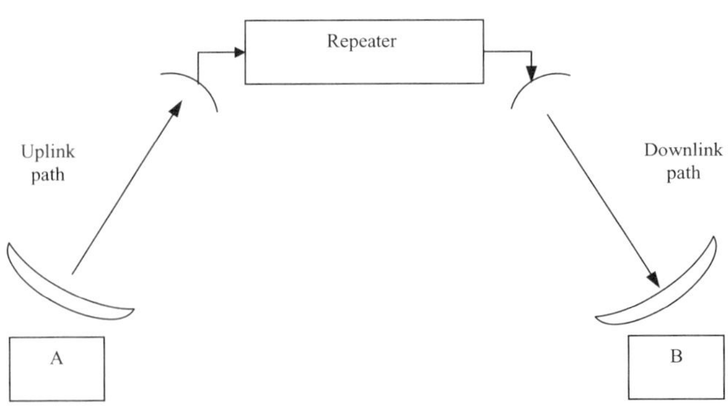

Figure 5.12 Synoptic diagram of satellite link.

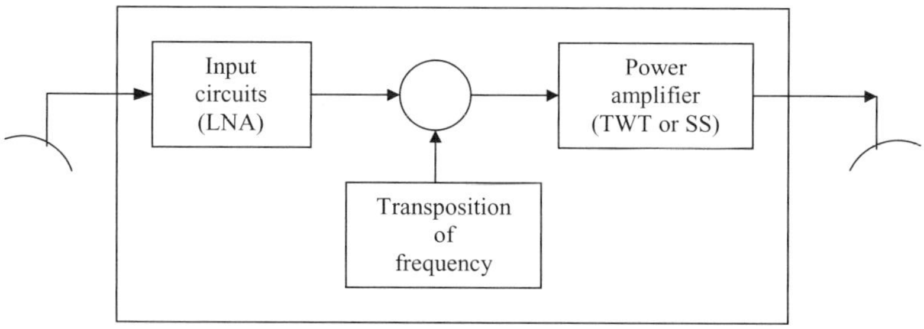

Figure 5.13 Schematic diagram of transparent repeater.

- A transparent, or bent-pipe, repeater is in general composed of an input circuit, a transposition in frequency, and an output power amplifier, as shown in Figure 5.13.
- A regenerative, or processing, repeater comprises, moreover, a demodulator and a modulator, making it possible to separate the uplink path from the downlink path.

The main features provided by the operator of the satellite which generally correspond to values guaranteed at the end of the lifetime at the edge of the coverage zones are as follows:

- Coverage geographical zones
- Repeater bandwidth
- Uplink and downlink frequency bands
- EIRP at saturation
- Input backoff
- Output backoff
- Figure of merit
- Power flux density at saturation
- EIRP of radio beacon

5.2.1.1 Zones of Coverage The zones of coverage correspond to the areas on Earth delimited by the radiation pattern of the radiating–receiving antennas of the repeater; these zones can be distinct, to connect two distant regions, or inside the same area and can take very diverse forms. The level contours between the edge of zone (EZ) and the center of zone (CZ) give the uplink and downlink geographical advantages according to the locus of the Earth station. Figure 5.14 presents a satellite system comprising two distinct zones.

5.2.1.2 Frequency Bandwidth The frequency bandwidth of the repeater corresponds to the total bandwidth of the various signals that share the

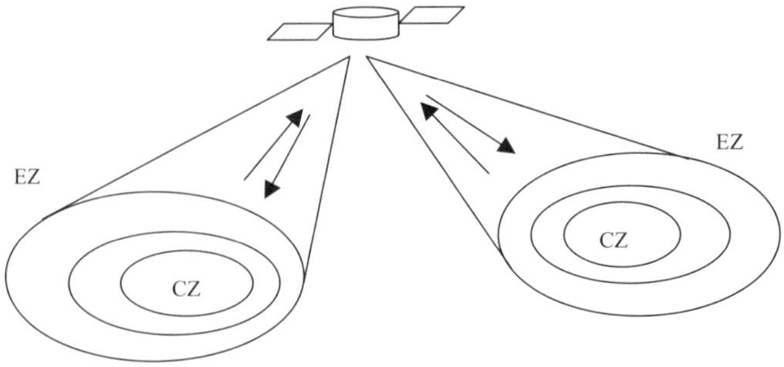

Figure 5.14 Uplink and downlink coverages.

resource. When the repeater is of the regenerative type, the uplink bandwidth can be different from the downlink bandwidth depending on their respective modulation type.

5.2.1.3 Figure of Merit The figure of merit is the ratio of the gain G of the antenna to the total noise temperature T of the repeater:

$$F_M = \frac{G}{T} \tag{5.16}$$

where G = antenna gain
T = total noise

The input circuits, which follow the reception antenna, comprise an LNA which, as we saw in Section 2.2, determines the equivalent noise temperature T_{eq} of the repeater; the total noise temperature of the repeater is then equal to

$$T = T_{eq} + T_0 \tag{5.17}$$

where $T_0 \approx 290$ K is the temperature of Earth that is seen by the receiving antenna. This characteristic is generally related to the uplink geographical advantage which depends on the position of the Earth station in the zone of coverage.

5.2.1.4 Power Amplifier The power amplifier is the most important element of the chain and is generally of two types, which are described in Section 2.3.2:

- Travelling-wave tube (TWT)
- Solid-state (transistors)

5.2.1.5 Power Flux Density at Saturation This corresponds to the value of the power flux density collected by the receiving antenna of the satellite for which the output power amplifier of the repeater is saturated; for the reasons mentioned above about the zone of coverage of the receiving antenna, we usually define the power flux density at saturation at the edge of the zone to which one associates the uplink geographical advantage, which refers to the location of the Earth station.

5.2.1.6 EIRP at Saturation The EIRP at saturation of the satellite can be expressed in dBW or in the form of a power flux density at the ground level expressed in dB(W m^{-2}); when this characteristic is given for the edge of the zone, we must associate the downlink geographical advantage to it in relation to the location of the Earth station.

5.2.1.7 Effective Input Noise Temperature of Repeater The noise generated by the repeater, expressed in dBW, is given by the relation

$$N_r = K + T + B \tag{5.18}$$

where K = Boltzmann's noise, $10\log(k) = -228.6\,\text{dB}(\text{W K}^{-1}\,\text{Hz}^{-1})$
T = Total noise temperature of repeater (dBK)
B = repeater bandwidth (dBHz)

We can also express it in terms of noise power flux density, in dB(W m^{-2}), using the relation

$$\Phi N_r = K + T + B + 10\log\left(\frac{4\pi}{\lambda^2}\right) + G_R \tag{5.19}$$

where G_R is the gain of the receiving antenna of the repeater (dBi), or in terms of the output backoff of the component of noise of the repeater,

$$\begin{aligned} \Phi N_r &= \Phi_{\text{sat}} - \text{IB} - (P_{\text{sat}} - \text{OB} - N_r) \\ \text{OB}(N_r) &= P_{\text{sat}} - N_r = \Phi_{\text{sat}} - \Phi N_r - \text{IB} + \text{OB} \end{aligned} \tag{5.20}$$

where Φ_{sat} = satellite power flux density at saturation [dB(W m^{-2})]
P_{sat} = output power at saturation (dBW)
IB,OB = input and output backoff of repeater (dB)

5.2.2 General Characteristics of Earth Station

The Earth station generally comprises an antenna associated with various modules for the emission part (e.g., modulation, transposition of frequency, power amplifier) and for the reception part (e.g., LNA, transposition of fre-

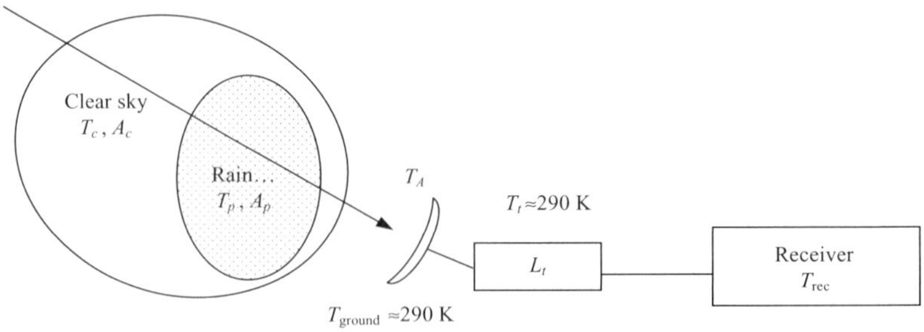

Figure 5.15 Noises source in chain of reception of Earth station.

quency, demodulation); among the elements which have an impact on dimensioning, such as emitted power and size of the antenna, the total noise temperature of the system of reception is the most determining.

5.2.2.1 *Sources of Noise of Reception System* Figure 5.15 presents the whole chain of reception between the antenna and the receiver as well as the natural sources of noise that affect the downlink path.

5.2.2.2 *Noise Temperature of Receiving Antenna* The noise temperature of the antenna is the sum of all the noises of various origins that are collected by the antenna at reception; it is expressed by the integral

$$T_A = \frac{1}{4\pi} \iint g(\Omega) T(\Omega) d\Omega \tag{5.21}$$

where $d\Omega$ = elementary solid angle in direction of Ω
$g(\Omega)$ = antenna gain
$T(\Omega)$ = equivalent temperature of source of noise in this direction

The contribution of each lobe is thus determined by the relation

$$T_i = G_i \left(\frac{\Omega_i}{4\pi} \right) T_G$$

where G_i = mean gain of lobe
Ω_i = solid angle of lobe
T_G = brightness temperature of ground and the brightness temperature has the following values for the lateral lobes (ITU-R Report 390):

Elevation angle	<–10°	$-10° < E < 0°$	$0° < E < 10°$	$10° < E < 90°$
T_G	290 K	150 K	50 K	10 K

The main sources of noise that contribute to determine the noise temperature of the antenna were treated in Section 2.2.

These sources of noise, which come from diverse origins, can be classified in the following way:

- Terrestrial noise:
  Ohmic losses of antenna and branching devices such as transmission lines
  Noise due to radiation from Earth, collected by side lobes
  Atmospheric noise, which is due to attenuation by oxygen, water vapour, rain, and so on
  Jamming produced by other operative systems in same frequency bands
- Extraterrestrial noise:
  Cosmic noise (2.7 K)
  Galactic noise
  Stellar radiosources, sun, moon, planets, and so on

Ground noise and atmospheric noise, which is characterized by its brightness temperature, vary in the inverse ratio of the elevation angle of the antenna while extraterrestrial noise depends primarily on the nature and the direction of the radio source. For example, the sun can cause an important disturbance when it passes in the main beam of the antenna, because of its very high temperature (6000 K) and great solid angle (0.5°), while the influence of remote stars is insignificant due to their small solid angle compared to the antenna beamwidth. We can thus define a characteristic T_{A0}, called typical noise temperature of the antenna per clear atmosphere, which includes the ground noise.

Consider now the contribution of noise brought by the clear atmosphere and an additional atmospheric attenuation, where A_c, T_c are the atmospheric attenuation and temperature due to the clear atmosphere and A_R, T_R the additional atmospheric attenuation and temperature. From relation (1.140), the total temperature of the antenna is given as[1]

$$T_A = \frac{T_c}{A_R}\left(1-\frac{1}{A_c}\right)+T_R\left(1-\frac{1}{A_R}\right)+T_{A0} \tag{5.22}$$

As the additional atmospheric loss is generally due to hydrometeors (e.g., rain[2]), the temperature T_R is about 290 K.

[1]It is advisable to check what the manufacturer of the antenna includes in the nominal characteristic T_{A0} in order to introduce the respective contributions of Earth, the clear sky, the atmosphere, and the rain.

[2]The additional attenuation by rain includes the atmospheric attenuation for the corresponding fraction $r_{0.01}$ of the path.

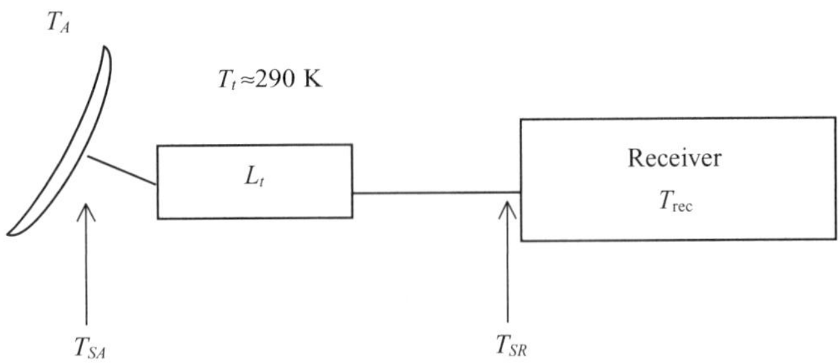

Figure 5.16 Total noise temperature of reception system.

5.2.2.3 Total Noise Temperature of Reception System

Let L_t, T_t be the loss and the physical temperature of the feeder line and T_{rec} the effective input noise temperature of the receiver. The total noise temperature of the system of reception, presented in Figure 5.16, is given by the following relations[3]:

- Input noise temperature at antenna port:

$$T_{SA} = T_A + (L_t - 1)T_t + L_t T_{rec} \tag{5.23}$$

- Input noise temperature at receiver port:

$$T_{SR} = T_{rec} + \left(1 - \frac{1}{L_t}\right)T_t + \frac{T_A}{L_t} \tag{5.24}$$

5.3 LINK BUDGET

Figure 5.17 illustrates the relations between the main characteristics of the repeater and the radiation pattern of the transmitting and receiving antennas.

5.3.1 Downlink Path

The EIRP that must be provided by the satellite, eirp_{SAT}, for a given BER, expressed in dBW, in the direction of the Earth station is given by the relation

$$\mathrm{eirp}_{SAT} = \left(\frac{C}{N_0}\right)_D + A_{FSD} + A_{aD} + M_D - \left(\frac{G}{T}\right)_{ST} + K \tag{5.25}$$

[3]The contribution of the ambient temperature at the entry of the two-port circuit in the relations in Section 2.2.6 is replaced here by the noise temperature of the antenna.

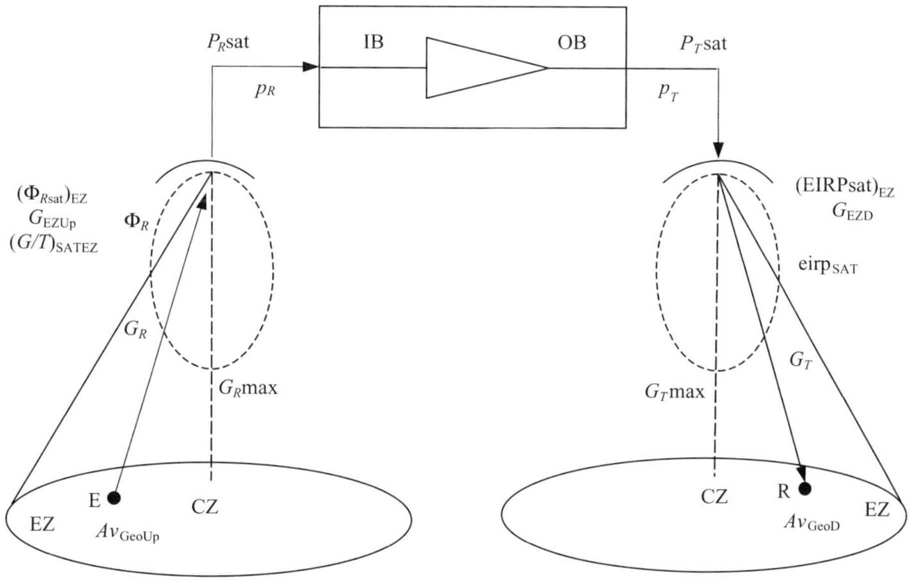

Figure 5.17 Main characteristics of repeater.

where $(C/N_0)_D$ = ratio of carrier-to-noise power spectral density (dBHz)
A_{FSD} = free-space downlink path propagation loss (dB)
A_{aD} = atmospheric attenuation (dB)
M_D = required margin (aging of equipment, misalignment of antenna, attenuation by hydrometeors, degradation of system temperature, jamming, etc.) (dB)
$(G/T)_{ST}$ = figure of merit of Earth station (dB K^{-1})

We can also write

$$\text{eirp}_{SAT} = p_T + G_T = p_T + G_{EZD} + Av_{GeoD} \tag{5.26}$$

where p_T = emitted power which must be allocated by repeater
G_T = gain of transmitting antenna toward Earth station
G_{EZD} = gain of transmitting antenna at edge of coverage zone
Av_{GeoD} = downlink geographic advantage according to location of Earth station

5.3.2 Resource Allocation by Repeater

As the output amplifier of the repeater is not linear and owing to the fact that its characteristics concern the total EIRP, the resource allocation of power for a single carrier is calculated by determining the identical number n of possible carriers for a given output backoff OB in the following way:

$$10\log n = P_{T_{\text{sat}}} - \text{OB} - p_T = (\text{EIRP}_{\text{sat}})_{\text{EZ}} - G_{\text{EZD}} - \text{OB} - \text{eirp}_{\text{SAT}} + G_{\text{EZD}} + Av_{\text{GeoD}}$$

where P_{Tsat} = the emitted power at saturation of repeater
$(\text{EIRP}_{\text{sat}})_{\text{EZ}}$ = repeater EIRP at saturation at edge of coverage zone

that is,

$$10\log n = (\text{EIRP}_{\text{sat}})_{\text{EZ}} - \text{OB} - \text{eirp}_{\text{SAT}} + Av_{\text{GeoD}} \tag{5.27}$$

5.3.3 Uplink Path

The relation between the received power and the power flux density at the level of the repeater is written as

$$\begin{aligned} p_R &= \Phi_R - 10\log\left(\frac{4\pi}{\lambda^2}\right) + G_R \\ &= \Phi_R - 10\log\left(\frac{4\pi}{\lambda^2}\right) + G_{\text{EZUp}} + Av_{\text{GeoUp}} \end{aligned} \tag{5.28}$$

where Φ_R = power flux density at reception antenna of repeater
G_R = repeater antenna gain in direction of Earth station
G_{EZUp} = repeater antenna gain at edge of coverage
Av_{GeoUp} = uplink geographic advantage according to location of Earth station

The relation which links together the power at saturation at the input of the repeater, $P_{R\text{sat}}$, and the characteristic of power flux density at saturation at the edge of the coverage zone, $(\Phi_{R\text{sat}})_{\text{EZ}}$, is given by

$$P_{R_{\text{sat}}} = (\Phi_{R_{\text{sat}}})_{\text{EZ}} - 10\log\left(\frac{4\pi}{\lambda^2}\right) + G_{\text{EZUp}} \tag{5.29}$$

The power required at the reception of the repeater to generate the resource necessary for the emission on the downlink path for a given input backoff IB is written as

$$p_R = P_{R_{\text{sat}}} - \text{IB} - 10\log n$$

That is,

$$p_R = (\Phi_{R_{\text{sat}}})_{\text{EZ}} - 10\log\left(\frac{4\pi}{\lambda^2}\right) + G_{\text{EZUp}} - \text{IB} - 10\log n \tag{5.30}$$

By solving the system of equations (5.28) and (5.30), we obtain the value of the required power flux density, expressed in dB(W m^{-2}), at the level of the satellite:

$$\Phi_R = (\Phi_{R_{\text{sat}}})_{\text{EZ}} - \text{IB} - 10\log n - Av_{\text{GeoUp}} \tag{5.31}$$

The necessary EIRP on behalf of the Earth station, EIRP_{ST}, expressed in dBW, is then given by the expression

$$\text{EIRP}_{\text{ST}} = \Phi_R + 10\log(4\pi d^2) + A_{a\text{Up}} + M_{\text{Up}} \tag{5.32}$$

where $A_{a\text{Up}}$ = uplink atmospheric attenuation (dB)
M_{Up} = uplink margin for hydrometeors, aging, misalignment, jamming, etc. (dB)

The carrier-to-noise ratio on the uplink path, in dBHz, is given by the relation

$$\left(\frac{C}{N_0}\right)_{\text{Up}} = \Phi_R + \left(\frac{G}{T}\right)_{\text{SATEZ}} + 10\log\left(\frac{\lambda^2}{4\pi}\right) + Av_{\text{GeoUp}} - K \tag{5.33}$$

5.3.4 Total Carrier-to-Noise Ratio

As with a transparent repeater, the noise on the uplink path is added to that of the downlink path, we can write that the level of total noise N_T at a given point of the chain of reception is equal to

$$N_T = N_{\text{Up}} + N_D$$

where N_{Up} = noise level on uplink path
N_D = noise level on downlink path

As the carrier-to-total-noise ratio corresponds to the same point of the chain of reception at the Earth station, this expression can be written by introducing the level of the carrier, C, at the considered measurement point:

$$\frac{1}{C/N_T} = \frac{1}{C/N_{\text{Up}}} + \frac{1}{C/N_D}$$

However, the levels of noise N_{Up} and N_D actually refer to various points of the transmission chain and, consequently, to various levels of the carrier. It is the same for the other components of noise, such as the intermodulation products I in the power amplifiers of the Earth station or the repeater and any other source of jamming J, which appear at any point of the transmission chain. Thus, from the preceding relation we write

$$\frac{1}{(C/N)_T}=\frac{1}{(C/N)_{\text{Up}}}+\frac{1}{(C/N)_D}+\frac{1}{(C/I)_R}+\frac{1}{(C/I)_{\text{PA}}}+\frac{1}{(C/J)_T}+\cdots \quad (5.34)$$

or, while referring to the noise power spectral density, since the signal occupies the same frequency bandwidth throughout the transmission chain,

$$\frac{1}{(C/N_0)_T}=\frac{1}{(C/N_0)_{\text{Up}}}+\frac{1}{(C/N_0)_D}+\frac{1}{(C/I_0)_R}+\frac{1}{(C/I_0)_{\text{PA}}}+\frac{1}{(C/J_0)_T}+\cdots$$

5.3.5 Power Spectral Density of Repeater

The power spectral density of the repeater, expressed in dBW Hz^{-1}, is the ratio of the EIRP at saturation at the edge of coverage, including the output backoff, to the total bandwidth:

$$\text{PSD}=(\text{EIRP}_{\text{sat}})_{\text{EZ}}-\text{OB}-B \quad (5.35)$$

where B is the total frequency bandwidth of the repeater (dBHz).

5.3.6 Occupancy Ratio of Repeater by Carrier

The occupancy ratio in power is written as

$$(\text{OR})_P=\text{eirp}_{\text{SAT}}-Av_{\text{GeoD}}-[(\text{EIRP}_{\text{sat}})_{\text{EZ}}-\text{OB}] \quad (5.36)$$

The occupancy ratio in frequency bandwidth is written the same way:

$$(\text{OR})_B=b-B \quad (5.37)$$

where b is the frequency bandwidth occupied by the carrier (dBHz).

5.3.7 Optimum Size for Receiving Antenna

In general, the point of operation of the repeater, that is, the value of the power flux density at saturation, is selected so as to privilege the downlink path; this condition makes it possible to avoid regeneration on board at the cost of a light oversizing of the Earth station and results in the following approximate relation except for physical units:

$$\left(\frac{C}{N_0}\right)_T\approx\left(\frac{C}{N_0}\right)_D\approx \text{eirp}_{\text{SAT}}-A_{\text{FSD}}-A_{aD}+\left(\frac{G}{T}\right)_{\text{ST}}-K-M_D$$

The condition of optimization is met when the occupancy in terms of frequency bandwidth and that in terms of power are equal, that is,

$$(\mathrm{EIRP_{sat}})_{\mathrm{EZ}} - \mathrm{OB} - \mathrm{eirp_{SAT}} + Av_{\mathrm{GeoD}} = B - b = 10\log n$$

and

$$\left(\frac{G_{\mathrm{OPT}}}{T}\right)_{\mathrm{ST}} = \left(\frac{C}{N_0}\right)_D + B - b - (\mathrm{EIRP_{sat}})_{\mathrm{EZ}} + \mathrm{OB} + A_{\mathrm{FSD}} + A_{aD} + K + M_D - Av_{\mathrm{GeoD}}$$

which yields the diameter of the optimal antenna:

$$D_{\mathrm{OPT}} = \frac{\lambda}{\pi}\sqrt{\frac{G_{\mathrm{OPT}}}{\eta}} \tag{5.38}$$

The optimal antenna thus corresponds to the minimal size that optimizes the resource of the repeater so that the power spectral density of the carrier is equal to that of the repeater. However, the size retained for the antenna of the Earth station is in general larger in order to respect, on the one hand, the rules of coordination between the services by satellite and the radio relay systems, as we saw in the Section 1.3.2, and, on the other hand, to reduce the power of emission on the uplink path.

5.3.8 General Characteristics of Signal

As we saw in Section 2.3, the modulation types are characterized by the number of states of the carrier and by an occupied bandwidth which also depends on the type of coding used for the correction of the errors.

5.3.9 Performance Objectives of Connection

The quality of the connection rests on the value of the characteristic E_b/N_0 corresponding to the probability of error considered. From the signal-to-total-noise power spectral density ratio $(C/N_0)_T$, given by relation (5.34), and from relation (2.42), we can calculate the value E_b/N_0 as well as the probability of error which results from it, writing, in decibels, for a bit rate R

$$\left(\frac{E_b}{N_0}\right) = \left(\frac{C}{N_0}\right)_T - 10\log R \tag{5.39}$$

In practice, we will thus adjust in relation (5.25) the value of the signal-to-noise power spectral density ratio $(C/N_0)_D$ on the downlink path, which is the most underprivileged, in order to obtain the required nominal quality of performance.

5.3.10 Multiple-Access Techniques

Multiple access is the possibility of several Earth stations sharing the same satellite repeater with their respective carrier. Any Earth station that is located in the zone of coverage can thus receive the carriers emitted by the other stations, and it is also possible to gather several carriers within the same station in order to serve several destinations. The modes of assignment of the channels are of two types:

- Multiple access using fixed assignment, in which the channels are assigned permanently to the various Earth stations for their exclusive use
- Multiple access using on-demand assignment, in which the channels are affected according to the needs for each communication as long as this one continues

The techniques of multiple access can be classified into three categories:

- Frequency-division multiple access (FDMA), where each station has its own assigned frequency
- Time-division multiple access (TDMA), where all the stations use the same carrier frequency and the same bandwidth according to a division in time
- Code-division multiple access (CDMA), where all the stations share simultaneously the same frequency bandwidth and recognize the signals by identification of a pseudorandom code specific to each carrier

5.3.10.1 Frequency-Division Multiple Access In an FDMA system, each carrier occupies a separate channel that should not overlap any other, as shown in Figure 5.18. However, products of intermodulation occur between the carrier frequencies, generated by the nonlinearity of the repeater, and each carrier is also affected by jamming from adjacent channels that is mainly due to the secondary lobes of modulation which could not be eliminated by filtering at the emission. Consequently, it is advisable to allot to each carrier a bandwidth guard according to the residual lateral bandwidth of each transmitted signal which must also take into account the drift of the local oscillators of the repeater and those of the Earth stations as well as the Doppler-Fizeau effect due to the relative movement between the satellite and the Earth stations, when those are mobile, as we will see. In addition to calculation of the effects of the intermodulation products, distortions, and jamming between carriers, it is also advisable to take into account the variations of the relative level of the signals owing to attenuation by the hydrometeors, misalignment of the antennas, and other losses.

5.3.10.2 Time-Division Multiple Access In TDMA system, various carriers emitted by the Earth stations can be received by the satellite in intervals

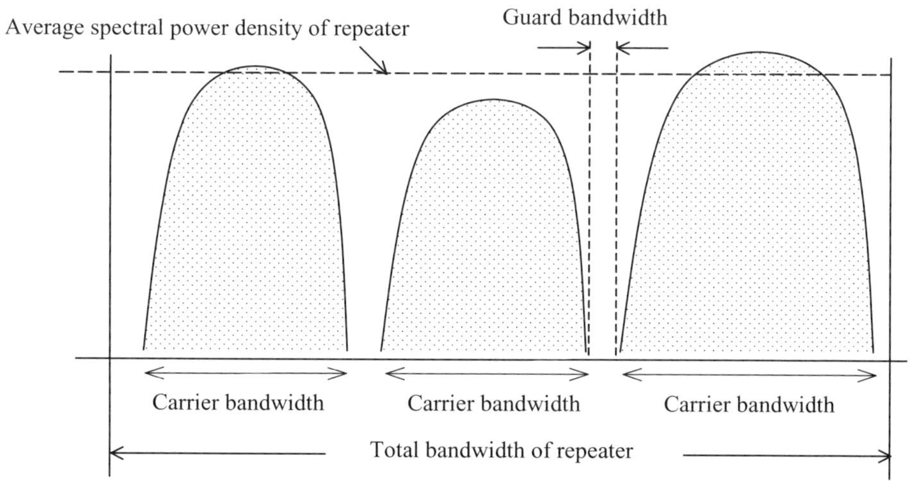

Figure 5.18 Frequency division multiple access.

of separated times per bursts and without overlapping. Insofar as the system occupies all the repeater resource, TDMA makes impossible the formation of intermodulation products between carriers since only one is transmitted at a time; there remains only the self-intermodulation products between all the components of the transmitted spectrum. Each terminal on Earth must be able to determine the time reference at the level of the satellite so that the emitted bursts are inserted within a periodic time structure of a given duration called a frame. The principal consequence of this access mode is that the binary rate is in general much higher than that of the continuous bit streams at the input of the terminal on Earth.

5.3.10.3 Code-Division Multiple Access In CDMA, the carriers simultaneously use the whole frequency bandwidth of the repeater, but each one is identified by a specific code; at reception, each terminal recognizes the signal that is intended to it by its code and extracts the basic information. This technique, which is also called multiple access by spread spectrum since it imposes a broad spread in frequency of the signal, has the following principal advantages:

- Useless coordination between transmitters
- Very low power spectral density
- Protection against jammers with narrow bandwidth
- Security of communications by code
- Resistance to propagation by multipath

On the other hand, the main disadvantage of CDMA systems lies in their low spectrum efficiency. The most used methods of coding are as follows:

- Direct sequence (DS-CDMA) or pseudonoise (PN)
- Frequency hopping (FH)
- Combination of both

The fundamental parameters necessary to define the performance of a CDMA system are the frequency bandwidth of the information b, roughly equal to the binary useful rate R, the frequency bandwidth B of the emitted signal, and the margin to jamming M_J. We can then deduce the gain g which results from the signal processing:

$$g = \frac{B}{b} \approx \frac{B}{R} \tag{5.40}$$

Calling J the power of the jammers, that is, the principal component of the total noise, we write

$$N_0 = \frac{J}{B}$$

If S is the power of the signal and E_b the average energy transmitted per bit of information, we have $S = E_b R$ and the margin to jamming becomes

$$M_J = \frac{J}{S} = \frac{BN_0}{E_b R} = \frac{g}{(E_b/N_0)} \tag{5.41}$$

5.3.11 Doppler–Fizeau Effect

The Doppler–Fizeau effect appears when the source of the emission, the receiver, or both at the same time move toward each other. This phenomenon thus occurs when the satellite is running on a low or medium orbit, on a circular geosynchronous orbit with nonzero inclination, describing then one "eight" on both sides of the equatorial plane, during the periodic correction of trajectory or when the Earth stations are mobile.

Figure 5.19 presents a satellites affected by a radial speed V_S being used as a relay for two Earth stations A and B which are respectively animated by a radial speed V_A and V_B; we will suppose that Earth station A emits a signal at the frequency f and that the propagation velocity is that of light in a vacuum, c.

Consider the uplink path of Earth station A toward the satellite; the Lorentz equations of transformation allow us to write that the relative radial speed V_{AS} from station A with regard to the satellite is equal to

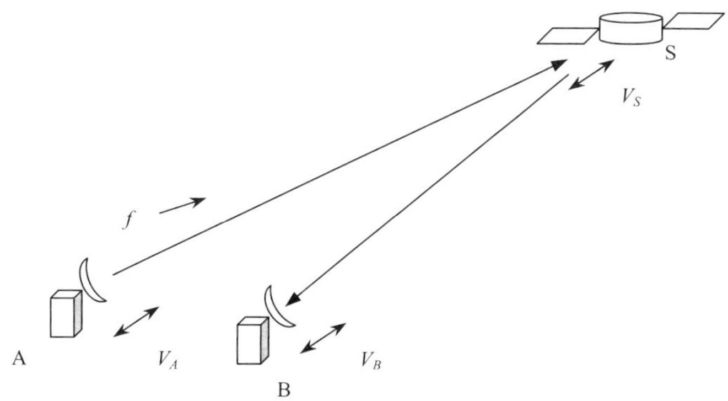

Figure 5.19 Doppler–Fizeau effect.

$$V_{AS} = \frac{V_S - V_A}{1 - V_S V_A / c^2} \tag{5.42}$$

that is,

$$V_{AS} \approx V_S - V_A$$

where V_S and V_A are small compared to c. The velocity of light being a universal constant, independent from the reference mark considered, only the wavelength can vary, and this one becomes

$$\lambda' = \lambda - \frac{V_{AS}}{f} \qquad f = \frac{c}{\lambda}$$

From

$$f' = \frac{c}{\lambda'} = f\left(\frac{1}{1 - V_{AS}/c}\right) \approx f\left(\frac{1}{1 - (V_S - V_A)/c}\right)$$

we obtain the relative frequency shift from Earth station A to the satellite:

$$\frac{\Delta f_{A \to S}}{f} = \frac{f' - f}{f} = \frac{V_S - V_A}{c - V_S - V_A} \approx \frac{V_S - V_A}{c} \tag{5.43}$$

In the same way, on the downlink path from the satellite toward Earth station B we have the relative frequency shift

$$\frac{\Delta f_{S \to B}}{f} \approx \frac{V_S - V_B}{c}$$

For a signal undergoing two successive shifts of frequency, we can write the following approximate relation for very slow speeds relative to the velocity of light which gives the total frequency shift for the whole connection between Earth stations A and B:

$$\frac{\Delta f_{A \to B}}{f} \approx \frac{\pm 2V_S \pm V_A \pm V_B}{c} \tag{5.44}$$

where the sign allocated to each component of speed depends on the direction of displacement with respect to the terrestrial reference mark.

As an example, Section 5.1.3.1 shows that, in the case of a LEO, the radial speed may reach up to $7.5\,\mathrm{km\,s^{-1}}$, to which corresponds, for a connection between two fixed Earth stations using the same coverage zone, a relative frequency shift of

$$\frac{\Delta f_{A \to B}}{f} \approx \pm \frac{2V_S}{c} = \pm \frac{1.5 \times 10^4}{3 \times 10^8} = \pm 5 \times 10^{-5}$$

5.4 METHOD OF PREDICTION

5.4.1 Earth–Space Propagation

We previously studied the various factors which affect the propagation of electromagnetic waves through:

- Nonionized atmosphere
  - (a) Atmospheric absorption and radiation
  - (b) Diffusion, depolarization, and radiation by hydrometeors
  - (c) Loss in gain of antenna due to divergence of beam caused by variation of refractivity of atmosphere with elevation angle, bending of rays which modifies their launch and arrival angles, and variation of phase caused by irregularities of refractivity
  - (d) Tropospheric scintillation and multipath
- Ionosphere
  - (a) Faraday's rotation,
  - (b) Dispersion, which results in group delay distortion
  - (c) Ionospheric scintillation, which affects amplitude and phase of signal
- External sources of noise of terrestrial and extraterrestrial origin

5.4.1.1 Atmospheric Attenuation We refer to Section 1.10 for the calculation of attenuation and to Section 2.2 for the calculation of the brightness temperature that influences the effective noise temperature of the system of reception.

5.4.1.2 Effects Due to Hydrometeors We refer to Section 1.11 for the calculation of attenuation and transpolarization and to Section 2.2 for the corresponding brightness temperature.

5.4.1.3 Antenna Gain Loss The antenna gain loss can be regarded as negligible for elevation angles higher than 5 °.

5.4.1.4 Effects Due to Tropospheric Scintillation and Multipath Recommendation ITU-R P.618 provides the methods of calculation of the distribution of fadings ascribable to tropospheric scintillation and multipath for:

- Elevation angles lower than 5 °, which are connected with that described in Section 3.4 for line-of-sight links
- Elevation angles higher than 4 °, which are referred in most cases to the field of Earth–space connections

Consequently we will limit calculation to the amplitude of the variations due to the tropospheric scintillation for elevation angles higher than 4 ° when using the method recommended by ITU-R which is described below. The parameters concerning the geographical location of the Earth station which come into play are as follows:

- Average ambient temperature T for one-month period or more
- Average humidity ratio HR for same period
- Frequency f between 4 and 20 GHz
- Elevation angle θ higher or equal to 4 °
- Diameter of antenna D
- Antenna efficiency η that we can take equal to 0.5

We start by calculating, for the temperature T and the humidity ratio HR considered and using the relations (1.78), the water vapor pressure at saturation p_S and the partial water vapor pressure p_v in hectopascals. We calculate the wet fraction N_{Wet} of relation (1.75):

$$N_{\text{Wet}} = 3.732 \times 10^5 \frac{p_v}{T^2} \tag{5.45}$$

Then, the reference standard deviation of the variation of the amplitude of the signal is given as

$$\sigma_{\text{ref}} = 3.6 \times 10^{-3} + 10^{-4} N_{\text{Wet}} \tag{5.46}$$

We introduce the parameters suitable for the Earth station by calculating the effective path length L:

$$L = \frac{2h_L}{\sqrt{(\sin\theta)^2 + 2.35\times10^{-4}} + \sin\theta} \tag{5.47}$$

where $h_L = 1000\,\text{m}$ is the height of the turbulent layer. We consider the effective antenna diameter D_{eff} in meters:

$$D_{\text{eff}} = \sqrt{\eta}D \tag{5.48}$$

Then the normative factor of the antenna is given as

$$g(x) = \sqrt{3.86(x^2+1)^{11/12}\sin\left[\frac{11}{6}\arctan\left(\frac{1}{x}\right)\right] - 7.08x^{5/6}} \tag{5.49}$$

where $x = 1.22D_{eff}^2(f/L)$. The standard deviation of the variation of the signal for the path considered then becomes

$$\sigma = \sigma_{\text{ref}} f^{7/12}\frac{g(x)}{(\sin\theta)^{1.2}} \tag{5.50}$$

The factor of percentage of time $a(p)$ for the percentage of time p ranging between 0.01 and 50% is given by the relation

$$a(p) = -0.061(\log p)^3 + 0.072(\log p)^2 - 1.71\log p + 3 \tag{5.51}$$

The fading depth of scintillation A_S according to the percentage of time p, expressed in decibels, is finally given by the relation

$$A_S(p) = a(p)\sigma \tag{5.52}$$

Figure 5.20 shows as an example the distribution of the magnitude of the atmospheric scintillation according to the percentage of time for various values of the elevation angle.

5.4.1.5 Effects Due to Ionosphere We refer to Section 1.12 for the calculation of Faraday's rotation and the time delay of propagation.

5.4.1.6 External Noise Sources We refer to Section 2.2.8 for the determination of the total external noise to take into account in the calculation of the effective noise temperature of the system of reception.

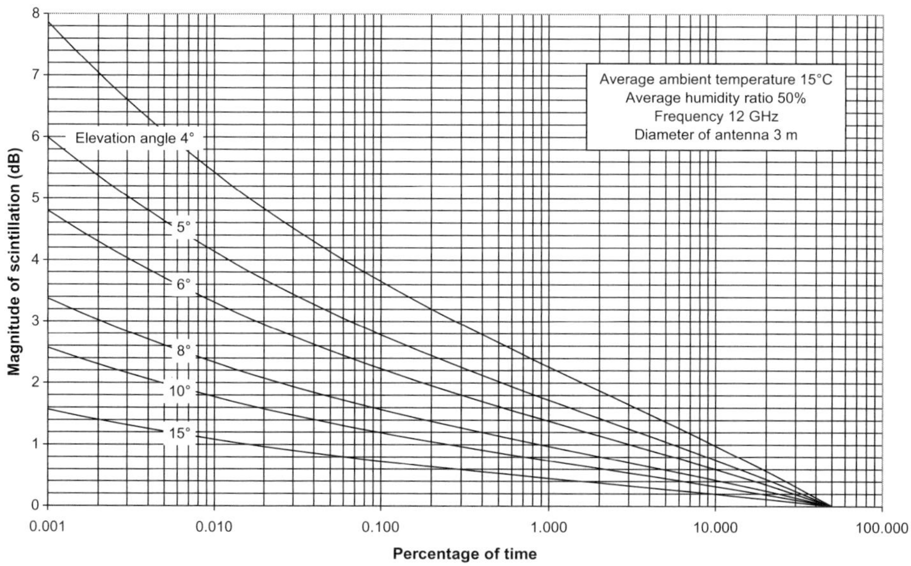

Figure 5.20 Magnitude of tropospheric scintillation.

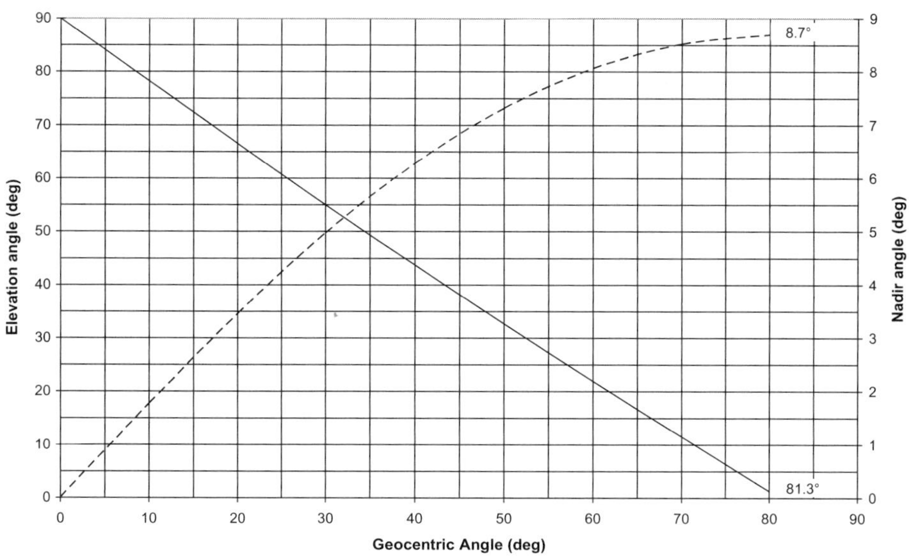

Figure 5.21 Elevation and nadir angles versus geocentric angle for global coverage.

TABLE 5.1 Satellite Connection: Station A to Station B

Longitude station A	0.0°	Uplink free-space loss	199.3 dB	Bit rate	256 kbits/s
Latitude station A	45.0°	Downlink free-space loss	196.2 dB	Modulation	4
Elevation angle	30.3°	Station A eirp	62.8 dBW	Coding	0.50
Azimuth	219.2°	SAT power flux density	–103.1 dBW/m^2	Required (E_b/N_o)	7.0 dB
Antenna diameter	4.5 m	SAT carrier eirp	10.4 dBW	Equipment margin	0.0 dB
Antenna efficiency	60%	Carrier OB	19.1 dB	Guard coefficient	1.4
Reception antenna gain	43.3 dBi	SAT Carrier EZ eirp	9.6 dBW	Spreading bandwidth	256.0 kHz
System temperature	150 K	Carrier spectral density	–44.5 dBW/Hz	Spreading gain	0.0 dB
Reception (G/T)	21.5 dB/K	Carrier-to-repeater SD	6.4 dB	Occupied bandwidth	358.4 kHz
Downlink clear sky	0.1 dB	$(C/N_o)_d$	61.4 dBHz	Transmitted bandwidth	256.0 kHz
Downlink rain	1.0 dB	$(C/N_o)_{up}$	79.0 dBHz	Effective bandwidth	256.0 kHz
Rain degradation C/N_0	2.4 dB	System $(C/l_o)_{total}$	73.9 dB	Power rate occupancy	3.1%
Antenna depointing	0.5 dB	$(E_b/N_o)_{total}$	7.0 dB	Rep. bandwidth rate occ.	1.0%
Downlink margin	3.0 dB	C/N	7.0 dB	Availability	99.99%
Downlink geo. adv.	0.9 dB	$C/P–C/N$	19.4 dB	Uplink rain	3.0 dB
Reception $(G/T)_{optimal}$	27.6 dB/K	Clear-sky uplink (B→A)	0.1 dB	Em. antenna diameter	6.0 m
Rec. ant. optimal gain	49.4 dBi	Uplink margin	3.6 dB	Em. antenna gain	49.3 dBi
Rec. ant. optimal diam.	9.0 m	Uplink geo. adv.	2.5 dB	Emitted power	22.3 W
Ambient temperature	15 °C	Humidity ratio	50%	Scintillation (0.01%)	0.2 dB
		Characteristics of Repeater: Global Coverage			
Longitude SAT	30.0°	Longitude CZ	30.0°	SAT antenna beamwidth	14.7°
Latitude SAT	0.0°	Latitude CZ	0.0°	G_{max}CZ-GainEZ	4.0 dB
Uplink frequency	6.0 GHz	Downlink frequency	4.0 GHz	Repeater bandwidth	36.0 MHz
SAT max. antenna gain	20.9 dBi	Gain to ST/reception	17.8 dBi	Distance ST/SAT	38,589.2 km
Elevation CZ/SAT	90.0°	Nadir angle SAT CZ	0.0°	SAT nadir angle ST/Rec.	7.5°
Distance CZ/SAT	35786.0 km	Distance ST/CZ	5815.1 km	Difference nadir angle	7.5°
Emitted power	15.0 W	Noise figure	4.2 dB	Repeater spectral density	–50.9 dBW/Hz
EIRP at saturation EZ	28.7 dBW	(G/T)EZ	–12.0 dBK	(C_r/I)repeater	16.0 dB
Output backoff	4.0 dB	Input backoff	5.5 dB	(C_r/I)network	20.0 dB
Noise OB	22.5 dB	SAT EZ flux at saturation	–80.0 dBW/m^2	(C_r/J)jamming	20.0 dB
Beacon EZ EIRP	4.0 dBW	Beacon $(C/N_o)_{min}$	32.0 dBHz	Beacon margin	26.2 dB

TABLE 5.2 Satellite Connection: Station A to Station A

Longitude station B	55.0°	Uplink free-space loss	199.7 dB	Bit rate	256 kbits/s
Latitude station B	–20.0°	Downlink free-space loss	195.8 dB	Modulation	4
Elevation angle	53.2°	Station A eirp	58.6 dBW	Coding	0.50
Azimuth	53.7°	SAT power flux density	–104.7 dBW/m^2	Required (E_b/N_o)	7.0 dB
Antenna diameter	6.0 m	SAT carrier eirp	8.9 dBW	Equipment margin	0.0 dB
Antenna efficiency	65%	Carrier OB	22.3 dB	Guard coefficient	1.4
Reception antenna gain	46.1 dBi	SAT carrier EZ eirp	6.4 dBW	Spreading bandwidth	256.0 kHz
System temperature	120 K	Carrier spectral density	–47.7 dBW/Hz	Spreading gain	0.0 dB
Reception (G/T)	25.3 dB/K	Carrier-to-repeater SD	3.2 dB	Occupied bandwidth	358.4 kHz
Downlink clear sky	0.1 dB	$(C/N_o)_d$	61.8 dBHz	Transmitted bandwidth	256.0 kHz
Downlink rain	2.0 dB	$(C/N_o)_{up}$	75.8 dBHz	Effective bandwidth	256.0 kHz
Rain degradation C/N_o	4.6 dB	System $(C/I_o)_{total}$	70.7 dB	Power rate occupancy	1.5%
Antenna depointing	0.5 dB	$(Eb/N_o)_{total}$	7.0 dB	Rep bandwidth rate occ.	1.0%
Downlink margin	5.2 dB	C/N	7.0 dB	Availability	99.99%
Downlink geo. adv.	2.5 dB	$C/P–C/N$	16.2 dB	Uplink rain	0.0 dB
Reception $(G/T)_{optimal}$	27.8 dB/K	Clear-sky uplink (A→B)	0.1 dB	Em. antenna diameter	4.5 m
Rec. ant. optimal gain	48.6 dBi	Uplink margin	0.6 dB	Em. antenna gain	47.2 dBi
Rec. ant. optimal diam	8.0 m	Uplink geo. adv.	0.9 dB	Emitted power	14.1 W
Ambient temperature	25 °C	Humidity ratio	80%	Scintillation (0.01%)	0.2 dB
		Characteristics of Repeater: Global Coverage			
Longitude SAT	30.0°	Longitude CZ	30.0°	SAT antenna beamwidth	14.7°
Latitude SAT	0.0°	latitude CZ	0.0°	G_{max}CZ-GainEZ	4.0 dB
Uplink frequency	6.0 GHz	Downlink frequency	4.0 GHz	Repeater bandwidth	36.0 MHz
SAT max. antenna gain	20.9 dBi	Gain to ST/Reception	19.4 dBi	Distance ST/SAT	36884.0 km
Elevation CZ/SAT	90.0°	Nadir angle SAT CZ	0.0°	SAT nadir angle ST/Rec.	5.2°
Distance CZ/SAT	35786.0 km	Distance ST/CZ	3518.5 km	Difference nadir angle	5.2°
Emitted power	15.0 W	Noise figure	4.2 dB	Repeater spectral density	–50.9 dBW/Hz
EIRP at saturation EZ	28.7 dBW	(G/T)EZ	–12.0 dB K	(C_r/I)repeater	16.0 dB
Output backoff	4.0 dB	Input backoff	5.5 dB	(C_r/I)network	20.0 dB
Noise OB	22.5 dB	SAT EZ flux at saturation	–80.0 dBW/m^2	(C_r/J)jamming	20.0 dB
Beacon EZ EIRP	4.0 dBW	Beacon $(C/N_o)_{min}$	32.0 dBHz	Beacon margin	32.1 dB

5.4.2 Prediction of Total Outage Probability

The total availability of the system of reception rests on the margins M_{Up} and M_D defined in Section 5.3 for the uplink and downlink paths with respect to the effects due to tropospheric scintillation and to hydrometeors, which are in general dominating. The Earth stations being distant from one another, these effects can be regarded as decorrelated on the uplink and downlink paths; indeed, their joint probability, which is then equal to the product of their individual probabilities so that the same fading occurs on both paths and at the same time, is of second order. The total probability of outage of a connection will thus be equal to the sum of the probabilities of outage on the uplink and downlink paths.

In the same way, the outage probability for each path is equal to the sum of the probabilities of interruption due to tropospheric scintillation and hydrometeors since these events can themselves be regarded as independent. It is thus enough to determine the following:

- A margin M_R for the rain composed of a given value for the attenuation as well as the consecutive degradation of the effective noise temperature of the system of reception, to which corresponds a probability of interruption P_R
- A margin M_S for scintillation, to which corresponds a probability of outage P_S that is given by relation (1.144) with $x = M_S/\sigma_{\text{ref}}$, where M_S and σ_{ref}, given by relation (5.46), are expressed in decibels

The total probability of interruption of the connection, or total unavailability due to propagation, then becomes

$$P_{\text{tot}} = (P_R + P_S)_{\text{UP}} + (P_R + P_S)_D \tag{5.53}$$

In case of site diversity, the outage probability for a system of two Earth stations in given in Section 1.11.1.3.

5.4.3 Example of Link Budget

Consider a link established under a global coverage zone where Figure 5.21 shows the geometric angle and the nadir angle. Tables 5.1 and 5.2 show the detailed calculations in the frequency band 4–6 GHz.

REFERENCES

G. Barué, *Foundations for line-of-sight microwave links: Foundations for troposcatter microwave links*, Thomson-CSF, 1982.

G. Barué, Process for stabilization of magnetic field by paramagnetic nuclear resonance synchro-cyclotron—CERN, Geneva; Patent E.S.I.E.E.–Scientific Electric, 1968. Electronique Industrielle et Microélectronique (Paris, April 1969).

G. Barué, Transverse directivity of the propagation medium in troposcatter mode. Process for automatic antenna alignment and optimization of the radio link. Patent, Thomson-CSF, 1982.

G. Barué, "Telecommunications and Shelters," Ellipses, 2003.

L. Boithias, Propagation of radioelectric waves in terrestrial environment. Technical and scientific collection for telecommunications, Bordas and Cnet-Enst, 1983.

L. Boithias, Reception in diversity in microwave radio links, L'Onde Electrique, Jan./Feb. 1990.

C. Collin, Empirical evaluation of the coherence bandwidth in troposcatter mode, Thomson-CSF, Sept. 1979.

R. E. Collin, *Antennas and Radiowave Propagation*, McGraw-Hill, New York, 1985.

R. L. Freeman, *Reference Manual for Telecommunications Engineering*, Wiley, New York, 1994.

J. D. Kraus, *Radio Astronomy*, Ohio State University, Cygnus-Quasar Books, 1986.

P. Lena, *Astrophysics—Physical methods for observation*, InterEditions, CNRS Editions, 1996.

J. Lavergnat and M. Sylvain, *Propagation of radioelectric waves*, Masson, 1997.

National Bureau of Standards (NBS), Transmission loss predictions for tropospheric communication circuits, technical note 101, Washington, DC, 1996.

Ph. F. Panter, *Communication Systems Design: Line-of-Sight and Tropo-Scatter Systems*, McGraw-Hill, New York, 1972.

J. G. Proakis, *Digital Communications*, McGraw-Hill, New York, 1995.

P. L. Rice, A. G. Longley, K. A. Norton, and A. P. Barsis, Institute for Telecommunications Sciences, and Astronomy, Environmental Science Services Administration, Boulder, Co.

S. Silver, *Microwave Antenna Theory and Design*, Peter Peregrinus, London, 1986.

B. Sklar, *Digital Communications Fundamentals and Applications*, Prentice-Hall International, New Jersey, 1988.

Microwave Engineering: Land & Space Radiocommunications, by Gérard Barué

J. L. Steinberg and J. Lequeux, *Radioastronomy*, Observatory of Paris Meudon, Dunod, 1960.

U.S. Department of Defense (DoD), *Military Handbook*, MIL-HDBK-217, DoD, Washington, DC, 1992.

INDEX